# ELASTIC PLATES

## Theory and Application

# ELASTIC PLATES

## Theory and Application

**HERBERT REISMANN**
State University of New York at Buffalo

A WILEY-INTERSCIENCE PUBLICATION
**JOHN WILEY & SONS**
**New York · Chichester · Brisbane · Toronto · Singapore**

*To Edith, Sandy, and Bobbie, who were frequently neglected while the work progressed. Their understanding was their contribution to the present undertaking.*





***Library of Congress Cataloging in Publication Data*:**

Reismann, Herbert.
Elastic plates: theory and application/Herbert Reismann.
p. cm.
"A Wiley-Interscience publication."
Bibliography: p.
Includes index.
ISBN 0-471-85601-0
1. Plates (Engineering). 2. Elastic plates and shells. I. Title.
TA660.P6R44 1988
624.1'7765--dc 19

Printed in the United States of America

10 9 8 7 6 5 4 3 2 1

*The scientist must put things in order; science is made of facts as a house is made of stones, but science is no more an accumulation of facts than a house is a pile of stones.*

*Henri Poincaré*

*It has been long understood that approximate solutions of problems in the ordinary branches of Natural Philosophy may be obtained by a species of abstractions, or rather limitations of the data, such as enables us easily to solve the modified form of the question, while we are well assured that the circumstances (so modified) affect the result only in a superficial manner.*

*Lord Kelvin and Peter Guthrie Tait,*
*"Treatise on Natural Philosophy"*

*Practice is best served by a good theory.*

*L. Boltzmann*

# PREFACE

The omnipresence of plates and platelike structures in modern technology is well known and needs no particular elaboration. Whether the concern is with aircraft and missile surface (skin) components, reinforced concrete floor slabs, glass-window panes, electric circuit boards, or certain layered geological formations, engineers and analysts are frequently called upon to predict deformations, stresses, natural frequencies, and stability of elastic plates. These topics require a thorough knowledge of the fundamental equations of plate theory as well as an understanding of the various methods of solution. Most of the existing textbooks treat this subject in an "ad hoc" manner—the usual "strength of materials" approach—and then proceed to provide the reader with a catalog of solutions of a large number of specific problems.

This book approaches the subject of elastic plates in a somewhat different manner. From the outset, the theory of plates is treated as a branch (or subspecies) of classical, three-dimensional elasticity theory. Thus, the first chapter covers the pertinent mathematical topics, and an appropriate notation (Cartesian tensors) is developed. Chapter 2 is a brief review of elasticity theory, and it includes a discussion of the variational principles employed in subsequent chapters (principle of virtual work, Hamilton's principle, etc.) Chapters 3 and 4 are concerned with the derivation of basic plate theory: definition of stress resultants, stress-strain, stress-displacement relations, equations of equilibrium, curvilinear coordinates, and so on. Both vectorial and scalar (or variational) methods are used, and the treatment is not restricted to classical plate theory. The more recently developed theory that accounts for shear as well as flexural deformations is also included. Chapter 5 presents a number of (analytical) solutions of the static plate equations (boundary value problems). The selection was based on the importance of the method of solution as well as the specific technical significance of the problem. Chapter 6 is concerned with dynamical problems within the context of plate theory: (a) the dynamics of plates of bounded extent and (b) wave motion in plates. The former considers the free vibrations and forced motion, while the latter examines the dynamics of plates of unbounded extent. This chapter also answers some basic questions raised in Chapters 3 and 4. Chapter 7 treats static and dynamic problems of plates in a (given) state of initial membrane stress. In a most natural manner, this topic leads to a discussion of the instability of elastic plates. Chapter 8 serves as an introduction to the approximate, numerical methods of solution of plate deformation problems. We discuss both finite element as well as finite difference methods, and applications are presented to problems of statics, dynamics, and stability. In Chapter 9, elastic plate problems where geometric, nonlinear effects predominate are formulated, and some solutions are presented. Chapter 10, written by Dr. Robert Wetherhold, serves as an introduction to the currently popular topic of laminated (composite) plates.

This book is designed to present, in concise form, the basic fundamentals of plate theory within a reasonably rigorous context and to explain the methods of solution applicable to the full spectrum of problems encountered by engineers and physical scientists. Thus, it is the author's hope that a well-prepared student who masters the content can proceed from fundamentals to the ability to treat new problems or research topics in the relatively short time span of one semester. To facilitate understanding and to fix the ideas contained in this text, the reader is urged to solve a substantial number of the problems appended to each chapter. These problems range, in difficulty, from trivial algebraic computations to minor research projects. They will be of special value to readers who are forced to study on their own and who are not associated with an educational institution. Some of the problems contain important results not covered in the text. For instance, thermo-elastic plate theory is relegated to exercises with suitable hints as an extension of the textual material.

The theory of plates has a colorful history. Classical plate theory was initiated by Mlle. Sophie Germain (1776–1831) in direct response to a prize offered by the French Academy (1811) for the explanation of the nodal curves of a vibrating plate, as demonstrated (experimentally) by E. Chladni (1756–1829) of Saxony (see the figure). After two attempts, Mlle. Germain received the prize in 1816 but only after Lagrange, a member of the examination committee, corrected her initially submitted paper. Subsequently, a controversy ensued about the appropriate, associated boundary conditions, and this was settled approximately 34 years after the correct partial differential equations were discovered. No less than the authorities G. R. Kirchhoff (1824–1887) and Lord Kelvin (William Thomson) (1824–1907) were responsible for this part of the theory. The names of Cauchy, Poisson, Navier, and Clebsch are also associated with numerous, early contributions. In modern times, major contributions were made by E. Reissner, R. D. Mindlin, and Y. S. Uflyand. A detailed study of the history of elastic plates is really a story of the evolution and development of a segment of classical mathematical (theoretical) physics. Many of the mathematical techniques, such as the Rayleigh-Ritz method, variational methods, double Fourier series, perturbation methods, eigenvalue-eigenvector problems, and so on, were initially developed to affect solutions of the static and dynamical problems of elastic plate theory.

At least three organizations influenced the present undertaking. The Space Vehicle Division of Martin-Marietta Corporation provided a variety of challenging problems within the context of "real world" applications. The Office of Scientific Research of the U.S. Air Force provided the financial assistance that, for many years, supported a number of research efforts during the course of which the idea for the present book was conceived. The author is also indebted to the State University of New York at Buffalo, where he has delivered many lectures on the theory of plates and shells, and to the many graduate students who wrote theses and dissertations in this area under the direction of the author. Without a doubt, they had an influence upon the present book.

The author acknowledges the specific contributions of Dr. D. P. Malone [Fig. 5.7(b)], Dr. M. W. Richman (Fig. 7.9), Dr. L. S. Segerlind (FORTRAN subroutines BDYVAL, DCMPBC, and SLVBD in Section 8.6), Mr. E. Dull (Figs. 5.14 and 5.16), and Mr. E. K. Marshman and Miss K. G. Solney (Table 8.1 and Fig. 8.17). Finally, the author wishes to thank Mrs. Ginger Adessa for her expertise and stamina in providing the initial typescript.

HERBERT REISMANN

*Williamsville, New York*
*May, 1988*

## Wetherhold's Preface

Contributor's Note: I was fortunate enough to be invited to add Chapter 10 on Laminated Plates to this book, and appreciate the invitation. The study of laminated plates is of great current interest in a variety of applications, including the aerospace and automotive fields, where high stiffness and strength to weight ratios are important. Several students assisted in the proofreading of this chapter, including Jerry Fritz, Dave Thomas, Lalit Jain, and John Bryant. I also thank Marty Fye for her excellent and patient typing efforts.

ROBERT C. WETHERHOLD

*Amherst, New York*
*May, 1988*

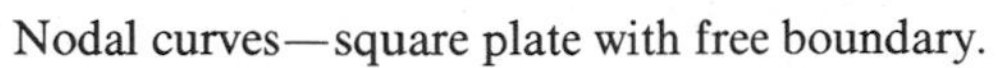

Nodal curves—square plate with free boundary.

# CONTENTS

# ELASTIC PLATES

**Theory and Application**

# 1

# MATHEMATICAL PRELIMINARIES

The present chapter is designed to acquaint the reader with rules and techniques of index notation and fundamentals of Cartesian tensors. The related topics of coordinate rotations and curvilinear coordinates are also covered. Chapter 1 establishes the "language" of this book. It may be skipped or used as a reference by those who have had previous exposure to this material.

## 1.1 INDEX NOTATION

The mathematical structure of continuum mechanics, elasticity theory, the theory of plates, shells, and so on, displays certain patterns and characteristics that make it convenient to introduce a special notation, often referred to as "index notation." Although these subjects can be developed using conventional notation, their development with the aid of index notation results in a more concise, transparent presentation and greatly assists in the discovery of important patterns and occasionally hidden symmetries. In addition, with the aid of this notation, many otherwise tedious and lengthy algebraic computations become trivial, and it often brings out the structural simplicity of formulas that look formidable and complex when expanded into conventional notation.

We begin by considering the expression

$$S = A_1B_1 + A_2B_2 + A_3B_3 \tag{1.1a}$$

where $A_i$, $B_i$, $i = 1, 2, 3$, are real numbers or integers. We can use a capital sigma as a summation symbol to write (1.1a) in the form

$$S = \sum_{i=1}^{3} A_iB_i \tag{1.1b}$$

In many physical applications it is tacitly understood that we need to sum over the repeated index $i$ from 1 to 3, and in such cases the sum (1.1b) is frequently written

$$S = A_i B_i \tag{1.1c}$$

In the case of (1.1c) and other, similar cases, the following rule is operative: If, in an expression, a subscript is repeated, then we must sum over all values of the subscript. In the case of (1.1c) we have

$$S \equiv A_i B_i \equiv \sum_{i=1}^{3} A_i B_i \equiv A_1 B_1 + A_2 B_2 + A_3 B_3 \tag{1.2}$$

The double-subscript summation convention was originally proposed by A. Einstein (1879–1955).

Let us next consider the expression

$$a_{ij} x_j = b_i, \qquad i = 1, 2, 3; \quad j = 1, 2, 3 \tag{1.3a}$$

In this case we have a repeated index $j$ and a free index $i$. If we apply the summation rule to the repeated index, we obtain (summing on the $j$'s)

$$a_{i1} x_1 + a_{i2} x_2 + a_{i3} x_3 = b_i, \qquad i = 1, 2, 3 \tag{1.3b}$$

and if we now successively assign the values $i = 1, 2, 3$ in (1.3b), we obtain

$$\begin{aligned} a_{11} x_1 + a_{12} x_2 + a_{13} x_3 &= b_1 \\ a_{21} x_1 + a_{22} x_2 + a_{23} x_3 &= b_2 \\ a_{31} x_1 + a_{32} x_2 + a_{33} x_3 &= b_3 \end{aligned} \tag{1.3c}$$

Thus, the system of three equations in the three unknowns $x_i$, $i = 1, 2, 3$, has the deceptively simple representation (1.3a) if the double-subscript summation convention is employed.

The summation convention can also be applied to the case of multiple sums, although more than one repetition of a subscript is not permitted. For example, consider the expression

$$R = A_{ij} B_{ij} \tag{1.4a}$$

First, we sum on $j$:

$$R = A_{i1} B_{i1} + A_{i2} B_{i2} + A_{i3} B_{i3} \tag{1.4b}$$

Next, we sum on $i$ for each of the three terms in (1.4b):

$$\begin{aligned} R = {} & A_{11} B_{11} + A_{21} B_{21} + A_{31} B_{31} + A_{12} B_{12} + A_{22} B_{22} + A_{32} B_{32} \\ & + A_{13} B_{13} + A_{23} B_{23} + A_{33} B_{33} \end{aligned} \tag{1.4c}$$

Equation (1.4c) is the fully expanded form of (1.4a), and it should be clear that

(1.4a) is a double sum that is frequently written as

$$R = \sum_{i=1}^{3} \sum_{j=1}^{3} A_{ij} B_{ij} \tag{1.4d}$$

This scheme is readily extended to other cases. For example, the reader is urged to show that the equation

$$A_{ij} = B_{ijkl} C_{kl} \begin{Bmatrix} i = 1, 2, 3 \\ j = 1, 2, 3 \\ k = 1, 2, 3 \\ l = 1, 2, 3 \end{Bmatrix} \tag{1.5}$$

requires summations over $k$ and $l$ (double sum), and in the fully expanded form there will be nine equations, the right-hand side of each equation consisting of the addition of nine product terms.

There are many other creative applications of the repeated index summation convention. For example, we can write

$$C_{ii} \equiv C_{11} + C_{22} + C_{33} \tag{1.6a}$$

$$\frac{\partial f_k}{\partial x_k} \equiv \frac{\partial f_1}{\partial x_1} + \frac{\partial f_2}{\partial x_2} + \frac{\partial f_3}{\partial x_3} \tag{1.6b}$$

$$\equiv f_{k,k} \equiv f_{1,1} + f_{2,2} + f_{3,3} \tag{1.6c}$$

The summation convention does not apply to expressions of the type $A_i \pm B_i$. It should also be noted that the summation index must not appear more than twice in any one expression. In the case of a repeated subscript, the actual symbol used is immaterial. For example,

$$A_i B_i \equiv A_k B_k \equiv A_j B_j, \quad \text{etc.}$$

$$C_{ii} \equiv C_{kk} \equiv C_{jj}, \quad \text{etc.}$$

$$f_{i,i} \equiv f_{k,k} \equiv f_{j,j}, \quad \text{etc.}$$

In the present book, we shall make the following distinction:

(a) A repeated, lowercase Latin subscript is summed from 1 to 3.
(b) A repeated, lowercase Greek subscript is summed from 1 to 2.

For example,

$$A_k B_k \equiv A_1 B_1 + A_2 B_2 + A_3 B_3$$

and

$$A_\alpha B_\alpha \equiv A_1 B_1 + A_2 B_2$$

We now introduce some notational devices that will enhance the convenience and dexterity of the repeated subscript summation convention. The Kronecker delta (Leopold Kronecker, 1827–1891) is defined as follows:

$$\delta_{ij} = 1 \quad \text{if } i = j, \qquad \delta_{ij} = 0 \quad \text{if } i \neq j \tag{1.7a}$$

The values of the Kronecker delta can be displayed by the matrix equation

$$\begin{bmatrix} \delta_{11} & \delta_{12} & \delta_{13} \\ \delta_{21} & \delta_{22} & \delta_{23} \\ \delta_{31} & \delta_{32} & \delta_{33} \end{bmatrix} = \begin{bmatrix} 1 & 0 & 0 \\ 0 & 1 & 0 \\ 0 & 0 & 1 \end{bmatrix} \tag{1.7b}$$

Let us consider the expression

$$\delta_{ij}A_i = \delta_{1j}A_1 + \delta_{2j}A_2 + \delta_{3j}A_3$$

According to (1.7), we have

$$\begin{aligned} \delta_{ij}A_i &= A_1 \quad \text{for } j = 1 \\ \delta_{ij}A_i &= A_2 \quad \text{for } j = 2 \\ \delta_{ij}A_i &= A_3 \quad \text{for } j = 3 \end{aligned}$$

and therefore we conclude that

$$\delta_{ij}A_i = A_j \tag{1.8}$$

Equation (1.8) reveals why the symbol $\delta_{ij}$ is frequently called the "substitution operator." Multiplication of $A_i$ by $\delta_{ij}$ and subsequent summation with respect to the repeated subscript $i$ results in the substitution (replacement) of the subscript $i$ by the subscript $j$ in $A_i$. This same property also leads to the relations

$$\delta_{ii} = \delta_{11} + \delta_{22} + \delta_{33} = 3 \tag{1.9a}$$

$$\delta_{ij}\delta_{jk} = \delta_{ik} \tag{1.9b}$$

Another notational device of great utility is provided by the alternator $e_{ijk}$:

$$e_{ijk} = \tfrac{1}{2}(j-k)(k-i)(i-j) \quad (\text{no sum on } i,\ j, \text{ and } k) \tag{1.10a}$$

By direct substitution in (1.10a), we obtain

$$e_{ijk} = \left\{ \begin{array}{rl} 1 & \text{if } (ijk) = (123), (312), (231) \\ -1 & \text{if } (ijk) = (321), (132), (213) \\ 0 & \text{if two or more subscripts are equal} \end{array} \right\} \tag{1.10b}$$

It can be shown that the following useful relations between the Kronecker delta and the alternator hold:

$$e_{ijk}e_{pqr} = \begin{vmatrix} \delta_{ip} & \delta_{iq} & \delta_{ir} \\ \delta_{jp} & \delta_{jq} & \delta_{jr} \\ \delta_{kp} & \delta_{kq} & \delta_{kr} \end{vmatrix} \tag{1.11a}$$

$$e_{ijk}e_{kqr} = \begin{vmatrix} \delta_{iq} & \delta_{ir} \\ \delta_{jq} & \delta_{jr} \end{vmatrix} = \delta_{iq}\delta_{jr} - \delta_{ir}\delta_{jq} \tag{1.11b}$$

For subsequent developments it will be convenient to define the two-dimensional Kronecker delta as

$$\delta_{\alpha\beta} = 1 \quad \text{for } \alpha = \beta, \qquad \delta_{\alpha\beta} = 0 \quad \text{for } \alpha \neq \beta \tag{1.12a}$$

so that

$$[\delta_{\alpha\beta}] = \begin{bmatrix} \delta_{11} & \delta_{12} \\ \delta_{21} & \delta_{22} \end{bmatrix} = \begin{bmatrix} 1 & 0 \\ 0 & 1 \end{bmatrix} \tag{1.12b}$$

Similarly, the two-dimensional alternator is defined by

$$e_{\alpha\beta} = \beta - \alpha \tag{1.13a}$$

or, equivalently,

$$\begin{bmatrix} e_{11} & e_{12} \\ e_{21} & e_{22} \end{bmatrix} = \begin{bmatrix} 0 & 1 \\ -1 & 0 \end{bmatrix} \tag{1.13b}$$

The following relations are readily established (see Exercises 1.11 and 1.12):

$$e_{\alpha\beta} = \begin{vmatrix} \delta_{\alpha 1} & \delta_{\alpha 2} \\ \delta_{\beta 1} & \delta_{\beta 2} \end{vmatrix} = \begin{vmatrix} \delta_{\alpha 1} & \delta_{\beta 1} \\ \delta_{\alpha 2} & \delta_{\beta 2} \end{vmatrix} \tag{1.14a}$$

$$e_{\alpha\beta}e_{\xi\eta} = \begin{vmatrix} \delta_{\alpha\xi} & \delta_{\alpha\eta} \\ \delta_{\beta\xi} & \delta_{\beta\eta} \end{vmatrix} \tag{1.14b}$$

$$e_{\alpha\gamma}e_{\beta\gamma} = \delta_{\alpha\beta}; \qquad e_{\alpha\beta}e_{\alpha\beta} = 2 \tag{1.14c}$$

## 1.2 COORDINATE ROTATIONS

We shall restrict the present discussion to right-handed, rectangular, Cartesian coordinate systems the axes of which are labeled $(x, y, z) \equiv (x_1, x_2, x_3)$. The origin $O$ is located at the intersection of the coordinate axes. The (unit) basis

**TABLE 1.1 Direction Cosines (Rotation of Cartesian Axes)**

| | $\hat{\mathbf{e}}_1$ | $\hat{\mathbf{e}}_2$ | $\hat{\mathbf{e}}_3$ |
|---|---|---|---|
| $\hat{\mathbf{e}}_{1'}$ | $a_{1'1}$ | $a_{1'2}$ | $a_{1'3}$ |
| $\hat{\mathbf{e}}_{2'}$ | $a_{2'1}$ | $a_{2'2}$ | $a_{2'3}$ |
| $\hat{\mathbf{e}}_{3'}$ | $a_{3'1}$ | $a_{3'2}$ | $a_{3'3}$ |

vectors associated with the axes $(x_1, x_2, x_3)$ are $(\hat{\mathbf{e}}_1, \hat{\mathbf{e}}_2, \hat{\mathbf{e}}_3)$, respectively. The position vector $\mathbf{r}$ characterizes the location of a generic point $P(x_1, x_2, x_3)$ relative to $O$. When referred to our axis system, we can write $\mathbf{r} = \hat{\mathbf{e}}_i x_i$. The numbers $x_i$, $i = 1, 2, 3$, are the scalar components of the vector $\mathbf{r}$. We now consider another, right-handed Cartesian coordinate system $x_{1'}, x_{2'}, x_{3'}$ (the "primed" system) with associated basis vectors $\hat{\mathbf{e}}_{1'}, \hat{\mathbf{e}}_{2'}, \hat{\mathbf{e}}_{3'}$. The primed and the unprimed coordinate systems share the same origin $O$. The relative orientation of the two coordinate systems is characterized by the nine direction cosines

$$a_{p'i} = \cos(\hat{\mathbf{e}}_{p'}, \hat{\mathbf{e}}_i) = \cos(x_{p'}, x_i) = \hat{\mathbf{e}}_{p'} \cdot \hat{\mathbf{e}}_i \tag{1.15}$$

and we can construct the table of direction cosines (Table 1.1). In view of (1.15) we have

$$a_{p'i} = \cos(\hat{\mathbf{e}}_{p'}, \hat{\mathbf{e}}_i) = \cos(\hat{\mathbf{e}}_i, \hat{\mathbf{e}}_{p'}) = a_{ip'} \tag{1.16}$$

We are now ready to frame the key question of the present section: How can we express the components $(x_{1'}, x_{2'}, x_{3'})$ of the position vector $\mathbf{r}$ relative to the primed system in terms of the components $(x_1, x_2, x_3)$ relative to the unprimed system? We have $\mathbf{r} = \hat{\mathbf{e}}_i x_i = \hat{\mathbf{e}}_{q'} x_{q'}$ (why?). Hence, $\hat{\mathbf{e}}_i x_i = \hat{\mathbf{e}}_{q'} x_{q'}$, and we dot multiply both sides of this equation by $\mathbf{e}_{p'}$. In view of (1.15) we obtain

$$x_{p'} = a_{p'i} x_i \tag{1.17a}$$

If we relabel the coordinate axes in (1.17a), we readily obtain

$$x_i = a_{ip'} x_{p'} \tag{1.17b}$$

We next wish to find formulas that express the unit vectors $\hat{\mathbf{e}}_{p'}$ in terms of the unit vectors $\hat{\mathbf{e}}_i$. In view of (1.17b), we have

$$\hat{\mathbf{e}}_i x_i = \hat{\mathbf{e}}_i a_{ip'} x_{p'} = \hat{\mathbf{e}}_{p'} x_{p'}$$

or

$$(\hat{\mathbf{e}}_{p'} - a_{p'i} \hat{\mathbf{e}}_i) x_{p'} = 0$$

However, the vector $x_{p'}$ is arbitrary, so that

$$\hat{\mathbf{e}}_{p'} = a_{p'i}\hat{\mathbf{e}}_i \tag{1.18a}$$

By relabeling the coordinates in (1.18a), we obtain

$$\hat{\mathbf{e}}_i = a_{ip'}\hat{\mathbf{e}}_{p'} \tag{1.18b}$$

In view of (1.7), (1.8), and (1.18a) we have

$$\begin{aligned}\delta_{p'q'} &= \hat{\mathbf{e}}_{p'} \cdot \hat{\mathbf{e}}_{q'} = \left(a_{p'i}\hat{\mathbf{e}}_i\right) \cdot \left(a_{q'j}\hat{\mathbf{e}}_j\right) \\ &= \delta_{ij}a_{p'i}a_{q'j} = a_{p'i}a_{q'i}\end{aligned} \tag{1.19a}$$

or, upon relabeling the coordinates,

$$\delta_{ij} = a_{ip'}a_{jp'} \tag{1.19b}$$

With reference to (1.19a), when $p' = q'$, this equation expresses the fact that the base vectors $\hat{\mathbf{e}}_{p'}$ are unit vectors, or $|\hat{\mathbf{e}}_{p'}| = 1$. When $p' \neq j'$, it expresses the fact that the unit vectors $\hat{\mathbf{e}}_{p'}$ and $\hat{\mathbf{e}}_{q'}$ are mutually perpendicular. For these reasons (1.19a) [or (1.19b)] are known as orthonormality relations. With reference to (1.19), it should be clear that the nine direction cosines $a_{ij'}$ are not independent but are related by the six equations (1.19). We also note the following useful relations, which follow directly from the definition of the dot and cross products and from (1.7) and (1.10):

$$\hat{\mathbf{e}}_i \cdot \hat{\mathbf{e}}_j = \delta_{ij} \tag{1.20}$$

$$\hat{\mathbf{e}}_i \times \hat{\mathbf{e}}_j = \hat{\mathbf{e}}_k e_{ijk} \tag{1.21}$$

We now consider the special case of a rotation about the $z$-axis:

$$x_{\xi'} = a_{\xi'\alpha}x_\alpha \tag{1.20a}$$

$$x_{z'} = x_z \tag{1.20b}$$

Expanding (1.20a), we obtain

$$\begin{aligned}x_{1'} &= a_{1'1}x_1 + a_{1'2}x_2 \\ x_{2'} &= a_{2'1}x_1 + a_{2'2}x_2\end{aligned} \tag{1.21a}$$

With reference to Fig. 1.1 we have

$$\begin{bmatrix} a_{1'1} & a_{1'2} \\ a_{2'1} & a_{2'2} \end{bmatrix} = \begin{bmatrix} \cos\theta & \sin\theta \\ -\sin\theta & \cos\theta \end{bmatrix} \tag{1.22}$$

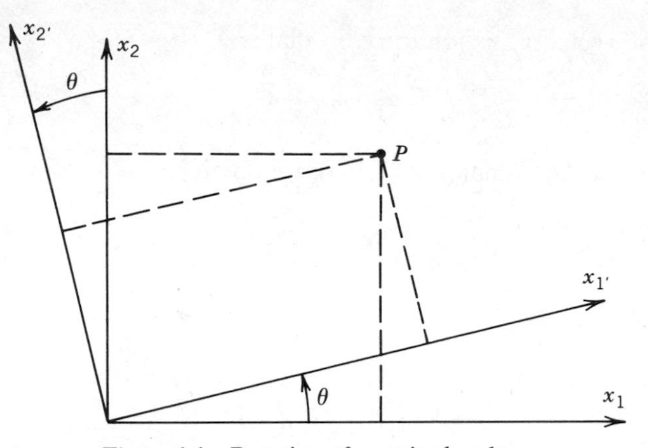


**Figure 1.1** Rotation of axes in the plane.

so that

$$\begin{aligned} x_{1'} &= x_1 \cos\theta + x_2 \sin\theta \\ x_{2'} &= -x_1 \sin\theta + x_2 \cos\theta \end{aligned} \tag{1.23}$$

Equations (1.23) are the usual relations between the coordinates $(x_{1'}, x_{2'})$ of the primed system and $(x_1, x_2)$ of the unprimed system, both characterizing the position of the point $P$ in the plane. The relative rotation of the two coordinate systems is specified by the angle $\theta$, as shown in Fig. 1.1.

## 1.3 ORTHOGONAL CURVILINEAR COORDINATES IN THE PLANE

Many boundary value problems of plate theory become (analytically) tractable only if the basic equations are referred to a curvilinear coordinate system. For this reason we now develop the basic notions of orthogonal curvilinear coordinates in the plane. In addition, we shall find a representation of the basic invariant differential operators when referred to orthogonal, curvilinear coordinates. To achieve these goals, let us express the Cartesian coordinates in the plane $(x, y)$ in terms of the curvilinear coordinates $(\alpha, \beta)$ by means of the equations

$$x = x(\alpha, \beta), \qquad y = y(\alpha, \beta) \tag{1.24}$$

where $x(\alpha, \beta)$ and $y(\alpha, \beta)$ denote single-valued functions of $\alpha, \beta$ with continuous partial derivatives of the first order in some given region $R$. The functions $x$ and $y$ must be independent, and the condition for this is that the

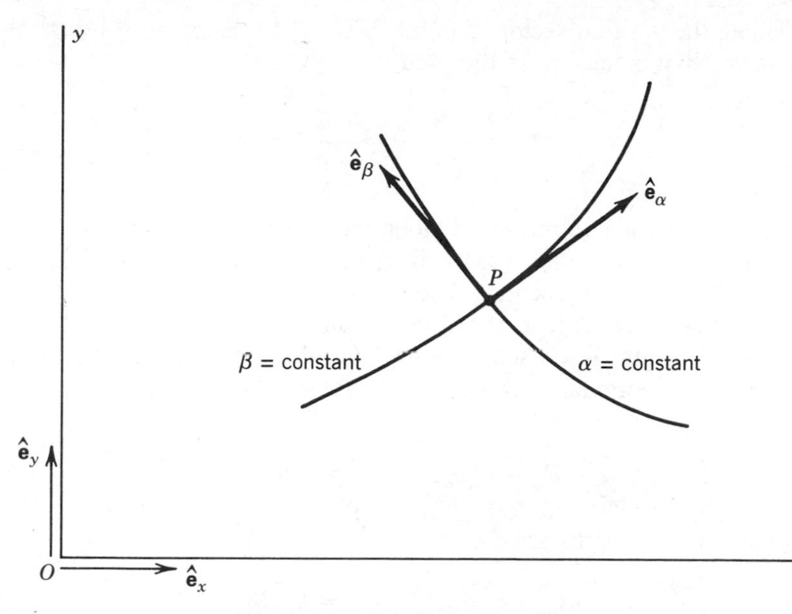


**Figure 1.2** Curvilinear coordinates.

Jacobian determinant

$$J = \begin{vmatrix} \dfrac{\partial x}{\partial \alpha} & \dfrac{\partial y}{\partial \alpha} \\ \dfrac{\partial x}{\partial \beta} & \dfrac{\partial y}{\partial \beta} \end{vmatrix} \neq 0 \text{ in } R \tag{1.25}$$

If these conditions prevail, it can be shown that $\alpha$ and $\beta$ can be expressed as single-valued functions of $(x, y)$, with continuous partial derivatives of the first order. Condition (1.25) may fail to be valid at isolated singular points, and there the analysis will require particular care.

A point $P$ in two-dimensional space can be characterized either by a pair of numbers $(x, y)$ or by $(\alpha, \beta)$, and we shall assume that this characterization is provided in a unique manner by (1.24) and its inverse. If $P$ moves such that only $\alpha$ varies and $\beta$ is held constant, it will describe a curve in the plane. This curve is known as the $\alpha$-curve. In a similar manner, we can define a $\beta$-curve, as shown in Fig. 1.2. In the following we shall limit our consideration to coordinate curves that are mutually perpendicular at each point $P$ of the plane. Such a coordinate system is called an orthogonal, curvilinear coordinate system. We now construct tangent unit vectors $\hat{\mathbf{e}}_\alpha$ and $\hat{\mathbf{e}}_\beta$ at $P$ such that $\hat{\mathbf{e}}_\alpha$ $(\hat{\mathbf{e}}_\beta)$ is tangent to the $\alpha$ $(\beta)$ curve at $P$, and we have

$$\hat{\mathbf{e}}_\alpha \cdot \hat{\mathbf{e}}_\beta = 0, \qquad |\hat{\mathbf{e}}_\alpha| = |\hat{\mathbf{e}}_\beta| = 1, \qquad \hat{\mathbf{e}}_\alpha \times \hat{\mathbf{e}}_\beta = \hat{\mathbf{k}} \tag{1.26}$$

If we denote the position vector from the origin $O$ to the point $P$ by $OP = \mathbf{R}$, then a vector that is tangent to the $\alpha$-curve is given by

$$\mathbf{E}_\alpha \equiv \frac{\partial \mathbf{R}}{\partial \alpha} = \frac{\partial \mathbf{R}}{\partial s_\alpha}\frac{ds_\alpha}{d\alpha} \tag{1.27}$$

where $s_\alpha$ is the arc length measured along the $\alpha$-curve. [*Note:* the double-subscript summation convention has been suspended in (1.27) as well as in the remainder of the present section.] The quantity $\hat{\mathbf{e}}_\alpha = \partial \mathbf{R}/\partial s_\alpha$ characterizes a unit vector that is tangent to the $\alpha$-curve, and it points in the direction of increasing arc length. If we set $h_\alpha \equiv ds_\alpha/d\alpha$, (1.27) assumes the form $\mathbf{E}_\alpha = h_\alpha \hat{\mathbf{e}}_\alpha$, and there are similar equations for the $\beta$-curve. Thus, the tangent vectors are

$$\mathbf{E}_\alpha = h_\alpha \hat{\mathbf{e}}_\alpha, \qquad \mathbf{E}_\beta = h_\beta \hat{\mathbf{e}}_\beta \tag{1.28}$$

the differential arc lengths are

$$ds_\alpha = h_\alpha\, d\alpha, \qquad ds_\beta = h_\beta\, d\beta \tag{1.29}$$

and it can be inferred that $h_\alpha$ and $h_\beta$ serve as scale factors associated with the $\alpha$- and $\beta$-curves, respectively. We have

$$\mathbf{R} = \hat{\mathbf{e}}_x x + \hat{\mathbf{e}}_y y \tag{1.30a}$$

$$\frac{\partial \mathbf{R}}{\partial \alpha} = \hat{\mathbf{e}}_x \frac{\partial x}{\partial \alpha} + \hat{\mathbf{e}}_y \frac{\partial y}{\partial \alpha} \tag{1.30b}$$

and in view of (1.28), (1.29), and (1.30b) we can write

$$|\mathbf{E}_\alpha| = h_\alpha = \left|\frac{\partial \mathbf{R}}{\partial \alpha}\right| = \frac{ds_\alpha}{d\alpha} = \sqrt{\left(\frac{\partial x}{\partial \alpha}\right)^2 + \left(\frac{\partial y}{\partial \alpha}\right)^2}$$

and a similar expression for $|\mathbf{E}_\beta|$. Consequently,

$$h_\alpha^2 = \left(\frac{\partial x}{\partial \alpha}\right)^2 + \left(\frac{\partial y}{\partial \alpha}\right)^2 \tag{1.31a}$$

$$h_\beta^2 = \left(\frac{\partial x}{\partial \beta}\right)^2 + \left(\frac{\partial y}{\partial \beta}\right)^2 \tag{1.31b}$$

Since

$$d\mathbf{R} = \frac{\partial \mathbf{R}}{\partial \alpha} d\alpha + \frac{\partial \mathbf{R}}{\partial \beta} d\beta = \hat{\mathbf{e}}_\alpha h_\alpha\, d\alpha + \hat{\mathbf{e}}_\beta h_\beta\, d\beta = \hat{\mathbf{e}}_x\, dx + \hat{\mathbf{e}}_y\, dy$$

**TABLE 1.2 Direction Cosines (Curvilinear Coordinates)**

| | $\hat{\mathbf{e}}_x$ | $\hat{\mathbf{e}}_y$ |
|---|---|---|
| $\hat{\mathbf{e}}_\alpha$ | $\dfrac{1}{h_\alpha}\dfrac{\partial x}{\partial \alpha}$ | $\dfrac{1}{h_\alpha}\dfrac{\partial y}{\partial \alpha}$ |
| $\hat{\mathbf{e}}_\beta$ | $\dfrac{1}{h_\beta}\dfrac{\partial x}{\partial \beta}$ | $\dfrac{1}{h_\beta}\dfrac{\partial y}{\partial \beta}$ |

the square of the distance $PQ \equiv d\mathbf{R}$ between two neighboring points $P$ and $Q$ in the plane is

$$ds^2 = d\mathbf{R} \cdot d\mathbf{R} = dx^2 + dy^2 \tag{1.32a}$$

or

$$ds^2 = h_\alpha^2\, d\alpha^2 + h_\beta^2\, d\beta^2 \tag{1.32b}$$

The direction cosine of the angle between the unit vectors $\hat{\mathbf{e}}_\alpha$ and $\hat{\mathbf{e}}_x$ is provided by

$$\cos(\hat{\mathbf{e}}_\alpha, \hat{\mathbf{e}}_x) = \hat{\mathbf{e}}_\alpha \cdot \hat{\mathbf{e}}_x = \frac{1}{h_\alpha}\frac{\partial \mathbf{R}}{\partial \alpha} \cdot \hat{\mathbf{e}}_x = \frac{1}{h_\alpha}\frac{\partial x}{\partial \alpha}$$

where we have used (1.30b). In this manner we can construct the set of four direction cosines relating the basis $(\hat{\mathbf{e}}_\alpha, \hat{\mathbf{e}}_\beta)$ to the basis $(\hat{\mathbf{e}}_x, \hat{\mathbf{e}}_y)$, and the results are presented in Table 1.2. With the aid of this table we obtain

$$h_\alpha \hat{\mathbf{e}}_\alpha = \frac{\partial x}{\partial \alpha}\hat{\mathbf{e}}_x + \frac{\partial y}{\partial \alpha}\hat{\mathbf{e}}_y \tag{1.33a}$$

$$h_\beta \hat{\mathbf{e}}_\beta = \frac{\partial x}{\partial \beta}\hat{\mathbf{e}}_x + \frac{\partial y}{\partial \beta}\hat{\mathbf{e}}_y \tag{1.33b}$$

or, inverting (1.33),

$$\hat{\mathbf{e}}_x = \frac{1}{h_\alpha}\frac{\partial x}{\partial \alpha}\hat{\mathbf{e}}_\alpha + \frac{1}{h_\beta}\frac{\partial x}{\partial \beta}\hat{\mathbf{e}}_\beta \tag{1.34a}$$

$$\hat{\mathbf{e}}_y = \frac{1}{h_\alpha}\frac{\partial y}{\partial \alpha}\hat{\mathbf{e}}_\alpha + \frac{1}{h_\beta}\frac{\partial y}{\partial \beta}\hat{\mathbf{e}}_\beta \tag{1.34b}$$

The derivatives of the basis vectors are linear combinations of the basis vectors. For example,

$$\frac{\partial \hat{\mathbf{e}}_\alpha}{\partial \beta} = c_\alpha \hat{\mathbf{e}}_\alpha + c_\beta \hat{\mathbf{e}}_\beta$$

where $c_\alpha$ and $c_\beta$ are functions of $\alpha$ and $\beta$. Hence, it can be shown (see Exercise 1.20) that

$$\begin{bmatrix} \dfrac{\partial \hat{\mathbf{e}}_\alpha}{\partial \alpha} & \dfrac{\partial \hat{\mathbf{e}}_\alpha}{\partial \beta} \\[2ex] \dfrac{\partial \hat{\mathbf{e}}_\beta}{\partial \alpha} & \dfrac{\partial \hat{\mathbf{e}}_\beta}{\partial \beta} \end{bmatrix} = \begin{bmatrix} -\dfrac{1}{h_\beta}\dfrac{\partial h_\alpha}{\partial \beta}\hat{\mathbf{e}}_\beta & \dfrac{1}{h_\alpha}\dfrac{\partial h_\beta}{\partial \alpha}\hat{\mathbf{e}}_\beta \\[2ex] \dfrac{1}{h_\beta}\dfrac{\partial h_\alpha}{\partial \beta}\hat{\mathbf{e}}_\alpha & -\dfrac{1}{h_\alpha}\dfrac{\partial h_\beta}{\partial \alpha}\hat{\mathbf{e}}_\alpha \end{bmatrix} \tag{1.35}$$

In view of (1.26), (1.33), (1.25) we have

$$(h_\alpha \hat{\mathbf{e}}_\alpha) \times (h_\beta \hat{\mathbf{e}}_\beta) = h_\alpha h_\beta \hat{\mathbf{k}} = \left(\hat{\mathbf{e}}_x \frac{\partial x}{\partial \alpha} + \hat{\mathbf{e}}_y \frac{\partial y}{\partial \alpha}\right) \times \left(\hat{\mathbf{e}}_x \frac{\partial x}{\partial \beta} + \hat{\mathbf{e}}_y \frac{\partial y}{\partial \beta}\right) = \hat{\mathbf{k}} J$$

and we conclude [see (1.25)] that

$$J = h_\alpha h_\beta = \begin{vmatrix} \dfrac{\partial x}{\partial \alpha} & \dfrac{\partial y}{\partial \alpha} \\[2ex] \dfrac{\partial x}{\partial \beta} & \dfrac{\partial y}{\partial \beta} \end{vmatrix} \neq 0 \tag{1.36}$$

Also, in view of (1.33), we have

$$(h_\alpha \hat{\mathbf{e}}_\alpha) \cdot (h_\beta \hat{\mathbf{e}}_\beta) = h_\alpha h_\beta (\hat{\mathbf{e}}_\alpha \cdot \hat{\mathbf{e}}_\beta) = \frac{\partial x}{\partial \alpha}\frac{\partial x}{\partial \beta} + \frac{\partial y}{\partial \alpha}\frac{\partial y}{\partial \beta} = 0 \tag{1.37}$$

and (1.37) serves as a check on the orthogonality of the $(\alpha, \beta)$ system of curvilinear coordinates. The vector area of an element is [see (1.28), (1.29), and (1.26)]

$$d\mathbf{A} = \mathbf{E}_\alpha \, d\alpha \times \mathbf{E}_\beta \, d\beta = (h_\alpha \hat{\mathbf{e}}_\alpha \, d\alpha) \times (h_\beta \hat{\mathbf{e}}_\beta \, d\beta) = h_\alpha h_\beta \, d\alpha \, d\beta \, \hat{\mathbf{k}}$$

so that the area element is

$$dA = |d\mathbf{A}| = h_\alpha h_\beta \, d\alpha \, d\beta = J \, d\alpha \, d\beta \tag{1.38}$$

where $J$ is the Jacobian of the transformation (1.24).

We now proceed to express the gradient operator in terms of orthogonal curvilinear coordinates $\alpha$ and $\beta$. The gradient vector **A** of the scalar field $\phi(x, y)$ is

$$\mathbf{A} = \mathbf{grad}\,\phi = \hat{\mathbf{e}}_x \frac{\partial \phi}{\partial x} + \hat{\mathbf{e}}_y \frac{\partial \phi}{\partial y} = \hat{\mathbf{e}}_\alpha A_\alpha + \hat{\mathbf{e}}_\beta A_\beta = \nabla \phi \tag{1.39}$$

and therefore,

$$A_\alpha = \hat{\mathbf{e}}_\alpha \cdot \left( \hat{\mathbf{e}}_x \frac{\partial \phi}{\partial x} + \hat{\mathbf{e}}_y \frac{\partial \phi}{\partial y} \right) = \frac{1}{h_\alpha} \left( \frac{\partial x}{\partial \alpha} \frac{\partial \phi}{\partial x} + \frac{\partial y}{\partial \alpha} \frac{\partial \phi}{\partial y} \right) = \frac{1}{h_\alpha} \frac{\partial \phi}{\partial x}$$

where we have used Table 1.2. A similar result is obtained for $A_\beta$, and we have

$$\mathbf{A} = \mathbf{grad}\,\phi = \frac{\hat{\mathbf{e}}_\alpha}{h_\alpha} \frac{\partial \phi}{\partial \alpha} + \frac{\hat{\mathbf{e}}_\beta}{h_\beta} \frac{\partial \phi}{\partial \beta} = \nabla \phi$$

and the (two-dimensional) gradient operator referred to orthogonal, curvilinear coordinates is

$$\nabla \equiv \frac{\hat{\mathbf{e}}_\alpha}{h_\alpha} \frac{\partial}{\partial \alpha} + \frac{\hat{\mathbf{e}}_\beta}{h_\beta} \frac{\partial}{\partial \beta} \tag{1.40}$$

In view of (1.39) and with the aid of (1.35) we obtain (see Exercise 1.20)

$$\nabla \cdot \mathbf{A} \equiv \mathbf{div}\,\mathbf{A} = \frac{1}{h_\alpha h_\beta} \left[ \frac{\partial}{\partial \alpha} (h_\beta A_\alpha) + \frac{\partial}{\partial \beta} (h_\alpha A_\beta) \right] \tag{1.41}$$

Using (1.40) and (1.35), we can readily show that

$$\nabla^2 \phi \equiv \nabla \cdot \nabla \phi = \frac{1}{h_\alpha h_\beta} \left[ \frac{\partial}{\partial \alpha} \left( \frac{h_\beta}{h_\alpha} \frac{\partial \phi}{\partial \alpha} \right) + \frac{\partial}{\partial \beta} \left( \frac{h_\alpha}{h_\beta} \frac{\partial \phi}{\partial \beta} \right) \right] \tag{1.42}$$

where $\nabla^2$ now is the Laplacian of $\phi$ referred to orthogonal, curvilinear coordinates $(\alpha, \beta)$.

We now present two specific examples of curvilinear coordinates.

**Example (a). Polar Coordinates:** In this case the coordinate transformation is given by $x + iy = \alpha e^{i\beta} \equiv re^{i\theta}$, $0 \le r < \infty$, $0 \le \theta < 2\pi$ $(\alpha = r,\ \beta = \theta)$, and $x = r\cos\theta$, $y = r\sin\theta$. The coordinate curves are circles $r =$ constant $(x^2 + y^2 = r^2)$ and radial lines $\theta =$ constant $[\theta = \tan^{-1}(y/x)]$, as shown in Fig. 1.3. Using (1.31), we obtain $h_r = 1$ and $h_\theta = r$, and from (1.32b) and (1.38) we

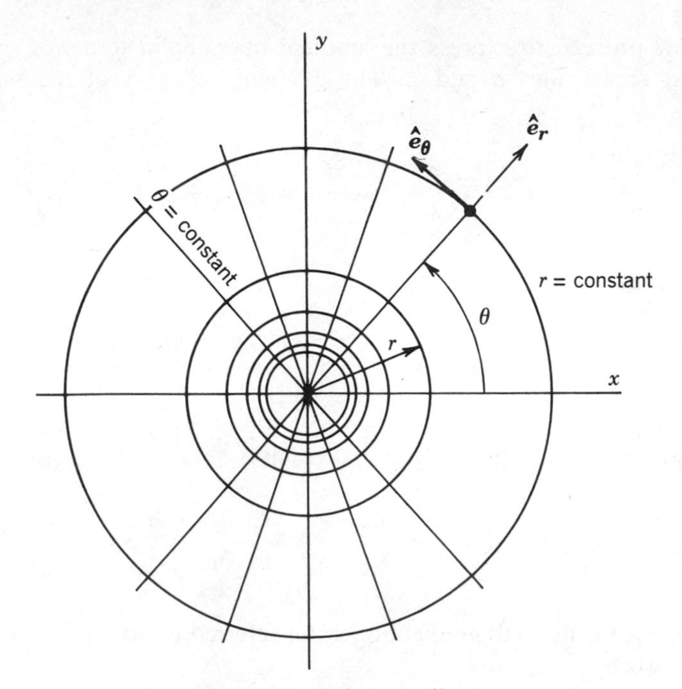


**Figure 1.3** Polar coordinates.

have $ds^2 = dr^2 + r^2\,d\theta^2$ and $dA = r\,dr\,d\theta$, respectively. In the present case, the direction cosines are listed in Table 1.3 (see Table 1.1).

We now employ (1.35) to calculate the derivatives of the unit vectors:

$$\begin{bmatrix} \dfrac{\partial \hat{\mathbf{e}}_r}{\partial r} & \dfrac{\partial \hat{\mathbf{e}}_r}{\partial \theta} \\ \dfrac{\partial \hat{\mathbf{e}}_\theta}{\partial r} & \dfrac{\partial \hat{\mathbf{e}}_\theta}{\partial \theta} \end{bmatrix} = \begin{bmatrix} 0 & \hat{\mathbf{e}}_\theta \\ 0 & -\hat{\mathbf{e}}_r \end{bmatrix} \tag{1.43}$$

and with the aid of (1.40)–(1.43) we obtain

$$\mathbf{grad}\, f = \nabla \cdot f = \hat{\mathbf{e}}_r \frac{\partial f}{\partial r} + \hat{\mathbf{e}}_\theta \frac{1}{r} \frac{\partial f}{\partial \theta} \tag{1.44a}$$

**TABLE 1.3 Direction Cosines (Polar Coordinates)**

| | $\hat{\mathbf{e}}_x$ | $\hat{\mathbf{e}}_y$ |
|---|---|---|
| $\hat{\mathbf{e}}_\alpha$ | $\cos\theta$ | $\sin\theta$ |
| $\hat{\mathbf{e}}_\beta$ | $-\sin\theta$ | $\cos\theta$ |

$$\operatorname{div} \mathbf{A} = \nabla \cdot \mathbf{A} = \frac{1}{r}\frac{\partial}{\partial r}(rA_r) + \frac{1}{r}\frac{\partial A_\theta}{\partial \theta} \tag{1.44b}$$

$$\nabla^2 f = \nabla \cdot \nabla f = \frac{1}{r}\frac{\partial}{\partial r}\left(r\frac{\partial f}{\partial r}\right) + \frac{1}{r^2}\frac{\partial^2 f}{\partial \theta^2} \tag{1.44c}$$

where $f(\alpha, \beta)$ and $\mathbf{A} = \hat{\mathbf{e}}_r A_r(\alpha, \beta) + \hat{\mathbf{e}}_\theta A_\theta(\alpha, \beta)$ are scalar and vector fields, respectively.

**Example (b). Elliptic Coordinates:** We follow the procedure used in Example (a). The reader is urged to verify the results. The coordinate transformation is $x + iy = c\cosh(\alpha + i\beta), 0 \leq \alpha < \infty, -\pi < \beta \leq \pi$. Hence, $x = c\cosh\alpha\cos\beta$, $y = c\sinh\alpha\sin\beta$, and the coordinate curves are

$$\frac{x^2}{c^2\cosh^2\alpha} + \frac{y^2}{c^2\sinh^2\alpha} = 1 = \cos^2\beta + \sin^2\beta$$

$$\frac{x^2}{c^2\cos^2\beta} - \frac{y^2}{c^2\sin^2\beta} = 1 = \cosh^2\alpha - \sinh^2\alpha$$

The ellipses corresponding to constant values of $\alpha$ are confocal, the distance between foci being $2c$. The curves corresponding to constant values of $\beta$ are hyperbolas confocal with one another and with the ellipses, as shown in Fig. 1.4. The semi-axes of the ellipses are $a, b$ where $a = c\cosh\alpha$, $b = c\sinh\alpha$, and $a^2 - b^2 = c^2$. We have

$$h_\alpha = h_\beta = c\sqrt{\cosh^2\alpha - \cos^2\beta} \tag{1.45}$$

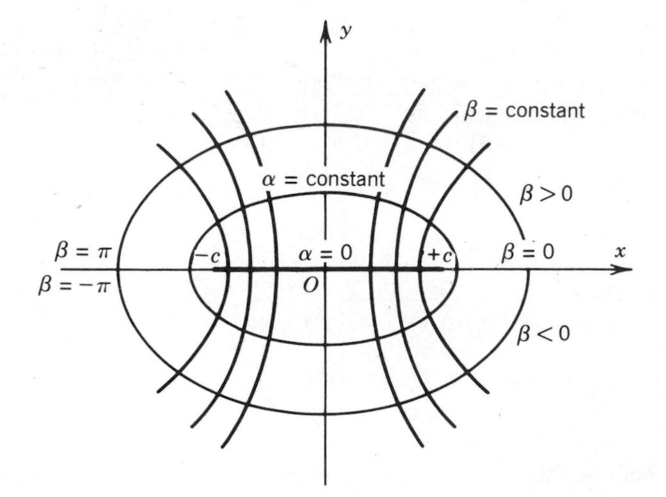


**Figure 1.4** Elliptic coordinates.

**TABLE 1.4 Direction Cosines (Elliptic Coordinates)**

| | $\hat{\mathbf{e}}_x$ | $\hat{\mathbf{e}}_y$ |
|---|---|---|
| $\hat{\mathbf{e}}_\alpha$ | $\dfrac{\sinh\alpha\cos\beta}{\sqrt{\cosh^2\alpha - \cos^2\beta}}$ | $\dfrac{\cosh\alpha\sin\beta}{\sqrt{\cosh^2\alpha - \cos^2\beta}}$ |
| $\hat{\mathbf{e}}_\beta$ | $-\dfrac{\cosh\alpha\sin\beta}{\sqrt{\cosh^2\alpha - \cos^2\beta}}$ | $\dfrac{\sinh\alpha\cos\beta}{\sqrt{\cosh^2\alpha - \cos^2\beta}}$ |

and the direction cosines are listed in Table 1.4. When $\alpha \to \infty$, Table 1.4 reverts to the values in Table 1.3, with $\alpha = r$ and $\beta = \theta$. The square of the arc length and element of area are

$$ds^2 = c^2(\cosh^2\alpha - \cos^2\beta)(d\alpha^2 + d\beta^2) \tag{1.46a}$$

$$dA = c^2(\cosh^2\alpha - \cos^2\beta)\, d\alpha\, d\beta \tag{1.46b}$$

respectively. The derivatives of the basis vectors are

$$\begin{bmatrix} \dfrac{\partial\hat{\mathbf{e}}_\alpha}{\partial\alpha} & \dfrac{\partial\hat{\mathbf{e}}_\beta}{\partial\beta} \\ \dfrac{\partial\hat{\mathbf{e}}_\beta}{\partial\alpha} & \dfrac{\partial\hat{\mathbf{e}}_\beta}{\partial\beta} \end{bmatrix} = \begin{bmatrix} \dfrac{\hat{\mathbf{e}}_\beta\sin\beta\cos\beta}{\cosh^2\alpha - \cos^2\beta} & \dfrac{\hat{\mathbf{e}}_\beta\sinh\alpha\cosh\alpha}{\cosh^2\alpha - \cos^2\beta} \\ \dfrac{\hat{\mathbf{e}}_\alpha\sin\alpha\cos\beta}{\cosh^2\alpha - \cos^2\beta} & -\dfrac{\hat{\mathbf{e}}_\alpha\sinh\alpha\cosh\alpha}{\cosh^2\alpha - \cos^2\beta} \end{bmatrix} \tag{1.47}$$

The invariant operators are

$$\textbf{grad f} = \nabla f = \frac{1}{c\sqrt{\cosh^2\alpha - \cos^2\beta}}\left(\hat{\mathbf{e}}_\alpha\frac{\partial f}{\partial\alpha} + \hat{\mathbf{e}}_\beta\frac{\partial f}{\partial\beta}\right) \tag{1.48a}$$

$$\begin{aligned}\textbf{div A} = \nabla\cdot\mathbf{A} = \frac{1}{c^2(\cosh^2\alpha - \cos^2\beta)}&\left[\frac{\partial}{\partial\alpha}\left(c\sqrt{\cosh^2\alpha - \cos^2\beta}\cdot A_\alpha\right)\right.\\ &\left.+\frac{\partial}{\partial\beta}\left(c\sqrt{\cosh^2\alpha - \cos^2\beta}\cdot A_\beta\right)\right]\end{aligned} \tag{1.48b}$$

$$\nabla^2 f \equiv \nabla\cdot\nabla f = \frac{1}{c^2(\cosh^2\alpha - \cos^2\beta)}\left(\frac{\partial^2 f}{\partial\alpha^2} + \frac{\partial^2 f}{\partial\beta^2}\right) \tag{1.48c}$$

where $f = f(\alpha, \beta)$ and $\mathbf{A} = \hat{\mathbf{e}}_\alpha A_\alpha(\alpha, \beta) + \hat{\mathbf{e}}_\beta A_\beta(\alpha, \beta)$ are scalar and vector fields, respectively.

We conclude this section with the following observation: In the limiting case of Cartesian coordinates, $\alpha = x$, $\beta = y$, and in this case we have

$h_\alpha = h_\beta = 1$, and Table 1.2 of the direction cosines becomes

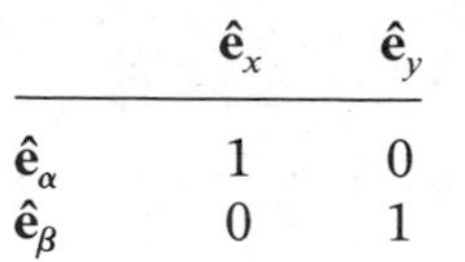

| | $\hat{\mathbf{e}}_x$ | $\hat{\mathbf{e}}_y$ |
|---|---|---|
| $\hat{\mathbf{e}}_\alpha$ | 1 | 0 |
| $\hat{\mathbf{e}}_\beta$ | 0 | 1 |

Thus, all basic expressions derived in this section revert to their (original) Cartesian form.

## 1.4 CARTESIAN TENSORS

Let us consider a vector $\mathbf{V}$ relative to the (Cartesian) basis vectors $\hat{\mathbf{e}}_i$, $i = 1, 2, 3$. In this case we can characterize $\mathbf{V}$ by the expression

$$\mathbf{V} = \hat{\mathbf{e}}_i V_i \tag{1.49}$$

where $V_i$, $i = 1, 2, 3$, are the scalar components of $\mathbf{V}$ relative to the basis $\hat{\mathbf{e}}_i$. Suppose we wish to find the scalar components $V_{p'}$ of the vector $\mathbf{V}$ relative to the basis $\hat{\mathbf{e}}_{p'}$, which is rotated with respect to the basis vectors $\hat{\mathbf{e}}_i$ in the sense of Section 1.2. Then we have

$$\hat{\mathbf{e}}_i V_i = \hat{\mathbf{e}}_{p'} V_{p'} \tag{1.50}$$

and since $\hat{\mathbf{e}}_i = a_{ip'} \hat{\mathbf{e}}_{p'}$, we have $V_i a_{ip'} \hat{\mathbf{e}}_{p'} = \hat{\mathbf{e}}_{p'} V_{p'}$, or

$$V_{p'} = a_{ip'} V_i = a_{p'i} V_i \tag{1.51}$$

Equation (1.51) will serve as the definition of a vector: A vector in three-dimensional space has three components relative to a Cartesian coordinate system, and for a rotation of the axis system characterized by the nine direction cosines $a_{p'i}$, the law of transformation is given by (1.51). Conversely, a set of three numbers (scalar components) that transform according to the law expressed by (1.51) are said to constitute a vector. This approach can be generalized, and it will be shown that a vector is a special case of an object called a tensor. Consider an expression of the type

$$\mathscr{T} = \hat{\mathbf{e}}_i \hat{\mathbf{e}}_j T_{ij} \tag{1.52}$$

Here, $T_{ij}$ are the (nine) tensor components, and $\hat{\mathbf{e}}_i$, $i = 1, 2, 3$, are again the basis vectors. However, the combination $\hat{\mathbf{e}}_i \hat{\mathbf{e}}_j$ is not a real product but a formal expression. For example, the expressions $\hat{\mathbf{e}}_1 \hat{\mathbf{e}}_2$ and $\hat{\mathbf{e}}_2 \hat{\mathbf{e}}_1$ are *not* dot products nor are they equal to one another. The symbol $\mathscr{T}$ in (1.52) serves as an invariant under rotation of axes and characterizes the tensor. Consider a new set of basis vectors $\hat{\mathbf{e}}_{p'}$ related to $\hat{\mathbf{e}}_i$, by the expression $\hat{\mathbf{e}}_i = a_{ip'}\hat{\mathbf{e}}_{p'}$. Then we have

$$\begin{aligned}\mathscr{T} &= T_{ij}\hat{\mathbf{e}}_i\hat{\mathbf{e}}_j = T_{p'q'}\hat{\mathbf{e}}_{p'}\hat{\mathbf{e}}_{q'} \\ &= T_{ij}a_{ip'}\hat{\mathbf{e}}_{p'}a_{jq'}\hat{\mathbf{e}}_{q'} = a_{p'i}a_{q'j}T_{ij}\hat{\mathbf{e}}_{p'}\hat{\mathbf{e}}_{q'}\end{aligned}$$

or

$$\left(T_{p'q'} - a_{p'i}a_{q'j}T_{ij}\right)\hat{\mathbf{e}}_{p'}\hat{\mathbf{e}}_{q'} = 0$$

But the basis $\hat{\mathbf{e}}_{p'}$ is arbitrary. Hence,

$$T_{p'q'} = a_{p'i}a_{q'j}T_{ij} \tag{1.53}$$

and this equation serves as the definition of a tensor of order two in a three-dimensional space. We note that such a tensor has $3 \times 3 = 9$ components. An example of a tensor of order two is provided by the Kronecker delta (see Section 1.1). In view of (1.18a) and (1.19a), we have

$$\delta_{p'q'} = \hat{\mathbf{e}}_{p'} \cdot \hat{\mathbf{e}}_{q'} = \left(a_{p'i}\hat{\mathbf{e}}_i\right) \cdot \left(a_{q'j}\hat{\mathbf{e}}_j\right) = a_{p'i}a_{q'j}\delta_{ij} \tag{1.54}$$

Comparison of (1.54) with (1.53) reveals that the nine scalar quantities $\delta_{ij}$ transform according to the rule expressed by (1.53). Hence, they are the components of a tensor of order two.

The generalization to tensors of higher order is straightforward. For example, a tensor of order three has the formal representation

$$\mathscr{U} = U_{ijk}\hat{\mathbf{e}}_i\hat{\mathbf{e}}_j\hat{\mathbf{e}}_k \tag{1.55}$$

and it can be shown that its components transform according to the rule

$$U_{p'q'r'} = a_{p'i}a_{q'j}a_{r'k}U_{ijk} \tag{1.56}$$

An example of a tensor of order three is provided by the alternator (1.10). It is easy to show (see Exercise 1.24b) that

$$e_{p'q'r'} = a_{p'i}a_{q'j}a_{r'k}e_{ijk} \tag{1.57}$$

In the present scheme of a "hierarchy of tensors," a vector is a tensor of order one, and a scalar is a tensor of order zero.

Whether or not a set of numbers are the components of a tensor can only be established by application of the laws of nature. For example, it can be shown that the nine stress components $\tau_{ij}$ constitute a tensor of order two. However, it is necessary to invoke the laws of statics to arrive at this conclusion. Furthermore, a law of nature must be form invariant and cannot depend on the particular choice of the coordinate system. For example, consider the stress equation of equilibrium

$$\tau_{ij,j} + F_i = 0 \tag{1.58a}$$

where $F_i$ is the body force (vector) per unit volume, and $\tau_{ij}$ is a tensor of order two, the stress tensor. With respect to rotated (primed) axes, we have

$$F_i = a_{ip'}F_{p'}, \qquad \tau_{ij} = a_{ip'}a_{jq'}\tau_{p'q'}$$

Furthermore, we have

$$\tau_{ij,j} = \frac{\partial \tau_{ij}}{\partial x_j} = \frac{\partial \tau_{ij}}{\partial x_{r'}}\frac{\partial x_{r'}}{\partial x_j} = a_{ip'}a_{jq'}a_{r'j}\tau_{p'q',r'} = a_{ip'}\tau_{p'q',q'}$$

and (1.58a) is transformed into

$$\left(\tau_{p'q',q'} + F_{p'}\right)a_{ip'} = 0$$

However, the (orthogonal) rotation of the coordinates is arbitrary. Consequently, we have

$$\tau_{p'q',q'} + F_{p'} = 0 \tag{1.58b}$$

A comparison of (1.58a) and (1.58b) reveals that both equations have the same structure. In fact, except for the primes, the two equations are identical. Hence, we can say that the stress equations of equilibrium are form invariant under arbitrary, orthogonal rotations of the coordinate system. This result should have been expected: The laws of nature [and (1.58) is such a law!] operate independently of the (man-made) coordinate system. Coordinate systems are introduced for analytical convenience. Thus, the form of an equation purporting to represent a law of nature must be the same for all (admissible) coordinate systems. In other words, the equation must display the property of form invariance with respect to any admissible set of coordinate transformations. It can be demonstrated that all Cartesian tensor equations possess this property. The form invariance of tensor equations is the inherent reason for their importance in the study of mechanics and, in a larger sense, in all of mathematical physics.

## 1.5 TENSOR ALGEBRA AND ANALYSIS

The definitions and considerations offered in Section 1.4 provide the motivation for the basic operations with tensors. The sum (or difference) of two tensors of the same order (and in the same space) is a tensor of that order. For example, $C_{ij} = A_{ij} \pm B_{ij}$. The addition or subtraction of tensors of different orders is meaningless. The outer product of a tensor of order $M$ with a tensor of order $N$ is a tensor of order $M + N$. For example, $H_{ijk} = A_i B_{jk}$ is the outer product of $A_i$ and $B_{jk}$. It is a tensor of order $1 + 2 = 3$. If two indices of a tensor are set equal and the result is summed according to the summation convention of Section 1.1, then we say that the indices have been contracted. A contraction performed on a tensor of order $N \geq 2$ results in a tensor of order $N - 2$. For example, $B_{ii}$ is the contraction of $B_{ij}$, and it is a tensor of order $2 - 2 = 0$, that is, a scalar. The inner product of a tensor is an outer product followed by a contraction. For example, $H_{iji} = A_i B_{ji}$ is an inner product of $A_i$ and $B_{jk}$.

A tensor of order two has $3^2 = 9$ components. If this tensor is symmetric with respect to the indices $i$ and $j$, it has only $9 - 3 = 6$ independent components in view of the relation $T_{ij} = T_{ji}$. Conversely, if $T_{ij}$ is antisymmetric (or skew symmetric) with respect to $i$ and $j$, we have $T_{ij} = -T_{ji}$, and there will be only $\frac{1}{2}(9 - 3) = 3$ independent components. Associated with every skew-symmetric second-order tensor $T_{ij}$ of order two is a dual vector $T_k$. If we set

$$T_{ij} = e_{ijk} T_k \tag{1.59}$$

then it follows from (1.59) and (1.10) that $T_{ij} = -T_{ji}$. We now multiply both sides of (1.59) by $e_{ijl}$ and obtain

$$e_{ijl} T_{ij} = e_{ijl} e_{ijk} T_k = 2\delta_{lk} T_k = 2T_l$$

where we have used (1.11b) and (1.8). We conclude that the components of the dual vector associated with the skew-symmetric tensor $T_{ij}$ are given by

$$T_k = \tfrac{1}{2} e_{ijk} T_{ij} \tag{1.60}$$

An arbitrary second-order tensor can always be decomposed into the sum of a symmetric and a skew-symmetric tensor. Let $T_{ij} = \varepsilon_{ij} + \omega_{ij}$, where $\varepsilon_{ij} = \varepsilon_{ji} = \frac{1}{2}(T_{ij} + T_{ji})$, and $\omega_{ij} = -\omega_{ji} = \frac{1}{2}(T_{ij} - T_{ji}) = e_{ijk} T_k$, where $T_k = \frac{1}{2} e_{ijk} T_{ij}$ by (1.60). It can be shown that this decomposition is unique (see Exercise 1.31).

The quotient laws for tensors are useful to establish "tensor character." As an example, consider the relation $A_{ij} B_i = C_j$, where it is known that $B_i$ and $C_j$ are vectors and that $B_i$ is independent of $A_{ij}$. Form invariance dictates that $A_{p'q'} B_{p'} = C_{q'}$ where primes denote a rotated coordinate system. But $B_i = a_{p'i} B_{p'}$ and $C_{q'} = a_{q'j} C_j$, so that $A_{p'q'} B_{p'} = a_{q'j} C_j = a_{q'j} A_{ij} B_i = a_{p'i} a_{q'j} A_{ij} B_{p'}$ and

$$\left( A_{p'q'} - a_{p'i} a_{q'j} A_{ij} \right) B_{p'} = 0$$

But the vector $B_p$ is arbitrary, so that

$$A_{p'q'} = a_{p'i}a_{q'j}A_{ij}$$

Thus, with reference to (1.53) we can conclude that $A_{ij}$ is a Cartesian tensor of order two. The method employed in this example is readily extended to other cases. For example, let us consider the expression $S = T_{ijk}A_iB_jC_k$, where $S$ is a scalar and $A_i$, $B_j$, and $C_k$ are independent, arbitrary vectors. The reader is urged to show that these conditions imply that $T_{ijk}$ is a Cartesian tensor of order three (see Exercise 1.24a).

In any continuum theory, such as elasticity or plate theory, we encounter scalar, vector, and tensor fields. The gradient of a scalar and the divergence of a vector or tensor are expressions that occur frequently. Their importance can be attributed to the fact that they are used to characterize physical concepts that are entirely independent of the coordinate system employed. For example, consider the gradient operator $\nabla \equiv \hat{\mathbf{e}}_i(\partial/\partial x_i)$ operating upon the vector $\mathbf{Q} = \hat{\mathbf{e}}_jQ_j$. In this case we have

$$\mathbf{div}\,\mathbf{Q} = \nabla \cdot \mathbf{Q} = \left(\hat{\mathbf{e}}_i\frac{\partial}{\partial x_i}\right)\cdot(\hat{\mathbf{e}}_jQ_j) = \delta_{ij}Q_{j,i} = Q_{i,i} \tag{1.61}$$

and this shows that the divergence of a vector is a scalar. Similarly, if $\mathscr{T} = \hat{\mathbf{e}}_i\hat{\mathbf{e}}_jT_{ij}$ is a tensor of order two, we obtain

$$\mathbf{div}\,\mathscr{T} = \nabla \cdot (\hat{\mathbf{e}}_i\hat{\mathbf{e}}_jT_{ij}) = \hat{\mathbf{e}}_k\frac{\partial}{\partial x_k}\cdot(\hat{\mathbf{e}}_i\hat{\mathbf{e}}_jT_{ij}) = \delta_{ik}\hat{\mathbf{e}}_jT_{ij,k} = \hat{\mathbf{e}}_jT_{ij,i} = \mathbf{T}_{i,i} \tag{1.62}$$

where $\mathbf{T}_i = \hat{\mathbf{e}}_jT_{ij}$, and it can be inferred that the divergence of a tensor of order two is a vector. We recall the divergence theorem, which in component form is given by

$$\int_V A_{i,i}\,dV = \int_S A_in_i\,dS \tag{1.63a}$$

where $\mathbf{A} = \hat{\mathbf{e}}_iA_i$ and $\hat{\mathbf{n}} = \hat{\mathbf{e}}_in_i$. Here $S$ is the surface that encloses the volume $V$ and $\hat{\mathbf{n}}$ is the outer normal unit vector to $S$. The invariant form of (1.63a) is

$$\int_V \mathbf{div}\,\mathbf{A}\,dV = \int_S \mathbf{A}\cdot\hat{\mathbf{n}}\,dS \tag{1.63b}$$

In view of (1.62), we can also write

$$\int_V \mathbf{div}\,\mathscr{T}\,dV = \int_S \mathscr{T}\cdot\hat{\mathbf{n}}\,dS \tag{1.64}$$

where $\mathscr{T} = \hat{\mathbf{e}}_i\hat{\mathbf{e}}_jT_{ij}$ and $\mathscr{T}\cdot\hat{\mathbf{n}} = n_i\mathbf{T}_i$, and where $\mathbf{T}_i = \hat{\mathbf{e}}_jT_{ij}$. The two-dimensional counterpart of the divergence theorem is known as Green's theorem in the plane (G. Green, 1793–1841):

$$\int_A B_{\alpha,\alpha}\, dA = \oint_C B_\alpha n_\alpha\, ds \tag{1.65}$$

where $\mathbf{B} = \hat{\mathbf{e}}_\alpha B_\alpha$, $\hat{\mathbf{n}} = \hat{\mathbf{e}}_\alpha n_\alpha$, and $A$ is the (plane) area with $C$ the boundary and $\hat{\mathbf{n}}$ the outer unit normal to $C$. The integrands and regions of integration in (1.63)–(1.65) must be suitably restricted, and this type of information is contained in the appropriate texts on mathematics.

## EXERCISES

**1.1** **(a)** Write the following equation in conventional notation:

$$T_{ij,j} = F_i = 0$$

**(b)** Show that these equations are form invariant under orthogonal rotations of the coordinate system.

**1.2** **(a)** Write the following equation in conventional notation:

$$2E_{ij} = U_{i,j} - U_{j,i} + U_{k,i}U_{k,j}$$

**(b)** Show that these equations are form invariant under orthogonal rotations of the coordinate system.

**1.3** Give the unabridged form of each of the following expressions. If the indicial notation is used incorrectly, explain why.

**(a)** $A_i = B_i$

**(b)** $F_i = G_i + H_{ji}A_j$

**(c)** $U_i = V_j$

**(d)** $A_i = B_i + C_i D_i$

**(e)** $F_i = A_i + B_{ij}C_j D_j$

**(f)** $\phi = \dfrac{\partial F_i}{\partial x_i}$

**(g)** $d = \sqrt{x_i x_i}$

**(h)** $K = \delta_{ij}A_i B_j$

**1.4** Prove the following identities:

**(a)** $\delta_{ii} = 3$

**(b)** $\delta_{ij}C_{ij} = C_{ii}$

**(c)** $A_{ij}e_{ijk} = 0$ if $A_{ij} = A_{ji}$

**(d)** $A_{ij}B_{kl}\delta_{ik} = A_{ij}B_{il}$

**1.5** **(a)** If $f = f(x_1, x_2, x_3)$, show that $df/dt = (\partial f/\partial x_i)(dx_i/dt)$.

**(b)** Expand the double sum $S = A_{ij}x_ix_j$.

**(c)** Show that $(a_ix_i)_{,p} = a_p$.

**(d)** Show that $\delta_{ip}\delta_{jq}a_{pq} = a_{ij}$.

**1.6** Let $S = a_{ij}x_ix_j \equiv 0$ for all values of the variables $x_1, x_2, x_3$. Show that $a_{ij} = -a_{ji}$.

**1.7** Let $S = a_{ijk}x_ix_jx_k \equiv 0$ for all values of the variables $x_1, x_2, x_3$. Show that

$$a_{ijk} + a_{kij} + a_{jki} + a_{jik} + a_{jki} + a_{ikj} = 0$$

**1.8** **(a)** Show that $e_{ijk}a_{1i}a_{2j}a_{3k}$ is the expansion of the determinant $|a_{ij}|$ by rows.

**(b)** Show that $e_{ijk}a_{i1}a_{j2}a_{k3}$ is the expansion of the determinant $|a_{ij}|$ by columns.

**1.9** Show that

$$e_{ijk}e_{pqr}a_{pqr} = a_{ijk} + a_{kij} + a_{jki} - a_{jik} - a_{kji} - a_{ikj}$$

**1.10** Write the following equations in conventional notation:

$$A_{ij} = B_{ijkl}C_{kl}, \qquad i, j, k, l = 1, 2, 3$$

**1.11** Show that

**(a)**

$$e_{\alpha\beta} = \begin{vmatrix} \delta_{\alpha 1} & \delta_{\alpha 2} \\ \delta_{\beta 1} & \delta_{\beta 2} \end{vmatrix} = \begin{vmatrix} \delta_{\alpha 1} & \delta_{\beta 1} \\ \delta_{\alpha 2} & \delta_{\beta 2} \end{vmatrix}$$

**(b)**

$$e_{\alpha\beta}e_{\xi\eta} = \begin{vmatrix} \delta_{\alpha\xi} & \delta_{\alpha\eta} \\ \delta_{\beta\xi} & \delta_{\beta\eta} \end{vmatrix}$$

**1.12** Show that

**(a)** $e_{ijk}e_{rjk} = 2\delta_{ir}$

**(b)** $e_{ijk}e_{ijk} = 6$

**(c)** $e_{\alpha\gamma}e_{\beta\gamma} = \delta_{\alpha\beta}$

**(d)** $e_{\alpha\beta}e_{\alpha\beta} = 2$

**1.13** Show that

$$a = \det(a_{\alpha\beta}) = e_{\alpha\beta} a_{1\alpha} a_{2\beta} = e_{\alpha\beta} a_{\alpha 1} a_{\beta 2}$$
$$a e_{\gamma\eta} = e_{\alpha\beta} a_{\alpha\gamma} a_{\beta\eta} = e_{\alpha\beta} a_{\gamma\alpha} a_{\eta\beta}$$
$$a = \tfrac{1}{2} e_{\alpha\beta} e_{\gamma\eta} a_{\alpha\gamma} a_{\beta\eta}$$

**1.14** Show that

$$\mathbf{A} \times (\mathbf{B} \times \mathbf{C}) = (\mathbf{C} \cdot \mathbf{A})\mathbf{B} - (\mathbf{B} \cdot \mathbf{A})\mathbf{C}$$

Use index notation in your demonstration.

**1.15** If $\mathbf{D} = \mathbf{A} \times \mathbf{B}$, show that

$$D_k = e_{ijk} A_i B_j = \begin{vmatrix} A_1 & A_2 & A_3 \\ B_1 & B_2 & B_3 \\ \delta_{k1} & \delta_{k2} & \delta_{k3} \end{vmatrix}$$

**1.16** Two coordinate systems are connected by the direction cosines $a_{i'j}$, where

$$[a_{i'j}] = \begin{bmatrix} \frac{1}{\sqrt{3}} & \frac{1}{\sqrt{3}} & \frac{1}{\sqrt{3}} \\ \frac{1}{\sqrt{2}} & 0 & -\frac{1}{\sqrt{2}} \\ -\frac{1}{\sqrt{6}} & \frac{2}{3} & -\frac{1}{\sqrt{6}} \end{bmatrix}$$

The unprimed reference is rectangular, Cartesian, and right-handed.

**(a)** Are the primed coordinate axes mutually orthogonal?

**(b)** Is the primed reference right-handed or left-handed?

**1.17** Use 1.19 to write all orthogonality relations and all normalization conditions on the direction cosines $a_{p'i}$ in conventional notation.

**1.18** Consider a rotation of axes in the $x$-$y$ plane as shown in Fig. 1.1. Show that $\hat{\mathbf{e}}_{\beta'} = a_{\alpha\beta'} \hat{\mathbf{e}}_{\alpha}$, $\hat{\mathbf{e}}_{\beta} = a_{\alpha'\beta} \hat{\mathbf{e}}_{\alpha'}$, where $a_{\alpha\beta'} = a_{\beta'\alpha} = \cos(\hat{\mathbf{e}}_{\alpha}, \hat{\mathbf{e}}_{\beta'})$. Show that $a_{11'} = a_{22'} = \cos\theta$, $a_{1'2} = -a_{2'1} = \sin\theta$, and $a = \det(a_{\alpha\beta'}) = 1$.

**1.19** Consider two rectangular Cartesian systems (RCSs) one primed and the other unprimed. If the direction cosines characterizing the relative rotation of the two coordinate systems are $a_{i'k} = \cos(\hat{\mathbf{e}}_{i'}, \hat{\mathbf{e}}_k)$ show that $x_k = x_{i'} a_{i'k}$. Also show that $\hat{\mathbf{e}}_j = a_{i'j} \hat{\mathbf{e}}_{i'}$.

**1.20** With reference to (1.35), show that

$$\frac{\partial \hat{\mathbf{e}}_\alpha}{\partial \alpha} = -\frac{1}{h_\beta}\frac{\partial h_\alpha}{\partial \beta}\hat{\mathbf{e}}_\beta, \qquad \frac{\partial \hat{\mathbf{e}}_\beta}{\partial \beta} = \frac{1}{h_\beta}\frac{\partial h_\alpha}{\partial \beta}\hat{\mathbf{e}}_\alpha$$

$$\frac{\partial \hat{\mathbf{e}}_\alpha}{\partial \beta} = \frac{1}{h_\alpha}\frac{\partial h_\beta}{\partial \alpha}\hat{\mathbf{e}}_\beta, \qquad \frac{\partial \hat{\mathbf{e}}_\beta}{\partial \beta} = -\frac{1}{h_\alpha}\frac{\partial h_\beta}{\partial \alpha}\hat{\mathbf{e}}_\alpha$$

*Hint:* Assume that the derivatives of the basis vectors are linear combinations of the basis vectors. Use the relations

$$\hat{\mathbf{e}}_\alpha \cdot \frac{\partial \hat{\mathbf{e}}_\alpha}{\partial \alpha} = 0, \qquad \hat{\mathbf{e}}_\alpha \frac{\partial \hat{\mathbf{e}}_\beta}{\partial \alpha} = -\hat{\mathbf{e}}_\beta \frac{\partial \hat{\mathbf{e}}_\alpha}{\partial \alpha}$$

$$\frac{\partial^2 \mathbf{R}}{\partial \alpha\, \partial \beta} = \frac{\partial}{\partial \alpha}(h_\beta \hat{\mathbf{e}}_\beta) = \frac{\partial}{\partial \beta}(h_\alpha \hat{\mathbf{e}}_\alpha), \quad \text{etc.}$$

In the next two exercises:

**(a)** Find the scale factors $h_\alpha$, $h_\beta$.

**(b)** Find the element of arc length $ds$ and the element of area $dA$.

**(c)** Construct a table of direction cosines (see Table 1.2).

**(d)** Find grad $f$, **div A**, and $\nabla^2 f$ if $f = f(\alpha, \beta)$ and $\mathbf{A} = \hat{\mathbf{e}}_\alpha A_\alpha(\alpha, \beta) + \hat{\mathbf{e}}_\beta A_\beta(\alpha, \beta)$.

**(e)** Draw the curves $\alpha$ = constant and $\beta$ = constant.

**1.21** Parabolic coordinates $x + iy = \frac{1}{2}c(\alpha + i\beta)^2$, $-\infty < \alpha < \infty$, $0 \le \beta < \infty$.

**1.22** Bipolar coordinates $x + iy = c \tanh \frac{1}{2}(\alpha + i\beta)$, $-\infty < \alpha < \beta$, $-\pi < \beta < \pi$.

**1.23** In the expression $\mathscr{T} = \hat{\mathbf{e}}_i\hat{\mathbf{e}}_j T_{ij}$ each term (such as $\hat{\mathbf{e}}_1\hat{\mathbf{e}}_2 T_{12}$) is called a dyad, and the sum of dyads $\hat{\mathbf{e}}_i\hat{\mathbf{e}}_j T_{ij}$ is called a dyadic. We now define the conjugate of the dyadic $\mathscr{T}$ as $\mathscr{T}_c = \hat{\mathbf{e}}_j\hat{\mathbf{e}}_i T_{ij}$. Show that the expression $\mathscr{T} = \mathscr{T}_c$ implies that $T_{ij} = T_{ji}$, that is, the tensor $T_{ij}$ is symmetric.

**1.24** **(a)** Consider the equation $S = T_{ijk}A_iB_jC_k$, where $S$ is a scalar and $A_i$, $B_j$, and $C_k$ are independent, arbitrary vectors. Show that $T_{ijk}$ is a tensor of order three.

**(b)** Show that the alternator $e_{ijk}$ is a tensor of order three.

**1.25** **(a)** Show that $\delta_{\alpha\beta} = \delta_{\beta\alpha}$ and $e_{\alpha\beta} = -e_{\beta\alpha}$ are Cartesian tensors of order two in a two-dimensional space.

**(b)** Show that in a two-dimensional space there is a unique scalar $T = \frac{1}{2}e_{\alpha\beta}T_{\alpha\beta}$ associated with every skew-symmetric tensor of order two so that $T_{\alpha\beta} = e_{\alpha\beta}T$.

**1.26** Given a vector $A_i$ and two second-order tensors $B_{ij}$ and $C_{ij}$, prove that:

**(a)** $F_{ij} = B_{ij} + C_{ij}$ is a second-order tensor.

**(b)** $H_{ijk} = A_i B_{jk}$ is a third-order tensor.

**(c)** $B_{ii}$ is a scalar.

**(d)** $H_{iji} = A_i B_{ji}$ is a vector.

**1.27** **(a)** Given that $T_{ij}N_iN_j = S$, where $S$ is a scalar and $N_i$ is an arbitrary vector, show that $T_{ij}$ is a symmetric, second-order tensor.

**(b)** Given that $T_{ij}A_iB_j = S$, where $S$ is a scalar and $A_i$ and $B_i$ are independent and arbitrary vectors. Show that $T_{ij}$ is a tensor of order two.

**(c)** Given the equation $T_{ijk}A_iB_jC_k = S$, where $S$ is a scalar and $A_i$, $B_j$, and $C_k$ are independent and arbitrary vectors. Prove that $T_{ijk}$ is a tensor of order three.

**1.28** Consider the nine scalar quantities $I_{ij} = \int_V (r^2\delta_{ij} - x_ix_j)\,dm$, where $x_i$ are Cartesian coordinates, $r^2 = x_ix_i$, and the integration extends over the volume $V$ of a rigid body with element of mass $dm$. Prove that $I_{ij}$ is a symmetric tensor of order two, known as the inertia tensor in rigid body mechanics.

**1.29** Show that the relation $T_{ij} = T_{ji}$ implies $T_{p'q'} = T_{q'p'}$ if $T_{ij}$ is a tensor of order two. The primed RCS is obtained from the unprimed RCS by an orthogonal rotation.

**1.30** Given the equation $T_{ij} = C_{ijkl}E_{kl}$, where $T_{ij} = T_{ji}$ and $E_{kl} = E_{lk}$ are symmetric tensors of order two, show that the 81 scalars $C_{ijkl}$ are the components of a tensor of order four with the symmetry properties $C_{ijkl} = C_{jikl} = C_{ijlk}$.

**1.31** An arbitrary second-order tensor can always be decomposed into the sum of a symmetric and a skew-symmetric tensor. Prove that this decomposition is unique.

**1.32** Consider the central quadric form $c_{\alpha\beta}x_\alpha x_\beta = K$, where $c_{\alpha\beta} = c_{\beta\alpha}$, $x_\alpha$ is a vector from the origin to the quadric, and $K$ is a scalar. Show that $c_{\alpha\beta}$ is a symmetric tensor of order two. Let $n_\alpha = x_\alpha/|\mathbf{r}|$, where $\mathbf{r} = \hat{\mathbf{e}}_\alpha x_\alpha$ and $n_\alpha n_\alpha = 1$. Then $c_{\alpha\beta}n_\alpha n_\beta = K/|\mathbf{r}|^2 = S$. Now extremize $S$ subject to the constraint $n_\alpha n_\alpha = 1$. To facilitate a symmetric calculation, use the method of Lagrange multipliers (see, e.g., C. R. Wylie, *Advanced Engineering Mathematics*, 4th ed., McGraw-Hill, New York, 1975, p. 595.) Form the function $F = S - \lambda n_\alpha n_\alpha = (c_{\alpha\beta} - \lambda\delta_{\alpha\beta})n_\alpha n_\beta$, where $\lambda$

is the Lagrange multiplier. Then

$$\frac{\partial F}{\partial n_\beta} = (c_{\alpha\beta} - \lambda\delta_{\alpha\beta})n_\alpha = 0$$

or

$$\begin{bmatrix} c_{11} - \lambda & c_{12} \\ c_{21} & c_{22} - \lambda \end{bmatrix}\begin{bmatrix} n_1 \\ n_2 \end{bmatrix} = \begin{bmatrix} 0 \\ 0 \end{bmatrix}$$

Show that

$$2\lambda_1 = (c_{11} + c_{22}) - \sqrt{(c_{11} - c_{22})^2 + 4c_{12}^2}$$

$$2\lambda_2 = (c_{11} + c_{22}) + \sqrt{(c_{11} - c_{22})^2 + 4c_{12}^2}$$

$$n_1^{(1)} = \frac{-c_{12}}{\sqrt{(\lambda_1 - c_{11})^2 + c_{12}^2}} = a_{1'1} = \cos\theta$$

$$n_2^{(1)} = \frac{-(\lambda_1 - c_{11})}{\sqrt{(\lambda_1 - c_{11})^2 + c_{12}^2}} = a_{1'2} = \sin\theta$$

$$n_1^{(2)} = \frac{+c_{12}}{\sqrt{(\lambda_2 - c_{11})^2 + c_{12}^2}} = a_{12'} = -\sin\theta$$

$$n_2^{(2)} = \frac{+(\lambda_2 - c_{11})}{\sqrt{(\lambda_2 - c_{11})^2 + c_{12}^2}} = a_{22'} = \cos\theta$$

Show that $\lambda = c_{\alpha\beta}n_\alpha n_\beta = c_{\alpha'\beta}n_{\alpha'}n_{\beta'}$ where the $x_{1'}$-axis has direction cosines $(n_1^{(1)}, n_2^{(2)})$ and the $x_{2'}$-axis has direction cosines $(n_1^{(2)}, n_2^{(2)})$. Show that $n_{1'}^{(1)} = 1$, $n_{2'}^{(1)} = 0$, $n_{1'}^{(2)} = 0$, $n_{2'}^{(2)} = 1$, and therefore $c_{1'1'} = \lambda_1$, $c_{2'2'} = \lambda_2$, and $c_{1'2'} = c_{2'1'} = 0$. Show that, with respect to primed axes, $K = c_{\alpha'\beta'}x_{\alpha'}x_{\beta'} = \lambda_1(x_{1'})^2 + \lambda_2(x_{2'})^2$, or

$$\frac{(x_{1'})^2}{K/\lambda_1} + \frac{(x_{2'})^2}{K/\lambda_2} = 1$$

The primed axes are the principal axes of the central quadric. Identify the angle $\theta$. Show that $\hat{\mathbf{n}}^{(1)} \times \hat{\mathbf{n}}^{(2)} = \hat{\mathbf{e}}_1 \times \hat{\mathbf{e}}_2$.

**1.33** Given the central quadric $8x^2 - 12xy + 17y^2 = 20$. With reference to Exercise 1.32, do the following:

**(a)** Show that the equation of the central quadric assumes the form $K = c_{\alpha\beta}x_\alpha x_\beta = 20$, where

$$[c_{\alpha\beta}] = \begin{bmatrix} 8 & -6 \\ -6 & 17 \end{bmatrix}$$

**(b)** Show that

$$\lambda_1 = 5, \qquad n_1^{(1)} = \frac{2}{\sqrt{5}}, \qquad n_2^{(1)} = \frac{1}{\sqrt{5}}$$

$$\lambda_2 = 20, \qquad n_1^{(2)} = \frac{1}{\sqrt{5}}, \qquad n_2^{(2)} = -\frac{2}{\sqrt{5}}$$

where

$$\hat{\mathbf{n}}^{(1)} = \hat{\mathbf{e}}_{1'} = \frac{2}{\sqrt{5}}\hat{\mathbf{e}}_1 + \frac{1}{\sqrt{5}}\hat{\mathbf{e}}_2$$

$$\hat{\mathbf{n}}^{(2)} = \hat{\mathbf{e}}_{2'} = \frac{1}{\sqrt{5}}\hat{\mathbf{e}}_1 - \frac{2}{\sqrt{5}}\hat{\mathbf{e}}_2$$

$\theta = 26.565°$.

**(c)** Show that with respect to rotated (primed) axes

$$\frac{(x_{1'})^2}{(2)^2} + \frac{(x_{2'})^2}{(1)^2} = 1$$

that is, the given quadric characterizes an ellipse with semi-axes of length 2 and 1.

**1.34** Find the point on the curve $y^2 = 4x$ that is closest to the point $(1, 0)$. Use the method of Lagrange multipliers (see Exercise 1.32).

**1.35** Maximize the volume of a rectangular box if the area of its six sides is $A$. Find the dimensions of the box. Use the method of Lagrange multipliers (see Exercise 1.32).

**1.36** Let $S$ be the surface of the sphere $r = a$ and define the vector field $\mathbf{A} = r\hat{\mathbf{e}}_r$. Show that the divergence theorem (1.63) is satisfied for this special case.

**1.37** Let $A$ be the area of the circle $r = a$ and define the two-dimensional, plane vector field $\mathbf{B} = r\hat{\mathbf{e}}_r$. Show that Green's theorem in the plane (1.65) is satisfied for this special case.

**1.38** Find the area $A$ enclosed by the ellipse $x = a\cos\theta$, $y = b\sin\theta$. Use (1.65) and set $B_1 = x$, $B_2 = y$. Show that

$$2A = \oint_C (x\,dy - y\,dx) = 2\pi ab$$

**1.39** Prove Green's theorem in the plane (see 1.65).

**1.40** Set $B_1 = f(\partial g/\partial x_1)$ and $B_2 = f(\partial g/\partial x_2)$ in Green's theorem (1.65).
**(a)** Show that

$$B_1 n_1 + B_2 n_2 = f(\hat{\mathbf{n}} \cdot \nabla g) = f\frac{dg}{dn}$$

**(b)** Show that

$$\int_A f\nabla^2 g\,dA + \int_A \frac{\partial g}{\partial x_1}\frac{\partial f}{\partial x_1} + \frac{\partial g}{\partial x_2}\frac{\partial f}{\partial x_2}\,dA = \oint_C f\frac{dg}{dn}\,ds$$

*Note:* $dg/dn = \hat{\mathbf{n}} \cdot \nabla g$ is the directional derivative of the function $g$ in the direction of the vector $\hat{\mathbf{n}}$.

# 2

# SURVEY OF ELASTICITY THEORY

The present chapter contains an outline and summary of those portions of elasticity theory that will prove to be useful for the development of plate theory in subsequent chapters. The reader is probably familiar with many of the topics covered, but the style and notation adopted will facilitate a proper "fit" with succeeding chapters. The reader who desires to review details is urged to consult any book on elasticity theory or advanced strength of materials (e.g., H. Reismann and P. S. Pawlik, *Elasticity*: *Theory and Applications*, Wiley, New York, 1980).

## 2.1 DEFORMATION AND STRAIN

We consider a solid body in its reference configuration $B_0$. Each material point in $B_0$ is located by its position vector $\mathbf{r} = \hat{\mathbf{e}}_i x_i$, where $x_i$, $i = 1, 2, 3$, are its Cartesian coordinates. The unit vectors $\hat{\mathbf{e}}_i$ form an orthogonal triad and are aligned with the coordinate axes $x_i$. After the solid deforms, it occupies the region $B$, and the point originally at $\mathbf{r}$ is now located by the position vector $\mathbf{R} = \hat{\mathbf{e}}_i X_i$. Hence, the displacement vector is $\mathbf{u} = \hat{\mathbf{e}}_i u_i = \mathbf{R} - \mathbf{r}$, or $u_i = X_i - x_i$. The measure of strain is taken as $dR^2 - dr^2$, where $dr$ and $dR$ are the lengths of material line elements before and after deformation, respectively. Thus, we have

$$dR^2 - dr^2 = dX_i\, dX_i - dx_i\, dx_i \tag{2.1}$$

It is assumed that there is a one-to-one mapping of the region $B_0$ onto $B$, and if we employ the Lagrangian description, this mapping is provided by $X_i = X_i(x_1, x_2, x_3)$. Consequently,

$$dX_k = X_{k,i}\, dx_i \tag{2.2}$$

and upon substitution of (2.2) into (2.1), we obtain

$$dR^2 - dr^2 = 2L_{ij}\,dx_i\,dx_j$$

where

$$2L_{ij} = X_{k,i}X_{k,j} - \delta_{ij} \tag{2.3}$$

However, $X_k = x_k + u_k$, and we have

$$X_{k,i} = u_{k,i} + \delta_{ki} \tag{2.4}$$

Upon substitution of (2.4) into (2.3) and subsequent simplification, we obtain

$$2L_{ij} = u_{i,j} + u_{j,i} + u_{k,i}u_{k,j} \tag{2.5}$$

The nine numbers $L_{ij}$ constitute the Lagrangian strain tensor, and it is easy to show that $L_{ij} = L_{ji}$. When referred to Cartesian coordinates, the Lagrangian strain tensor has the representation

$$\mathscr{L} = L_{ij}\hat{\mathbf{e}}_i\hat{\mathbf{e}}_j \tag{2.6}$$

By direct computation, it can be shown that

$$2L_{ij}\,dx_i\,dx_j = 2\,d\mathbf{u}\cdot d\mathbf{r} + d\mathbf{u}\cdot d\mathbf{u} \tag{2.7}$$

Associated with the strain tensor $L_{ij}$ is a rotation tensor $\Omega_{ij}$ (formally) obtained as follows: Since $X_i = x_i + u_i$, we have

$$dX_i = dx_i + u_{i,j}\,dx_j = \left(\delta_{ij} + L_{ij} + \Omega_{ij}\right)dx_j$$

where

$$2\Omega_{ij} = u_{i,j} - u_{j,i} - u_{k,i}u_{k,j} \tag{2.8}$$

The set of numbers $\Omega_{ij}$ characterizes the Lagrangian rotation tensor. If the solid body moves without deformation, $L_{ij} \equiv 0$, and in this case the quantities $\Omega_{ij} + \delta_{ij}$ characterize rigid body motion. For many useful applications, the (geometrically) linear theory of elasticity is adequate. In this case we assume conditions of linear strain so that $|u_{i,j}| \ll 1$. In this case we have [see (2.5) and (2.8)]

$$2L_{ij} \cong u_{i,j} + u_{j,i} \equiv 2\varepsilon_{ij} \tag{2.9}$$

and

$$2\Omega_{ij} \cong u_{i,j} - u_{j,i} \equiv 2\omega_{ij} = -2\omega_{ji} \tag{2.10}$$

The "infinitesimal" rotation tensor $\omega_{ij}$ is antisymmetric and of order two. Thus, according to (1.60), for the case of sufficiently small rotations we have the (infinitesimal) rotation vector

$$\omega_k = \tfrac{1}{2} e_{ijk} \omega_{ij} \tag{2.11}$$

In the case of infinitesimal rigid body motion, we have the displacement $\mathbf{C} = \hat{\mathbf{e}}_i C_i$ as well as the rotation $\boldsymbol{\omega} = \hat{\mathbf{e}}_k \omega_k$, and therefore the displacement field is characterized by

$$\mathbf{u}(x, t) = \mathbf{C} + \mathbf{r} \times \boldsymbol{\omega} \tag{2.12a}$$

or

$$u_i(x, t) = C_i + e_{ijk} x_j \omega_k \tag{2.12b}$$

We note that the rigid body displacement field (2.12b) satisfies the equation $2\varepsilon_{ij} = u_{i,j} + u_{j,i} = 0$ identically; that is, the strain field vanishes for arbitrary rigid body motion of the solid. As an example, consider the infinitesimal rigid body rotation $\theta$ of a prism with rectangular cross-section about the $z$-axis (see Fig. 2.1). In this case $\mathbf{C} = \mathbf{0}$, $\mathbf{r} = \hat{\mathbf{e}}_x x + \hat{\mathbf{e}}_y y$, and $\boldsymbol{\omega} = \hat{\mathbf{e}}_z \theta$. Therefore, the displacement field is $\mathbf{u} = \mathbf{r} \times \boldsymbol{\omega} = \hat{\mathbf{e}}_x \cdot x \cdot \theta - \hat{\mathbf{e}}_y \cdot y \cdot \theta$, and we have $u_x = \theta x$, $u_y = -\theta y$, $u_z \equiv 0$. Note that $u_{x,x} = -u_{y,y} = \theta$ and therefore $|u_{x,x}| = |u_{y,y}| = |\theta|$, and this rotation is small provided $|\theta| \ll 1$.

Equation (2.9) relates the linearized strain field $\varepsilon_{ij}$ to the displacement field $u_i$. The following question is now posed: Given the (six) sufficiently well-behaved functions $\varepsilon_{ij}$, does the system of equations (2.9) characterize three equally well-behaved functions $u_i$? The answer is yes, if the following six

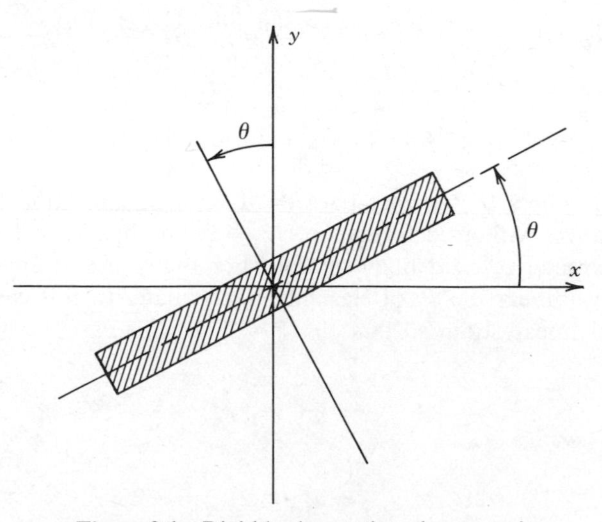


**Figure 2.1** Rigid body rotation about $z$-axis.

compatibility equations are satisfied:

$$e_{ipr}e_{jqs}\varepsilon_{pq,rs} = 0 \tag{2.13a}$$

If we characterize the strain tensor by its invariant form $\mathscr{E} = \hat{\mathbf{e}}_i\hat{\mathbf{e}}_j\varepsilon_{ij}$, then (2.13a) has the representation

$$\nabla \times (\nabla \times \mathscr{E})_c = \mathbf{0} \tag{2.13b}$$

where the subscript $c$ denotes the conjugate of the dyadic $\nabla \times \mathscr{E}$ (see Exercise 1.23). Equation (2.13a) can also be written as

$$\varepsilon_{il,jk} + \varepsilon_{jk,il} - \varepsilon_{ik,jl} - \varepsilon_{jl,ik} = 0 \tag{2.13c}$$

We note that (2.13c) are the six St. Venant equations of compatibility. They are necessary and sufficient conditions that ensure the existence of a single-valued, continuous displacement field for a simply connected region, resulting from a solution of (2.9). However, satisfaction of (2.13) does not ensure uniqueness of the displacement field because any infinitesimal rigid body motion (2.12) can be added to the displacement field without affecting the strain field.

## 2.2 STRESS AND EQUILIBRIUM

If we wish to specify the stress vector $\mathbf{T}$ at an interior point $P$ of a solid, we pass an imaginary plane $\Pi$ through $P$, and the orientation of the plane is characterized by its outer unit normal vector $\hat{\mathbf{n}} = \hat{\mathbf{e}}_i n_i$. The stress vector at $P$ is denoted by the symbol $\mathbf{T}(n)$, and it is formally defined by

$$\mathbf{T}(n) = \lim_{\Delta s \to 0} \frac{\Delta \mathbf{F}}{\Delta s} = \frac{d\mathbf{F}}{ds} \tag{2.14}$$

where $\Delta\mathbf{F}$ is the force resultant acting on the area $\Delta s$ in $\Pi$ (see Fig. 2.2). Cauchy's theorem states that

$$\mathbf{T}(\hat{\mathbf{n}}) = \mathbf{T}_i n_i \tag{2.15}$$

where $\mathbf{T}_i$ is the stress vector acting on a plane with normal $\hat{\mathbf{e}}_i$, that is,

$$\mathbf{T}_i \equiv \mathbf{T}(\hat{\mathbf{e}}_i) \tag{2.16}$$

If a solid occupies a region $B$ and is in equilibrium, then at every point we require

$$\left.\begin{aligned} \mathbf{T}_{i,i} + \mathbf{F} &= \mathbf{0} \quad &(2.17a)\\ \hat{\mathbf{e}}_i \times \mathbf{T}_i &= \mathbf{0} \quad &(2.17b) \end{aligned}\right\} \text{ in } B$$

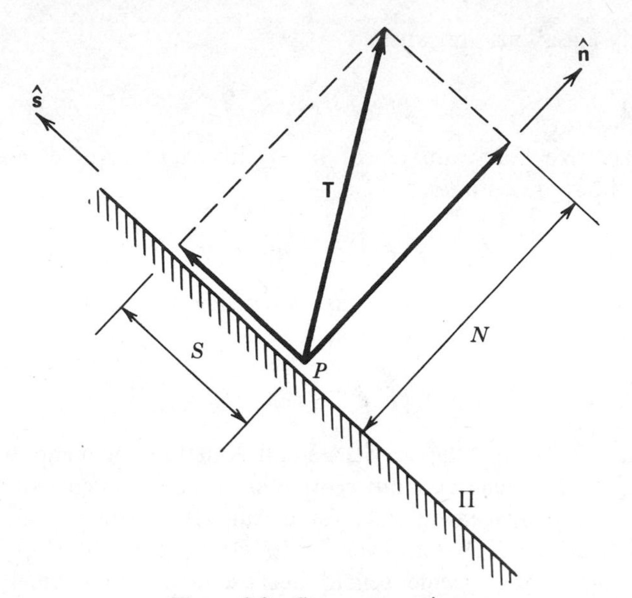


**Figure 2.2** Stress at a point.

where **F** is the body force per unit of volume. Equations (2.17) have the following invariant representation:

$$\left.\begin{aligned} \mathbf{div}\,\mathscr{T} + \mathbf{F} &= \mathbf{0} \\ \mathscr{T} &= \mathscr{T}_c \end{aligned}\right\} \quad \text{in } B \qquad \begin{aligned} &(2.18\text{a}) \\ &(2.18\text{b}) \end{aligned}$$

where

$$\mathscr{T} = \mathbf{T}_i \hat{\mathbf{e}}_i \tag{2.19}$$

is the stress tensor. We can also write

$$\mathbf{T}_i = \hat{\mathbf{e}}_j \tau_{ij} \tag{2.20}$$

where $\tau_{ij}$ are the (Cartesian) components of the stress vector $\mathbf{T}_i$. In view of (2.19) and (2.20), we have

$$\mathscr{T} = \mathbf{T}_i \hat{\mathbf{e}}_i = \hat{\mathbf{e}}_i \hat{\mathbf{e}}_j \tau_{ij} \tag{2.21}$$

and with reference to (2.21), we note that the nine numbers $\tau_{ij}$ are the components of the stress tensor referred to Cartesian coordinates. Upon substitution of (2.21) into (2.18), we obtain

$$\left.\begin{aligned} \tau_{ji,j} + F_i &= 0 \\ \tau_{ij} &= \tau_{ji} \end{aligned}\right\} \quad \text{in } B \qquad \begin{aligned} &(2.22\text{a}) \\ &(2.22\text{b}) \end{aligned}$$

We also note that the components of the stress vector relative to Cartesian axes are

$$\hat{\mathbf{e}}_k \cdot \mathbf{T} = n_i \tau_{ij} \delta_{kj} = n_i \tau_{ik} = T_k \tag{2.23}$$

The components of the stress vector normal and parallel to the plane Π (see Fig. 2.2) are

$$N = \mathbf{T} \cdot \hat{\mathbf{n}} = T_i n_i = n_i n_j \tau_{ij} \tag{2.24a}$$

$$S = \mathbf{T} \cdot \hat{\mathbf{s}} = T_i s_i = n_i s_j \tau_{ij} \tag{2.24b}$$

respectively.

Since $\tau_{ij}$ are the components of a tensor of order two, we have the law of transformation

$$\tau_{p'q'} = a_{p'i} a_{q'j} \tau_{ij} \tag{2.25}$$

(see 1.53), and it can be shown that there is at least one set of three mutually perpendicular axes at $P$ for which the terms $\tau_{p'q'}$, $p' \neq q'$, vanish. These axes are called principal axes, and we have

$$[\tau_{p'q'}] = \begin{bmatrix} \tau_{1'1'} & 0 & 0 \\ 0 & \tau_{2'2'} & 0 \\ 0 & 0 & \tau_{3'3'} \end{bmatrix} = \begin{bmatrix} \sigma_1 & 0 & 0 \\ 0 & \sigma_2 & 0 \\ 0 & 0 & \sigma_3 \end{bmatrix} \tag{2.26}$$

that is, the off-diagonal terms vanish for principal axes. The stress and strain tensor components can be combined to result in certain tensor invariants. These combinations have the same value for all possible coordinate rotations. They are, for the stress tensor,

$$\begin{aligned} I_1 &= \tau_{ii}, \qquad I_2 = \tfrac{1}{2}\left(I_1^2 - \tau_{ij}\tau_{ij}\right) \\ I_3 &= \tfrac{1}{3}\left(3I_1 I_2 - I_1^3 + \tau_{ij}\tau_{jk}\tau_{ki}\right) \end{aligned} \tag{2.27}$$

and for the strain tensor

$$\begin{aligned} J_1 &= \varepsilon_{ii}, \qquad J_2 = \tfrac{1}{2}\left(J_1^2 - \varepsilon_{ij}\varepsilon_{ij}\right) \\ J_3 &= \tfrac{1}{3}\left(3J_1 J_2 - J_1^3 + \varepsilon_{ij}\varepsilon_{jk}\varepsilon_{ki}\right) \end{aligned} \tag{2.28}$$

For example, we have $I_1 = \tau_{kk} = \tau_{p'p'}$, where the primed axes are any set of rotated axes.

## 2.3 HOOKE'S LAW AND STRAIN ENERGY

The most general *linear* relationship between stress tensor and strain tensor components at a point in a solid is provided by the equations

$$\tau_{ij} = C_{ijkl} \varepsilon_{kl} \tag{2.29}$$

In the case of a homogeneous, elastic solid, the symbols $C_{ijkl}$ are constants that characterize its elastic behavior. Since $\tau_{ij}$ and $\varepsilon_{ij}$ are tensors of order two, the $3^4 = 81$ numbers $C_{ijkl}$ are the components of a tensor of order four, according to the quotient law. The 81 numbers $C_{ijkl}$ are not independent. They satisfy the symmetry relations

$$C_{ijkl} = C_{ijlk}, \qquad C_{ijkl} = C_{jikl}, \qquad C_{ijkl} = C_{klij} \tag{2.30}$$

and in view of these symmetries there will be $\frac{1}{2}(6^2 - 6) + 6 = 21$ independent elastic constants in the case of a generally anisotropic, elastic solid.

If we stipulate the existence of a strain energy density function $W = W(\varepsilon_{ij})$, we have

$$\tau_{ij} = \frac{\partial W(\varepsilon_{ij})}{\partial \varepsilon_{ij}} \tag{2.31}$$

We also assume that stress and strain always vanish simultaneously in the "natural" state of the solid. In this case the relation (2.29) is obtained if we set

$$W(\varepsilon_{ij}) = \tfrac{1}{2} C_{ijkl} \varepsilon_{ij} \varepsilon_{kl} \tag{2.32}$$

and in view of (2.30), (2.29), and (2.31), we have

$$W = \tfrac{1}{2} \tau_{ij} \varepsilon_{ij} \tag{2.33}$$

We now assume that the solid is isotropic; that is, its elastic properties are independent of orientation. In this case the strain energy density function $W$ must be unaffected by a rotation of the reference axes, and in the case of linear elasticity $W$ must be a homogeneous quadratic function of the strain tensor components. Both conditions are fulfilled if we set [see (2.28)]

$$W = AJ_1^2 - BJ_2 \tag{2.34}$$

where $A$ and $B$ are constants. Upon substitution of (2.34) into (2.31), we obtain

$$\tau_{ij} = \frac{\partial W}{\partial \varepsilon_{ij}} = (2A - B)\varepsilon_{kk}\delta_{ij} + B\varepsilon_{ij}$$

and in order to conform to the usual notation, we set

$$2A - B = \lambda = \frac{E\nu}{(1+\nu)(1-2\nu)}, \qquad B = 2G = \frac{E}{1+\nu}$$

where $E$ is Young's modulus, $G$ is the shear modulus, $\nu$ is Poisson's ratio, and $\lambda$ is Lamé's constant. Thus, in the case of an isotropic, elastic solid we have only two independent, elastic constants, and Hooke's law assumes the form

$$\tau_{ij} = \lambda \varepsilon_{kk} \delta_{ij} + 2G\varepsilon_{ij} \tag{2.35}$$

If we invert (2.35), we obtain

$$2G\varepsilon_{ij} = \tau_{ij} - \frac{\lambda \delta_{ij}}{2G + 3\lambda} \tau_{kk} \tag{2.36}$$

and the expanded form of (2.36) can be written as

$$\begin{aligned} E\varepsilon_{xx} &= \tau_{xx} - \nu(\tau_{yy} + \tau_{zz}) \\ E\varepsilon_{yy} &= \tau_{yy} - \nu(\tau_{zz} + \tau_{xx}) \\ E\varepsilon_{zz} &= \tau_{zz} - \nu(\tau_{xx} + \tau_{yy}) \\ 2G\varepsilon_{xy} &= \tau_{xy}, 2G\varepsilon_{yz} = \tau_{yz}, 2G\varepsilon_{zx} = \tau_{zx} \end{aligned} \tag{2.37}$$

## 2.4 PRINCIPLE OF VIRTUAL WORK (LINEAR STRAIN)

The principle of virtual work relates the work done by external forces during a virtual displacement to the corresponding change in strain energy for a solid in equilibrium:

$$\delta U = \delta \mathscr{W} \tag{2.38a}$$

where

$$\delta U = \int_B \delta W \, dV \tag{2.38b}$$

is the change in strain energy, and

$$\delta \mathscr{W} = \int_B F_i \, \delta u_i \, dV + \int_{S_1} T_i^* \, \delta u_i \, dS \tag{2.38c}$$

is the work done by external forces during a virtual displacement $\delta u_i$. Here $T_i^*$ is the surface traction vector, and $F_i$ is the body force vector per unit volume. The vector $T_i^*$ is prescribed on $S_1$, and the displacement $u_i$ is prescribed on $S_2$, where $S = S_1 + S_2$ is the surface that encloses the volume $B$. The symbol $\delta u_i$ denotes a virtual displacement that is arbitrary, except that it must not violate the kinematic boundary condition, that is, $\delta u_i = 0$ on $S_2$.

We can use (2.38) to derive the stress equations of equilibrium for the solid. With reference to (2.31) and (2.9) we have

$$\delta W = \frac{\partial W}{\partial \varepsilon_{ij}} \delta \varepsilon_{ij} = \tau_{ij} \, \delta \varepsilon_{ij} = \tau_{ij} \, \delta u_{i,j}$$

and with reference to (2.38b) we have

$$\delta U = \int_B \tau_{ij}\, \delta u_{i,j}\, dV = \int_{S_1} \tau_{ij}\, \delta u_i\, n_j\, dS - \int_B \tau_{ij,j}\, \delta u_i\, dV \tag{2.39}$$

where we have used the divergence theorem. Thus (see 2.38a), we obtain

$$\int_B \left(\tau_{ij,j} + F_i\right) \delta u_i\, dV + \int_{S_1} \left(T_i^* - \tau_{ij} n_j\right) \delta u_i\, dS = 0$$

The region $B$ is arbitrary, and therefore,

$$\left(\tau_{ij,j} + F_i\right) \delta u_i = 0 \quad \text{in } B$$

$$\left(T_i^* - \tau_{ij} n_j\right) \delta u_i = 0 \quad \text{on } S_1$$

The vector $\delta u_i$ is arbitrary. Hence,

$$\tau_{ij,j} + F_i = 0 \quad \text{in } B \tag{2.40a}$$

$$T_i^* - \tau_{ij} n_j = 0 \quad \text{on } S_1 \tag{2.40b}$$

[see (2.22)]. We note that the application of the principle of virtual work (2.38) results not only in the equilibrium equations (2.40a) but also in the admissible (traction) boundary condition (2.40b).

## 2.5 HAMILTON'S PRINCIPLE

Hamilton's principle is the dynamical counterpart of the principle of virtual work in statics (see Section 2.4). A formal statement of Hamilton's principle will now be given.

Define the Lagrangian function

$$L = T - U \tag{2.41}$$

where $T$ is the kinetic energy and $U$ is the strain energy in the solid. Then the motion of the deformable solid is characterized by

$$\int_{t_1}^{t_2} \delta L\, dt = -\int_{t_1}^{t_2} \delta \mathscr{W}\, dt \tag{2.42}$$

where $\delta L$ is the first variation of $L$ and $\delta \mathscr{W}$ is the virtual work of external forces during a virtual displacement $u_i$ of the solid [see (2.38c)]. In words: The time integral of the variation of the Lagrangian function $L$ over the time interval $t_1 \le t \le t_2$ is equal to the time integral of the virtual work of external forces over the same time interval for the "actual" motion. The following proviso must be imposed: Only such variations are admitted as vanish, first, at times $t_1$ and $t_2$ at all points of the body and, second, throughout the entire time interval wherever the displacements are prescribed. The application of Hamilton's principle to deformable solids results in the equations of motion as well as the associated, admissible boundary conditions. We shall use Hamilton's principle in Chapter 6 to derive the equations of motion and associated admissible boundary conditions for elastic plates.

To demonstrate its application, we shall now employ (2.42) to derive the equations of motion and associated admissible boundary conditions of a hyperelastic solid. If $\rho$ is the specific mass, the kinetic energy of the solid is

$$T = \frac{1}{2}\int_B \rho \dot{u}_i \dot{u}_i \, dV \tag{2.43}$$

where $\dot{u}_i$ is the velocity vector. The total strain energy of the solid is

$$U = \int_B W \, dV \tag{2.44}$$

where $W = W(\varepsilon_{ij})$ is the strain energy density [see (2.31)]. The virtual work of the applied forces is given by (2.38b), where $T_i{}^*$ is the surface traction vector on $S_1$ and $F_i$ is the body force vector per unit volume. We have

$$\delta T = \int_B \rho \dot{u}_i \, \delta \dot{u}_i \, dV$$

and

$$\int_{t_1}^{t_2} \delta T \, dt = \int_{t_1}^{t_2} \int_B \rho \dot{u}_i \, \delta \dot{u}_i \, dV \, dt = -\int_{t_1}^{t_2} \int_B \rho \ddot{u}_i \, \delta u_i \, dV \tag{2.45}$$

where we have integrated by parts (with respect to $t$) and utilized the condition $\delta u_i = 0$ in $B$ for $t = t_1$ and $t = t_2$. Upon substitution of (2.45) and (2.39) into (2.42), we obtain

$$\int_{t_1}^{t_2} \int_B (\rho \ddot{u}_i - \tau_{ij,j} - F_i) \, \delta u_i \, dV \, dt + \int_{t_1}^{t_2} \int_{S_1} (\tau_{ij} n_j - T_i{}^*) \, \delta u_i \, dS \, dt = 0$$

The time interval $t_2 - t_1$ and the region $B$ are arbitrary. Consequently, we have

$$(\rho \ddot{u}_i - \tau_{ij,j} - F_i) \, \delta u_i = 0 \quad \text{in } B$$

$$(\tau_{ij} n_j - T_i{}^*) \, \delta u_i = 0 \quad \text{on } S_1$$

The vector $\delta u_i$ is arbitrary. Hence, we have

$$\rho \ddot{u}_i - \tau_{ij,j} - F_i = 0 \quad \text{in } B \tag{2.46a}$$

$$\tau_{ij} n_j - T_i^* = 0 \quad \text{on } S_1 \tag{2.46b}$$

Equations (2.46a) are the equations of motion of the solid, and (2.46b) are the associated, admissible (traction) boundary conditions.

## 2.6 PRINCIPLE OF VIRTUAL WORK FOR FINITE DEFORMATIONS

In Chapter 9 it will be necessary to employ the principle of virtual work to derive the (geometrically) nonlinear plate bending and stretching equations of von Kármán. We shall now derive this principle for the case of a three-dimensional, hyperelastic solid subjected to arbitrary, finite deformations.

Consider a solid that is in static equilibrium under the action of prescribed body forces $F_i$ in $B$, surface tractions $T_i^*$ on $S_1$, and prescribed displacements $u_i^*$ on $S_2$, where $S = S_1 + S_2$ is the surface that encloses the region $B$ occupied by the solid. The equilibrium condition for this solid is characterized as follows:

$$s_{ki,i} + F_k = 0 \quad \text{in } B \tag{2.47a}$$

$$s_{ki} n_i - T_k^* = 0 \quad \text{on } S_1 \tag{2.47b}$$

$$u_i - u_i^* = 0 \quad \text{on } S_2 \tag{2.47c}$$

where

$$s_{ki} = \sigma_{ij}\left(\delta_{k,j} + u_{k,j}\right) \tag{2.48}$$

The six independent numbers $\sigma_{ij} = \sigma_{ji}$ constitute the (second) Piola-Kirchhoff stress tensor, and the symbols $B$, $S_1$, and $S_2$ now refer to the reference (undeformed) configuration of the solid. We note that the quantities $\sigma_{ij}$ are not stress components in a physical sense. In fact, it can be shown that the physical (or engineering) stress components per unit area of the undeformed body are given by

$$t_{ij} = \sigma_{ij}\sqrt{1 + L_{(jj)}} \tag{2.49}$$

However, the numbers $t_{ij}$ do not constitute a tensor, in contradistinction to the $\sigma_{ij}$. Equations (2.47) and (2.48) are referred to Lagrangian coordinates $x_i$; that is, $x_i$ is the position of the material particle in the reference configuration. The displacement field is $u_i = X_i - x_i$ and $u_i = u_i(x_1, x_2, x_3)$. In the present development, commas denote differentiation with respect to $x_i$, that is, $(\ )_{,i} \equiv \partial(\ )/\partial x_i$.

We now imagine that the solid is (slightly) displaced from its equilibrium configuration by the amount $\delta u_i$. This "virtual displacement" is arbitrary, except that it must not violate the kinematic boundary condition; that is,

$$\delta u_i = 0 \quad \text{on } S_2 \tag{2.50}$$

The principle of virtual work relates the work performed by the external forces during a virtual displacement to the corresponding change in strain energy. It can be derived as follows: As a consequence of (2.47a) and (2.47b), we can write

$$-\int_B (s_{ki,i} + F_k)\, \delta u_k \, dV + \int_{S_1} (s_{ik} n_i - T_k^*)\, \delta u_k \, dS = 0 \tag{2.51}$$

where the region $B$ is associated with the *undeformed* configuration of the solid.

However, with the aid of the divergence theorem, we have

$$\int_{S_1} s_{ki} n_i \, \delta u_k \, ds = \int_B (s_{ki}\, \delta u_k)_{,i} \, dV = \int_B s_{ki}\, \delta u_{k,i} \, dV + \int_B s_{ki,i}\, \delta u_k \, dV$$

and therefore,

$$\int_B s_{ki}\, \delta u_{k,i} \, dV - \int_B F_k \, \delta u_k \, dV - \int_{S_1} T_k^* \, \delta u_k \, dS = 0$$

However,

$$\begin{aligned}
s_{ki}\, \delta u_{k,i} &= \tfrac{1}{2} s_{ki}\, \delta u_{k,i} + \tfrac{1}{2} s_{kj}\, \delta u_{k,j} \\
&= \tfrac{1}{2}\sigma_{ij}(\delta_{ki} + u_{k,i})\, \delta u_{k,j} + \tfrac{1}{2}\sigma_{ij}(\delta_{k,j} + u_{k,j})\, \delta u_{k,i} \\
&= \tfrac{1}{2}\sigma_{ij}\left[(\delta_{ki} + u_{k,i})\, \delta u_{k,j} + (\delta_{kj} + u_{k,j})\, \delta u_{k,i}\right] \\
&= \tfrac{1}{2}\sigma_{ij}\left[\delta u_{i,j} + \delta u_{j,i} + u_{k,i}\, \delta u_{k,j} + u_{k,j}\, \delta u_{k,i}\right] \\
&= \sigma_{ij}\, \delta L_{ij}
\end{aligned} \tag{2.52}$$

where $L_{ij} = \frac{1}{2}(u_{i,j} + u_{j,i} + u_{k,i} u_{k,j})$ is the Lagrangian strain tensor [see (2.5)]. Consequently, we can write

$$\int_B \sigma_{ij}\, \delta L_{ij} \, dV = \int_B F_i \, \delta u_i \, dV + \int_{S_1} T_i^* \, \delta u_i \, dA \tag{2.53}$$

If the solid is hyperelastic, $W = W(L_{ij})$, and we have

$$\sigma_{ij}\, \delta L_{ij} = \frac{\partial W}{\partial L_{ij}} \delta L_{ij} = \delta W \tag{2.54}$$

and (2.53) can now be written in the form

$$\delta U = \delta \mathscr{W} \tag{2.55}$$

where $\delta \mathscr{W}$ is the work of external forces during a virtual displacement $\delta u_i$ (the "virtual work") as characterized by (2.38c) and

$$\delta U = \int_B \sigma_{ij}\, \delta L_{ij}\, dV \tag{2.56}$$

is the change in strain energy due to virtual displacements $\delta u_i$. Equation (2.55) in conjunction with (2.38c) and (2.56) is a statement of the principle of virtual work that is valid for solids subjected to *finite* deformations.

## EXERCISES

**2.1** Write (2.5) in expanded, conventional notation.

**2.2** Demonstrate the validity of (2.7).

**2.3** Show that (2.13a), (2.13b), and (2.13c) are equivalent.

**2.4** Write (2.13) in expanded, conventional notation.

**2.5** Use (2.21) and (2.18) to demonstrate the validity of (2.22).

**2.6** Consider the stress tensor with components

$$[\tau_{ij}] = \begin{bmatrix} \tau_{11} & \tau_{12} & 0 \\ \tau_{21} & \tau_{22} & 0 \\ 0 & 0 & \sigma_3 \end{bmatrix}$$

(the $x_3$-axis is principal, and $\tau_{33} = \sigma_3$). The stress vector $\mathbf{T} = \hat{\mathbf{n}}N + \hat{\mathbf{s}}S$ acts on the plane with outer unit normal vector $\hat{\mathbf{n}} = n_1\hat{\mathbf{e}}_1 + n_2\hat{\mathbf{e}}_2$ and is parallel to the $x_1$-$x_2$ plane. Show that

$$\begin{aligned} n_1 &= \cos\theta, & n_2 &= \sin\theta, & n_3 &= 0 \\ s_1 &= -\sin\theta, & s_2 &= \cos\theta, & s_3 &= 0 \end{aligned}$$

Show that

$$\begin{aligned} N &= \tfrac{1}{2}(\tau_{11} + \tau_{22}) + \tfrac{1}{2}(\tau_{11} - \tau_{22})\cos 2\theta + \tau_{12}\sin 2\theta \\ S &= -\tfrac{1}{2}(\tau_{11} - \tau_{22})\sin 2\theta + \tau_{12}\cos 2\theta \end{aligned}$$

How is the Mohr circle related to these equations?

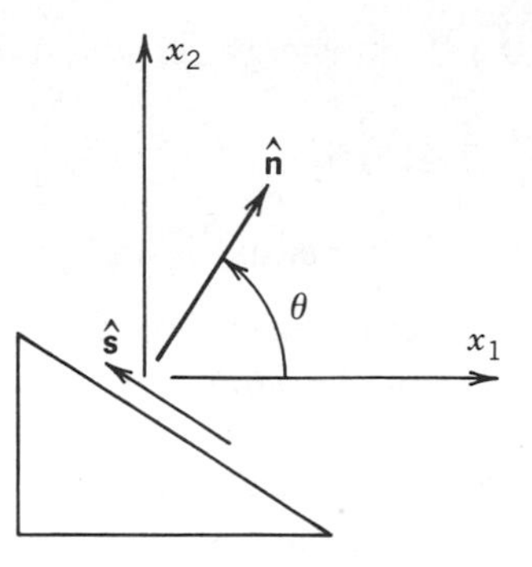


**2.7** With reference to (2.27), show that

$$I_2 = \begin{vmatrix} \tau_{11} & \tau_{12} \\ \tau_{21} & \tau_{22} \end{vmatrix} + \begin{vmatrix} \tau_{22} & \tau_{23} \\ \tau_{32} & \tau_{33} \end{vmatrix} + \begin{vmatrix} \tau_{33} & \tau_{31} \\ \tau_{13} & \tau_{11} \end{vmatrix}$$

$$I_3 = \begin{vmatrix} \tau_{11} & \tau_{12} & \tau_{13} \\ \tau_{21} & \tau_{22} & \tau_{23} \\ \tau_{31} & \tau_{32} & \tau_{33} \end{vmatrix}$$

**2.8** Derive (2.37) from (2.36). Note that $G = E/2(1 + \nu)$.

**2.9** Invert (2.35). Show that the result is (2.36).

**2.10** Use the method of Section 2.6 to derive (2.38). Assume $|u_{i,j}| \ll 1$.

**2.11** Use Hamilton's principle to derive the equation of the vibrating string.

**2.12** Use Hamilton's principle to derive the equation of the vibrating beam.

**2.13** Use (2.55) to derive (2.47).

**2.14** Consider a point $P$ in a stressed solid. Construct principal axes $x_i$ at $P$. Construct a (small) rectangular parallelepiped about $P$, with sides $a_1, a_2, a_3$ parallel to $x_1, x_2, x_3$, respectively. Let $A_3 = a_1 a_2$, $A_1 = a_2 a_3$, and $A_2 = a_3 a_1$ and note that the volume of the parallelepiped is $V = a_1 a_2 a_3$. If $W$ denotes the potential energy per unit of volume in a linearly elastic solid, show that the potential energy stored in the parallelepiped is

$$\begin{aligned} WV &= \tfrac{1}{2}(\sigma_1 A_1)(a_1 \varepsilon_1) + \tfrac{1}{2}(\sigma_2 A_2)(a_2 \varepsilon_2) + \tfrac{1}{2}(\sigma_3 A_3)(a_3 \varepsilon_3) \\ &= \tfrac{1}{2}(\sigma_1 \varepsilon_1 + \sigma_2 \varepsilon_2 + \sigma_3 \varepsilon_3) V \end{aligned}$$

Hence, show that the strain energy density at $P$ is

$$W = \tfrac{1}{2}(\sigma_1\varepsilon_1 + \sigma_2\varepsilon_2 + \sigma_3\varepsilon_3) = \tfrac{1}{2}\tau_{ij}\varepsilon_{ij}$$

Show that the strain energy density is an invariant.

**2.15** **(a)** Assume that the strain energy density $W(\varepsilon_{ij})$ of a solid is a quadratic function of the strain tensor components $\varepsilon_{ij}$ and that $\tau_{ij} = \partial W/\partial\varepsilon_{ij}$. Use Euler's theorem for homogeneous functions to show that $\varepsilon_{ij}(\partial W/\partial\varepsilon_{ij}) = 2W$, and show that $W = \frac{1}{2}\tau_{ij}\varepsilon_{ij}$.

**(b)** Express $\varepsilon_{ij}$ as a linear function of the $\tau_{ij}$ (Hooke's law). In this case $W(\tau_{ij})$ is a quadratic function of the $\tau_{ij}$. Now use Euler's theorem for homogeneous functions to show that $\tau_{ij}(\partial W/\partial\tau_{ij}) = 2W = \tau_{ij}\varepsilon_{ij}$ by (a) and deduce that $\varepsilon_{ij} = \partial W/\partial\tau_{ij}$.

**2.16** Hamilton's principle (2.42) can be applied to the case of finite deformations with appropriate modifications. We have [see (2.52)]

$$\delta U = \int_B \sigma_{ij}\,\delta L_{ij}\,dV = \int_B s_{ki}\,\delta u_{k,i}\,dV$$

$$= \int_S s_{ik}n_k\,\delta u_i\,dS - \int_B s_{ik,k}\,\delta u_i\,dV \qquad (\alpha)$$

where $s_{ki} = \sigma_{ij}(\delta_{k,j} + u_{k,j})$, $\sigma_{ij} = \sigma_{ji}$, is the Piola-Kirchhoff stress tensor and $L_{ij}$ is the Lagrangian strain tensor (2.5). Use (2.42) in conjunction with $(\alpha)$, (2.38c), and (2.45) to show that

$$s_{ik,k} + F_i = \rho\ddot{u}_i \quad \text{in } B$$

$$s_{ik}n_k = T_i{}^* \quad \text{on } S_1$$

$$u_i = u_i^* \quad \text{on } S_2$$

It should be noted that for the present case, the symbols $B$, $S_1$, and $S_2$ are associated with the reference (or undeformed) configuration of the solid (see Section 2.5).

# 3

# FUNDAMENTALS OF PLATE THEORY

## 3.1 PRELIMINARIES

For the present, a plate is considered to be a load-carrying structural member bounded by two parallel planes, called its faces, and a cylindrical surface, called its edge. The generators of the cylindrical surface are perpendicular to the plane faces. The distance between the plane faces is called the thickness of the plate, and it will be assumed that the thickness is small compared to the characteristic dimensions of the faces (length, width, diameter, etc.). The plane parallel to and equidistant from the faces is called the midplane, or median plane, of the plate. Its intersection with the cylindrical boundary is assumed to be a piecewise, smooth, simple, closed curve $C$. The plane region enclosed by $C$ is called the area of the plate and is denoted by the symbol $A$ (see Figs. 3.1 and 3.4). If the region $A$ is oriented in space, then a sense can be assigned to the curve $C$ as follows: The region $A$ always lies to the left of an observer who traverses the curve $C$ in the positive direction of the positive side of $A$. Plates that contain holes, or cut-outs, are readily included in the preceding definition by permitting the region $A$ to be multiply connected. In this case the symbol $C$ is understood to represent the sum of all curves that define the boundary of $A$.

In the following it will be assumed that the plate is subjected to two types of loading: (a) distributed surface loads, which act normal to the faces of the plate, and (b) distributed line loads in the form of moments and shearing forces that act upon the edges of the plate.

## 3.2 GEOMETRY OF DEFORMATION (DISPLACEMENTS)

The present section is concerned with the "small" deformation of a plate of thickness $h$. The material plate is referred to an orthogonal, Cartesian coordinate system $(x_1, x_2, z)$, and the $x_1 - x_2$ plane coincides with the median plane of the plate in the reference configuration, as shown in Fig. 3.1. It is now

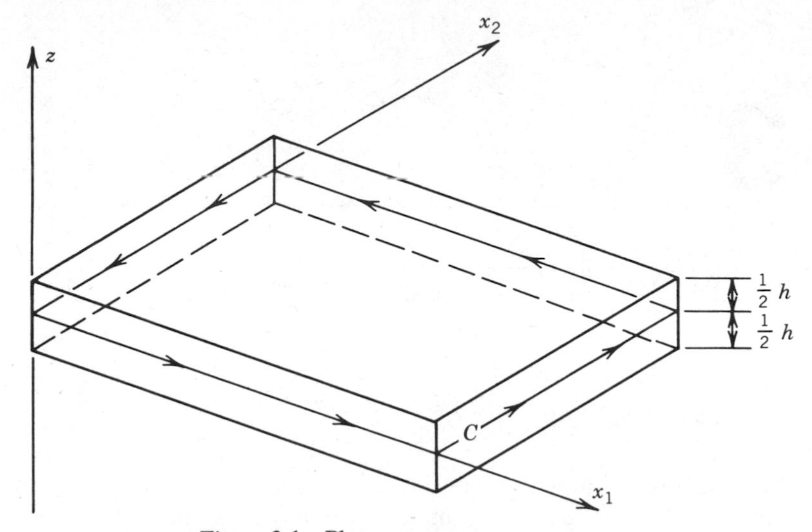


**Figure 3.1** Plate geometry.

assumed that (a) Material points in the midplane displace in the transverse (or $z$) direction only, so that points that lie in the midplane have no displacement components parallel to the $x_1 - x_2$ plane, that is, the midplane is a neutral surface and no membrane forces are induced when the plate deforms. (b) Straight material line elements that are perpendicular to the midplane in the reference configuration remain straight in the deformed state. Figure 3.2 shows a portion of the plate above the neutral surface. All angles are referred to $(x', y', z')$ axes taken to be parallel to the reference axes $x$, $y$, $z$ or $(x_1, x_2, z)$, respectively. The point $O$ lies in the median surface of the plate. In the undeformed state of the plate, the material line element $OP$ is coincident with the $z'$-axis and normal to the $x'$-$y'$ plane. In the deformed state, the line element $OP$ subtends the component angles $\psi_x$ and $\psi_y$, measured in the planes $x'$-$z'$ and $y'$-$z'$, respectively. The angle between the $z'$-axis and the rotated line element $OP$ is denoted by $\psi$. Thus, for arbitrary, finite rotations we have

$$\tan \psi_x = \frac{L \sin \psi \cos \phi}{L \cos \psi} = \tan \psi \cdot \cos \phi \tag{3.1a}$$

$$\tan \psi_y = \frac{L \sin \psi \sin \phi}{L \cos \psi} = \tan \psi \cdot \sin \phi \tag{3.1b}$$

We now linearize (3.1) such that $\psi$, $\psi_x$, and $\psi_y$ are taken as "small" angles in the sense that, for example, $\sin \psi \cong \tan \psi \cong \psi$. Thus,

$$\psi_x = \psi \cos \phi \tag{3.2a}$$

$$\psi_y = \psi \sin \phi \tag{3.2b}$$

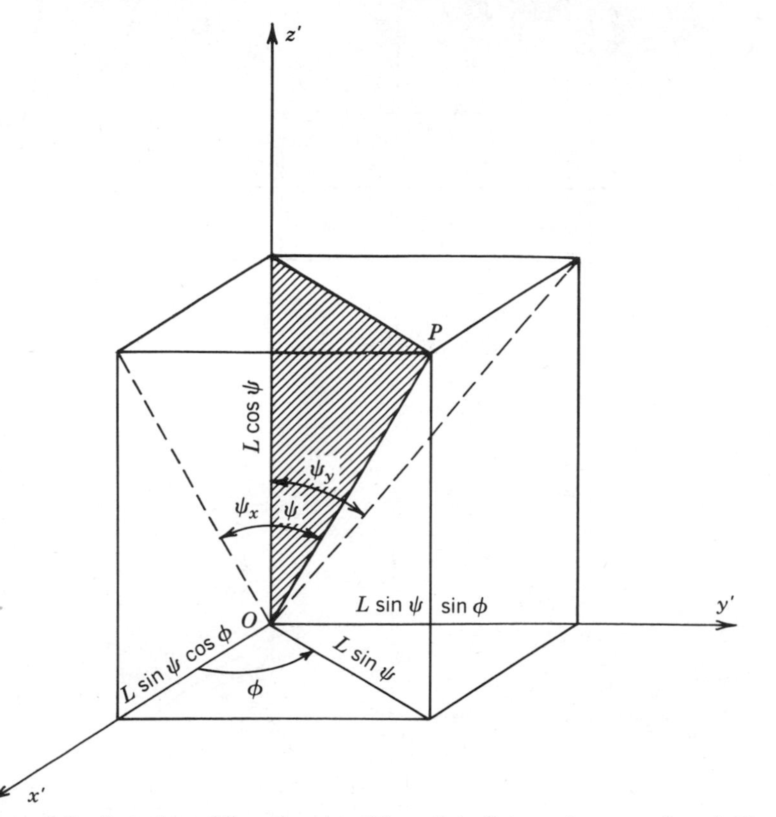


**Figure 3.2** Rotation of line element. (*Note:* $L$ is distance between $O$ and $P$).

and with the aid of (3.2) it can be shown (see Exercise 3.1) that for small rotations of $OP$, the quantities $\psi_x$ and $\psi_y$ are Cartesian components of a vector $\boldsymbol{\psi} = \hat{\mathbf{e}}_x\psi_x + \hat{\mathbf{e}}_y\psi_y$ in a space of two dimensions. Because $\psi$ is a small angle, $\cos\psi \cong 1$, and therefore the transverse displacement of a point in the plate is given by

$$u_z = w - z + z\cos\psi \cong w$$

where $w$ is the local transverse displacement of the median surface, as shown in Fig. 3.3. Thus, it can be inferred that for small rotations of the normal to the median surface, all points on the line element $OP$ will experience the same transverse displacement $u_z = w(x_1, x_2)$; that is, there is no "thickness stretch" of the plate. In view of the preceding considerations, we can write the displacement vector of a material point in the plate $-\frac{1}{2}h \leqslant z \leqslant \frac{1}{2}h$ in the form

$$\mathbf{u} = \hat{\mathbf{e}}_i u_i = \hat{\mathbf{e}}_z w(x_1, x_2) + z\boldsymbol{\psi}(x_1, x_2) \tag{3.3}$$

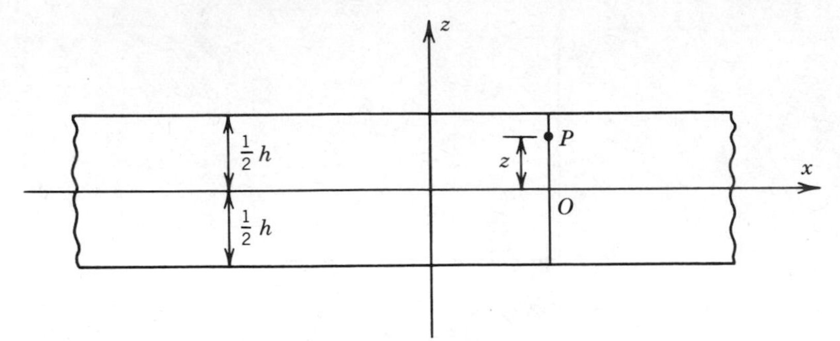


(a) Plate in reference configuration

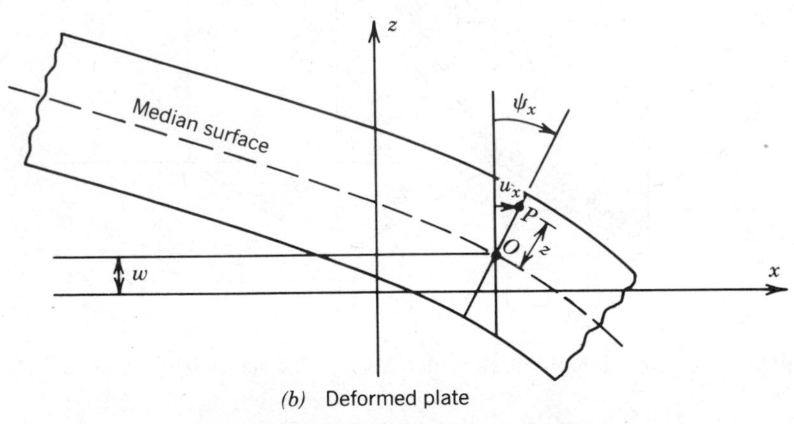


(b) Deformed plate

**Figure 3.3** Deformation of plate.

With reference to (3.3) it is to be noted that the plate displacement field is completely characterized by the three functions

$$w = w(x, y), \qquad \psi_x = \psi_x(x, y), \qquad \psi_y = \psi_y(x, y) \tag{3.4}$$

We now consider the directed curve $C$, which characterizes the edge (boundary) of the plate. With reference to Fig. 3.4, $\hat{\mathbf{n}}$ and $\hat{\mathbf{s}}$ are unit normal and tangent vectors, respectively, where

$$\hat{\mathbf{n}} = n_\alpha \hat{\mathbf{e}}_\alpha, \qquad \hat{\mathbf{s}} = s_\alpha \hat{\mathbf{e}}_\alpha, \qquad \hat{\mathbf{n}} \cdot \hat{\mathbf{s}} = 0, \qquad \hat{\mathbf{n}} \times \hat{\mathbf{s}} = \hat{\mathbf{e}}_z \tag{3.5}$$

The following relations are readily established along $C$ (see Exercise 3.2):

$$\boldsymbol{\psi} = \psi_\alpha \hat{\mathbf{e}}_\alpha = \psi_n \hat{\mathbf{n}} + \psi_s \hat{\mathbf{s}} \tag{3.6}$$

$$\psi_n = \boldsymbol{\psi} \cdot \hat{\mathbf{n}} = \psi_\alpha n_\alpha \tag{3.7a}$$

$$\psi_s = \boldsymbol{\psi} \cdot \hat{\mathbf{s}} = \psi_\alpha s_\alpha \tag{3.7b}$$

$$\psi_\alpha = \psi_n n_\alpha + \psi_s s_\alpha \tag{3.8}$$

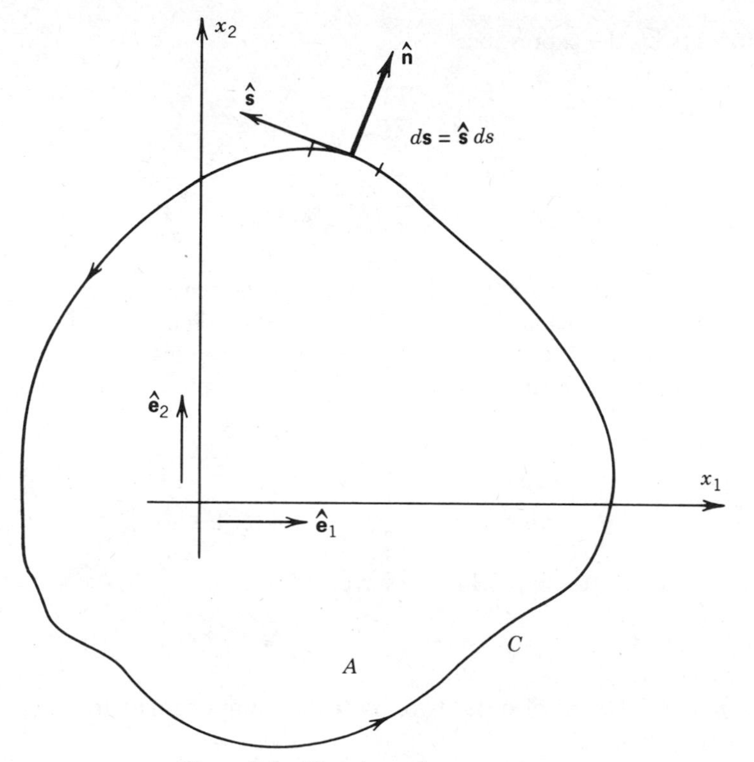


**Figure 3.4** Plate boundary geometry.

Furthermore, it can be shown (see Exercise 3.3) that

$$e_{\alpha\beta} = n_\alpha s_\beta - n_\beta s_\alpha \tag{3.9}$$

where $e_{\alpha\beta}$ is the two-dimensional alternating tensor (see Chapter 1).

## 3.3 PLATE STRAINS

Upon substitution of the displacement assumptions (3.3) into the strain-displacement relations of the three-dimensional theory (2.9), we readily obtain

$$\varepsilon_{\alpha\beta} = \tfrac{1}{2}z(\psi_{\alpha,\beta} + \psi_{\beta,\alpha}) \tag{3.10a}$$

$$\varepsilon_{\alpha z} = \tfrac{1}{2}(\psi_\alpha + w_{,\alpha}) \tag{3.10b}$$

$$\varepsilon_{zz} = 0 \tag{3.10c}$$

In subsequent developments it will be convenient to define the plate strain components by the expressions

$$m_{\alpha\beta} = \frac{12}{h^3}\int_{-h/2}^{h/2} z\varepsilon_{\alpha\beta}\,dz \tag{3.11a}$$

$$q_{\alpha} = \frac{2}{h}\int_{-h/2}^{h/2} \varepsilon_{\alpha z}\,dz \tag{3.11b}$$

where $m_{\alpha\beta}$ are the flexural (bending) strain components and $q_{\alpha}$ are the plate shear strain components. It can be shown (see Exercise 3.4) that $m_{\alpha\beta}$ is a tensor of order two and $q_{\alpha}$ is a vector, both in a two-dimensional space. Upon substitution of (3.10) into (3.11), we obtain the plate strain-displacement relations

$$m_{\alpha\beta} = \tfrac{1}{2}(\psi_{\alpha,\beta} + \psi_{\beta,\alpha}) \tag{3.12a}$$

$$q_{\alpha} = \psi_{\alpha} + w_{,\alpha} \quad \text{or} \quad \mathbf{q} = \boldsymbol{\psi} + \nabla w \tag{3.12b}$$

and in view of (3.10) and (3.12), we have

$$\varepsilon_{\alpha\beta} = zm_{\alpha\beta}, \qquad \varepsilon_{\alpha z} = \tfrac{1}{2}q_{\alpha} \tag{3.13}$$

Since $m_{\alpha\beta}$ is a tensor of order two, its scalar components with respect to $\hat{\mathbf{n}}\hat{\mathbf{n}}$ and $\hat{\mathbf{n}}\hat{\mathbf{s}}$ are, respectively,

$$\begin{aligned} m_{nn} &= m_{\alpha\beta}n_{\alpha}n_{\beta} = \tfrac{1}{2}(\psi_{\alpha,\beta} + \psi_{\beta,\alpha})n_{\alpha}n_{\beta} = \psi_{\alpha,\beta}n_{\alpha}n_{\beta} \\ m_{ns} &= m_{\alpha\beta}n_{\alpha}s_{\beta} = \tfrac{1}{2}(\psi_{\alpha,\beta} + \psi_{\beta,\alpha})n_{\alpha}s_{\beta} \end{aligned} \tag{3.14a}$$

or

$$2m_{ns} = \psi_{\alpha,\beta}n_{\alpha}s_{\beta} + \psi_{\beta,\alpha}s_{\beta}n_{\alpha} \tag{3.14b}$$

However, we have

$$d\psi = \psi_{,\alpha}\,dx_{\alpha} = \left(dx_{\alpha}\frac{\partial}{\partial x_{\alpha}}\right)\psi = (d\mathbf{r}\cdot\nabla)\psi$$

where $d\mathbf{r} = \hat{\mathbf{e}}_{\alpha}\,dx_{\alpha}$ and $\nabla = \hat{\mathbf{e}}_{\alpha}(\partial/\partial x_{\alpha})$. Therefore,

$$\frac{d\psi}{|d\mathbf{r}|} = \left(\frac{d\mathbf{r}}{|d\mathbf{r}|}\cdot\nabla\right)\psi = (\hat{\mathbf{n}}\cdot\nabla)\psi \equiv \mathbf{D}_n \tag{3.15a}$$

where $\hat{\mathbf{n}} = d\mathbf{r}/|d\mathbf{r}| = \hat{\mathbf{e}}_{\alpha}n_{\alpha}$, and $\mathbf{D}_n$ is recognized as the directional derivative of $\psi$ in the direction of the unit normal vector $\hat{\mathbf{n}}$. Similarly,

$$\frac{d\psi}{|ds|} = (\mathbf{s}\cdot\nabla)\psi \equiv \mathbf{D}_s \tag{3.15b}$$

where $\hat{\mathbf{s}} = \hat{\mathbf{e}}_\alpha s_\alpha$ and $\mathbf{D}_s$ is the directional derivative of $\boldsymbol{\psi}$ in the direction of the unit tangent vector $\hat{\mathbf{s}}$. In view of (3.14) and (3.15), we have

$$m_{nn} = \psi_{\alpha,\beta} n_\alpha n_\beta = \hat{\mathbf{n}} \cdot [(\hat{\mathbf{n}} \cdot \nabla)\boldsymbol{\psi}] = \hat{\mathbf{n}} \cdot \mathbf{D}_n \tag{3.16a}$$

$$2m_{ns} = \hat{\mathbf{n}} \cdot [(\hat{\mathbf{s}} \cdot \nabla)\boldsymbol{\psi}] + \hat{\mathbf{s}} \cdot [(\hat{\mathbf{n}} \cdot \nabla)\boldsymbol{\psi}] = \hat{\mathbf{n}} \cdot \mathbf{D}_s + \hat{\mathbf{s}} \cdot \mathbf{D}_n \tag{3.16b}$$

Equations (3.16) are invariant representations of the flexural strain tensor components, and they can be used with advantage to calculate the strain components relative to any orthogonal, curvilinear coordinate system.

**Example:** We calculate the plate strain-displacement relations referred to a polar coordinate system. In this case we have [see (1.44a), (3.15), and (3.16)]

$$\nabla = \hat{\mathbf{e}}_r \frac{\partial}{\partial r} + \hat{\mathbf{e}}_\theta \frac{1}{r}\frac{\partial}{\partial \theta}$$

$$\boldsymbol{\psi} = \hat{\mathbf{e}}_r \psi_r + \hat{\mathbf{e}}_\theta \psi_\theta$$

$$\mathbf{D}_r = (\hat{\mathbf{e}}_r \cdot \nabla)\boldsymbol{\psi} = \frac{\partial \boldsymbol{\psi}}{\partial r} = \hat{\mathbf{e}}_r \frac{\partial \psi_r}{\partial r} + \hat{\mathbf{e}}_\theta \frac{\partial \psi_\theta}{\partial r}$$

$$\mathbf{D}_\theta = (\hat{\mathbf{e}}_\theta \cdot \nabla)\boldsymbol{\psi} = \frac{1}{r}\frac{\partial}{\partial \theta}\boldsymbol{\psi} = \hat{\mathbf{e}}_r \left(\frac{1}{r}\frac{\partial \psi_r}{\partial \theta} - \frac{1}{r}\psi_\theta\right) + \hat{\mathbf{e}}_\theta \left(\frac{1}{r}\frac{\partial \psi_\theta}{\partial \theta} + \frac{\psi_r}{r}\right)$$

$$m_{rr} = \hat{\mathbf{e}}_r \cdot \mathbf{D}_r = \frac{\partial}{\partial r}\psi_r \tag{3.17a}$$

$$m_{\theta\theta} = \hat{\mathbf{e}}_\theta \cdot \mathbf{D}_\theta = \frac{1}{r}\frac{\partial \psi_\theta}{\partial \theta} + \frac{\psi_r}{r} \tag{3.17b}$$

$$2m_{r\theta} = \hat{\mathbf{e}}_r \cdot \mathbf{D}_\theta + \hat{\mathbf{e}}_\theta \cdot \mathbf{D}_r = \frac{1}{r}\frac{\partial}{\partial \theta}\psi_r - \frac{1}{r}\psi_\theta + \frac{\partial \psi_\theta}{\partial r} \tag{3.17c}$$

$$\begin{gathered} \mathbf{q} = \boldsymbol{\psi} + \nabla w = q_r \hat{\mathbf{e}}_r + q_\theta \hat{\mathbf{e}}_\theta = \hat{\mathbf{e}}_r \left(\psi_r + \frac{\partial w}{\partial r}\right) + \hat{\mathbf{e}}_\theta \left(\psi_\theta + \frac{1}{r}\frac{\partial w}{\partial \theta}\right) \\ q_r = \psi_r + \frac{\partial w}{\partial r}, \qquad q_\theta = \psi_\theta + \frac{1}{r}\frac{\partial w}{\partial \theta} \end{gathered} \tag{3.18}$$

## 3.4 RIGID BODY MOTION

Analogous to three-dimensional theory, we can find the components of plate rotation in the following manner:

$$d\psi_\alpha = \psi_{\alpha,\beta}\, dx_\beta = (m_{\alpha\beta} + \Omega_{\alpha\beta})\, dx_\beta \tag{3.19}$$

where [see (3.12a)]

$$\Omega_{\alpha\beta} = \tfrac{1}{2}(\psi_{\alpha,\beta} - \psi_{\beta,\alpha}) \tag{3.20}$$

are the components of the rotation tensor due to flexure alone. Similarly,

$$dw = w_{,\alpha}\, dx_\alpha = \tfrac{1}{2}(q_\alpha + \Omega_\alpha)\, dx_\alpha$$

where [see (3.12b)]

$$\Omega_\alpha = w_{,\alpha} - \psi_\alpha$$

is the rotation vector due to shear strains. In the case of (allowable) rigid body motions we have a vanishing strain field (i.e., $m_{\alpha\beta} \equiv 0$ and $q_\alpha \equiv 0$), so that $\psi_{\alpha,\beta} = -\psi_{\beta,\alpha}$ and $w_{,\alpha} = -\psi_\alpha$. Consequently,

$$\Omega_\alpha = -2\psi_\alpha, \qquad \Omega_{\alpha\beta} = \psi_{\alpha,\beta}$$

and

$$w = w_0 + \tfrac{1}{2}\Omega_\alpha x_\alpha \tag{3.21a}$$

$$\psi_\alpha = -\tfrac{1}{2}\Omega_\alpha \tag{3.21b}$$

Equations (3.21) characterize a permissible rigid body motion of the plate: $w_0$ is a pure translation in the $z$-direction, and $\Omega_\alpha$ characterizes an "infinitesimal" rigid rotation of the undeformed plate median surface. From this analysis we can conclude that a given field of plate strain will not generate a unique displacement field (if it exists) unless we also impose three independent conditions that fix the values of $w_0$, $\Omega_1$, and $\Omega_2$, and it is assumed in subsequent discussions that this has been done.

## 3.5 EQUATIONS OF COMPATIBILITY

An inspection of the strain-displacement relations (3.12) reveals that there are three components of displacement $(w, \psi_1, \psi_2)$ and five components of strain $(m_{11}, m_{12}, m_{22}, q_1, q_2)$. If the displacement field is given, then the plate strain components are readily computed with the aid of (3.12). However, if the functions $(m_{11}, m_{12}, m_{22}, q_1, q_2)$ are assigned arbitrarily, (3.12) will not, in general, have a single-valued solution because we have five equations to find three unknown functions. The required integrability conditions for (3.12) are given by

$$S_{12} = 2m_{12,12} - m_{11,22} - m_{22,11} = 0 \tag{3.22a}$$

$$R_1 = 2(m_{11,2} - m_{12,1}) + q_{2,11} - q_{1,21} = 0 \tag{3.22b}$$

$$R_2 = 2(m_{22,1} - m_{21,2}) + q_{1,22} - q_{2,12} = 0 \tag{3.22c}$$

It can be shown, by direct substitution, that (3.12) satisfy (3.22) identically (see Exercise 3.6). From this fact we can conclude that (3.22) are necessary conditions on the plate strain components to ensure the generation of a single-valued displacement field. Equations (3.22) can also be shown to be sufficient conditions to ensure the existence of a single-valued displacement field provided the region of the plate is simply connected. In the case of a multiply connected region, (3.22) are still necessary for the generation of a single-valued displacement field, but in this case sufficient conditions are provided by (3.22) augmented by the satisfaction of certain conditions involving line integrals (see Exercise 3.8). We also note that (3.22) are not independent because

$$R_{1,2} + R_{2,1} = -2S_{12} \tag{3.23}$$

## 3.6 PLATE STRESSES

In the analysis of plates, it is convenient to define the following plate stress resultants:

$$Q_\alpha = \int_{-h/2}^{h/2} \tau_{\alpha z}\, dz \tag{3.24a}$$

$$M_{\alpha\beta} = \int_{-h/2}^{h/2} \tau_{\alpha\beta} z\, dz \tag{3.24b}$$

where $\tau_{ij} = \tau_{ji}$ are the components of the stress tensor. In view of definition (3.24b), it can be shown that $M_{\alpha\beta} = M_{\beta\alpha}$. With reference to (3.24b), we speak of plate bending moments per unit length when $\alpha = \beta$ and of plate twisting moments per unit length when $\alpha \neq \beta$. The four numbers $M_{\alpha\beta}$ constitute the components of a second-order tensor, whereas the numbers $Q_\alpha$ are the components of a vector $\mathbf{Q} = \hat{\mathbf{e}}_\alpha Q_\alpha$ in a two-dimensional space. We adopt a sign convention for moments and shear forces in the plate that is consistent with the sign conventions of elasticity theory, and positive resultants acting upon the plate are shown in Fig. 3.5.

Just as the stress vector is a convenient device in three-dimensional elasticity theory, an analogous development will show that the concept of a moment vector will prove useful in subsequent developments. We shall use Cauchy's formulas (2.23) as a point of departure, and these are now written in the form

$$T_\alpha = \tau_{\beta\alpha} n_\beta + \tau_{z\alpha} n_z \tag{3.25a}$$

$$T_z = \tau_{\alpha z} n_\alpha + \tau_{zz} n_z \tag{3.25b}$$

where $(T_1, T_2, T_z)$ are the components of the surface traction vector acting on the edge (boundary) of the plate. Consider a point $P$ on the boundary curve $C$ of the plate, as shown in Fig. 3.6. Then $\hat{\mathbf{n}} \times \hat{\mathbf{s}} = \hat{\mathbf{e}}_z$, and on the edge of the plate $n_z \equiv 0$. Therefore, Cauchy's formulas (3.25) reduce to

$$T_\alpha = \tau_{\beta\alpha} n_\beta \tag{3.26a}$$

$$T_z = \tau_{\alpha z} n_\alpha \tag{3.26b}$$

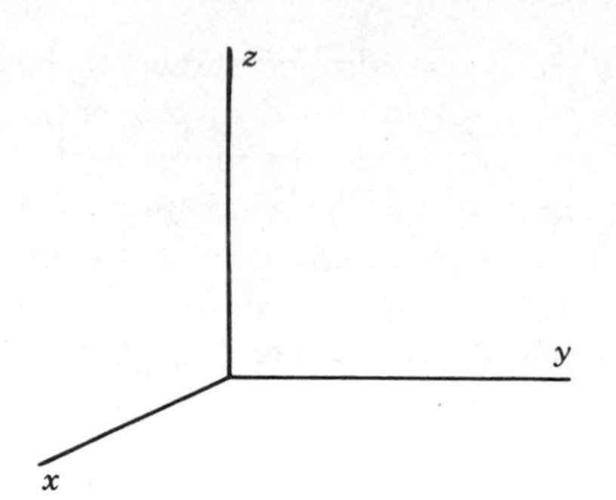


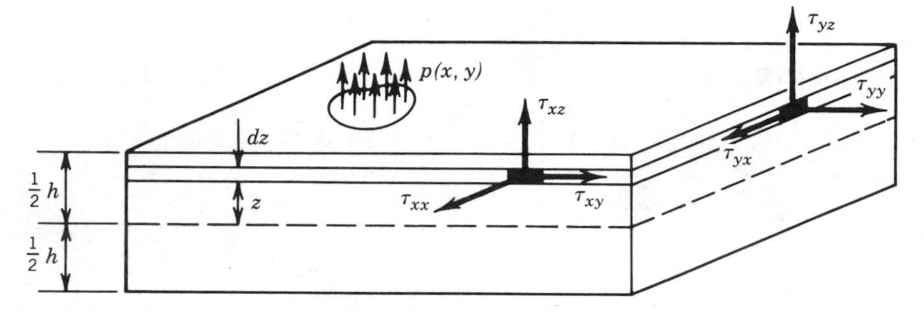


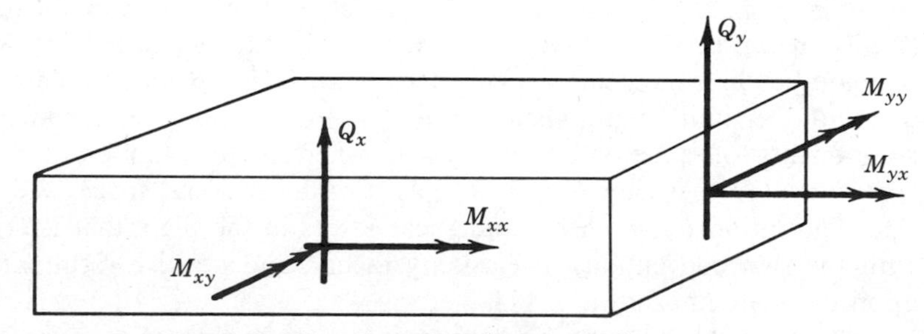


**Figure 3.5** Plate element and stress resultants.

We now apply the operations $\int_{-h/2}^{h/2} \ldots z\,dz$ to (3.26a) and $\int_{-h/2}^{h/2} \ldots dz$ to (3.26b) and define the moment vector as

$$M_\alpha = \int_{-h/2}^{h/2} T_\alpha z\,dz \tag{3.27}$$

and the transverse shear force by

$$Q_n = \int_{-h/2}^{h/2} T_z\,dz \tag{3.28}$$

In view of (3.24), (3.26), and (3.27), we readily obtain

$$M_\alpha = M_{\alpha\beta} n_\beta \tag{3.29}$$

$$Q_n = Q_\alpha n_\alpha = \mathbf{Q} \cdot \hat{\mathbf{n}} \tag{3.30a}$$

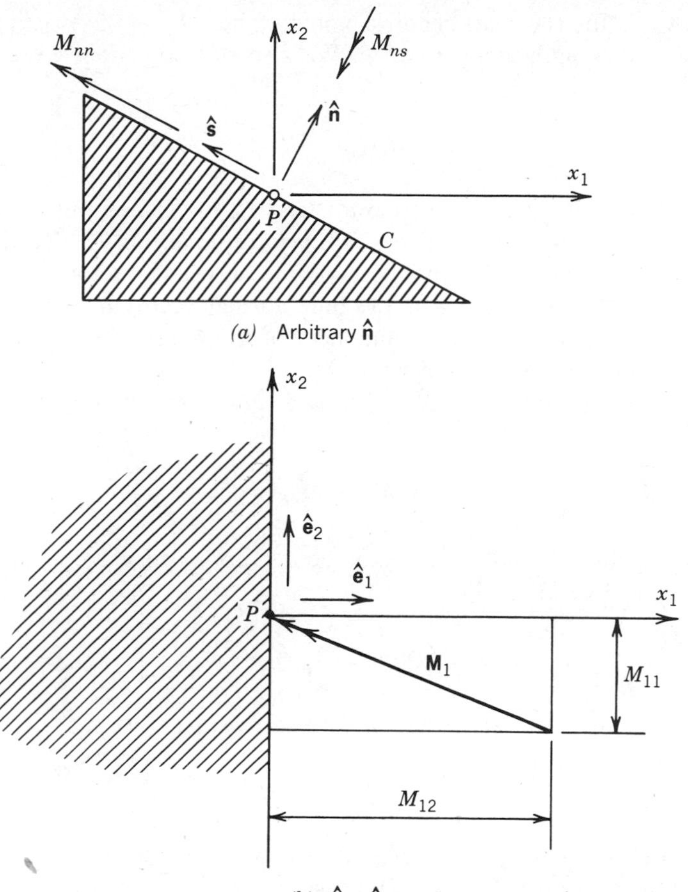


**Figure 3.6** Moment vector components.

where

$$\mathbf{Q} = Q_\alpha \hat{\mathbf{e}}_\alpha = Q_n \hat{\mathbf{n}} \tag{3.30b}$$

In order to maintain notational uniformity and consistency, we take as the formal representation of the moment vector at $P$ the expression

$$\mathbf{M}(\hat{\mathbf{n}}) = M_{nn}\hat{\mathbf{n}} + M_{ns}\hat{\mathbf{s}} = \hat{\mathbf{e}}_\alpha M_\alpha \tag{3.31}$$

although the (positive) bending and twisting moments $M_{nn}$ and $M_{ns}$, respectively, act in the manner shown by the double arrows in Fig. 3.6 and the right-hand rule applies. Consequently,

$$M_{nn} = \mathbf{M} \cdot \hat{\mathbf{n}} = M_\alpha n_\alpha = n_\alpha n_\beta M_{\alpha\beta} \tag{3.32a}$$

$$M_{ns} = \mathbf{M} \cdot \hat{\mathbf{s}} = M_\alpha s_\alpha = s_\alpha n_\beta M_{\alpha\beta} \tag{3.32b}$$

where $M_{nn}$ is the (normal) bending moment and $M_{ns}$ is the twisting moment, both per unit length along $C$ at $P$. We also note that in view of (3.6) and (3.31),

$$\mathbf{M} \cdot \boldsymbol{\psi} = M_\alpha \psi_\alpha = M_{nn}\psi_n + M_{ns}\psi_s \tag{3.33}$$

By application of the quotient rule for tensors, it can be inferred from (3.32) that $M_{\alpha\beta}$ is a tensor of order two. We also observe that (3.32a) displays the scalar bending moment component of the moment vector as a quadratic function of the components of the unit normal vector $\hat{\mathbf{n}}$. Similarly, (3.32b) displays the twisting moment component of the moment vector as a bilinear function of the components of $\hat{\mathbf{n}}$ and $\hat{\mathbf{s}}$.

If we multiply both sides of (3.29) by $\hat{\mathbf{e}}_\alpha$ and subsequently sum over $\alpha$, we obtain

$$\hat{\mathbf{e}}_\alpha M_\alpha = M_{\alpha\beta}\hat{\mathbf{e}}_\alpha n_\beta \tag{3.34}$$

We now define the moment vector acting upon the edge with unit outer normal $\hat{\mathbf{e}}_\beta$ as [see (3.31)]

$$\mathbf{M}(\hat{\mathbf{e}}_\beta) \equiv \mathbf{M}_\beta = M_{\alpha\beta}\hat{\mathbf{e}}_\alpha \tag{3.35}$$

and with the aid of (3.31) and (3.35), we can write (3.34) in the form

$$\mathbf{M} = \mathbf{M}_\alpha n_\alpha \tag{3.36}$$

Equation (3.36) is the plate analog of Cauchy's theorem in three-dimensional continuum theory (see Chapter 2). Equation (3.35) can be inverted to read

$$M_{\alpha\beta} = \mathbf{M}_\alpha \cdot \hat{\mathbf{e}}_\beta \tag{3.37}$$

We also note that

$$\mathbf{M}_{\alpha,\alpha} = \mathbf{div}\, \mathcal{M} \equiv \nabla \cdot \mathcal{M} = \hat{\mathbf{e}}_\beta M_{\alpha\beta,\alpha} \tag{3.38}$$

because

$$\mathcal{M} = \hat{\mathbf{e}}_\alpha \hat{\mathbf{e}}_\beta M_{\alpha\beta} \tag{3.39}$$

and

$$\nabla \cdot \mathcal{M} = \hat{\mathbf{e}}_\gamma \cdot \frac{\partial}{\partial x_\gamma}\left(M_{\alpha\beta}\hat{\mathbf{e}}_\alpha \hat{\mathbf{e}}_\beta\right) = \delta_{\gamma\alpha} M_{\alpha\beta,\gamma}\hat{\mathbf{e}}_\beta = M_{\alpha\beta,\alpha}\hat{\mathbf{e}}_\beta = \mathbf{M}_{\alpha,\alpha}$$

[see (3.36), (3.34), and (3.39)].

## 3.7 PRINCIPAL AXES OF THE MOMENT TENSOR

With reference to (3.31) and Fig. 3.6($a$), we write the square of the twisting moment as

$$M_{ns}^2 = |\mathbf{M}|^2 - M_{nn}^2 = M_1^2 + M_2^2 - M_{nn}^2 \tag{3.40}$$

We now pose the following problem: With reference to (3.31), does there exist an orientation $\hat{\mathbf{n}}$ of the plate edge for which the twisting moment $M_{ns}$ vanishes? In this case $M_{nn} = |\mathbf{M}| = \mu$, and $\mathbf{M}$ is normal to $\hat{\mathbf{n}}$. In view of (3.29), we have $M_\alpha = \mu n_\alpha = M_{\alpha\beta} n_\beta$, where $\mu$ is the scalar principal bending moment. Consequently,

$$(M_{\alpha\beta} - \mu\delta_{\alpha\beta}) n_\beta = 0 \tag{3.41}$$

Condition (3.41) can also be obtained by posing the present problem in a somewhat different manner. With reference to (3.32a), what is the direction of $\hat{\mathbf{n}}$ for which $M_{nn}$ assumes stationary values subject to the constraint $|\hat{\mathbf{n}}| = 1$? Using the Lagrange multiplier technique with multiplier $\mu$, we define (See Exercise 1.32)

$$F = M_{\alpha\beta} n_\alpha n_\beta - \mu(n_\gamma n_\gamma - 1) = (M_{\alpha\beta} - \mu\delta_{\alpha\beta}) n_\alpha n_\beta + \mu$$

and require

$$\frac{\partial F}{\partial n_\beta} = (M_{\alpha\beta} - \mu\delta_{\alpha\beta}) n_\alpha = 0$$

These conditions are identical to (3.41). In unabridged form, (3.41) becomes

$$(M_{11} - \mu) n_1 + M_{12} n_2 = 0 \tag{3.42a}$$

$$M_{12} n_1 + (M_{22} - \mu) n_2 = 0 \tag{3.42b}$$

Because $|\hat{\mathbf{n}}| = 1$, we have

$$\hat{\mathbf{n}} \cdot \hat{\mathbf{n}} = n_1^2 + n_2^2 = 1 \tag{3.43}$$

and it can be inferred that at least one of the components $n_\alpha \neq 0$. Consequently, a solution of (3.42) requires that

$$\begin{vmatrix} M_{11} - \mu & M_{12} \\ M_{21} & M_{22} - \mu \end{vmatrix} = 0 \tag{3.44}$$

Upon expanding (3.44), we obtain

$$\mu^2 - I_1 \mu + I_2 = 0 \tag{3.45}$$

where

$$I_1 = M_{11} + M_{22} \tag{3.46a}$$

$$I_2 = \begin{vmatrix} M_{11} & M_{12} \\ M_{21} & M_{22} \end{vmatrix} \tag{3.46b}$$

The roots of (3.45) are

$$\mu_1 = \tfrac{1}{2}(M_{11} + M_{22}) + \tfrac{1}{2}\sqrt{(M_{11} - M_{22})^2 + 4M_{12}^2} \tag{3.47a}$$

$$\mu_2 = \tfrac{1}{2}(M_{11} + M_{22}) - \tfrac{1}{2}\sqrt{(M_{11} - M_{22})^2 + 4M_{12}^2} \tag{3.47b}$$

or, in view of (3.46),

$$2\mu_1 = I_1 + \sqrt{I_1^2 - 4I_2} \tag{3.48a}$$

$$2\mu_2 = I_1 - \sqrt{I_1^2 - 4I_2} \tag{3.48b}$$

The principal bending moments $\mu_1$ and $\mu_2$ are scalar quantities that are independent of the choice of the (Cartesian) coordinate system. Equation (3.45), which characterizes the principal stresses, will be obtained regardless of the orientation of the coordinate system selected. For these reasons the coefficients $I_1$ and $I_2$ must be scalar quantities. These quantities are said to be invariants of the tensor $M_{\alpha\beta}$, that is, they will not change as a result of coordinate rotations. If the tensor components are referred to principal axes, the invariants assume the simple forms [see (3.46)]:

$$I_1 = \mu_1 + \mu_2 \tag{3.49a}$$

$$I_2 = \mu_1 \mu_2 \tag{3.49b}$$

Equations (3.49) can also be obtained by direct calculation using (3.48) (see Exercise 3.12).

The radicands in (3.47) are always nonnegative. Hence, the roots $\mu_1$ and $\mu_2$ will always be real. Let us now suppose that $M_{12} \neq 0$. Then $\mu_1 \neq \mu_2$, and there are two distinct principal values $\mu_1$ and $\mu_2$ associated with the unit vectors $\hat{\mathbf{n}}^{(1)}$ and $\hat{\mathbf{n}}^{(2)}$, respectively. Using (3.42), we find that the vectors $\hat{\mathbf{n}}^{(1)}$ and $\hat{\mathbf{n}}^{(2)}$ have slopes

$$\tan\phi_1 = \frac{n_2^{(1)}}{n_1^{(1)}} = \frac{\mu_1 - M_{11}}{M_{12}} = \frac{M_{12}}{\mu_1 - M_{22}} \tag{3.50a}$$

$$\tan\phi_2 = \frac{n_2^{(2)}}{n_1^{(2)}} = \frac{\mu_2 - M_{11}}{M_{12}} = \frac{M_{12}}{\mu_2 - M_{22}} \tag{3.50b}$$

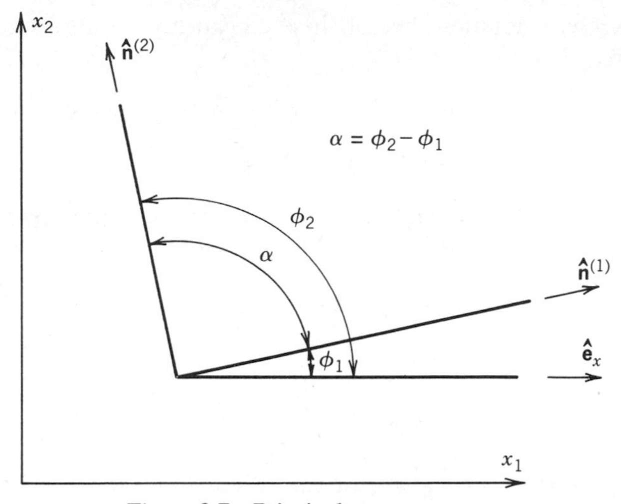


**Figure 3.7** Principal axes.

respectively, where $\phi_1$ and $\phi_2$ are the angles between the $x_1$-axis and the principal axes of the tensor $\mathscr{M}$. By a simple manipulation of (3.50), we obtain

$$\tan\phi_1 \tan\phi_2 = \frac{\mu_2 - M_{11}}{\mu_1 - M_{22}} \tag{3.51}$$

But from trigonometry, we have, with reference to Fig. 3.7,

$$\tan\alpha = \tan(\phi_2 - \phi_1) = \frac{\tan\phi_2 - \tan\phi_1}{1 + \tan\phi_1 \tan\phi_2}$$

and if $\alpha = \frac{1}{2}\pi$, then $1 + \tan\phi_1 \tan\phi_2 = 0$, and in view of (3.46a) and (3.49a), (3.51) will be satisfied. Hence we conclude that $\phi_2 = \phi_1 + \frac{1}{2}\pi$, and therefore $\hat{\mathbf{n}}^{(1)}$ is perpendicular to $\hat{\mathbf{n}}^{(2)}$. Furthermore, since $\hat{\mathbf{n}}^{(1)}$ and $\hat{\mathbf{n}}^{(2)}$ are unit vectors, we have

$$n_1^{(1)} = \cos\phi_1, \qquad n_2^{(1)} = \sin\phi_1 \tag{3.52a}$$

$$n_1^{(2)} = \cos\phi_2 = -\sin\phi_1, \qquad n_2^{(2)} = \sin\phi_2 = \cos\phi_1 \tag{3.52b}$$

## 3.8 PLATE STRESS-STRAIN RELATIONS

We now assume that the plate is a three-dimensional, linearly elastic, isotropic medium. In this case the relationship between stress and strain is expressed by Hooke's law (2.37), which is here written in the form

$$E\varepsilon_{ij} = (1 + \nu)\tau_{ij} - \nu\tau_{kk}\delta_{ij} \tag{3.53}$$

In the present notation, five of these constitutive relations can be written in the form

$$E\varepsilon_{\alpha\beta} = (1+\nu)\tau_{\alpha\beta} - \nu\tau_{\gamma\gamma}\delta_{\alpha\beta} - \nu\tau_{zz}\delta_{\alpha\beta} \tag{3.54a}$$

$$\tau_{z\alpha} = 2G\varepsilon_{z\alpha} = Gq_\alpha \tag{3.54b}$$

where $\tau_{\gamma\gamma} = \tau_{11} + \tau_{22} = \tau_{\alpha\alpha}$ and where $\delta_{\alpha\beta}$ is the two-dimensional Kronecker delta. By setting $\alpha = \beta$ in (3.54a), we obtain

$$E\varepsilon_{\alpha\alpha} = (1-\nu)\tau_{\alpha\alpha} - 2\nu\tau_{zz}$$

or

$$\tau_{\alpha\alpha} = \frac{E}{1-\nu}\varepsilon_{\alpha\alpha} + \frac{2\nu}{1-\nu}\tau_{zz}$$

With the aid of this equation, we can solve (3.54a) for $\tau_{\alpha\beta}$ in terms of $\varepsilon_{\alpha\beta}$ and $\tau_{zz}$. The result is

$$\tau_{\alpha\beta} = \frac{E}{1+\nu}\varepsilon_{\alpha\beta} + \frac{\nu E}{1-\nu^2}\varepsilon_{\gamma\gamma}\delta_{\alpha\beta} + \frac{\nu}{1-\nu}\tau_{zz}\delta_{\alpha\beta}$$

or, in view of (3.13),

$$\tau_{\alpha\beta} = \frac{Ez}{1-\nu^2}\left[(1-\nu)m_{\alpha\beta} + \nu m\delta_{\alpha\beta}\right] + \frac{\nu}{1-\nu}\tau_{zz}\delta_{\alpha\beta} \tag{3.55}$$

where the plate area dilatation $m$ is defined by

$$m = m_{\alpha\alpha} = \psi_{\alpha,\alpha} = \nabla\cdot\psi = \frac{12}{h^3}\int_{-h/2}^{h/2}\vartheta\,dz \tag{3.56}$$

and $\vartheta = u_{i,i}$ is the volume dilatation. In this case $\nabla = \hat{\mathbf{e}}_\alpha(\partial/\partial x_\alpha)$ is the two-dimensional gradient operator. Upon substitution of (3.55) into (3.24b), we readily obtain

$$M_{\alpha\beta} = D\left[(1-\nu)m_{\alpha\beta} + \nu m\delta_{\alpha\beta}\right] + \frac{\nu\delta_{\alpha\beta}}{1-\nu}\int_{-h/2}^{h/2} z\tau_{zz}\,dz \tag{3.57a}$$

where

$$D = \frac{Eh^3}{12(1-\nu^2)}$$

is referred to as the plate flexural rigidity. Similarly, the substitution of (3.54b) into (3.24a) yields

$$Q_\alpha = Ghq_\alpha \tag{3.57b}$$

These results will now be altered in two respects: (1) we drop the terms $\nu/(1-\nu)\int_{-h/2}^{h/2} z\tau_{zz}\,dz$ from (3.57a) and (2) we replace $G$ by $\kappa^2 G$ in (3.57b), where $\kappa^2$ is a "shear coefficient," a correction factor that can be determined by dynamical considerations (See Section 6.10). It is shown (R. D. Mindlin, "Influence of Rotatory Inertia and Shear on Flexural Motions of Isotropic Elastic Plates," *Journal of Applied Mechanics*, Vol. 18, No. 1, 1951, pp. 31–38) that $\kappa$ is a nearly linear function of Poisson's ratio $\nu$, such that $0.874 < \kappa < 0.955$ for $0 < \nu < 0.5$. For example, if $\nu = 0.3$, then $\kappa = 0.9225$, or $\kappa^2 = 0.86$. In view of these considerations, the (approximate) plate stress-strain relations to be used are

$$M_{\alpha\beta} = D\left[(1-\nu)m_{\alpha\beta} + \nu m\delta_{\alpha\beta}\right] \tag{3.58a}$$

$$Q_\alpha = \kappa^2 Ghq_\alpha \tag{3.58b}$$

Upon substitution of (3.12) into (3.58), we obtain the pertinent plate stress-displacement relations

$$M_{\alpha\beta} = \tfrac{1}{2}D\left[(1-\nu)(\psi_{\alpha,\beta} + \psi_{\beta,\alpha}) + 2\nu\psi_{\gamma,\gamma}\delta_{\alpha\beta}\right] \tag{3.59a}$$

$$Q_\alpha = \kappa^2 Gh(\psi_\alpha + w_{,\alpha}) \quad \text{or} \quad \mathbf{Q} = \kappa^2 Gh(\boldsymbol{\psi} + \nabla w) \tag{3.59b}$$

In the case of polar coordinates $(r, \theta)$, the stress-strain relations (3.58) assume the form

$$\begin{aligned} M_{rr} &= D(m_{rr} + \nu m_{\theta\theta}) \\ M_{\theta\theta} &= D(m_{\theta\theta} + \nu m_{rr}) \\ M_{r\theta} &= D(1-\nu)m_{r\theta} \end{aligned} \tag{3.60a}$$

$$Q_r = \kappa^2 Ghq_r, \qquad Q_\theta = \kappa^2 Ghq_\theta \tag{3.60b}$$

and substitution of (3.17) and (3.18) into (3.60) results in the stress-displacement relations

$$\begin{aligned} M_{rr} &= D\left[\frac{\partial\psi_r}{\partial r} + \frac{\nu}{r}\left(\psi_r + \frac{\partial\psi_\theta}{\partial\theta}\right)\right] \\ M_{\theta\theta} &= D\left[\nu\frac{\partial\psi_r}{\partial r} + \frac{1}{r}\left(\psi_r + \frac{\partial\psi_\theta}{\partial\theta}\right)\right] \\ M_{r\theta} = M_{\theta r} &= \frac{D}{2}(1-\nu)\left[\frac{\partial\psi_\theta}{\partial r} + \frac{1}{r}\frac{\partial\psi_r}{\partial\theta} - \frac{1}{r}\psi_\theta\right] \end{aligned} \tag{3.61a}$$

$$Q_r = \kappa^2 Gh\left(\psi_r + \frac{\partial w}{\partial r}\right), \qquad Q_\theta = \kappa^2 Gh\left(\psi_\theta + \frac{1}{r}\frac{\partial w}{\partial\theta}\right) \tag{3.61b}$$

## 3.9 STRAIN ENERGY

From classical, three-dimensional elasticity theory, the expression for strain energy density in a linear elastic, isotropic solid is given by [see (2.33)]

$$U = \tfrac{1}{2}\tau_{ij}\varepsilon_{ij} = \tfrac{1}{2}(\tau_{\alpha\beta}\varepsilon_{\alpha\beta} + \tau_{\alpha z}\varepsilon_{\alpha z} + \tau_{z\alpha}\varepsilon_{z\alpha} + \tau_{zz}\varepsilon_{zz})$$

However, in view of (3.10c) and (3.13), we have for a plate

$$U = \tfrac{1}{2}(z\tau_{\alpha\beta}m_{\alpha\beta} + \tau_{\alpha z}q_{\alpha})$$

We now define the plate strain energy density (energy per unit area) as

$$W_p = \int_{-h/2}^{h/2} U\,dz = \frac{1}{2}\int_{-h/2}^{h/2} (z\tau_{\alpha\beta}m_{\alpha\beta} + \tau_{\alpha z}q_{\alpha})\,dz$$

so that

$$W_p = \tfrac{1}{2}(M_{\alpha\beta}m_{\alpha\beta} + Q_{\alpha}q_{\alpha}) \tag{3.62}$$

If we substitute (3.58) into (3.62), we obtain the plate strain energy density in terms of plate strain components:

$$W_p = \tfrac{1}{2}D\left[(1-\nu)m_{\alpha\beta}m_{\alpha\beta} + \nu m^2\right] + \tfrac{1}{2}\kappa^2 Ghq_{\alpha}q_{\alpha} \tag{3.63}$$

where $m = m_{\alpha\alpha}$. Conversely, since $M_{\alpha\alpha} = D(1+\nu)m = M$, or $m = M/(1+\nu)D$, we can write $W_p$ as a function of plate stress components:

$$W_p = \frac{1}{2(1-\nu^2)D}\left[(1+\nu)M_{\alpha\beta}M_{\alpha\beta} - \nu M^2\right] + \frac{Q_{\alpha}Q_{\alpha}}{2\kappa^2 Gh} \tag{3.64}$$

We can use (3.63) to obtain the stress-strain relation (3.58):

$$M_{\alpha\beta} = \frac{\partial W_p}{\partial m_{\alpha\beta}} = D\left[(1-\nu)m_{\alpha\beta} + \nu m\delta_{\alpha\beta}\right] \tag{3.65a}$$

$$Q_{\alpha} = \frac{\partial W_p}{\partial q_{\alpha}} = \kappa^2 Ghq_{\alpha} \tag{3.65b}$$

Conversely, we can use (3.64) to obtain the strain-stress relations (see Exercise 3.21):

$$m_{\alpha\beta} = \frac{\partial W_p}{\partial M_{\alpha\beta}} = \frac{1}{(1-\nu^2)D}\left[(1+\nu)M_{\alpha\beta} - \nu M\delta_{\alpha\beta}\right] \tag{3.66a}$$

$$q_{\alpha} = \frac{\partial W_p}{\partial Q_{\alpha}} = \frac{Q_{\alpha}}{\kappa^2 Gh} \tag{3.66b}$$

The existence of the stress potential (or strain energy density function) $W_p$ for plates is discussed in Exercise 6.34.

## 3.10 EQUILIBRIUM

We shall now derive the stress equations of equilibrium of the plate. In the following, body forces are neglected, and as a point of departure, we rewrite the equations of equilibrium (2.22a) of the three-dimensional theory in the form

$$\tau_{\alpha\beta,\beta} + \tau_{\alpha z,z} = 0 \tag{3.67a}$$

$$\tau_{\alpha z,\alpha} + \tau_{zz,z} = 0 \tag{3.67b}$$

The operator $\int_{-h/2}^{h/2} \dots dz$ is now applied to each term of (3.67b), so that

$$\int_{-h/2}^{h/2} \tau_{\alpha z,\alpha}\, dz + \int_{-h/2}^{h/2} \tau_{zz,z}\, dz = 0 \tag{3.68a}$$

However,

$$\int_{-h/2}^{h/2} \tau_{zz,z}\, dz = \tau_{zz}\left(\frac{h}{2}\right) - \tau_{zz}\left(-\frac{h}{2}\right) = p(x, y) \tag{3.69}$$

where $p$ is readily identified as the intensity of the transverse force (i.e., pressure) acting upon the plate. In view of (3.69) and (3.24a), (3.68a) can now be written in the form

$$Q_{\alpha,\alpha} + p = 0 \tag{3.70a}$$

Next, we apply the operation $\int_{-h/2}^{h/2} \dots z\, dz$ to each term of (3.67a), resulting in

$$\int_{-h/2}^{h/2} \tau_{\alpha\beta,\beta} z\, dz + \int_{-h/2}^{h/2} \tau_{\alpha z,z} z\, dz = 0 \tag{3.68b}$$

The second integral in (3.68b) can be integrated by parts, and we obtain

$$\int_{-h/2}^{h/2} \tau_{\alpha z,z} z\, dz = \left(\tau_{\alpha z} z\right)_{z=-h/2}^{z=h/2} - \int_{-h/2}^{h/2} \tau_{\alpha z}\, dz = -Q_\alpha \tag{3.71}$$

In view of (3.24a) and because the plate faces are free from shear stresses, we have $\tau_{\alpha z}(\pm \frac{1}{2}h) = 0$. With the aid of (3.71) and (3.24b), we can write (3.68b) in the form

$$M_{\alpha\beta,\beta} - Q_\alpha = 0 \tag{3.70b}$$

In view of (1.62), we can write the plate stress equations of equilibrium in invariant form (see Exercise 3.18):

$$\mathbf{div}\, \mathcal{M} = \nabla \cdot \mathcal{M} = \mathbf{Q} \tag{3.72a}$$

$$\mathbf{div}\, \mathbf{Q} = \nabla \cdot \mathbf{Q} = -p \tag{3.72b}$$

where $\mathcal{M} = \hat{\mathbf{e}}_\alpha \hat{\mathbf{e}}_\beta M_{\alpha\beta}$ is the moment tensor [see (3.39)]. We can utilize (3.72) with advantage to calculate the plate stress equations of equilibrium referred to any given orthogonal curvilinear coordinate system.

**Example:** We calculate the plate stress equations of equilibrium referred to a polar coordinate system. In this case we have [see (3.39) and Fig. 1.3]

$$\mathcal{M} = M_{rr}\hat{\mathbf{e}}_r\hat{\mathbf{e}}_r + M_{r\theta}\hat{\mathbf{e}}_r\hat{\mathbf{e}}_\theta + M_{\theta r}\hat{\mathbf{e}}_\theta\hat{\mathbf{e}}_r + M_{\theta\theta}\hat{\mathbf{e}}_\theta\hat{\mathbf{e}}_\theta$$

$$\mathbf{Q} = \hat{\mathbf{e}}_r Q_r + \hat{\mathbf{e}}_\theta Q_\theta \qquad \text{[see (3.30)]}$$

$$\nabla = \hat{\mathbf{e}}_r \frac{\partial}{\partial r} + \hat{\mathbf{e}}_\theta \frac{1}{r}\frac{\partial}{\partial \theta} \qquad \text{[see (1.44a)]}$$

and we have

$$\begin{aligned}
\operatorname{div} \mathcal{M} &= \left(\hat{\mathbf{e}}_r \frac{\partial}{\partial r} + \hat{\mathbf{e}}_\theta \frac{1}{r}\frac{\partial}{\partial \theta}\right) \cdot \left(M_{rr}\hat{\mathbf{e}}_r\hat{\mathbf{e}}_r + M_{r\theta}\hat{\mathbf{e}}_r\hat{\mathbf{e}}_\theta + M_{\theta r}\hat{\mathbf{e}}_\theta\hat{\mathbf{e}}_r + M_{\theta\theta}\hat{\mathbf{e}}_\theta\hat{\mathbf{e}}_\theta\right) \\
&= \hat{\mathbf{e}}_r \left[\frac{\partial M_{rr}}{\partial r} + \frac{1}{r}\frac{\partial M_{\theta r}}{\partial \theta} + \frac{1}{r}(M_{rr} - M_{\theta\theta})\right] \\
&\quad + \hat{\mathbf{e}}_\theta \left(\frac{\partial M_{r\theta}}{\partial r} + \frac{M_{r\theta}}{r} + \frac{M_{\theta R}}{r} + \frac{1}{r}\frac{\partial M_{\theta\theta}}{\partial \theta}\right) \\
\operatorname{div} \mathbf{Q} &= \left(\hat{\mathbf{e}}_r \frac{\partial}{\partial r} + \hat{\mathbf{e}}_\theta \frac{1}{r}\frac{\partial}{\partial \theta}\right) \cdot (\hat{\mathbf{e}}_r Q_r + \hat{\mathbf{e}}_\theta Q_\theta) \\
&= \frac{\partial Q_r}{\partial r} + \frac{Q_r}{r} + \frac{1}{r}\frac{\partial Q_\theta}{\partial \theta}
\end{aligned}$$

where we have used (1.43). Thus, upon substitution into (3.72) and subsequent decomposition, we obtain

$$\frac{\partial Q_r}{\partial r} + \frac{1}{r}Q_r + \frac{1}{r}\frac{\partial Q_\theta}{\partial \theta} + p = 0 \tag{3.73a}$$

$$\frac{\partial M_{rr}}{\partial r} + \frac{1}{r}\frac{\partial M_{\theta r}}{\partial \theta} + \frac{1}{r}(M_{rr} - M_{\theta\theta}) = Q_r \tag{3.73b}$$

$$\frac{\partial M_{r\theta}}{\partial r} + \frac{1}{r}\frac{\partial M_{\theta\theta}}{\partial \theta} + \frac{2M_{r\theta}}{r} = Q_\theta \tag{3.73c}$$

Equations (3.73) are the plate stress equations of equilibrium referred to polar coordinates.

The equations of equilibrium can also be expressed in terms of displacements. If we substitute (3.59) into (3.70a) and (3.70b), we obtain

$$\tfrac{1}{2}D\left[(1-\nu)\psi_{\alpha,\beta\beta}+(1+\nu)\psi_{\beta,\beta\alpha}\right]-\kappa^2Gh(\psi_\alpha+w_{,\alpha})=0 \quad (3.74a)$$

$$\kappa^2Gh(\psi_{\alpha,\alpha}+w_{,\alpha\alpha})+p=0 \quad (3.74b)$$

or, in (invariant) vector notation,

$$\tfrac{1}{2}D\left[(1-\nu)\nabla^2\boldsymbol{\psi}+(1+\nu)\nabla(\nabla\cdot\boldsymbol{\psi})\right]-\kappa^2Gh(\boldsymbol{\psi}+\nabla w)=\mathbf{0} \quad (3.75a)$$

$$\kappa^2Gh(\nabla\cdot\boldsymbol{\psi}+\nabla^2 w)+p=0 \quad (3.75b)$$

Equations (3.74) or (3.75) are the displacement equations of equilibrium for the present (shear-deformable) plate model (improved theory). We note that these equations are coupled with respect to the displacement variables $w$ and $\boldsymbol{\psi}=\hat{\mathbf{e}}_\alpha\psi_\alpha$.

A set of equilibrium equations can also be written in uncoupled form. Toward this end we use (3.70b), (3.56), and (3.58a) to show

$$M_{\alpha\beta,\beta\alpha}=Q_{\alpha,\alpha}=-p=D\nabla^2 m \quad (3.76)$$

where $m=\psi_{\alpha,\alpha}$ [see (3.56)]. From (3.70a), (3.56), (3.59b), and (3.76) we have $-p=Q_{\alpha,\alpha}=\kappa^2Gh(m+\nabla^2 w)$, or

$$\nabla^2 w=-\frac{p}{\kappa^2Gh}-m \quad (3.77a)$$

Using (3.70b), (3.56), and (3.59), we can show that

$$Q_\alpha=M_{\alpha\beta,\beta}=\tfrac{1}{2}D\left[(1-\nu)\nabla^2\psi_\alpha+(1+\nu)m_{,\alpha}\right]=\kappa^2Gh(\psi_\alpha+w_{,\alpha})$$

or

$$\left(\frac{h^2}{12\kappa^2}\nabla^2-1\right)\psi_\alpha=\frac{\partial}{\partial x_\alpha}\left(w-\frac{1+\nu}{1-\nu}\frac{h^2}{12\kappa^2}m\right) \quad (3.77b)$$

Further manipulation of (3.77) results in

$$\nabla^4 w=\frac{p}{D}-\frac{\nabla^2 p}{\kappa^2Gh} \quad (3.78a)$$

$$\left(\frac{h^2}{12\kappa^2}\nabla^2-1\right)\left(D\nabla^4\psi_\alpha+\frac{\partial p}{\partial x_\alpha}\right)=0 \quad (3.78b)$$

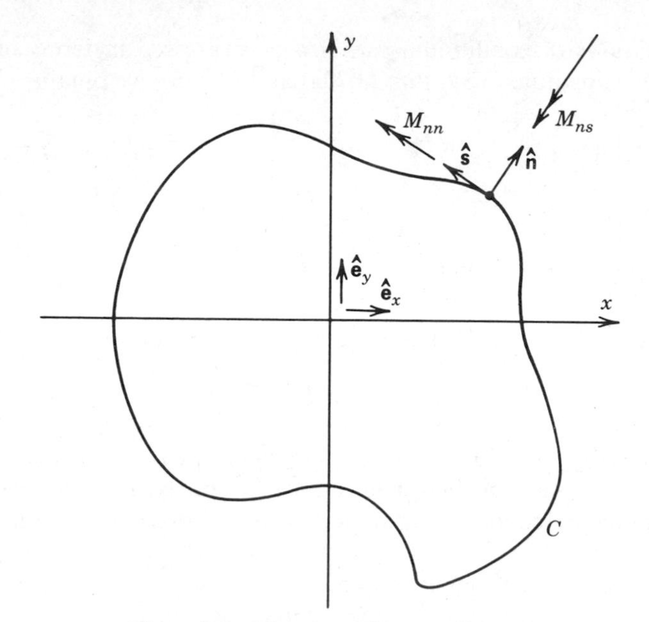


**Figure 3.8** Plate boundary conditions.

Each of the equilibrium equations (3.78) contains only a single displacement variable, that is, these equations are completely decoupled. We also note that the two equations (3.78b) can be written in (invariant) vector form

$$\left(\frac{h^2}{12\kappa^2}\nabla^2 - 1\right)\left(D\nabla^4\boldsymbol{\psi} + \nabla p\right) = \mathbf{0} \tag{3.78c}$$

It can be shown (see Section 4.5) that unique solutions of (3.74) or (3.78) result if we impose the following boundary conditions upon the solution along the plate boundary $C$: We specify

$$\text{Either } w \text{ or } Q_n = Q_\alpha n_\alpha \tag{3.79a}$$

and

$$\text{Either } \psi_n \text{ or } M_{nn} = M_{\alpha\beta} n_\alpha n_\beta \tag{3.79b}$$

and

$$\text{Either } \psi_s \text{ or } M_{ns} = M_{\alpha\beta} n_\alpha s_\beta \tag{3.79c}$$

where (see Fig. 3.8) $Q_n$ is the transverse shear force/unit length along $C$, $M_{nn}$ is the bending moment/unit length along $C$, $M_{ns}$ is the twisting moment/unit length along $C$ and

$$\boldsymbol{\psi} = \hat{\mathbf{e}}_x\psi_x + \hat{\mathbf{e}}_y\psi_y = \hat{\mathbf{n}}\psi_n + \hat{\mathbf{s}}\psi_s$$

where $\hat{\mathbf{n}}$ and $\hat{\mathbf{s}}$ are unit normal and tangent vectors to $C$, respectively. It should be clear that $\psi_n$ and $\psi_s$ are components of the angular rotation vector $\psi$ that are normal and tangential, respectively, to the boundary curve $C$.

## EXERCISES

**3.1** With reference to (3.2), show that $\psi = \mathbf{e}_x\psi_x + \mathbf{e}_y\psi_y$ is a vector in a two-dimensional space.

**3.2** Establish the relations (3.6), (3.7), and (3.8).

**3.3**

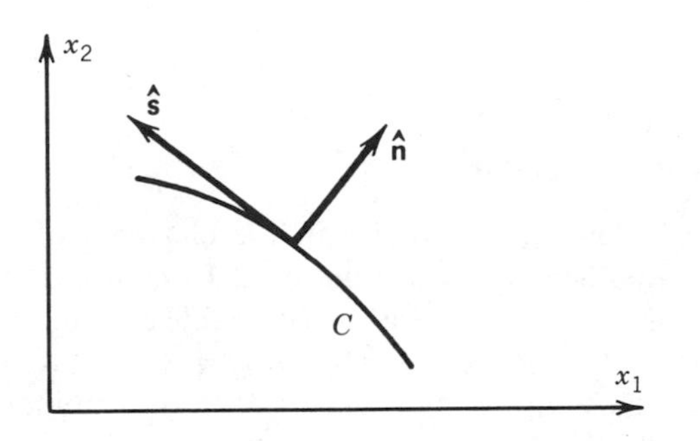


Let $\hat{\mathbf{n}}$ and $\hat{\mathbf{s}}$ be unit normal and tangent vectors to $C$, respectively, as shown. Show that $e_{\alpha\beta} = n_\alpha s_\beta - n_\beta s_\alpha$, where $e_{\alpha\beta} = \beta - \alpha$.

**3.4** **(a)** Show that $m_{\alpha\beta}$ is a tensor of order two.
**(b)** Show that $q_\alpha$ is a vector.

**3.5** Consider the case of a simply connected region in the median plane. Show that (3.22) are sufficient conditions for the existence of a single-valued plate displacement field.

**3.6** Substitute (3.12) into (3.22), and show that $S_{12} = R_1 = R_2 = 0$.

**3.7** Show that $R_{1,2} + R_{2,1} = -2S_{12}$ [see (3.23)].

**3.8**

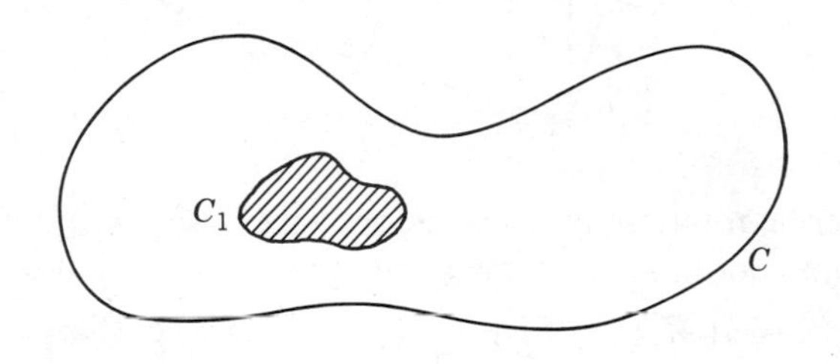


**(a)** A plate strain field ($q_\alpha$, $m_{\alpha\beta}$) is specified for a plate with an internal cut-out as shown. Show that a sufficient condition for the existence of a single-valued plate displacement field $(w, \psi_x, \psi_y)$ is given by (3.22) augmented by the requirements

$$\oint_{C_1} dw = 0 \quad \text{and} \quad \oint_{C_1} d\psi_\alpha = 0$$

**(b)** Generalize (a) to the case of a plate with $n$ internal cut-outs.

**3.9** Use (3.16) to derive the plate bending strain tensor components referred to Cartesian coordinates.

**3.10** **(a)** Given that $\varepsilon_{ij}$ are the (three-dimensional) strain tensor components, show that $m_{\gamma'\delta'} = a_{\gamma'\alpha} a_{\delta'\beta} m_{\alpha\beta}$, where $a_{\gamma'\alpha} = \cos(\gamma', \alpha)$.

**(b)** Show that $q_{\beta'} = a_{\beta'\alpha} q_\alpha$.

**3.11** Provide a detailed derivation of (3.17) and (3.18).

**3.12** Derive (3.49) from (3.48).

**3.13** In classical plate theory, material line elements of the plate that are normal to the median surface before deformation remain normal after deformation: $\psi_\alpha = -w,_\alpha$. Show that in this case $q_\alpha = 0$, and [see (3.22) and (3.12)] $R_1 = R_2 = S_{12} = 0$, $m_{\alpha\beta} = w,_{\alpha\beta}$, provided the median surface is sufficiently smooth, that is, there is a unique tangent plane at every point of the deformed median surface.

**3.14** Show that the moment tensor is symmetric; that is, show that $M_{\alpha\beta} = M_{\beta\alpha}$.

**3.15**

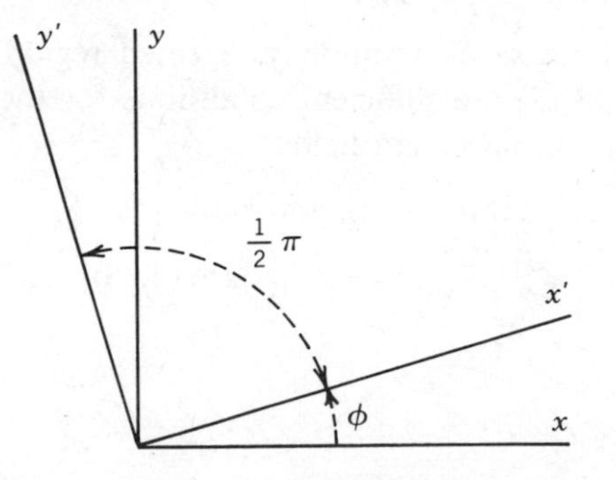


Consider a rotation of axes as shown. Both $(x, y)$ and $(x', y')$ are Cartesian coordinate systems in the plane. Show that

$$M_{x'x'} = \tfrac{1}{2}(M_{xx} + M_{yy}) + \tfrac{1}{2}(M_{xx} - M_{yy})\cos 2\phi + M_{xy}\sin 2\phi$$

$$M_{y'y'} = \tfrac{1}{2}(M_{xx} + M_{yy}) - \tfrac{1}{2}(M_{xx} - M_{yy})\cos 2\phi - M_{xy}\sin 2\phi$$

$$M_{x'y'} = -\tfrac{1}{2}(M_{xx} - M_{yy})\sin 2\phi + M_{xy}\cos 2\phi$$

**3.16** Consider a rotation of axes as in Exercise 3.15.

**(a)** Show that

$$M_{x'x'} - M_{y'y'} + 2iM_{x'y'} = \left(M_{xx} - M_{yy} + 2iM_{xy}\right)e^{2i\phi}$$
$$M_{x'x'} + M_{y'y'} = M_{xx} + M_{yy}$$

where $i = \sqrt{-1}$.

**(b)** Show that the results in (a) are equivalent to the results of Exercise 3.15.

**3.17** Use the results of Exercise 3.15 to establish the Mohr circle construction for the moment tensor. Show that

$$(M_{xx} - H)^2 + (M_{xy})^2 = R^2$$

where

$$H = \tfrac{1}{2}(M_{xx} + M_{yy}) = \tfrac{1}{2}I_1$$

and

$$R^2 = \left(\frac{M_{xx} - M_{yy}}{2}\right)^2 + M_{xy}^2 = \left(\frac{\mu_1 - \mu_2}{2}\right)^2 = \left(\frac{I_1}{2}\right)^2 - I_2$$

Note that the coordinates of the center of the circle are $(H, 0)$, and its radius is $R$.

**3.18** Write (3.72) relative to Cartesian coordinate axes.

**3.19** At a point in the plate, the stress tensor components are $M_{xx} = 10{,}000$ N, $M_{yy} = -15{,}000$ N, $M_{xy} = -12{,}000$ N. Find the extreme values of the bending moment and the associated principal axes. Draw the Mohr circle.

**3.20** At a point in the plate, the components of the shear force vector are $Q_x = 4500$ N/m and $Q_y = -3000$ N/m. Find the extreme values of the shear force and specify the planes upon which they act.

**3.21** Derive (3.66) from (3.64).

**3.22** Consider a point $P$ on a smooth surface $w = w(x, y)$, where $x$, $y$ are Cartesian axes in the plane. Perform a rotation of axes $x_{\beta'} = a_{\beta'\alpha}x_\alpha$ where $a_{\beta'\alpha} = \cos(x_{\beta'}, x_\alpha)$.

**(a)** Show that the slope of the surface at $P$ is a vector; that is, prove that $w_{,\beta'} = a_{\beta'\alpha}w_{,\alpha}$.

**(b)** Prove $w,_{\gamma'\delta'} = a_{\gamma'\alpha} a_{\delta'\beta} w,_{\alpha\beta}$. Now assume that the square of the extreme value of the slope at $P$ with respect to the $x$-$y$ plane is small compared to unity.

**(c)** Show that in this case $1/R_x \cong -\partial^2 w/\partial x^2$ and $1/R_y \cong -\partial^2 w/\partial y$, and define the "twist" by $1/R_{xy} = -\partial^2 w/\partial x\, \partial y = 1/R_{yx}$. Hence, show that the four numbers

$$\begin{bmatrix} 1/R_x & 1/R_{xy} \\ 1/R_{yx} & 1/R_y \end{bmatrix}$$

are the components of a tensor of order two in a two-dimensional space.

**3.23** Consider a (median) surface whose equation is $w = w(x, y)$, where $(x, y)$ are Cartesian coordinates in the reference plane. The partial derivatives $\partial w/\partial x$ and $\partial w/\partial y$ are the slopes of the curves created by the intersection of the surface with the planes $y =$ constant and $x =$ constant, respectively. The extreme value of the slope is given by $[(\partial w/\partial x)^2 + (\partial w/\partial y)^2]^{1/2} = |\text{grad } w|$. Construct a tangent plane to the surface at point 0. Now take $x$-$y$ axes in the tangent plane with origin at 0. The MacLaurin expansion of the function $w(x, y)$ for sufficiently small values of $x$ and $y$ is

$$w = \frac{1}{2}\left[\left(\frac{\partial^2 w}{\partial x^2}\right)_0 x^2 + 2\left(\frac{\partial^2 w}{\partial x\, \partial y}\right)_0 xy + \left(\frac{\partial^2 w}{\partial y^2}\right)_0 y^2\right]$$

because, for the present case, we have $w = \partial w/\partial x = \partial w/\partial y = 0$ at $(x, y) = (0, 0)$. If we take sections of the (shallow) surface by the coordinate planes $x0z$ and $y0z$, the coefficients $\kappa_{xx} = (\partial^2 w/\partial x^2)_0$ and $\kappa_{yy} = (\partial^2 w/\partial y^2)_0$ can be interpreted as curvatures of the section at 0. If we also set $\kappa_{xy} = (\partial^2 w/\partial x\, \partial y)_0$, the equation of the intersection of the surface with the plane $w = c$ takes the form

$$2c = \kappa_{xx} x^2 + 2\kappa_{xy} xy + \kappa_{yy} y^2$$

This is the equation of a conic and will therefore have two principal axes. Referred to these axes, the equation of the surface becomes $2w = \kappa_{11}(x')^2 + \kappa_{22}(y')^2$, where $\kappa_{11}$ and $\kappa_{22}$ are the principal curvatures (see Exercise 3.25).

**3.24**

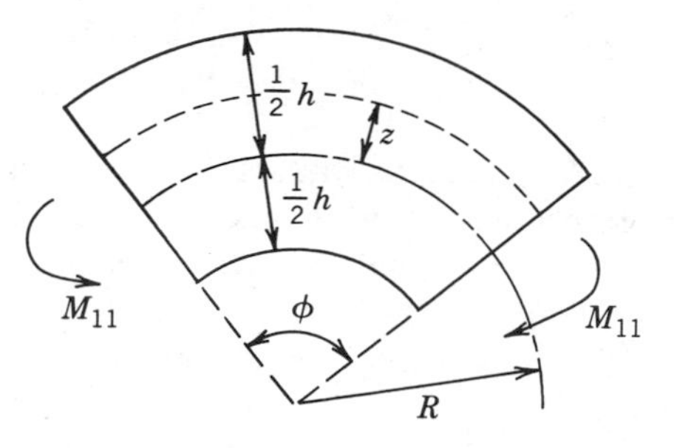


Consider cylindrical, pure bending of a plate resulting in the (principal) curvature $1/R$, as shown.

**(a)** Show that $\varepsilon_{11} = z/R$ is the change in length per unit of original length of a longitudinal fiber.

**(b)** Show that

$$M_{11} = \int_{-h/2}^{h/2} \tau_{11} z \, dz = \frac{D}{R}, \qquad M_{22} = \int_{-h/2}^{h/2} \tau_{22} z \, dz = \frac{\nu D}{R}$$

where $D = Eh^3/12(1 - \nu^2)$. Assume $\tau_{33} \equiv 0$. Note that $h$ and $R$ are arbitrary, with the proviso $h < 2R$.

**3.25**

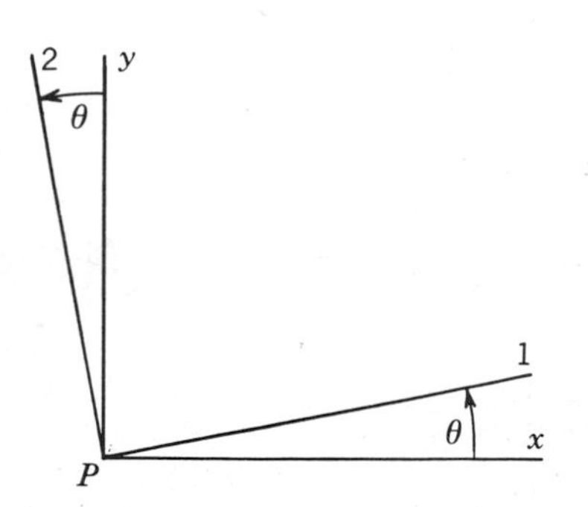


Consider a curved surface and the associated tangent plane $\Pi$ to this surface at $P$. Let $1/R$ be the normal curvature of the surface curve lying in the normal plane through $P$. There exist two mutually perpendicular, normal planes at $P$ for which $1/R$ attains its extreme values $1/R_1$ and $1/R_2$. We now establish a Cartesian axis system $x$-$y$ in $\Pi$ with origin at $P$. If 1 and 2 denote principal axes of curvature, Euler's theorem states that

$$\frac{1}{R_x} = \frac{\cos^2\theta}{R_1} + \frac{\sin^2\theta}{R_2}, \qquad \frac{1}{R_y} = \frac{\sin^2\theta}{R_1} + \frac{\cos^2\theta}{R_2}$$

where $1/R_x$ and $1/R_y$ are curvatures in the normal planes through the $x$ and $y$ axes, respectively. For details, see Erwin Kreyszig, *Introduction to Differential Geometry and Riemannian Geometry*, University of Toronto Press, Toronto, 1968, pp. 89–95. The quantities $\frac{1}{2}(1/R_1 + 1/R_2)$ and $1/R_1R_2$ are known as the mean and Gaussian curvatures, respectively.

**3.26** Let $R_1$ and $R_2$ be principal radii of curvature, as described in Exercise 3.25. Show that $m_{xx} + m_{yy} = 1/R_1 + 1/R_2$ and $m_{xx}m_{yy} - m_{xy}^2 = 1/R_1R_2$ are invariants with respect to orthogonal rotations of the $x$-$y$ coordinate system.

**3.27** Let $R_1$ and $R_2$ be principal radii of curvature of the plate median surface (see Exercise 3.25) and neglect shearing strains.

**(a)** Show that the strain energy density of the plate is given by

$$W_p = \tfrac{1}{2}D\left[\left(\frac{1}{R_1} + \frac{1}{R_2}\right)^2 - 2(1-\nu)\frac{1}{R_1}\frac{1}{R_2}\right]$$
$$= \tfrac{1}{2}D\left(\frac{1}{R_1^2} + \frac{1}{R_2^2} + \frac{2\nu}{R_1R_2}\right)$$

**(b)** Using the results of (a), show that $W_p \geq 0$.

**3.28** Take axes 1 and 2 as principal axes for curvature. Use (3.14) and Exercise 3.24 to show that

$$m_{11} = \frac{1}{R_1} = \tfrac{1}{2}(m_{xx} + m_{yy}) + \tfrac{1}{2}(m_{xx} - m_{yy})\cos 2\theta + m_{xy}\sin 2\theta$$

$$m_{22} = \frac{1}{R_2} = \tfrac{1}{2}(m_{xx} + m_{yy}) - \tfrac{1}{2}(m_{xx} - m_{yy})\cos 2\theta - m_{xy}\sin 2\theta$$

$$m_{12} = 0 = -\tfrac{1}{2}(m_{xx} - m_{yy})\sin 2\theta + m_{xy}\cos 2\theta$$

Show that

$$m_{xx} + m_{yy} = \frac{1}{R_1} + \frac{1}{R_2}, \qquad m_{xx} - m_{yy} = \left(\frac{1}{R_1} - \frac{1}{R_2}\right)\cos 2\theta,$$

$$2m_{xy} = \left(\frac{1}{R_1} - \frac{1}{R_2}\right)\sin 2\theta$$

Consequently, using Euler's theorem (Exercise 3.25) show that

$$m_{xx} = \frac{\cos^2\theta}{R_1} + \frac{\sin^2\theta}{R_2} = \frac{1}{R_x}, \qquad m_{yy} = \frac{\sin^2\theta}{R_1} + \frac{\cos^2\theta}{R_2} = \frac{1}{R_y}$$

**3.29** Assume that $|w,_{\alpha}| \ll 1$. Show that in this case $M_{\alpha\beta} = -w,_{\alpha\beta}$, $1/R_1 + 1/R_2 = -\nabla^2 w$, and

$$\frac{1}{R_1}\frac{1}{R_2} = \begin{vmatrix} -w,_{xx} & -w,_{xy} \\ -w,_{yx} & -w,_{yy} \end{vmatrix} = w,_{xx}w,_{yy} - w^2,_{xy}$$

and therefore, neglecting shearing strains,

$$W_p = \tfrac{1}{2}D\left[(\nabla^2 w)^2 - 2(1-\nu)\left(w,_{xx}w,_{yy} - w^2,_{xy}\right)\right]$$

**3.30** *Thermo-Elastic Plate Stress-Strain Relations*. Hooke's law can be modified to account for (linear) thermo-elastic effects:

$$E\varepsilon_{ij} = (1+\nu)\tau_{ij} - \nu\tau_{kk}\delta_{ij} + E\alpha T\delta_{ij}$$

where $T$ is temperature change and $\alpha$ is the coefficient of the linear (thermal) expansion of the material. Use this equation in place of (3.53) and repeat the analysis in Section 3.8. Show that in this case

$$N_{\alpha\beta} = C\left[(1-\nu)n_{\alpha\beta} + \nu n\delta_{\alpha\beta}\right] - \frac{N_T}{1-\nu}\delta_{\alpha\beta}$$

$$M_{\alpha\beta} = D\left[(1-\nu)m_{\alpha\beta} + \nu m\delta_{\alpha\beta}\right] - \frac{M_T}{1-\nu}\delta_{\alpha\beta}$$

$$Q_\alpha = Ghq_\alpha$$

where

$$C = \frac{Eh}{1-\nu^2}$$

$$N_{\alpha\beta} = \int_{-h/2}^{h/2} \tau_{\alpha\beta}\,dz, \qquad n_{\alpha\beta} = \int_{-h/2}^{h/2} \varepsilon_{\alpha\beta}\,dz$$

$$n = n_{\alpha\alpha}, \qquad M_T = \alpha E\int_{-h/2}^{h/2} zT\,dz$$

$$N_T = \alpha E\int_{-h/2}^{h/2} T\,dz$$

For details and an application see H. Reismann, D.P. Malone, and P.S. Pawlik, "Laser Induced Thermoelastic Response of Circular Plates", Solid Mechanics Archives, Vol. 5, No. 3, Aug. 1980, pp. 253–323.

**3.31** *Hooke's Law for Anisotropic Plates*. From three-dimensional elasticity theory, the strain energy density $W$ of a material with general anisotropy is $W = \frac{1}{2}C_{ijkl}\varepsilon_{ij}\varepsilon_{kl}$, where $C_{ijkl} = C_{jikl} = C_{ijlk} = C_{klij}$ are the components of the elasticity tensor. Set $\varepsilon_{zz} = 0$. Show that, for plates,

$$W_p = \int_{-h/2}^{h/2} W\,dz = \tfrac{1}{2}D_{\alpha\beta\gamma\delta}m_{\alpha\beta}m_{\gamma\delta} + \tfrac{1}{2}D_{\alpha\beta}q_\alpha q_\beta$$

where $D_{\alpha\beta\gamma\delta} = D_{\beta\alpha\gamma\delta} = D_{\alpha\beta\delta\gamma} = D_{\gamma\delta\alpha\beta}$ (six independent constants) and $D_{\alpha\beta} = D_{\beta\alpha}$ (three independent constants). Show that $W_p$ has $6 + 3 = 9$ independent elastic constants, and

$$M_{\alpha\beta} = \frac{\partial W_p}{\partial m_{\alpha\beta}} = D_{\alpha\beta\gamma\delta}m_{\gamma\delta}, \qquad Q_\alpha = 2D_{\alpha\beta}q_\beta$$

**3.32** *Isotropic Elastic Plates*. In the case of isotropy, $W_p(m_{\alpha\beta}, q_\alpha)$ must be quadratic in the strain components and expressible in terms of strain invariants (why?). Hence, we can write $W_p = \frac{1}{2}(Am^2 + Bm_{\alpha\beta}m_{\alpha\beta} + Cq_\alpha q_\alpha)$, where $m = m_{\alpha\alpha}$ and

$$M_{\alpha\beta} = \frac{\partial W_p}{\partial m_{\alpha\beta}} = Am\delta_{\alpha\beta} + Bm_{\alpha\beta}, \qquad Q_\alpha = \frac{\partial W_p}{\partial q_\alpha} = Cq_\alpha$$

Now use the results of Exercise 3.35 and Section 6.10 to determine the constants $A$, $B$, and $C$. Show that $A = \nu D$, $B = D(1 - \nu)$, and $C = \kappa^2 Gh$.

**3.33** **(a)** Write the plate stress equations of equilibrium relative to elliptical coordinates $(u, v)$, where $x = a\cosh u \cos v$, $y = a \sinh u \sin v$, and $u \geq 0$ and $0 \leq v < 2\pi$ [see Example (b), Section 1.3].

**(b)** Write the plate stress equations of equilibrium relative to parabolic coordinates $(u, v)$, where $x = \frac{1}{2}(u^2 - v^2)$, $y = uv$, and $-\infty < u < \infty$ and $v \geq 0$ (see Exercise 1.21).

**3.34** Show that the displacement field

$$u = (a - \nu b)xz, \qquad v = (b - \nu a)yz$$
$$w = -\tfrac{1}{2}(a - \nu b)x^2 - \tfrac{1}{2}(b - \nu a)y^2 - \tfrac{1}{2}\nu(a + b)z^2$$

generates the strain field

$$\varepsilon_{xx} = (a - \nu b)z = \frac{z}{R_x}, \qquad \varepsilon_{yy} = (b - \nu a)z = \frac{z}{R_y}$$

$$\varepsilon_{zz} = -\nu(a + b)z = -\frac{\nu}{1 - \nu}\left(\frac{1}{R_x} + \frac{1}{R_y}\right)z$$

Here $\varepsilon_{ij} \equiv 0$ for $i \neq j$, and the stress field $\tau_{xx} = Eaz$, $\tau_{yy} = Ebz$, $\tau_{zz} = 0$, and $\tau_{ij} = 0$ for $i \neq j$. Show that the stress field satisfies the stress equations of equilibrium (2.22). Show that this three-dimensional elasticity solution corresponds to the case of a rectangular plate of thickness $h$ bounded by planes $z = \pm \frac{1}{2}h$ and bent by moments $M_{xx}$ and $M_{yy}$ uniformly distributed along its edges. Show that

$$\frac{\partial^2 w}{\partial x^2} = \nu b - a = -\frac{1}{R_x}, \qquad \frac{\partial^2 w}{\partial y^2} = \nu a - b = -\frac{1}{R_y}$$

$$a = \frac{1}{1-\nu^2}\left(\frac{1}{R_x} + \frac{\nu}{R_y}\right), \qquad b = \frac{1}{1-\nu^2}\left(\frac{1}{R_y} + \frac{\nu}{R_x}\right)$$

$$m_{xx} = \frac{12}{h^3}\int_{-h/2}^{h/2} z\varepsilon_{xx}\, dz = \frac{1}{R_x}, \qquad m_{yy} = \frac{1}{R_y}, \qquad m_{xy} = q_x = q_y = 0$$

$$M_{xx} = \int_{-h/2}^{h/2} \tau_{xx} z\, dz = D\left(\frac{1}{R_x} + \frac{\nu}{R_y}\right)$$

$$M_{yy} = D\left(\frac{1}{R_y} + \frac{\nu}{R_x}\right), \qquad M_{xy} = Q_x = Q_y = 0$$

Each plane $z =$ constant is bent with curvatures $1/R_x$ and $1/R_y$ in the planes $x =$ constant and $y =$ constant, respectively. The centers of curvature lie in the negative direction of the $z$-axis for positive curvatures. Note the implied approximation $(\partial w/\partial x)^2 \ll 1$, $(\partial w/\partial y)^2 \ll 1$. When $M_{xx} = M_{yy} = M$, we have $R_x = R_y = R$ and $M = D(1-\nu)(1/R)$. In this case the plate is bent into a "spherical" surface.

**3.35** *Pure Bending of Plates*. Consider a rectangular plate referred to (principal) axes 1, 2 that coincide with a pair of intersecting edges. The plate is bent by uniformly distributed bending moments $M_{11}$ and $M_{22}$ along the edges. Let $1/R_1$ and $1/R_2$ be the curvatures of the median surface parallel to coordinates axes 1 and 2, respectively.

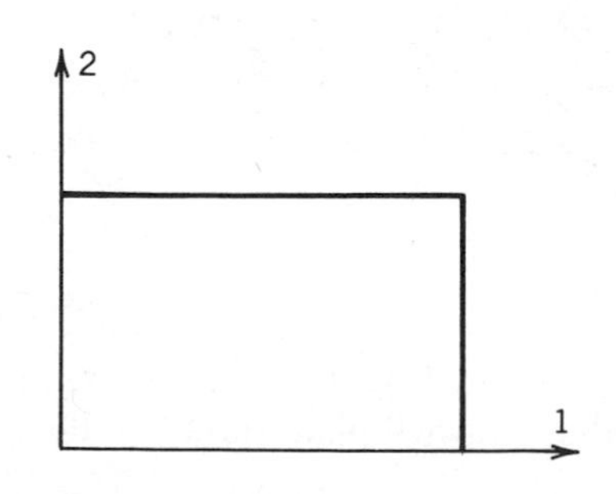

**(a)** Show that

$$M_{11} = D(m_{11} + \nu m_{22}) = D\left(\frac{1}{R_1} + \frac{\nu}{R_2}\right)$$

$$M_{22} = D(m_{22} + \nu m_{11}) = D\left(\frac{1}{R_2} + \frac{\nu}{R_1}\right)$$

$$M_{12} = 0$$

**(b)** Show that

$$\frac{1}{R_1} = \frac{12}{Eh^3}(M_{11} - \nu M_{22}), \qquad \frac{1}{R_2} = \frac{12}{Eh^3}(M_{22} - \nu M_{11})$$

Note that when $M_{22} = 0$, we have $R_2 = -\nu R_1$. From this result we conclude that when a rectangular plate is bent by moments acting on opposite (parallel) edges, an opposite, or "anticlastic," curvature is produced in the plane of the cross-section. The magnitude of the ratio of $R_2$ to $R_1$ is equal to Poisson's ratio $\nu$ (see the figure that follows).

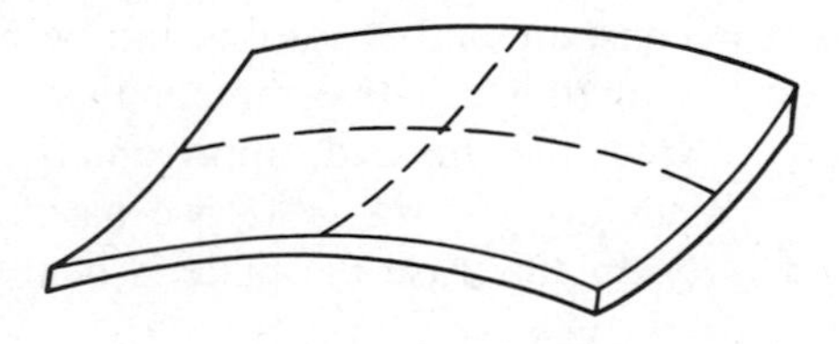

**3.36**

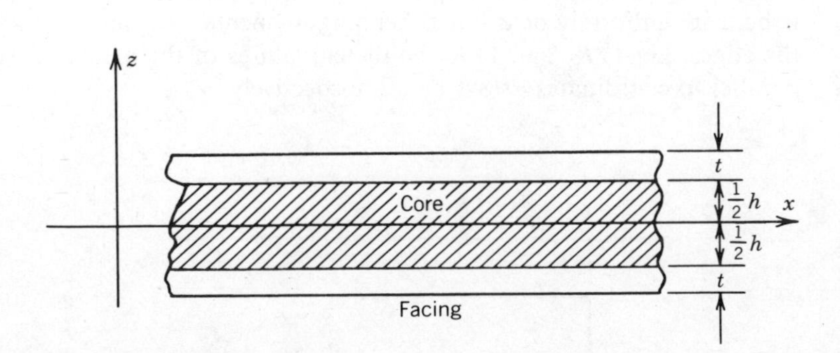


The theory developed in Chapter 3 can be readily modified to predict the behavior of the sandwich plate shown in the figure. We shall use the letters $f$ and $c$ to denote quantities pertaining to facing and core, respectively. Assuming that deformations are continuous, show that appropriate stress-displacement relations are given by

$$M_{xx} = M_{xx}^{(c)} + M_{xx}^{(f)} = \left( \frac{E_c I_c}{1 - \nu_c^2} + \frac{E_f I_f}{1 - \nu_f^2} \right) \frac{\partial \psi_x}{\partial x}$$

$$+ \left( \frac{\nu_c E_c I_c}{1 - \nu_c^2} + \frac{\nu_f E_f I_f}{1 - \nu_f^2} \right) \frac{\partial \psi_y}{\partial y}$$

$$M_{yy} = M_{yy}^{(c)} + M_{yy}^{(f)} = \left( \frac{E_c I_c}{1 - \nu_c^2} + \frac{E_f I_f}{1 - \nu_f^2} \right) \frac{\partial \psi_y}{\partial y}$$

$$+ \left( \frac{\nu_c E_c I_c}{1 - \nu_c^2} + \frac{\nu_f E_f I_f}{1 - \nu_f^2} \right) \frac{\partial \psi_x}{\partial x}$$

$$M_{xy} = M_{xy}^{(c)} + M_{xy}^{(f)} = \left( G_c I_c + G_f I_f \right) \left( \frac{\partial \psi_x}{\partial y} + \frac{\partial \psi_y}{\partial x} \right)$$

$$Q_x = Q_x^{(c)} + Q_x^{(f)} = \kappa^2 \left( G_c A_c + G_f A_f \right) \left( \psi_x + \frac{\partial w}{\partial x} \right)$$

$$Q_y = Q_y^{(c)} + Q_y^{(f)} = \kappa^2 \left( G_c A_c + G_f A_f \right) \left( \psi_y + \frac{\partial w}{\partial y} \right)$$

where

$$M_{xx}^{(c)} = \int_{-h/2}^{h/2} z \tau_{zz} \, Dz = \frac{E_c I_c}{1 - \nu_c^2} \left( \frac{\partial \psi_x}{\partial x} + \nu_c \frac{\partial \psi_y}{\partial y} \right)$$

$$M_{xx}^{(f)} = \int_{-(h/2+t)}^{-h/2} z \tau_{zz} \, dz + \int_{h/2}^{h/2+t} z \tau_{zz} \, dz = \frac{E_f I_f}{1 - \nu_f^2} \left( \frac{\partial \psi_x}{\partial x} + \nu_f \frac{\partial \psi_y}{\partial y} \right)$$

$$A_c = h, \qquad A_f = 2t$$

$$I_c = \frac{h^3}{12}, \qquad I_f = \tfrac{2}{3} t \left( \tfrac{3}{4} h^2 + \tfrac{3}{2} h t + t^2 \right)$$

and so on.

**3.37** Assume axial symmetry, and write the pertinent displacement equations of equilibrium referred to polar coordinates. Assume the plate has the shape of a circle. Use improved (shear-deformable) plate theory.

**3.38** Derive thermo-elastic plate stress-displacement relations. Use the results of Exercise 3.30 to show that

$$N_{\alpha\beta} = \tfrac{1}{2}C\left[(1-\nu)(u_{\alpha,\beta} + u_{\beta,\alpha}) + 2\nu u_{\gamma,\gamma}\delta_{\alpha\beta}\right] - \frac{N_T\delta_{\alpha\beta}}{(1-\nu)}$$

$$M_{\alpha\beta} = \tfrac{1}{2}D\left[(1-\nu)(\psi_{\alpha,\beta} + \psi_{\beta,\alpha}) + 2\nu\psi_{\gamma,\gamma}\delta_{\alpha\beta}\right] - \frac{M_T\delta_{\alpha\beta}}{(1-\nu)}$$

$$Q_\alpha = \kappa^2 Gh(\psi_\alpha + w_{,\alpha})$$

See the reference in Exercise 3.30.

**3.39** *Improved Plate Theory*. Consider a material point in the plate. In the (unstrained) reference configuration, its position vector is $\mathbf{r} = \hat{\mathbf{e}}_x x + \hat{\mathbf{e}}_y y + \hat{\mathbf{e}}_z z$. When the plate is deformed, its position is characterized by $\mathbf{R} = \hat{\mathbf{e}}_x x + \hat{\mathbf{e}}_y y + \hat{\mathbf{e}}_z w(x, y) + \hat{\boldsymbol{\eta}} z$, where the unit vector $\hat{\boldsymbol{\eta}} = \hat{\mathbf{e}}_x\psi_x + \hat{\mathbf{e}}_y\psi_y + \hat{\mathbf{e}}_z$ and $\psi_x^2 + \psi_y^2 \ll 1$. Show that the displacement vector of the material point is

$$\mathbf{u} = \mathbf{R} - \mathbf{r} = \hat{\mathbf{e}}_x z\psi_x + \hat{\mathbf{e}}_y z\psi_y + \hat{\mathbf{e}}_z w$$

and therefore $u_x = z\psi_x$, $u_y = z\psi_y$, and $u_z = w$ [see (3.4)]. Note that

$$\hat{\mathbf{e}}_x \cdot \hat{\boldsymbol{\eta}} = \cos\left(\tfrac{1}{2}\pi - \psi_x\right) = \sin\psi_x \approx \psi_x$$

$$\hat{\mathbf{e}}_y \cdot \hat{\boldsymbol{\eta}} = \cos\left(\tfrac{1}{2}\pi - \psi_y\right) = \sin\psi_y \approx \psi_y$$

**3.40** *Improved Plate Theory*. Within the framework of improved plate theory, show that relative to Cartesian coordinates

$$M_{xx} = D\left(\frac{\partial\psi_x}{\partial x} + \nu\frac{\partial\psi_y}{\partial y}\right)$$

$$M_{yy} = D\left(\frac{\partial\psi_y}{\partial y} + \nu\frac{\partial\psi_x}{\partial x}\right) \qquad (3.59a)$$

$$M_{xy} = \tfrac{1}{2}D(1-\nu)\left(\frac{\partial\psi_x}{\partial y} + \frac{\partial\psi_y}{\partial x}\right)$$

$$Q_x = \kappa^2 Gh\left(\psi_x + \frac{\partial w}{\partial x}\right)$$

$$Q_y = \kappa^2 Gh\left(\psi_y + \frac{\partial w}{\partial y}\right) \qquad (3.59b)$$

**3.41** Assume that flexural stresses in the plate are linearly proportional to $z$ for $-\frac{1}{2}h \leq z \leq \frac{1}{2}h$. Show that when the expression $\tau_{\alpha\beta} = (12M_{\alpha\beta}/h^3)z$ is substituted into (3.24b), an identity is obtained.

**3.42** Assume that shearing stresses in the plate are distributed according to a parabolic law for $-\frac{1}{2}h \leq z \leq \frac{1}{2}h$. Show that when the expression $\tau_{z\alpha} = (6Q_\alpha/h^3)[(h/2)^2 - z^2]$ is substituted into (3.24a), an identity is obtained.

# 4

# BASIC EQUATIONS AND ASSOCIATED BOUNDARY CONDITIONS

## 4.1 IMPROVED PLATE THEORY (IPT)

As a point of departure, we use the principle of virtual displacements (see Section 2.4). In the present case, the virtual work of all external forces acting upon the plate is given by

$$\delta \mathscr{W} = \int_A p\, \delta w\, dA + \oint_{C_1} (\mathbf{M}^* \cdot \delta \boldsymbol{\psi} + Q_n^* \,\delta w)\, ds \tag{4.1}$$

where $p$ is the intensity of the transverse, distributed load and $\mathbf{M}^*$ and $Q_n^*$ are prescribed stress resultants acting on the boundary $C_1$ of the plate. The quantities $\delta \boldsymbol{\psi}$ and $\delta w$ are virtual displacements. Furthermore, the variation of strain energy in the plate is $\int_A \delta W_p \, dA$, where (see Section 3.9)

$$\delta W_p = \frac{\partial W_p}{\partial m_{\alpha\beta}} \delta m_{\alpha\beta} + \frac{\partial W_p}{\partial q_\alpha} \delta q_\alpha = M_{\alpha\beta}\, \delta m_{\alpha\beta} + Q_\alpha\, \delta q_\alpha$$

In view of (3.13), (3.37), and (3.6), we have

$$M_{\alpha\beta}\, \delta m_{\alpha\beta} = M_{\alpha\beta}\, \delta \psi_{\alpha,\beta} = \mathbf{M}_\alpha \cdot \delta \boldsymbol{\psi}_{,\alpha}$$

and in view of (3.12b) and (3.30b), we have

$$Q_\alpha\, \delta q_\alpha = \mathbf{Q} \cdot \delta(\boldsymbol{\psi} + \nabla w)$$

so that

$$\int_A \delta W_p \, dA = \int_A [\mathbf{M}_\alpha \cdot \delta \boldsymbol{\psi}_{,\alpha} + \mathbf{Q} \cdot \delta(\boldsymbol{\psi} + \nabla w)]\, dA \tag{4.2}$$

However,

$$\int_A \mathbf{M}_\alpha \cdot \delta\boldsymbol{\psi}_{,\alpha}\, dA = \int_A (\mathbf{M}_\alpha \cdot \delta\boldsymbol{\psi})_{,\alpha}\, dA - \int_A \mathbf{M}_{\alpha,\alpha} \cdot \delta\boldsymbol{\psi}\, dA$$

$$= \oint_C (n_\alpha \mathbf{M}_\alpha) \cdot \delta\boldsymbol{\psi}\, ds - \int_A \mathbf{M}_{\alpha,\alpha} \cdot \delta\boldsymbol{\psi}\, dA$$

and $n_\alpha \mathbf{M}_\alpha = \mathbf{M}$ [see (3.36)] and $\mathbf{M}_{\alpha,\alpha} = \mathbf{div}\, \mathcal{M}$ [see (3.38)], so that

$$\int_A \mathbf{M}_\alpha \cdot \delta\boldsymbol{\psi}_{,\alpha}\, dA = \oint_C \mathbf{M} \cdot \delta\boldsymbol{\psi}\, ds - \int_A \mathbf{div}\, \mathcal{M} \cdot \delta\boldsymbol{\psi}\, dA \tag{4.3}$$

Also,

$$\int_A \mathbf{Q} \cdot \delta(\nabla w)\, dA = \int_A \mathbf{Q} \cdot \nabla(\delta w)\, dA = \int_A [\nabla \cdot (\mathbf{Q}\, \delta w) - \delta w\, (\nabla \cdot \mathbf{Q})]\, dA$$

$$= \oint_C (\mathbf{Q} \cdot \hat{\mathbf{n}})\, \delta w\, ds - \int_A (\nabla \cdot \mathbf{Q})\, \delta w\, dA$$

$$= \oint_C Q_n\, \delta w\, ds - \int_A (\nabla \cdot \mathbf{Q})\, \delta w\, dA \tag{4.4}$$

where we have used Green's theorem and (3.30b). Upon substitution of (4.3) and (4.4) into (4.2), we obtain

$$\int_A \delta W_p\, dA = \oint_C (\mathbf{M} \cdot \delta\boldsymbol{\psi} + Q_n\, \delta w)\, ds$$

$$+ \int_A [(\mathbf{Q} - \mathbf{div}\, \mathcal{M}) \cdot \delta\boldsymbol{\psi} - (\nabla \cdot \mathbf{Q})\, \delta w]\, dA \tag{4.5}$$

The principle of virtual work (Section 2.4) requires that

$$\delta U = \int_A \delta W_p\, dA = \delta \mathcal{W} \tag{4.6}$$

Upon substitution of (4.1) and (4.5) into (4.6) and after rearrangement, we obtain

$$\int_A [(\mathbf{Q} - \mathbf{div}\, \mathcal{M}) \cdot \delta\boldsymbol{\psi} - (\nabla \cdot \mathbf{Q} + p)\, \delta w]\, dA$$

$$= \oint_C [(\mathbf{M}^* - \mathbf{M}) \cdot \delta\boldsymbol{\psi} + (Q_n^* - Q_n)\, \delta w]\, ds \tag{4.7}$$

where $C$ is the boundary curve of the plate enclosing the area $A$. The region $A$

of the plate is arbitrary. Therefore, the two integrands in (4.7) must vanish independently. Hence,

$$(\mathbf{Q} - \mathbf{div}\, \mathscr{M}) \cdot \delta\boldsymbol{\psi} - (\nabla \cdot \mathbf{Q} + p)\, \delta w = 0 \quad \text{in } A \tag{4.8a}$$

$$(\mathbf{M}^* - \mathbf{M}) \cdot \delta\boldsymbol{\psi} + (Q_n^* - Q_n)\, \delta w = 0 \quad \text{on } C \tag{4.8b}$$

The virtual displacements $\delta\boldsymbol{\psi}$ and $\delta w$ are arbitrary in $A$ and on $C$. Hence,

$$\left.\begin{aligned} \mathbf{Q} - \mathbf{div}\, \mathscr{M} &= \mathbf{0} \\ \nabla \cdot \mathbf{Q} + p &= 0 \end{aligned}\right\} \quad \text{in } A \tag{4.9}$$

and

$$\left.\begin{aligned} \mathbf{M} = \mathbf{M}^* &\text{ or } \delta\boldsymbol{\psi} = \mathbf{0} \\ Q_n = Q_n^* &\text{ or } \delta w = 0 \end{aligned}\right\} \quad \text{on } C \tag{4.10}$$

Equations (4.9) are the stress equations of equilibrium (3.70) obtained by the conventional methods of statics in Section 3.10. Equations (4.10) are the admissible boundary conditions on $C$ associated with the equations of equilibrium (4.9). If we now substitute (3.59) into (4.9), we obtain the displacement equations of equilibrium referred to Cartesian coordinates:

$$\begin{aligned} \tfrac{1}{2}D\left[(1-\nu)\psi_{\alpha,\beta\beta} + (1+\nu)\psi_{\beta,\beta\alpha}\right] - \kappa^2 Gh(\psi_\alpha + w_{,\alpha}) &= 0 \\ \kappa^2 Gh(\psi_{\alpha,\alpha} + w_{,\alpha\alpha}) + p &= 0 \end{aligned} \tag{4.11}$$

Equations (4.11) have the following invariant (vector) representation:

$$\begin{aligned} \tfrac{1}{2}D\left[(1-\nu)\nabla^2\boldsymbol{\psi} + (1+\nu)\nabla(\nabla \cdot \boldsymbol{\psi})\right] - \kappa^2 Gh(\boldsymbol{\psi} + \nabla w) &= \mathbf{0} \\ \kappa^2 Gh(\nabla \cdot \boldsymbol{\psi} + \nabla^2 w) + p &= 0 \end{aligned} \tag{4.12}$$

In view of (4.10), in conjunction with the proper interpretation of the principle of virtual displacements (see Section 2.4), we conclude that a set of admissible boundary conditions associated with (4.12) is provided by the statement

$$\text{Either } \mathbf{M} \text{ or } \boldsymbol{\psi} \qquad \text{and} \qquad \text{either } Q_n \text{ or } w \tag{4.13a}$$

must be specified along the boundary $C$ of the plate. Equivalently, a set of admissible boundary conditions associated with (4.11) is provided by the statement

$$\text{Either } M_{nn} \text{ or } \psi_n \qquad \text{and} \qquad \text{either } M_{ns} \text{ or } \psi_s \qquad \text{and} \qquad \text{either } Q_n \text{ or } w \tag{4.13b}$$

must be specified along $C$.

## 4.2 CLASSICAL PLATE THEORY (CPT)

The most popular and most commonly used plate theory in present-day technical applications was originally developed by Germain, Lagrange, and Cauchy in the nineteenth century. In order to derive this theory, we need to add the following assumptions to the developments in Sections 3.2–3.6, 3.8, and 3.9.

Material line elements that are originally straight and normal to the plate median surface in the undeformed (or natural) state remain straight and normal to this surface when deformed, that is,

$$\psi_\alpha = -w,_\alpha \tag{4.14}$$

Equation (4.14) implies the neglect of transverse shear deformation in the plate because in view of (4.14) and (4.10b), we have

$$\varepsilon_{\alpha z} = \tfrac{1}{2}(\psi_\alpha + w,_\alpha) = 0.$$

Upon substitution of (4.14) into (3.12), the plate strain-displacement relations now assume the form

$$m_{\alpha\beta} = -w,_{\alpha\beta}, \qquad q_\alpha \equiv 0 \tag{4.15}$$

Further substitution of (4.15) into (3.58), with the aid of (3.56), results in the stress-displacement relations pertaining to CPT:

$$M_{\alpha\beta} = -D\left[(1-\nu)w,_{\alpha\beta} + \nu(\nabla^2 w)\delta_{\alpha\beta}\right] \tag{4.16a}$$

$$Q_\alpha = M_{\alpha\beta,\beta} = -D(\nabla^2 w),_\alpha \tag{4.16b}$$

or, in vector notation,

$$\mathbf{Q} = -D\nabla(\nabla^2 w). \tag{4.16c}$$

We note that the plate is now rigid with respect to transverse shear deformation, and (3.58b) yields

$$q_\alpha = \lim_{\kappa^2 G \to \infty}\left(\frac{Q_\alpha}{\kappa^2 G h}\right) = 0$$

for arbitrary (but finite) values of $Q_\alpha$. By differentiating (4.16b), we find that

$$Q_{\alpha,\alpha} = -D(\nabla^2 w),_{\alpha\alpha} = -D\nabla^2(\nabla^2 w) = -D\nabla^4 w$$

and therefore, (3.68a) can now be written as

$$D\nabla^4 w = p \tag{4.17}$$

where $\nabla^4 \equiv \nabla^2(\nabla^2)$ is the bi-harmonic operator. Equation (4.17) is the displacement equation of equilibrium of CPT.

To derive the admissible boundary conditions associated with CPT, we again employ the principle of virtual work as our point of departure, and we proceed as in Section 4.1. However, we systematically introduce the relations (4.14). Using (3.33), we have

$$\mathbf{M} \cdot \delta\boldsymbol{\psi} = M_{nn}\,\delta\psi_n + M_{ns}\,\delta\psi_s$$

and because the quantity $M_{ns}\,\delta w$ is single valued and continuous along the plate boundary $C$, we have

$$0 = \oint_C \frac{\partial}{\partial s}(M_{ns}\,\delta w)\,ds = \oint_C M_{ns}\frac{\partial}{\partial s}\,\delta w\,ds + \oint_C \frac{\partial M_{ns}}{\partial s}\,\delta w\,ds.$$

Consequently,

$$\oint_C M_{ns}\,\delta\psi_s\,ds = -\oint_C M_{ns}\frac{\partial}{\partial s}\,\delta w\,ds = \oint_C \frac{\partial M_{ns}}{\partial s}\,\delta w\,ds$$

$$\oint_C M_{nn}\,\delta\psi_n\,ds = -\oint_C M_{nn}\frac{\partial}{\partial n}\,\delta w\,ds = -\oint_C M_{nn}\,\delta\left(\frac{\partial w}{\partial n}\right)ds$$

and therefore, for the case of CPT

$$\oint_C \mathbf{M} \cdot \delta\boldsymbol{\psi}\,ds = \oint_C \left[\frac{\partial M_{ns}}{\partial s}\,\delta w - M_{nn}\,\delta\left(\frac{\partial w}{\partial n}\right)\right]ds.$$

Upon substitution on the right-hand side of (4.7), we obtain

$$\oint_{C_1} \left[(\mathbf{M}^* - \mathbf{M}) \cdot \delta\boldsymbol{\psi} + (Q_n^* - Q_n)\,\delta w\right]ds$$

$$= \oint_{C_1}\left[(V_n^* - V_n)\,\delta w + (M_{nn} - M_{nn}^*)\,\delta\left(\frac{\partial w}{\partial n}\right)\right]ds$$

where

$$V_n = Q_n + \frac{\partial M_{ns}}{\partial s} \tag{4.18}$$

Consequently, (4.7) now assumes the form

$$\int_A \left[(\mathbf{M}_{\alpha,\alpha} - \mathbf{Q}) \cdot \nabla\,\delta w - (\nabla \cdot \mathbf{Q} + p)\,\delta w\right]dA$$

$$= \oint_C \left[(V_n^* - V_n)\,\delta w + (M_{nn} - M_{nn}^*)\,\delta\left(\frac{\partial w}{\partial n}\right)\right]ds$$

and by arguments similar to the ones employed in Section 4.1, we obtain the equilibrium equations (4.9) and the admissible boundary conditions associated with CPT:

$$\text{Either } V_n \text{ or } w \qquad \text{and} \qquad \text{either } M_{nn} \text{ or } \frac{\partial w}{\partial n} \tag{4.19}$$

must be specified along $C$.

Two important observations are in order: (1) The boundary characterized by $C$ in the preceding derivation has a continuously turning tangent; that is, there are no "corners." (2) Only two conditions are to be specified along the plate edge in the case of CPT. Three conditions are required in the case of IPT. The quantity $V_n$ in (4.18) is usually referred to as the Kirchhoff shear. The originators of CPT erroneously believed that three independent boundary

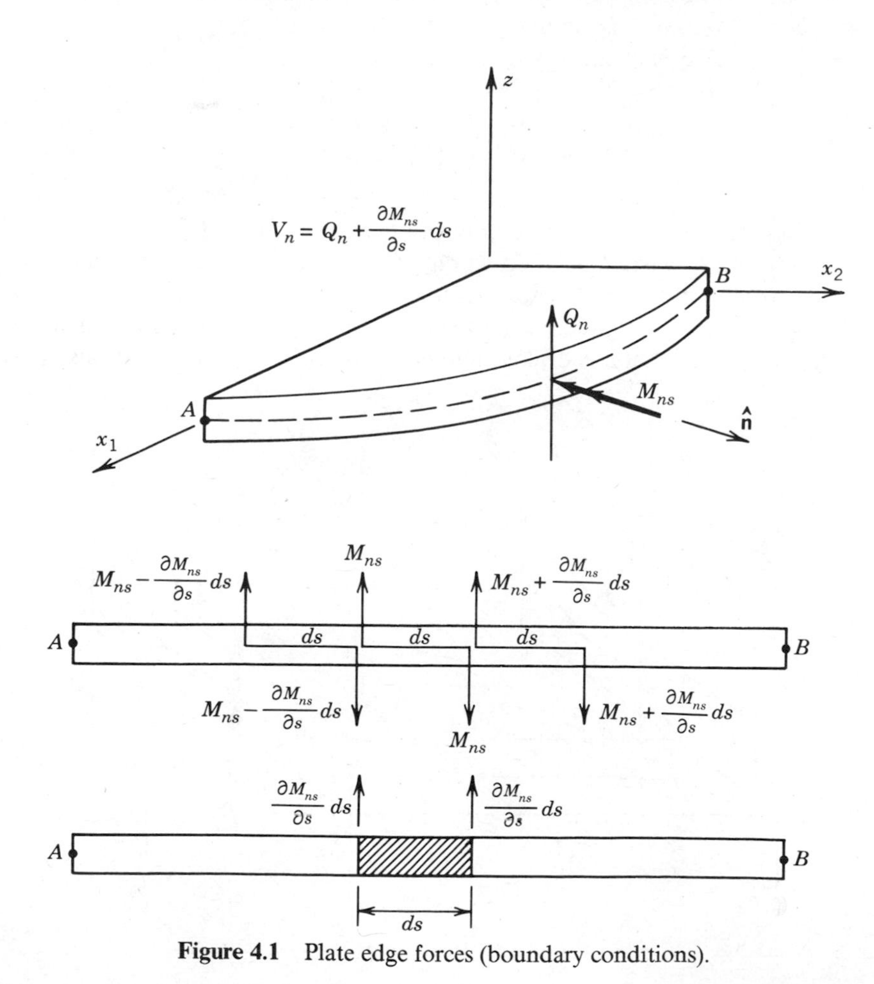


**Figure 4.1** Plate edge forces (boundary conditions).

conditions must be specified along each edge of the plate. For example, they assumed that at a free edge with outer normal vector $\hat{\mathbf{n}}$, $M_{nn} = Q_n = M_{ns} = 0$. The specification of three conditions along a plate edge resulted in an overdetermination, and it was Kirchhoff who first employed a variational derivation of CPT, resulting in the correct formulation of the associated admissible boundary conditions (see *J. Mathematik* (*Crelle*), Vol. 40, 1850, p. 51, and also *Vorlesungen über Mathematische Physik* (*Mechanik*), 1877, p. 450). He showed that along a free edge with outer normal $\hat{\mathbf{n}}$ it is necessary to set $M_{nn} = 0$ and $V_n = Q_n + \partial M_{ns}/\partial s = 0$. A heuristic, physical interpretation of Kirchhoff's result was provided by Lord Kelvin (W. Thomson) and P. G. Tait in their *Treatise on Natural Philosophy*, Cambridge University Press, London, 1879. With reference to Fig. 4.1, acting on an element of length $ds$, we have the twisting moment $M_{ns}\, ds$, which is (statically) replaced by two parallel vertical forces of magnitude $M_{ns}$ separated by the distance $ds$ as shown. This statically equivalent force system results in a net vertical force of magnitude $(\partial M_{ns}/\partial s)\, ds$ acting upon each element. Thus, the total vertical force acting on an element of length $ds$ will be

$$V_n\, ds = Q_n\, ds + \frac{\partial M_{ns}}{\partial s}\, ds$$

or the "replacement" shear force intensity is $V_n = Q_n + \partial M_{ns}/\partial s$.

The preceding variational derivation of CPT requires that the boundary curve $C$ be smooth; that is, it was assumed that there is a unique normal $\hat{\mathbf{n}}$ at each point of $C$. If the plate has corners, that is, if $C$ has points at which the direction of the normal $\hat{\mathbf{n}}$ is discontinuous, additional "corner conditions" may

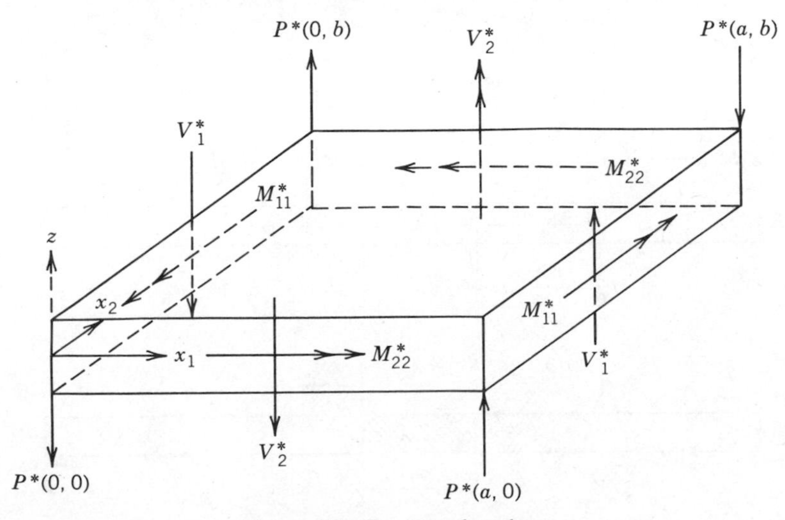


**Figure 4.2** Rectangular plate.

result that must be adjoined to the boundary conditions (4.19). For this reason we shall present a variational derivation of the frequently encountered case of a rectangular plate, as shown in Fig. 4.2. In this case the virtual work of the "internal forces" is readily shown to be

$$\delta U = \int_A \delta W_p \, dA = \int_A M_{\alpha\beta} \, \delta w,_{\alpha\beta} \, dA = \int_A M_{\alpha\beta,\alpha\beta} \, \delta w \, dA$$
$$+ \int_0^a (M_{22} \, \delta w,_2 - V_2 \, \delta w)_0^b \, dx_1 + \int_0^b (M_{11} \, \delta w,_1 - V_1 \, \delta w)_0^a \, dx_2$$
$$+ \left[ (M_{12} \, \delta w)_0^a \right]_0^b + \left[ (M_{21} \, \delta w)_0^b \right]_0^a$$

where we have integrated by parts and utilized (4.16b) and (4.18) and set $dA = dx_1 \, dx_2$. The virtual work of external forces is

$$\delta \mathscr{W} = \int_A p \, \delta w \, dA - \oint_{C_1} M_{\alpha\beta}^* n_\beta \, \delta w,_\alpha \, ds + \oint_{C_1} Q_\alpha^* n_\alpha \, \delta w \, ds$$
$$= \int_A p \, \delta w \, dA + \int_0^a (V_2^* \, \delta w - M_{22}^* \, \delta w,_2)_0^b \, dx_1$$
$$+ \int_0^b (V_1^* \, \delta w - M_{11}^* \, \delta w,_1)_0^a \, dx_2 - \left[ (M_{12}^* \, \delta w)_0^a \right]_0^b - \left[ (M_{21}^* \, \delta w)_0^b \right]_0^a$$

where $V_1^*$, $M_{12}^*$, ... denote boundary (and corner) values.

The preceding expressions for $\delta U$ and $\delta \mathscr{W}$ in conjunction with the principle of virtual work again result in the equation of equilibrium (4.17). However, for the present case, the boundary conditions assume the following form: On the edges $x_1 = 0, a$, we must specify

$$\text{Either } M_{11} \text{ or } w,_1 \qquad \text{and} \qquad \text{either } V_1 = (Q_1 + M_{12,2}) \text{ or } w \tag{4.20a}$$

On the edges $x_2 = 0, b$, we must specify

$$\text{Either } M_{22} \text{ or } w,_2 \qquad \text{and} \qquad \text{either } V_2 = (Q_2 + M_{21,1}) \text{ or } w \tag{4.20b}$$

At the corners $(x_1, x_2) = (0,0), (a,0), (a, b), (0, b)$, we must specify

$$\text{Either } M_{12} \text{ or } w \tag{4.20c}$$

The boundary conditions along the straight edges (4.20a) and (4.20b) correspond to (4.19). The corner conditions are given by (4.20c). A heuristic, physical interpretation of (4.20c) is obtained as follows: As in Fig. 4.1, we replace the twisting moments $M_{12}$ and $M_{21}$ by the statically equivalent force intensities $M_{12,2}$ and $M_{21,1}$, respectively, and this situation is shown in Fig. 4.3. It is clear from this figure that at the plate corners the "forces" $M_{12}$ and

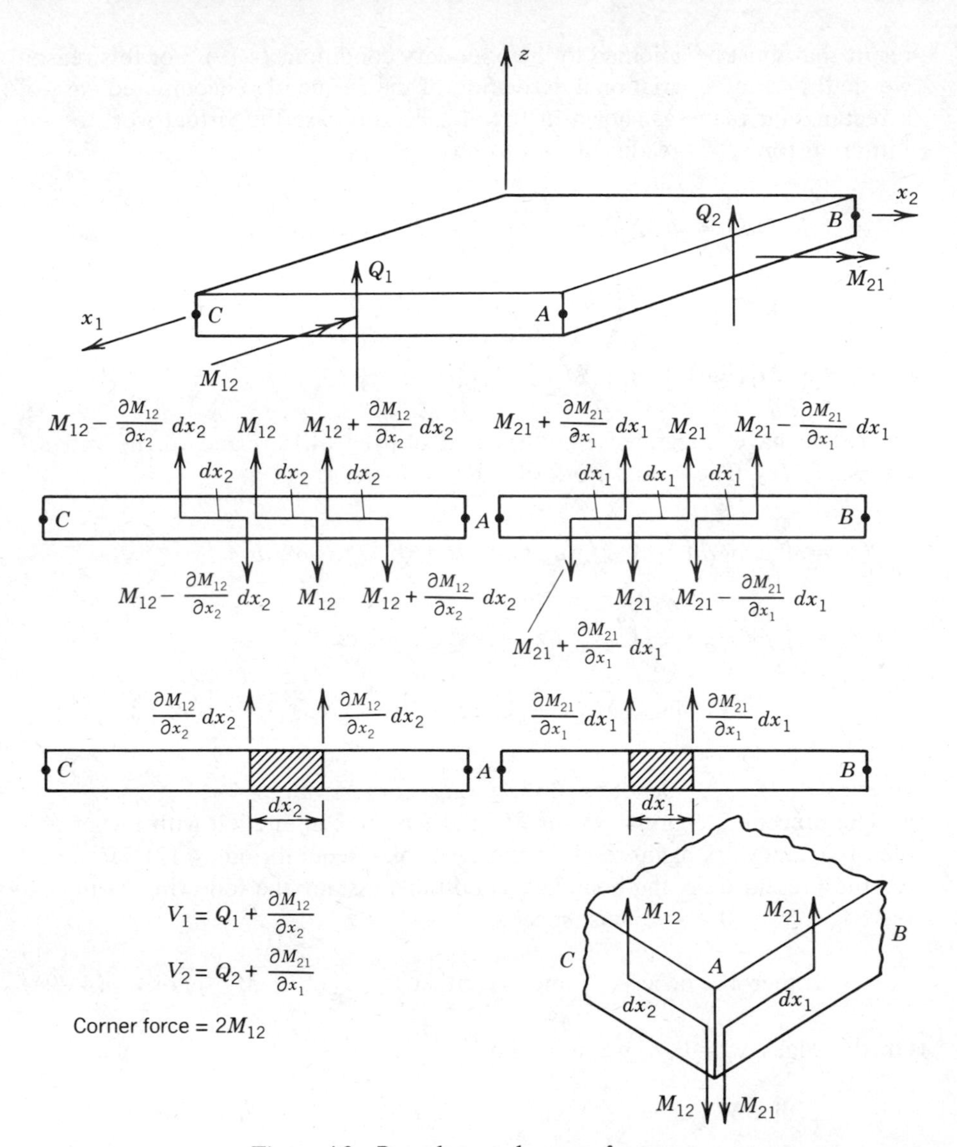


**Figure 4.3** Boundary and corner forces.

$M_{21}$ combine so as to result in a corner force of magnitude $M_{12} + M_{21} = 2M_{12} = P$.

## 4.3 THE RECIPROCAL THEOREM

Consider a linearly elastic plate, and let $m_{\alpha\beta}^{(1)}, q_{\alpha}^{(1)}$ and $m_{\alpha\beta}^{(2)}, q_{\alpha}^{(2)}$ represent two different states of plate strain this plate can experience. Let $M_{\alpha\beta}^{(1)}, Q_{\alpha}^{(1)}$ and

$M_{\alpha\beta}^{(2)}, Q_\alpha^{(2)}$ be the corresponding plate stresses. Then, in view of (3.58), we have

$$M_{\alpha\beta}^{(1)} m_{\alpha\beta}^{(2)} = D\left[(1-\nu) m_{\alpha\beta}^{(1)} m_{\alpha\beta}^{(2)} + \nu m^{(1)} m^{(2)}\right] = M_{\alpha\beta}^{(2)} m_{\alpha\beta}^{(1)}$$
$$Q_\alpha^{(1)} q_\alpha^{(2)} = \kappa^2 G h q_\alpha^{(1)} q_\alpha^{(2)} = Q_\alpha^{(2)} q_\alpha^{(1)}$$

Consequently, with reference to (3.63), we have

$$\begin{aligned} W_p^{(1,2)} &= \tfrac{1}{2}\left(M_{\alpha\beta}^{(1)} m_{\alpha\beta}^{(2)} + Q_\alpha^{(1)} q_\alpha^{(2)}\right) \\ &= \tfrac{1}{2}\left(M_{\alpha\beta}^{(2)} m_{\alpha\beta}^{(1)} + Q_\alpha^{(2)} q_\alpha^{(1)}\right) = W_p^{(2,1)} \end{aligned}$$

and therefore, the total strain energy is

$$U^{(1,2)} = \int_A W_p^{(1,2)}\, dA = \int_A W_p^{(2,1)}\, dA = U^{(2,1)} \tag{4.21}$$

In view of (3.12),

$$\begin{aligned} U^{(1,2)} &= \frac{1}{2}\int_A \left(M_{\alpha\beta}^{(1)} m_{\alpha\beta}^{(2)} + Q_\alpha^{(1)} q_\alpha^{(2)}\right) dA \\ &= \frac{1}{2}\int_A \left(M_{\alpha\beta}^{(1)} \psi_{\alpha,\beta}^{(2)} + Q_\alpha^{(1)} \psi_\alpha^{(2)} + Q_\alpha^{(1)} w_{,\alpha}^{(2)}\right) dA \\ &= \frac{1}{2}\int_A \left[\left(M_{\alpha\beta}^{(1)} \psi_\alpha^{(2)}\right)_{,\beta} + \left(Q_\alpha^{(1)} w^{(2)}\right)_{,\alpha}\right] dA \\ &\quad - \frac{1}{2}\int_A \left[\left(M_{\alpha\beta,\beta}^{(1)} - Q_\alpha^{(1)}\right)\psi_\alpha^{(2)} + Q_{\alpha,\alpha}^{(1)} w^{(2)}\right] dA \end{aligned}$$

We now apply Green's theorem and invoke the conditions of equilibrium (3.70). With the aid of (3.7) and (3.32), we obtain

$$U^{(1,2)} = \frac{1}{2}\int_A p^{(1)} w^{(2)}\, dA + \frac{1}{2}\oint_C \left(M_{nn}^{(1)} \psi_n^{(2)} + M_{ns}^{(1)} \psi_s^{(2)} + Q_n^{(1)} w^{(2)}\right) ds \tag{4.22a}$$

and in a similar manner,

$$U^{(2,1)} = \frac{1}{2}\int_A p^{(2)} w^{(1)}\, dA + \frac{1}{2}\oint_C \left(M_{nn}^{(2)} \psi_n^{(1)} + M_{ns}^{(2)} \psi_s^{(1)} + Q_n^{(2)} w^{(1)}\right) ds \tag{4.22b}$$

Upon substitution of (4.22) into (4.21), we obtain

$$\begin{aligned} &\frac{1}{2}\int_A p^{(1)} w^{(2)}\, dA + \frac{1}{2}\oint_C \left(M_{nn}^{(1)} \psi_n^{(2)} + M_{ns}^{(1)} \psi_s^{(2)} + Q_n^{(1)} w^{(2)}\right) ds \\ &\quad = \frac{1}{2}\int_A p^{(2)} w^{(1)}\, dA + \frac{1}{2}\oint_C \left(M_{nn}^{(2)} \psi_n^{(1)} + M_{ns}^{(2)} \psi_s^{(1)} + Q_n^{(2)} w^{(1)}\right) ds \end{aligned} \tag{4.23a}$$

This is the usual form in which the reciprocal theorem appears. Equation (4.23a) states that the work performed by the first set of loads acting through the displacements produced by the second set of loads is equal to the work performed by the second set of loads acting through the displacements produced by the first set of loads. Equation (4.23a) is the Betti-Rayleigh reciprocal relation applicable to IPT.

In the case of CPT, it can be shown (see Exercise 4.8) that the Betti-Rayleigh reciprocal relation assumes the form

$$\frac{1}{2}\int_A p^{(1)}w^{(2)}\,dA + \frac{1}{2}\oint_C\left[V_n^{(1)}w^{(2)} + M_{nn}^{(1)}\left(-\frac{\partial w^{(2)}}{\partial n}\right)\right]ds$$

$$= \frac{1}{2}\int_A p^{(2)}w^{(1)}\,dA + \frac{1}{2}\oint_C\left[V_n^{(2)}w^{(1)} + M_{nn}^{(2)}\left(-\frac{\partial w^{(1)}}{\partial n}\right)\right]ds \quad (4.23b)$$

In the case of a plate with admissible, homogeneous boundary conditions, the reciprocal theorem (4.23) reduces to

$$\int_A p^{(1)}w^{(2)}\,dA = \int_A p^{(2)}w^{(1)}\,dA \quad (4.24)$$

We now consider the case of concentrated loads acting in $A$:

$$p^{(1)}(x, y) = P^{(1)}\delta(x - x_1, y - y_1) \quad (4.25a)$$

$$p^{(2)}(x, y) = P^{(2)}\delta(x - x_2, y - y_2) \quad (4.25b)$$

where $P^{(i)}$ is the magnitude of the concentrated load acting at $(x_i, y_i)$, $i = 1, 2$, and $\delta(x - x_i, y - y_i)$ denotes the Dirac delta function at $(x_i, y_i)$, $i = 1, 2$. We note the following sifting property of the Dirac delta function:

$$\int_{-\infty}^{\infty}\int_{-\infty}^{\infty} f(x, y)\delta(x - x_1, y - y_1)\,dx\,dy = f(x_1, y_1) \quad (4.26)$$

Upon substitution of (4.25) into (4.24) and utilization of (4.26), we obtain the relation

$$P^{(1)}w^{(2)}(x_1, y_1) = P^{(2)}w^{(1)}(x_2, y_2) \quad (4.27)$$

Let $w(x, y; \xi, \eta)$ be the Green function of the plate, and set

$$P^{(1)} = P^{(2)} = 1$$

$$w^{(1)}(x, y) = w(x, y; \xi_1, \eta_1), \qquad w^{(2)}(x, y) = w(x, y; \xi_2, \eta_2)$$

Then (4.27) implies that

$$w(\xi_2, \eta_2; \xi_1, \eta_1) = w(\xi_1, \eta_1; \xi_2, \eta_2) \quad (4.28)$$

Equation (4.28) states that the source point and the field point are interchangeable, and this is a well-known symmetry property of the Green's function. Equation (4.28) states that the deflection at point 1 due to a unit load at point 2 is equal to the deflection at point 2 due to a unit load at point 1.

## 4.4 WORK AND ENERGY RELATIONS

In this section we shall derive expressions that relate the work of surface pressure forces and boundary forces to the change in strain energy in the plate.

### 4.4.1 Relations for Improved Plate Theory

Starting with the expression for plate strain energy density (3.62), it is readily shown that

$$2W_p = M_{\alpha\beta} m_{\alpha\beta} + Q_\alpha q_\alpha = M_{\alpha\beta} \psi_{\alpha,\beta} + Q_\alpha \psi_\alpha + Q_\alpha w_{,\alpha}$$

However,

$$(M_{\alpha\beta}\psi_\alpha)_{,\beta} = M_{\alpha\beta}\psi_{\alpha,\beta} + M_{\alpha\beta,\beta}\psi_\alpha = M_{\alpha\beta}\psi_{\alpha,\beta} + Q_\alpha \psi_\alpha$$
$$(Q_\alpha w)_{,\alpha} = Q_\alpha w_{,\alpha} + Q_{\alpha,\alpha} w = Q_\alpha w_{,\alpha} - pw$$

so that

$$2W_p = (M_{\alpha\beta}\psi_\alpha)_{,\beta} + (Q_\alpha w)_{,\alpha} + pw$$

Consequently, twice the total strain energy in the plate is given by

$$2U = \int_A \left[ (M_{\alpha\beta}\psi_\alpha)_{,\beta} + (Q_\alpha w)_{,\alpha} + pw \right] dA$$
$$= \oint_C (M_\alpha \psi_\alpha + Q_n w)\, ds + \int_A pw\, dA$$
$$= \oint_C (\mathbf{M} \cdot \boldsymbol{\psi} + Q_n w)\, ds + \int_A pw\, dA$$

where we used Green's theorem and employed (3.33). Thus, we obtain

$$U = \frac{1}{2}\int_A pw\, dA + \frac{1}{2}\oint_C (\mathbf{M} \cdot \boldsymbol{\psi})\, ds + \frac{1}{2}\oint_C (Q_n w)\, ds \tag{4.29}$$

Equation (4.29) states that the strain energy stored in the plate is equal to the work performed by all external forces acting upon the plate. The "forces" are $p\,dA$, $\mathbf{M}\,ds$, and $Q_n\,ds$, and the corresponding "displacements" are $w(x, y)$, $\boldsymbol{\psi}(s)$, and $w(s)$, respectively.

### 4.4.2 Relations for Classical Plate Theory

We shall now modify the analysis in Section 4.4.1 so as to make it applicable to CPT. In this case

$$\begin{aligned}\mathbf{M} \cdot \boldsymbol{\psi} &= M_\alpha \psi_\alpha = -M_{\alpha\beta} w,_\alpha n_\beta \\ &= -M_{nn} \frac{\partial w}{\partial n} - M_{ns} \frac{\partial w}{\partial s}\end{aligned}$$

However, we have $\oint_C (\partial/\partial s)(M_{ns} w)\, ds = 0$ because $M_{ns} w$ is a continuous and single-valued function on $C$. Consequently,

$$\oint_C M_{ns} \frac{\partial w}{\partial s}\, ds = -\oint_C w \frac{\partial M_{ns}}{\partial s}\, ds$$

and therefore, in the case of CPT we have

$$\oint_C \mathbf{M} \cdot \boldsymbol{\psi}\, ds = \oint_C M_{nn} \frac{\partial w}{\partial n}\, ds - \oint_C w \frac{\partial M_{ns}}{\partial s}\, ds.$$

Consequently, (4.29) now assumes the form

$$U = \frac{1}{2} \int_A pw\, dA + \frac{1}{2} \oint_C M_{nn} \left( -\frac{\partial w}{\partial n} \right) ds + \frac{1}{2} \oint_C V_n w\, ds \tag{4.30}$$

where we have used (4.18). Equation (4.30) relates the (elastic) strain energy stored in the plate to the work of external forces acting upon the plate. In the present case of CPT, the boundary "forces" on $C$ are $M_{nn}\, ds$ and $V_n\, ds$, and these correspond to the "displacements" $-\partial w/\partial n$ and $w$, respectively. It should be noted that (4.29) and (4.30) are characterizations of the first law of thermodynamics applied to elastic plate theory where thermal (or dissipative) effects have been neglected. Equations (4.29) and (4.30) are the equivalent of Clapeyron's theorem in classical elasticity theory.

## 4.5 UNIQUENESS OF SOLUTION

In order to establish the uniqueness of the solution of the boundary value problem characterized by (3.70) and (4.13), we now assume that it is possible to obtain two different solutions:

$$\left.\begin{aligned} &w^{(1)}, \psi_\alpha^{(1)}, \qquad \alpha = 1, 2 \\ &w^{(2)}, \psi_\alpha^{(2)}, \qquad \alpha = 1, 2 \end{aligned}\right\} \quad \text{in } A$$

Because of the linear character of the differential equations, it is clear that the set of functions defined by the formulas

$$w = w^{(1)} - w^{(2)}, \qquad \psi_\alpha = \psi_\alpha^{(1)} - \psi_\alpha^{(2)}$$

will satisfy (3.70) [in conjunction with (3.59)] with $p \equiv 0$. Thus, for the difference $w, \psi_\alpha$ of the two solutions, we have, from (4.29),

$$2U = \oint_C (\mathbf{M} \cdot \boldsymbol{\psi})\, ds + \frac{1}{2} \oint_C (Q_n w)\, ds.$$

But because the two different solutions satisfy the same boundary conditions, it follows that $\mathbf{M} = \mathbf{M}^{(1)} - \mathbf{M}^{(2)}$ or $\boldsymbol{\psi} = \boldsymbol{\psi}^{(1)} - \boldsymbol{\psi}^{(2)}$ and $Q_n = Q_n^{(1)} - Q_n^{(2)}$ or $w = w^{(1)} - w^{(2)}$ vanish on the plate boundary $C$. Thus, we conclude that

$$U = \int_A W_p\, dV = 0.$$

But with reference to (3.63), $W_p$ is a positive definite quadratic form in the plate strain components, and this integral can vanish only for $W_p = 0$, that is, for $m_{\alpha\beta} = q_\alpha = 0$. In this case $m_{\alpha\beta}^{(1)} = m_{\alpha\beta}^{(2)}$ and $q_\alpha^{(1)} = q_\alpha^{(2)}$, and we conclude that the plate strain components for the two solutions must be identical. Therefore, the components of the plate stresses are also identical. Regarding the uniqueness of displacements, we recall from Section 3.4 that they are determined to within the quantities that characterize rigid body motions. These can be made to vanish, for example, by fixing a part of the boundary $C$. A uniqueness proof for solutions obtained within the framework of CPT can be constructed along similar lines. (See Exercise 4.9.)

## EXERCISES

**4.1** Show that

$$M_{\alpha\beta}\delta m_{\alpha\beta} = M_{\alpha\beta}\delta\psi_{\alpha,\beta} = \mathbf{M}_\alpha \cdot \delta\boldsymbol{\psi}_{,\alpha}$$

*Hint:* Use (3.12), (3.37), and (3.6).

**4.2** Show that

$$\mathbf{Q} \cdot \delta\mathbf{q} = Q_\alpha \delta q_\alpha = \mathbf{Q} \cdot \delta(\boldsymbol{\psi} + \nabla w)$$

*Hint:* Use (3.12b) and (3.30b).

**4.3** Specialize (4.12) to Cartesian coordinates; that is, show that when referred to $(x, y)$ axes, (4.12) reduces to (4.11).

**4.4** Show that

$$\mathbf{M} \cdot \delta\boldsymbol{\psi} = M_{\alpha\beta} n_{\alpha} \delta\psi_{\beta} \qquad \text{for IPT}$$
$$\mathbf{M} \cdot \delta\boldsymbol{\psi} = -M_{\alpha\beta} n_{\beta} \delta w_{,\alpha} \qquad \text{for CPT}$$

**4.5**

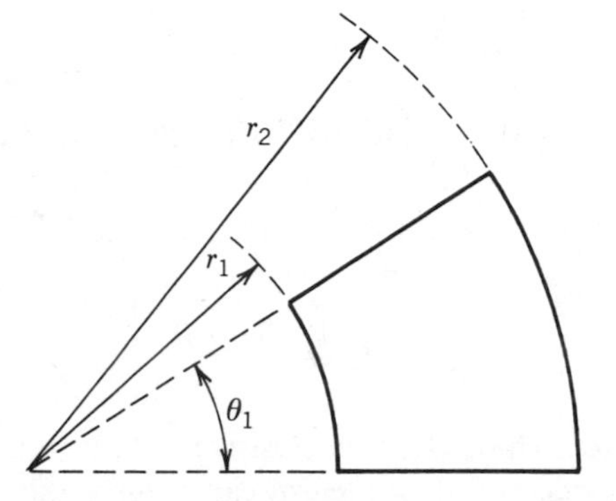


Following the procedure of Sections 4.1 and 4.2, apply the principle of virtual work to the plate element $(0 \leq r_1 < r < r_2, 0 < \theta < \theta_1 \leq 2\pi)$ shown in the figure to obtain the stress equations of equilibrium referred to polar coordinates $(r, \theta)$. What are the associated, admissible boundary conditions for **(a)** IPT and **(b)** CPT. Discuss the corner conditions for (b).

**4.6** Show that, in the case of CPT,

$$Q_{\alpha,\alpha} = M_{\alpha\beta,\alpha\beta} = -D(\nabla^2 w)_{,\alpha\alpha} = -D\nabla^4 w = -p$$

**4.7** Consider a linearly elastic plate (IPT) subjected to two different sets of forces and boundary conditions:
**(a)** $p^{(1)}$ in $A$ and either $\mathbf{M}^{(1)}$ or $\boldsymbol{\psi}^{(1)}$ and either $Q_n^{(1)}$ or $w^{(1)}$ on $C$.
**(b)** $p^{(2)}$ in $A$ and either $\mathbf{M}^{(2)}$ or $\boldsymbol{\psi}^{(2)}$ and either $Q_n^{(2)}$ or $w^{(2)}$ on $C$.
Show that, in $A$,

$$M_{\alpha\beta}^{(1)} m_{\alpha\beta}^{(2)} + Q_{\alpha}^{(1)} q_{\alpha}^{(2)} = M_{\alpha\beta}^{(2)} m_{\alpha\beta}^{(1)} + Q_{\alpha}^{(2)} q_{\alpha}^{(1)}$$

where superscripts (1) and (2) denote effects due to loading conditions (a) and (b), respectively.

**4.8** Derive the Betti-Rayleigh reciprocal relation (4.23b) applicable to CPT. *Hint:* Show that

$$\oint_C M_{ns}^{(1)} \frac{\partial w^{(2)}}{\partial s} = -\oint_C w^{(2)} \frac{\partial M_{ns}^{(1)}}{\partial s}\, ds$$

because $\oint_C (\partial/\partial s)(M_{ns}^{(1)} w^{(2)})\, ds = 0$ if the function $M_{ns}^{(1)} w^{(2)}$ is single valued on the boundary $C$.

**4.9** Using the method of Section 4.5, demonstrate that the solution of the problem posed by CPT is unique.

**4.10** Derive the thermo-elastic equations of equilibrium in terms of displacements. Use the results of Exercise 3.38 to show that [see (3.12)]

$$\tfrac{1}{2}C\left[(1-\nu)\nabla^2 \mathbf{u} + (1+\nu)\nabla(\nabla \cdot \mathbf{u})\right] = \frac{\nabla(N_T)}{1-\nu}$$

$$\tfrac{1}{2}D\left[(1-\nu)\nabla^2 \boldsymbol{\psi} + (1+\nu)\nabla(\nabla \cdot \boldsymbol{\psi})\right] - \kappa^2 Gh(\boldsymbol{\psi} + \nabla w) = \frac{\nabla(M_T)}{1-\nu}$$

$$\kappa^2 Gh(\nabla \cdot \boldsymbol{\psi} + \nabla^2 w) + p = 0$$

**4.11** *Classical Plate Theory.* Consider a material point in the plate. In the (unstrained) reference configuration, its position vector is $\mathbf{r} = \hat{\mathbf{e}}_x x + \hat{\mathbf{e}}_y y + \hat{\mathbf{e}}_z z$. When the plate is deformed, its position vector is characterized by $\mathbf{R} = \hat{\mathbf{e}}_x x + \hat{\mathbf{e}}_y y + \hat{\mathbf{e}}_z w(x, y) + \hat{\boldsymbol{\varepsilon}} z$, where the unit vector $\hat{\boldsymbol{\varepsilon}}$ is normal to the plate median surface in the reference and deformed configuration.

**(a)** Show that the displacement vector of the material point is $\mathbf{u} = \mathbf{R} - \mathbf{r} = (w - z)\hat{\mathbf{e}}_z + \hat{\boldsymbol{\varepsilon}} z$.

**(b)** Points on the (deformed) median surface are characterized by $z = 0$, and their position vector is given by $\mathbf{R}_0 = \hat{\mathbf{e}}_x x + \hat{\mathbf{e}}_y y + \hat{\mathbf{e}}_z w(x, y)$. Show that

$$\hat{\boldsymbol{\varepsilon}} = \frac{(\partial \mathbf{R}_0/\partial x) \times (\partial \mathbf{R}_0/\partial y)}{|(\partial \mathbf{R}_0/\partial x) \times (\partial \mathbf{R}_0/\partial y)|} \cong \hat{\mathbf{e}}_x\left(-\frac{\partial w}{\partial x}\right) + \hat{\mathbf{e}}_y\left(-\frac{\partial w}{\partial y}\right) + \hat{\mathbf{e}}_z$$

provided $|\text{grad}\, w| \ll 1$ (see E. Kreyszig, *Introduction to Differential and Riemannian Geometry*, University of Toronto Press, 1975, p. 64).

**(c)** Using (a) and (b), show that the components of the displacement vector $\mathbf{u} = \mathbf{R} - \mathbf{r}$ are given by [see (4.14)]

$$u_x = -\frac{\partial w}{\partial x} z, \qquad u_y = -\frac{\partial w}{\partial y} z, \qquad u_z = w(x, y)$$

**4.12** *Classical Plate Theory*. Use (4.16a) to show that $M_{\alpha\alpha} = -D(1+\nu)\nabla^2 w$.

**4.13** *Classical Plate Theory*. Within the framework of CPT, show that, relative to Cartesian coordinates,

$$M_{xx} = -D\left(\frac{\partial^2 w}{\partial x^2} + \nu\frac{\partial^2 w}{\partial y^2}\right)$$

$$M_{yy} = -D\left(\frac{\partial^2 w}{\partial y^2} + \nu\frac{\partial^2 w}{\partial x^2}\right) \tag{4.16a}$$

$$M_{xy} = -D(1-\nu)\frac{\partial^2 w}{\partial x\,\partial y}$$

$$Q_x = -D\frac{\partial}{\partial x}(\nabla^2 w), \qquad Q_y = -D\frac{\partial}{\partial y}(\nabla^2 w) \tag{4.16b}$$

**4.14** *Classical Plate Theory*. Within the framework of CPT, show that, relative to polar coordinates,

$$M_{rr} = -D\left[\frac{\partial^2 w}{\partial r^2} + \nu\left(\frac{1}{r}\frac{\partial w}{\partial r} + \frac{1}{r^2}\frac{\partial^2 w}{\partial \theta^2}\right)\right]$$

$$M_{\theta\theta} = -D\left(\frac{1}{r}\frac{\partial w}{\partial r} + \frac{1}{r^2}\frac{\partial^2 w}{\partial \theta^2} + \nu\frac{\partial^2 w}{\partial r^2}\right)$$

$$M_{r\theta} = -(1-\nu)D\frac{1}{r}\frac{\partial}{\partial \theta}\left(\frac{\partial w}{\partial r} - \frac{w}{r}\right)$$

$$Q_r = -D\frac{\partial}{\partial r}(\nabla^2 w), \qquad Q_\theta = -D\frac{1}{r}\frac{\partial}{\partial \theta}(\nabla^2 w)$$

where

$$\nabla^2 w = \frac{\partial^2 w}{\partial r^2} + \frac{1}{r}\frac{\partial w}{\partial r} + \frac{1}{r^2}\frac{\partial^2 w}{\partial \theta^2}$$

**4.15** For CPT, write the equilibrium equation in terms of displacement referred to polar coordinates $(r, \theta)$. Show that

$$D\nabla^4 w = p(r, \theta)$$

where

$$\nabla^2 w = \frac{\partial^2 w}{\partial r^2} + \frac{1}{r}\frac{\partial w}{\partial r} + \frac{1}{r^2}\frac{\partial^2 w}{\partial \theta^2}$$

**4.16** In the case of CPT, show that (infinitesimal) rigid body motion is characterized by $q_\alpha \equiv 0$, $m_{\alpha\beta} = -w,_{\alpha\beta} = 0$, and therefore $w = ax_1 + bx_2 + c$, where $a$, $b$, and $c$ are constants. (Refer to Exercise 3.13.)

**4.17** The plate theory of Section 4.1 accounts for flexural and shear deformations, whereas the theory of Section 4.2 accounts only for flexural deformations. Now consider the model of a "shear plate," that is, a plate that deforms in shear only but remains rigid with respect to other types of deformation.

**(a)** Show that the stress equation of equilibrium is

$$\nabla \cdot \mathbf{Q} + p = 0$$

**(b)** Assume (and justify) the displacement field $u_1 = u_2 = 0$, $u_3 = w(x_1, x_2)$, and show that the appropriate stress-displacement relations are

$$Q_\alpha = \int_{-h/2}^{h/2} \tau_{z\alpha}\, dz = \kappa^2 G h w,_\alpha$$

where $\kappa^2$ is the shear coefficient.

**(c)** Formulate a well-posed problem within the framework of this theory and discuss its solution.

# 5

# STATIC DEFORMATION OF TRANSVERSELY LOADED PLATES

## 5.1 CLASSICAL PLATE THEORY

In CPT we assume that

(a) The midsurface does not stretch or contract.

(b) Material line elements originally normal to the midsurface in the undeformed configuration remain normal in the deformed configuration.

(c) There is no thickness stretch.

The basic equations of CPT were derived in Section 4.2. When referred to Cartesian axes, the equilibrium equation is (see 4.17)

$$D \nabla^4 w = p(x, y) \tag{5.1}$$

where

$$\nabla^2 w = \frac{\partial^2 w}{\partial x^2} + \frac{\partial^2 w}{\partial y^2}$$

$$\nabla^4 w = \nabla^2(\nabla^2 w) = \frac{\partial^2 w}{\partial x^4} + 2\frac{\partial^4 w}{\partial x^2\, \partial y^2} + \frac{\partial^4 w}{\partial y^4}$$

The stress resultants are [see (4.16)]

$$M_{xx} = -D\left(\frac{\partial^2 w}{\partial x^2} + \nu\frac{\partial^2 w}{\partial y^2}\right) \tag{5.2a}$$

$$M_{yy} = -D\left(\frac{\partial^2 w}{\partial y^2} + \nu\frac{\partial^2 w}{\partial x^2}\right) \tag{5.2b}$$

$$M_{xy} = -D(1-\nu)\frac{\partial^2 w}{\partial x\,\partial y} \tag{5.2c}$$

$$Q_x = -D\frac{\partial}{\partial x}(\nabla^2 w) \tag{5.3a}$$

$$Q_y = -D\frac{\partial}{\partial y}(\nabla^2 w) \tag{5.3b}$$

$$V_x = Q_x + \frac{\partial M_{xy}}{\partial y}, \qquad V_y = Q_y + \frac{\partial M_{yx}}{\partial x} \tag{5.4}$$

When referred to polar coordinates, the equation of equilibrium is

$$D\nabla^4 w = p(r, \theta) \tag{5.5}$$

where

$$\nabla^2 w = \frac{\partial^2 w}{\partial r^2} + \frac{1}{r}\frac{\partial w}{\partial r} + \frac{1}{r^2}\frac{\partial^2 w}{\partial \theta^2}$$

The stress resultants are (see Exercise 4.14)

$$M_{rr} = -D\left[\frac{\partial^2 w}{\partial r^2} + \nu\left(\frac{1}{r}\frac{\partial w}{\partial r} + \frac{1}{r^2}\frac{\partial^2 w}{\partial \theta^2}\right)\right] \tag{5.6a}$$

$$M_{\theta\theta} = -D\left[\nu\frac{\partial^2 w}{\partial r^2} + \frac{1}{r}\frac{\partial w}{\partial r} + \frac{1}{r^2}\frac{\partial^2 w}{\partial \theta^2}\right] \tag{5.6b}$$

$$M_{r\theta} = -(1-\nu)D\frac{1}{r}\frac{\partial}{\partial \theta}\left(\frac{\partial w}{\partial r} - \frac{w}{r}\right) \tag{5.6c}$$

$$Q_r = -D\frac{\partial}{\partial r}(\nabla^2 w) \tag{5.7a}$$

$$Q_\theta = -D\frac{1}{r}\frac{\partial}{\partial \theta}(\nabla^2 w) \tag{5.7b}$$

$$V_r = Q_r + \frac{1}{r}\frac{\partial M_{r\theta}}{\partial \theta} \tag{5.8a}$$

$$V_\theta = Q_\theta + \frac{\partial M_{\theta r}}{\partial r} \tag{5.8b}$$

In the Sections 5.1.1–5.1.9 we shall formulate and solve a series of problems using (5.1)–(5.4) and (5.5)–(5.8). These problems were selected (a) because they answer questions that frequently arise in modern technology and (b) because they can be solved by relatively elementary, classical mathematical techniques that are accessible to all students and graduates of engineering and/or applied science programs.

### 5.1.1 The Simply Supported Rectangular Plate [Solution of Navier (1785 – 1836)]

As a point of departure, we consider the simply supported plate shown in Fig. 5.1. The plate is subjected to the distributed, transverse pressure load characterized by

$$p(x, y) = p_0 \sin\frac{m\pi x}{a}\sin\frac{n\pi y}{b}, \qquad m = 1, 2, 3, \ldots, \quad n = 1, 2, 3, \ldots \tag{5.9}$$

In view of the required simple support conditions, we need to specify [see (4.20)]

$$w(0, y) = M_{xx}(0, y) = 0 \tag{5.10a}$$

$$w(x, 0) = M_{yy}(x, 0) = 0 \tag{5.10b}$$

$$w(a, y) = M_{xx}(a, y) = 0 \tag{5.10c}$$

$$w(x, b) = M_{yy}(x, b) = 0 \tag{5.10d}$$

In view of (5.1) and (5.9), we require a solution $w = w(x, y)$ of

$$D\nabla^4 w = p(x, y) = p_0 \sin\frac{m\pi x}{a}\sin\frac{n\pi y}{b} \tag{5.11}$$

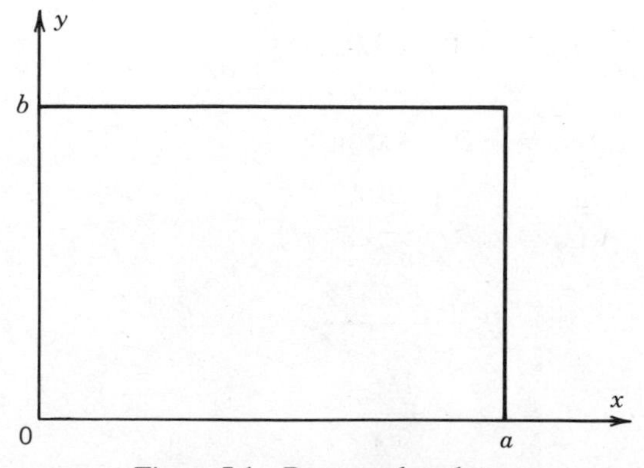


**Figure 5.1** Rectangular plate.

subject to the boundary conditions (5.10). We assume a solution of the form

$$w = C \sin\frac{m\pi x}{a}\sin\frac{n\pi y}{b} \tag{5.12}$$

and it is noted that (5.12) satisfies the required boundary conditions (5.10) for positive integer values of $m$ and $n$. Upon substitution of (5.12) into (5.11), we find that

$$C = \frac{p_0}{D\left[(m\pi/a)^2 + (n\pi/b)^2\right]^2}$$

and therefore, the solution of (5.11), subject to the boundary conditions (5.10), is given by

$$w = \frac{p_0 \sin(m\pi x/a)\sin(n\pi y/b)}{D\left[(m\pi/a)^2 + (n\pi/b)^2\right]^2} \tag{5.13}$$

It will be instructive to perform a (static) check of equilibrium in the $z$-direction for the case $m = n = 1$. The total transverse load is

$$W = \int_0^b\int_0^a p_0 \sin\frac{\pi x}{a}\sin\frac{\pi y}{b}\,dx\,dy = \frac{4p_0 ab}{\pi^2} \tag{5.14}$$

We calculate [see (5.4)]

$$V_x = Q_x + \frac{\partial M_{xy}}{\partial y}$$

$$= \left\{\frac{p_0(\pi/a)}{(\pi/a)^2 + (\pi/b)^2} + \frac{p_0(1-\nu)(\pi/a)(\pi/b)^2}{\left[(\pi/a)^2 + (\pi/b)^2\right]^2}\right\}\cos\frac{\pi x}{a}\sin\frac{\pi y}{b}$$

$$V_y = Q_y + \frac{\partial M_{yx}}{\partial x}$$

$$= \left\{\frac{p_0(\pi/b)}{(\pi/a)^2 + (\pi/b)^2} + \frac{p_0(1-\nu)(\pi/a)^2(\pi/b)}{\left[(\pi/a)^2 + (\pi/b)^2\right]^2}\right\}\sin\frac{\pi x}{a}\cos\frac{\pi y}{b}$$

Consequently, the integrated Kirchhoff shear forces along the four edges are

$$L(x,0) = \int_0^a V_y(x,0)\,dx, \qquad L(x,b) = \int_0^a V_y(x,b)\,dx$$

$$L(0,y) = \int_0^b V_x(0,y)\,dy, \qquad L(a,y) = \int_0^b V_x(a,y)\,dy$$

where

$$L(x,0) = -L(x,b) = \frac{2p_0(a/b)}{(\pi/a)^2 + (\pi/b)^2} + \frac{2p_0(1-\nu)(\pi/a)(\pi/b)}{\left[(\pi/a)^2 + (\pi/b)^2\right]^2} \tag{5.15a}$$

$$L(0,y) = -L(a,y) = \frac{2p_0(b/a)}{(\pi/a)^2 + (\pi/b)^2} + \frac{2p_0(1-\nu)(\pi/a)(\pi/b)}{\left[(\pi/a)^2 + (\pi/b)^2\right]^2} \tag{5.15b}$$

According to (4.20c) (see also Figs. 4.2 and 4.3), the force at the corner $(x, y)$ is

$$P(x,y) = 2M_{xy} = -\frac{2p_0(1-\nu)(\pi/a)(\pi/b)}{\left[(\pi/a)^2 + (\pi/b)^2\right]^2}\cos\frac{\pi x}{a}\cos\frac{\pi y}{b}$$

Consequently, at the four corners of the plate we have

$$\begin{aligned} P(0,0) &= -P_0, \qquad P(a,0) = P_0 \\ P(a,b) &= -P_0, \qquad P(0,b) = P_0 \end{aligned} \tag{5.16a}$$

where

$$P_0 = \frac{2p_0(1-\nu)(\pi/a)(\pi/b)}{\left[(\pi/a)^2 + (\pi/b)^2\right]^2} \tag{5.16b}$$

In view of (5.14) and (5.16), the forces acting in the positive $z$-direction are $W + 4P_0$, while the forces acting in the negative $z$-direction are $2L(x,0) + 2L(0,y)$. Thus, equilibrium requires that

$$W + 4P_0 = 2L(x,0) + 2L(0,y) \tag{5.17}$$

and upon substitution of (5.14), (5.16b), and (5.15) into (5.17), we obtain an identity in $a$ and $b$. These considerations show that the applied distributed forces $p(x, y)$ are equilibrated by the reactions consisting of the Kirchhoff shear forces along the four plate edges plus the four corner forces.

Because of the linearity of the present plate theory, the principle of superposition is applicable. Solution (5.13) in conjunction with the concept of a double Fourier series (see Exercise 5.3) can be used to generate solutions for simply supported rectangular plates subjected to arbitrarily distributed, transverse loads. Following the method of Navier, the distributed load is characterized by

$$p(x, y) = \sum_{m=1}^{\infty} \sum_{n=1}^{\infty} p_{mn} \sin \frac{m\pi x}{a} \sin \frac{n\pi y}{b} \tag{5.18}$$

where, according to Exercise 5.3, we have

$$p_{mn} = \frac{4}{ab} \int_0^b \int_0^a p(x, y) \sin \frac{m\pi x}{a} \sin \frac{n\pi y}{b}\, dx\, dy \tag{5.19}$$

We proceed as in the previous case and note that (5.13) may be looked upon as a single harmonic component of the desired solution. Consequently, for the present case we obtain (by superposition)

$$w(x, y) = \sum_{m=1}^{\infty} \sum_{n=1}^{\infty} \frac{p_{mn} \sin(m\pi x/a) \sin(n\pi y/b)}{D\left[(m\pi/a)^2 + (n\pi/b)^2\right]^2} \tag{5.20}$$

It should be noted that each term of the series (5.20) satisfies the simple support boundary conditions (5.10). We can utilize (5.2) and (5.3) in conjunction with (5.13) to calculate the moment tensor and shear force vector in the plate (see Exercise 5.6).

**Example:** We consider a simply supported rectangular plate subjected to a uniformly distributed load of intensity $p(x, y) = p_0 =$ constant. With the aid of (5.19), we have

$$\begin{aligned} p_{mn} = \frac{4p_0}{ab} \int_0^b \int_0^a \sin \frac{m\pi x}{a} \sin \frac{n\pi y}{b}\, dx\, dy &= \frac{16 p_0}{\pi^2 mn}, \qquad m, n = 1, 3, 5, \ldots \\ &= 0, \qquad m, n = 2, 4, 6, \ldots \end{aligned}$$

and upon substitution into (5.20), we obtain the equation characterizing the midplane deflection of the plate:

$$w(x, y) = \frac{16 p_0}{\pi^6 D} \sum_{m=1}^{\infty} \sum_{n=1}^{\infty} \frac{\sin(m\pi x/a) \sin(n\pi y/b)}{mn\left(m^2/a^2 + n^2/b^2\right)^2} \tag{5.21}$$

$$m, n = 1, 3, 5, \ldots$$

The maximum deflection occurs at the center of the plate, $(x, y) = (\frac{1}{2}a, \frac{1}{2}b)$, and

$$w_{\max} = w\left(\tfrac{1}{2}a, \tfrac{1}{2}b\right) = \frac{16 p_0}{\pi^6 D} \sum_{m=1}^{\infty} \sum_{n=1}^{\infty} \frac{(-1)^{[(m+n)/2]-1}}{mn\left(m^2/a^2 + n^2/b^2\right)^2} \tag{5.22}$$

$$m, n = 1, 3, 5, \ldots$$

For example, when $a = b$ (square plate) and $\nu = 0.3$, we have

$$w_{\max} = 0.00406\left(p_0 a^4/D\right) \text{ and } \left(M_{xx}\right)_{\max} = \left(M_{yy}\right)_{\max} = 0.0479 a^2 p_0.$$

### 5.1.2 Rectangular Plate with Two Simply Supported Opposite Edges [Method of M. Lévy (1838 – 1910)]

We next consider a rectangular plate (Fig. 5.1) for which the edges $y = 0$ and $y = b$ are simply supported and for which the load can be characterized by a series of terms, each of which can be written as

$$p = p(x, y) = p_n \sin\frac{n\pi y}{b}, \qquad n = 1, 2, 3, \ldots \tag{5.23}$$

Hence, we shall seek a solution of

$$D\nabla^4 w = p(x, y) = p_n \sin\frac{n\pi y}{b} \tag{5.24}$$

subject to the boundary conditions

$$w(x, 0) = M_{yy}(x, 0) = 0 \tag{5.25a}$$

$$w(x, b) = M_{yy}(x, b) = 0 \tag{5.25b}$$

Assuming a product solution of the form

$$w(x, y) = f_n(x) \sin\frac{n\pi y}{b} \tag{5.26}$$

we note that (5.26) satisfies the simple support boundary conditions (5.25) provided $n = 1, 2, 3, \ldots$ . Upon substitution of (5.26) into (5.23), we obtain the ordinary differential equation

$$\frac{d^4 f_n}{dx^4} - 2\frac{n^2\pi^2}{b^2}\frac{d^2 f_n}{dx^2} + \frac{n^4\pi^4}{b^4} f_n = \frac{p_n}{D} \tag{5.27}$$

The complete solution of (5.27) is

$$f_n(x) = (A_n + B_n x)\cosh\frac{n\pi x}{b} + (C_n + D_n x)\sinh\frac{n\pi x}{b} + \frac{p_n b^4}{n^4\pi^4 D} \tag{5.28a}$$

or, in the alternate but equivalent form,

$$f_n(x) = (A'_n + B'_n x)e^{-n\pi x/b} + (C'_n + D'_n x)e^{n\pi x/b} + \frac{p_n b^4}{n^4\pi^4 D} \tag{5.28b}$$

where $A_n$, $B_n$, $C_n$, $D_n$ and $A'_n$, $B'_n$, $C'_n$, $D'_n$ are constants that must be determined from boundary conditions along the edges $x = 0$ and $x = a$. The form (5.28a) is convenient for a plate of bounded extent, while the representation (5.28b) is useful when we consider a semi-infinite plate or a plate that is unbounded in the sense $-\infty < x < \infty$.

When the load is characterized by the Fourier series

$$p(x, y) = p(y) = \sum_{n=1}^{\infty} p_n \sin \frac{n\pi y}{b} \tag{5.29}$$

where (see Exercise 5.1)

$$p_n = \frac{2}{b} \int_0^b p(y) \sin \frac{n\pi y}{b}\, dy \tag{5.30}$$

then by superposition, in conjunction with the preceding analysis, we readily obtain the solution [see (5.26)]

$$w(x, y) = \sum_{n=1}^{\infty} f_n(x) \sin \frac{n\pi y}{b} \tag{5.31}$$

where $f_n(x)$ is given by (5.28).

**Example (a):** We consider the case of a rectangular plate with simply supported (ss) edges $y = 0$ and $y = b$ and clamped (cl) edges at $x = -\frac{1}{2}a$ and $x = \frac{1}{2}a$, as shown in Fig. 5.2. The plate is acted upon by a uniformly distributed load $p(x, y) = p_0 = \text{constant}$, and in view of (5.30), we have

$$p_n = \frac{2}{b} p_0 \int_0^b \sin \frac{n\pi y}{b}\, dy = \begin{cases} \dfrac{4p_0}{\pi} \dfrac{1}{n}, & n = 1, 3, 5, \ldots \\ 0, & n = 2, 4, 6, \ldots \end{cases} \tag{5.32}$$

If we select the coordinate system shown in Fig. 5.2, then we can take

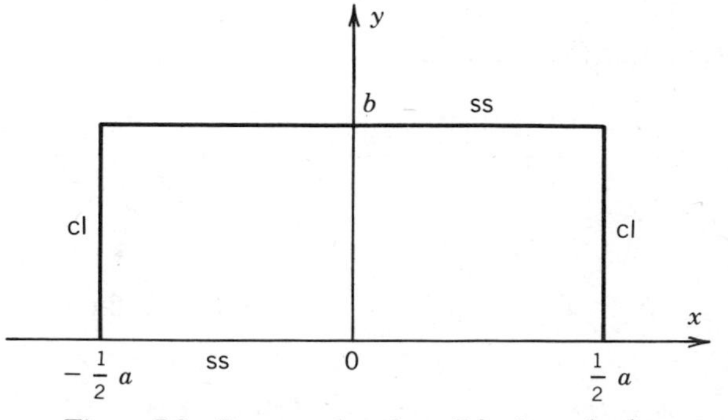


**Figure 5.2** Rectangular plate (Lévy's method).

advantage of the symmetry of the deflected median surface with respect to the $y$-$z$ plane: $f_n(x, y) = f_n(-x, y)$. Consequently, $B_n = C_n = 0$ in (5.28a), and we have

$$f_n(x) = A_n \cosh\frac{n\pi x}{b} + D_n x \sinh\frac{n\pi x}{b} + \frac{p_n b^4}{n^4\pi^4 D} \tag{5.33}$$

The boundary conditions at $x = \pm\frac{1}{2}a$ are

$$w\left(\pm\tfrac{1}{2}a, y\right) = \left(\frac{dw}{dx}\right)_{(\pm a/2,\, y)} = 0 \tag{5.34}$$

and upon substitution of (5.31) into (5.34), we obtain

$$f_n\left(\pm\tfrac{1}{2}a\right) = \left(\frac{df_n}{dx}\right)_{(\pm a/2)} = 0 \tag{5.35}$$

Further substitution of (5.33) into (5.35) results in the linear algebraic equations

$$\begin{bmatrix} \cosh N & \frac{1}{2}a \sinh N \\ \dfrac{n\pi}{b}\sinh N & N\cosh N + \sinh N \end{bmatrix} \begin{Bmatrix} A_n \\ D_n \end{Bmatrix} = \begin{Bmatrix} -\dfrac{p_n b^4}{n^4\pi^4 D} \\ 0 \end{Bmatrix}$$

where

$$N = \frac{n\pi a}{2b}$$

$$A_n = -\frac{p_n b^4}{n^4\pi^4 D}\,\frac{N\cosh N + \sinh N}{N + \sinh N\cosh N} \tag{5.36a}$$

$$D_n = -\frac{p_n b^4}{n^4\pi^4 D}\,\frac{(n\pi/b)\sinh N}{N + \sinh N\cosh N} \tag{5.36b}$$

Upon substitution of (5.36) into (5.33), we obtain

$$f_n(x) = \frac{p_n b^4}{n^4\pi^4 D}\left[1 - \frac{(N\cosh N + \sinh N)\cosh X - (\sinh N)(X\sinh X)}{N + \sinh N\cosh N}\right] \tag{5.37}$$

where $X = n\pi x/b$. In view of (5.31), (5.32), and (5.37), the required solution is

$$w = \sum_{n=1,3,5,\ldots}^{\infty} \frac{4p_0 b^4}{\pi^5 n^5 D}\left[1 - g(N, X)\right] \sin\frac{n\pi y}{b} \tag{5.38}$$

where

$$g(N, X) = \frac{(N\cosh N + \sinh N)\cosh X - (\sinh N)(X\sinh X)}{N + \sinh N\cosh N}$$

and

$$N = n\pi a/2b \text{ and } X = n\pi x/b.$$

**Example (b):** In this case we consider a semi-infinite plate strip $0 \le x \le \infty$ and $0 \le y \le b$ (see Fig. 5.3) that is simply supported along $y = 0$ and $y = b$. The plate strip is clamped along $x = 0$, and it is acted upon by a uniformly distributed, transverse load of intensity $p(x, y) = p_0 =$ constant. In the present case we shall use (5.28b) with the proviso that $\lim_{x\to\infty} w(x, y)$ remain bounded. Hence, we set $C_n' = D_n' = 0$ in (5.28b) and obtain

$$f_n(x) = (A_n' + B_n'x)e^{-n\pi x/b} + \frac{p_n b^4}{n^4\pi^4 D} \tag{5.39}$$

where $p_n$ is again defined by (5.32). In view of the clamped boundary condition along $x = 0$, we require

$$w(0, y) = \left(\frac{\partial w}{\partial x}\right)_{(0, y)} = 0$$

or, equivalently,

$$f_n(0) = \left(\frac{df_n}{dx}\right)_{x=0} = 0 \tag{5.40}$$

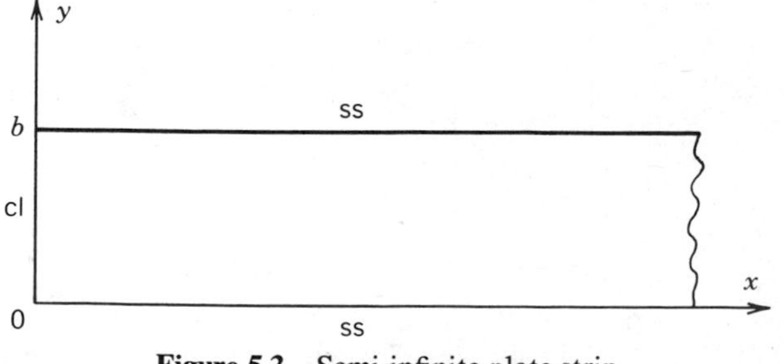


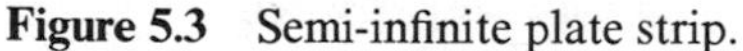
**Figure 5.3** Semi-infinite plate strip.

Upon substitution of (5.39) into (5.40), we readily obtain

$$A'_n = -\frac{p_n b^4}{n^4 \pi^4 D} \tag{5.41a}$$

$$B'_n = -\frac{p_n b^3}{n^3 \pi^3 D} \tag{5.41b}$$

and further substitution of (5.41) into (5.39) results in

$$f_n(x) = \frac{p_n b^4}{n^4 \pi^4 D}(1 - e^{-X} - Xe^{-X}) \qquad \text{where} \quad X = \frac{n\pi x}{b} \tag{5.42}$$

With the aid of (5.41), (5.42), and (5.32), we finally obtain the solution

$$w = \sum_{n=1,3,5,\ldots}^{\infty} \frac{4p_0 b^4}{\pi^5 n^5 D}(1 - e^{-X} - Xe^{-X}) \sin\frac{n\pi y}{b} \tag{5.43}$$

### 5.1.3 The Simply Supported Triangular Plate

In the present section we consider the case of a plate in the shape of an equilateral triangle subjected to a uniformly distributed load of intensity $p(x, y) = p_0 =$ constant, as shown in Fig. 5.4. The plate is simply supported along its edges, each of which is of length $2a/\sqrt{3}$. The origin of our coordinate system coincides with the centroid of our triangle, and the $x$-axis bisects the vertex of the triangle.

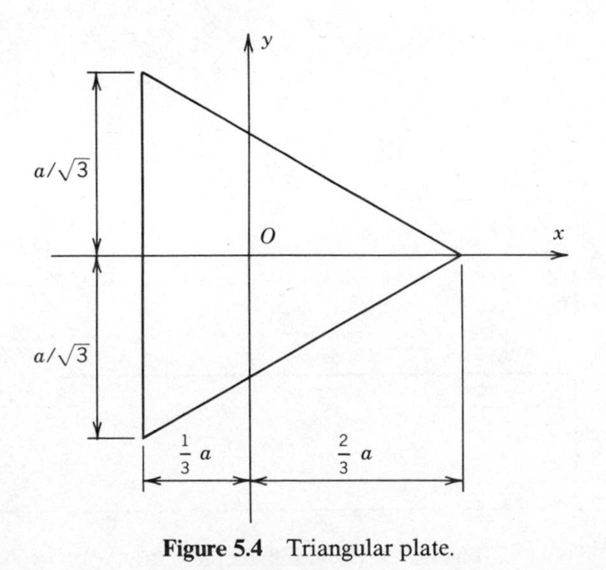


**Figure 5.4** Triangular plate.

The function

$$w = \frac{p_0}{64aD}\left(x + \tfrac{1}{3}a\right)\left(x - \tfrac{2}{3}a + y\sqrt{3}\right)\left(x - \tfrac{2}{3}a - y\sqrt{3}\right)\left(\tfrac{4}{9}a^2 - x^2 - y^2\right) \tag{5.44}$$

satisfies the equation $\nabla^4 w = p_0 =$ constant and vanishes along the plate boundaries

$$x = -\tfrac{1}{3}a \tag{5.45a}$$

$$x = \tfrac{2}{3}a - y\sqrt{3} \tag{5.45b}$$

$$x = \tfrac{2}{3}a + y\sqrt{3} \tag{5.45c}$$

According to Exercise 4.12, we have

$$M_{xx} + M_{yy} = M_{ss} + M_{tt} = -D(1+\nu)\nabla^2 w \tag{5.46}$$

and for the present case we calculate

$$\nabla^2 w = -\frac{p_0}{4aD}\left(x + \tfrac{1}{3}a\right)\left(x - \tfrac{2}{3}a + y\sqrt{3}\right)\left(x - \tfrac{2}{3}a - y\sqrt{3}\right) \tag{5.47}$$

With reference to (5.46), (5.47), and (5.45), it is clear that the normal bending moment vanishes along the plate edges. Thus, we conclude that (5.44) characterizes the deflection of the midplane of an equilateral, triangular plate that is simply supported and subjected to a uniformly distributed load of intensity $p_0$.

With the aid of (5.4) and (5.44), we calculate the Kirchhoff shear force as

$$V_x = \frac{p_0}{4a}\left(3x^2 - 3y^2 - 2ax\right) - \frac{(1-\nu)p_0}{16a}\left(3x^2 + 9y^2 + 2ax - \tfrac{2}{3}a^2\right)$$

and along the edge $x = -\frac{1}{3}a$, we have

$$V_x\left(-\tfrac{1}{3}a, y\right) = \frac{p_0 a}{4}\left[\left(1 - 3\frac{y^2}{a^2}\right) + \tfrac{1}{4}(1-\nu)\left(1 - 9\frac{y^2}{a^2}\right)\right] \tag{5.48}$$

The total reaction along the edge $x = -\frac{1}{3}a$ is

$$\int_{-a/\sqrt{3}}^{a/\sqrt{3}} V_x\left(-\tfrac{1}{3}a, y\right) dy = \frac{p_0 a^2}{3\sqrt{3}}$$

In view of the threefold symmetry, the total reactive force is $p_0 a^2/\sqrt{3}$, acting in the direction of the negative $z$-axis. The total applied force is $p_0(\frac{1}{2})(2a/\sqrt{3})a = p_0a^2/\sqrt{3}$, acting in the direction of the positive $z$-axis. Thus, we conclude that the plate is in a state of equilibrium and there are no concentrated forces acting at the vertices of the triangle. The bending moments are

$$M_{xx} = \frac{p_0}{16a}\Big[(5-\nu)x^3 - (3+\nu)ax^2 - \tfrac{2}{3}a^2(1-\nu)x - 3(1+3\nu)xy^2 - (1+3\nu)ay^2 + \tfrac{8}{27}(1+\nu)a^3\Big]$$

$$M_{yy} = \frac{p_0}{16a}\Big[(5\nu-1)x^3 - (1+3\nu)ax^2 + \tfrac{2}{3}(1-\nu)a^2x - 3(3+\nu)xy^2 - (3+\nu)ay^2 + \tfrac{8}{27}(1+\nu)a^3\Big]$$

Along the line of symmetry $y = 0$, we have

$$M_{xx} = \frac{p_0}{16a}\Big[(5-\nu)x^3 - (3+\nu)ax^2 - \tfrac{2}{3}a^2(1-\nu)x + \tfrac{8}{27}(1+\nu)a^3\Big] \tag{5.49a}$$

$$M_{yy} = \frac{p_0}{16a}\Big[(5\nu-1)x^3 - (1+3\nu)ax^2 + \tfrac{2}{3}a^2(1-\nu)x + \tfrac{8}{27}(1+\nu)a^3\Big] \tag{5.49b}$$

and the values of bending moments along $y = 0$ are shown in Table 5.1. Table 5.2 gives the values of $V_x$ along $x = -\frac{1}{3}a$ [see (5.48)]. We note that there is a reversal (change in direction) of $V_x(-\frac{1}{3}a, y)$ in the vicinity of the vertices. We also note that the twisting moment $M_{xy}$ vanishes along $y = 0$ and on the

**TABLE 5.1 Bending Moments along $y = 0$**

| $x/a$ | $M_{xx}/p_0a^2$ | $M_{yy}/p_0a^2$ |
|---|---|---|
| −0.333 | 0.00000 | 0.00000 |
| −0.300 | 0.00633 | 0.00378 |
| −0.200 | 0.0193 | 0.01323 |
| −0.100 | 0.0246 | 0.0199 |
| 0.000 | 0.0241 | 0.0241 |
| 0.100 | 0.0194 | 0.0258 |
| 0.200 | 0.0123 | 0.0254 |
| 0.300 | 0.00469 | 0.0230 |
| 0.400 | −0.00179 | 0.0188 |
| 0.500 | −0.00533 | 0.0129 |
| 0.600 | −0.00424 | 0.00557 |
| 0.667 | 0.00000 | 0.00000 |

**TABLE 5.2 Kirchhoff Shear Force along $x = -\frac{1}{3}a$**

| $y/a$ | $V_x$ |
|---|---|
| 0.0 | 0.294 |
| 0.1 | 0.282 |
| 0.2 | 0.248 |
| 0.3 | 0.191 |
| 0.4 | 0.111 |
| 0.5 | 0.008 |
| $1/\sqrt{3}$ | $-0.088$ |

circle $(x + \frac{1}{3}a)^2 + y^2 = (a/\sqrt{3})^2$. With the aid of (5.49) we find the following extreme values ($\nu = 0.3$):

$$(M_{xx})_{\max} = M_{xx}(0.062a, 0) = 0.0248 p_0 a^2$$

$$(M_{yy})_{\min} = M_{xx}(-0.53a, 0) = 0.0056 p_0 a^2$$

$$(M_{yy})_{\max} = M_{yy}(0.12a, 0) = 0.0259 p_0 a^2$$

At the centroid of the plate $(x, y) = (\frac{1}{3}a, 0)$ we have the values

$$M_{xx} = M_{yy} = (1 + \nu)\frac{p_0 a^2}{54} = 0.0241 p_0 a^2$$

$$w = \frac{p_0 a^4}{973D} = 0.001029 \frac{p_0 a^4}{D}$$

### 5.1.4 Twisting of a Rectangular Plate by Corner Forces

In this section we consider the flexure of a rectangular plate induced by four transverse corner forces, each of magnitude $P$, as shown in Fig. 5.5. The forces acting at $A$ and $C$ are in the direction of the positive $z$-axis (up), and the forces acting at $B$ and $D$ are in the direction of the negative $z$-axis (down). The four edges of the plate are free, so that

$$M_{xx}\left(\tfrac{1}{2}a, y\right) = V_x\left(\tfrac{1}{2}a, y\right) = 0 \tag{5.50a}$$

$$M_{yy}\left(x, \tfrac{1}{2}b\right) = V_y\left(x, \tfrac{1}{2}b\right) = 0 \tag{5.50b}$$

$$M_{xx}\left(-\tfrac{1}{2}a, y\right) = V_x\left(-\tfrac{1}{2}a, y\right) = 0 \tag{5.50c}$$

$$M_{yy}\left(x, -\tfrac{1}{2}b\right) = V_y\left(x, -\tfrac{1}{2}b\right) = 0 \tag{5.50d}$$

In the present case, the plate is free from surface pressures, and therefore,

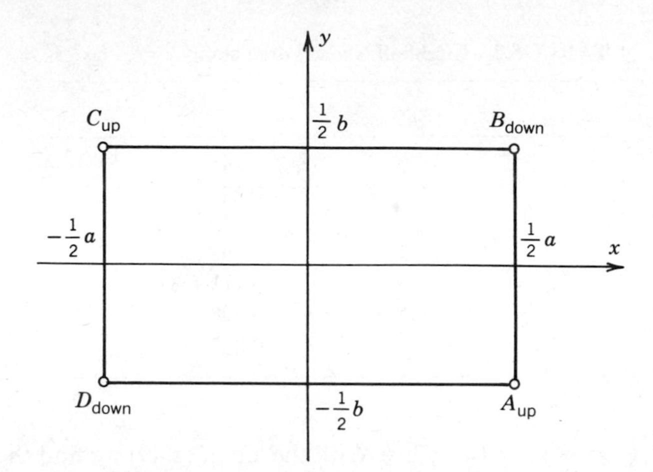


**Figure 5.5** Rectangular plate with corner forces.

$p \equiv 0$. Consequently, we seek a solution of [see (5.1)]

$$\nabla^4 w = 0 \tag{5.51}$$

subject to the boundary conditions (5.50) and the corner conditions [see (4.20c)]

$$P = 2M_{xy} \quad \text{at } \left(\tfrac{1}{2}a, -\tfrac{1}{2}b\right), \left(\tfrac{1}{2}a, \tfrac{1}{2}b\right), \left(-\tfrac{1}{2}a, \tfrac{1}{2}b\right), \left(-\tfrac{1}{2}a, -\tfrac{1}{2}b\right) \tag{5.52}$$

A solution of (5.51) is

$$w = \theta xy \tag{5.53}$$

where $\theta$ is a constant (the angle of twist per unit length). With the aid of Exercise 4.13, it is readily shown that

$$M_{xx} = M_{yy} = 0, \qquad M_{xy} = -D(1-\nu)\theta$$
$$Q_x = Q_y = 0, \qquad V_x = V_y = 0$$

and

$$P = 2M_{xy} = -\frac{Eh^3\theta}{6(1+\nu)} = -\frac{Gh^3\theta}{3}$$

so that

$$\theta = -\frac{3P}{Gh^3}$$

We conclude that our solution is

$$w = -\frac{3P}{Gh^3}xy \tag{5.54}$$

With reference to (5.54) and Fig. 5.5, the deflection of the plate at $A$ and $C$ is $\frac{3}{4}(abF/h^3G)$ (up), and at $B$ and $D$ it is $-\frac{3}{4}(abF/h^3G)$ (down). This solution is attributed to Sir Horace Lamb (1849–1934).

### 5.1.5 Pure Bending of a Rectangular Plate

This section is concerned with the deformation of a rectangular plate by the action of distributed couples along two opposite edges. With reference to Fig. 5.6 we consider a plate with dimensions $a:b:h$ that is acted upon by distributed moments $M_{xx} = M_0$ along the edges $x = \pm\frac{1}{2}a$. The edges $y = \pm\frac{1}{2}b$ are free, and the plate is free from transverse pressure forces. Consequently, we require a solution of the equation

$$\nabla^4 w = 0 \tag{5.55}$$

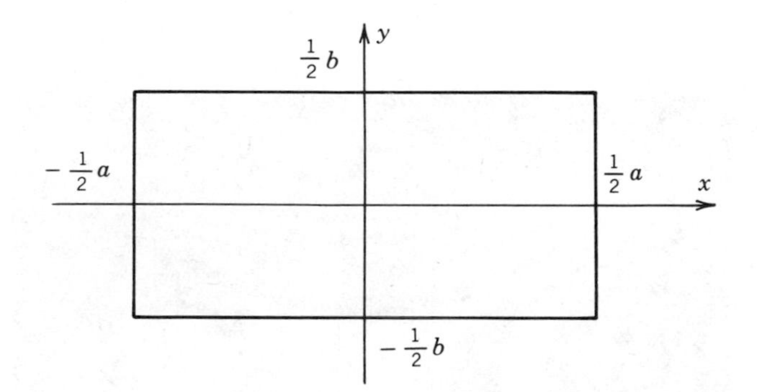


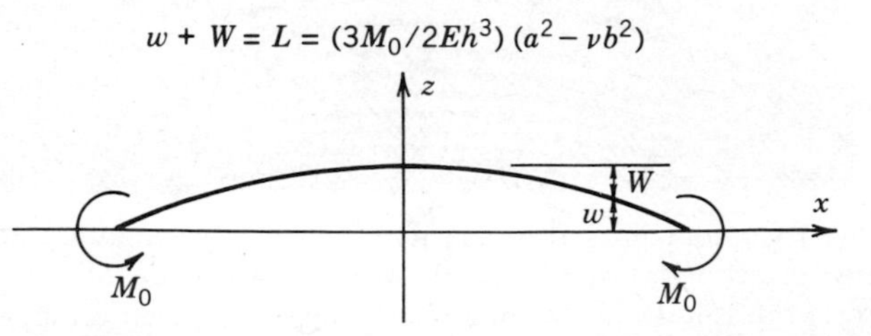


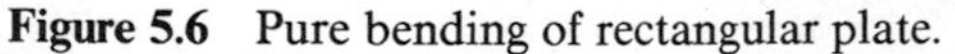
**Figure 5.6** Pure bending of rectangular plate.

subject to the boundary conditions

$$
\begin{aligned}
w\left(\pm\tfrac{1}{2}a, \pm\tfrac{1}{2}b\right) &= 0, & M_{xx}\left(\pm\tfrac{1}{2}a, y\right) &= M_0 \\
M_{yy}\left(x, \pm\tfrac{1}{2}b\right) &= 0, & V_y\left(x, \pm\tfrac{1}{2}b\right) &= 0 \\
& & V_x\left(\pm\tfrac{1}{2}a, y\right) &= 0
\end{aligned}
\tag{5.56}
$$

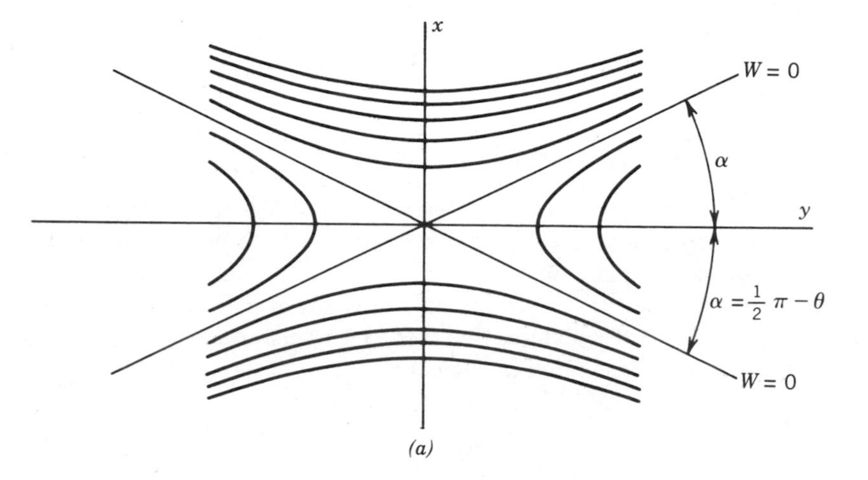


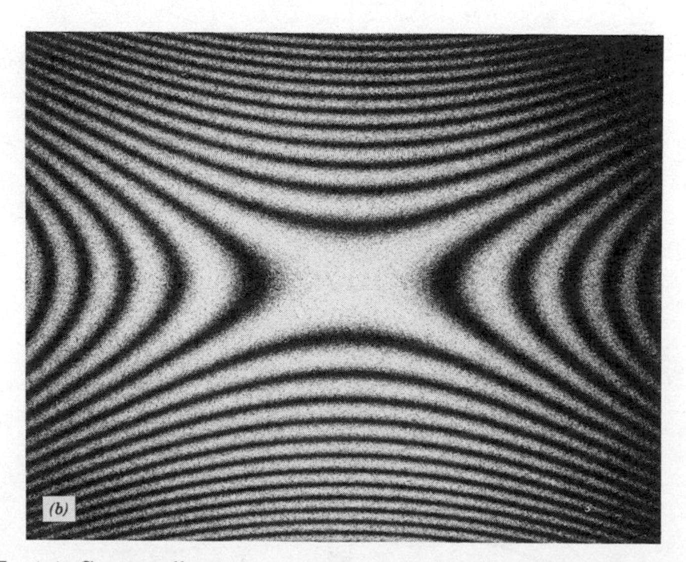


**Figure 5.7** ($a$) Contour lines $W =$ constant. ($b$) Double-exposure hologram of deformed plate surface. (Holographic work was performed by Mr. P. Malyak in the laboratory of Professor D. P. Malone, Department of Electrical Engineering, State University of New York at Buffalo.)

A solution of (5.55) that also satisfies the conditions (5.56) is

$$w = -\frac{6M_0}{Eh^3}(x^2 - \nu y^2) + \frac{6M_0}{Eh^3}\left(\frac{a^2}{4} - \nu\frac{b^2}{4}\right) \tag{5.57}$$

For subsequent purposes, it will be convenient to translate (shift) the reference in the $z$-direction and to measure plate deflection from a plane that is tangent to the deformed surface at the point $(x, y) = (0, 0)$ (see Fig. 5.6)

$$W = -w + L = \frac{6M_0}{Eh^3}(x^2 - \nu y^2) \tag{5.58}$$

where $L = \frac{3}{2}M_0(a^2 - \nu b^2)$. The contour map corresponding to (5.58) is shown in Fig. 5.7($a$). We note that the contour lines consist of two families of hyperbolas, each having two branches. The asymptotes are straight lines characterized by $\tan\theta = y/x = 1/\sqrt{\nu}$ and $\alpha = \frac{1}{2}\pi - \theta$. An experimental technique called holographic interferometry is uniquely suited to measure sufficiently small deformations of a plate. In Fig. 5.7($b$) we show a double-exposure hologram of the anticlastic deformation of a plate loaded as shown in Fig. 5.6. This hologram was taken using a two-beam technique, utilizing Kodak Holographic 120-02 plates. The laser was a 10-mW He–Ne laser, 632.8 nm, with beam ratio 4 : 1. The fringe lines in Fig. 5.7($b$) correspond to the contour lines of Fig. 5.7($a$). The close correspondence between theory and experiment is readily observed. We also note that this technique results in the nondestructive experimental determination of Poisson's ratio $\nu$ of the plate.

### 5.1.6 Circular Plates (Axial Symmetry)

In this section we consider the case of a circular plate (or a plate in the shape of an annulus) subjected to axially symmetric loads and boundary conditions. It will be convenient to refer all pertinent basic equations to polar coordinates, omitting all terms of the type $\partial(\quad)/\partial\theta$. Thus, we require that

$$D\nabla^4 w = p(r) \tag{5.59}$$

where [see (5.5)]

$$\nabla^2 w = \frac{d^2w}{dr^2} + \frac{1}{r}\frac{dw}{dr} = \frac{1}{r}\frac{d}{dr}\left(r\frac{dw}{dr}\right) \tag{5.60}$$

and

$$\nabla^4 w = \nabla^2(\nabla^2 w) = \frac{1}{r}\frac{d}{dr}\left\{r\frac{d}{dr}\left[\frac{1}{r}\frac{d}{dr}\left(r\frac{dw}{dr}\right)\right]\right\} \tag{5.61}$$

For the present applications, the stress-displacement relations are

$$M_{rr} = -D\left(\frac{d^2w}{dr^2} + \nu\frac{1}{r}\frac{dw}{dr}\right) \tag{5.62a}$$

$$M_{\theta\theta} = -D\left(\frac{1}{r}\frac{dw}{dr} + \nu\frac{d^2w}{dr^2}\right) \tag{5.62b}$$

$$M_{r\theta} \equiv 0 \tag{5.62c}$$

$$Q_r = -D\frac{d}{dr}(\nabla^2 w) \tag{5.63a}$$

$$Q_\theta \equiv 0 \tag{5.63b}$$

For a detailed derivation of (5.59)–(5.63), the reader is referred to Exercises 4.14 and 4.15.

We shall consider the case of a uniformly distributed load $p(r) = p_0 =$ const. Then (5.61) assumes the form

$$\frac{d}{dr}\left[r\frac{d}{dr}\nabla^2 w\right] = \frac{rp_0}{D}$$

and

$$\frac{d}{dr}(\nabla^2 w) = \frac{p_0}{D}\frac{r}{2} + \frac{A'}{r}$$

$$\nabla^2 w = \frac{1}{r}\frac{d}{dr}\left(r\frac{dw}{dr}\right) = \frac{p_0}{D}\frac{r^2}{4} + A'\ln r + B'$$

Further integration results in

$$\frac{d}{dr}\left(r\frac{dw}{dr}\right) = \frac{p_0}{D}\frac{r^3}{4} + A'r\ln r + B'r$$

$$r\frac{dw}{dr} = \frac{p_0}{D}\frac{r^4}{16} + A'\left(\frac{r^2}{2}\ln r - \frac{r^2}{4}\right) + B'\frac{r^2}{2} + C'$$

$$w = \frac{p_0}{D}\frac{r^4}{64} + A'\left(\frac{r^2}{4}\ln r - \frac{r^2}{4}\right) + \frac{B'r^2}{4} + C'\ln r + H'$$

and by redefining the constants of integration, we obtain the complete solution of (5.59):

$$w = \frac{p_0 r^4}{64D} + Ar^2\ln r + Br^2 + C\ln r + H \tag{5.64}$$

The constants of integration $A$, $B$, $C$, and $H$ can be determined by the application of suitable boundary conditions at $r = a$ and at $r = b$ or, in the

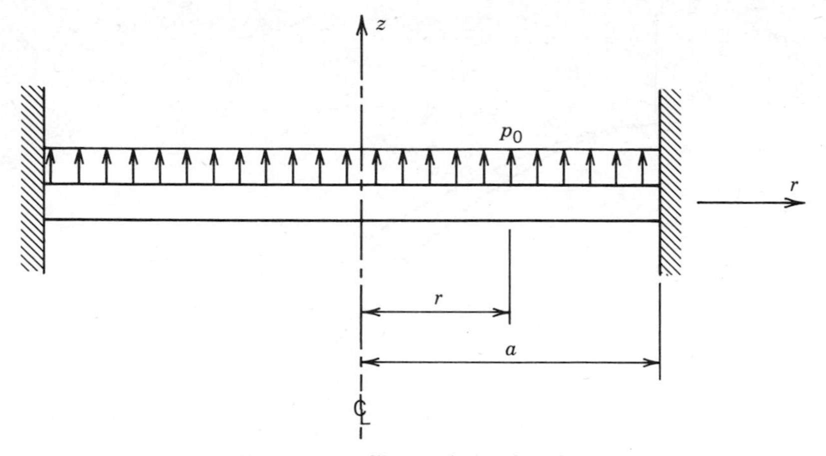


**Figure 5.8** Clamped circular plate.

case of a solid circular plate, by the application of regularity conditions at $r = 0$ and boundary conditions at $r = a$.

**Example (a):** We consider a solid circular plate clamped along its boundary $r = a$, as shown in Fig. 5.8. The plate is acted upon by a uniformly distributed load of intensity $p(r) = p_0 =$ constant. In this case we require that $w$, $M_{rr}$, $M_{\theta\theta}$, and $Q_r$ remain bounded at $r = 0$. This results in $A = C = 0$. Consequently, the solution can be written in the form

$$w = \frac{p_0 r^4}{64D} + Br^2 + H \tag{5.65}$$

At the clamped edge we have

$$w(a) = 0 \quad \text{and} \quad \left(\frac{dw}{dr}\right)_{r=a} = 0 \tag{5.66}$$

Upon substitution of (5.65) into (5.66), we obtain

$$B = -\frac{p_0 a^2}{32D}, \qquad H = \frac{p_0 a^4}{64D} \tag{5.67}$$

and insertion of (5.67) into (5.65) provides the solution

$$w = \frac{p_0}{64D}(a^2 - r^2)^2 \tag{5.68}$$

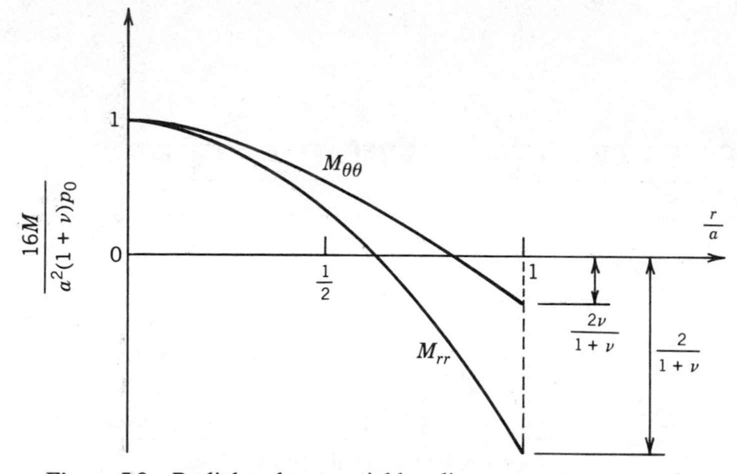


**Figure 5.9** Radial and tangential bending moments versus $r/a$.

With the aid of (5.62), (5.63), and (5.68), we obtain

$$M_{rr} = \frac{p_0}{16}\left[a^2(1+\nu) - r^2(3+\nu)\right] \tag{5.69a}$$

$$M_{\theta\theta} = \frac{p_0}{16}\left[a^2(1+\nu) - r^2(1+3\nu)\right] \tag{5.69b}$$

$$\nabla^2 w = \frac{p_0}{8D}(2r^2 - a^2) \tag{5.70a}$$

$$Q_r = -D\frac{\partial}{\partial r}\nabla^2 w = -\frac{p_0 r}{2} \tag{5.70b}$$

The maximum deflection occurs at the center and is

$$w_{\text{max}} = w(0) = \frac{p_0 a^4}{64D}$$

A plot of $M_{rr}$ and $M_{\theta\theta}$ versus $r$ is shown in Fig. 5.9. We note that

$$(M_{rr})_{\text{max}} = M_{rr}(0) = \tfrac{1}{16}(1+\nu)a^2 p_0$$
$$(M_{\theta\theta})_{\text{max}} = M_{\theta\theta}(0) = \tfrac{1}{16}(1+\nu)a^2 p_0$$
$$(M_{rr})_{\text{min}} = M_{rr}(a) = -\tfrac{1}{8}a^2 p_0$$
$$(M_{\theta\theta})_{\text{min}} = M_{\theta\theta}(a) = -\tfrac{1}{8}\nu a^2 p_0$$

**Example (b):** With reference to Fig. 5.10, we consider a plate annulus $a \leq r \leq b$ that is clamped at its outer boundary $r = b$. The inner boundary is welded to a

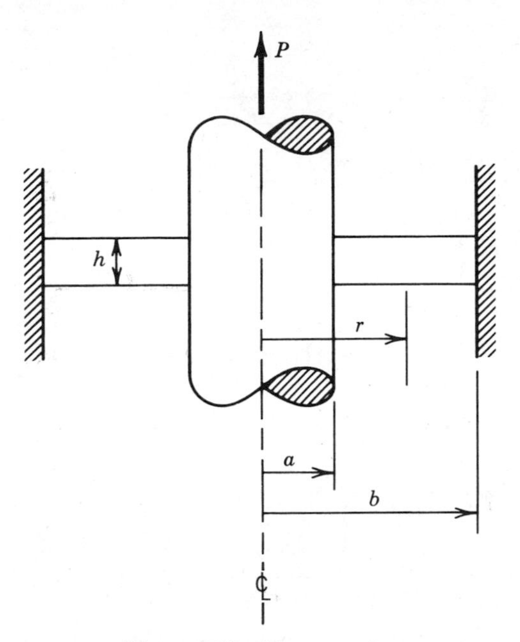


**Figure 5.10** Plate annulus.

cylindrical shaft at $r = a$, and an axial force $P$ is applied to the shaft. We wish to find the equation of the deformed plate median surface. In the present case $p \equiv 0$, and with reference to (5.64), the solution is

$$w = Ar^2 \ln r + Br^2 + C \ln r + H \tag{5.71}$$

The boundary conditions are

$$w(b) = 0, \qquad \left(\frac{dw}{dr}\right)_{r=b} = 0 \tag{5.72a}$$

$$\left(\frac{dw}{dr}\right)_{r=a} = 0, \qquad Q_r(a) = -\frac{P}{2\pi a} \tag{5.72b}$$

Using (5.7a) and (5.71), we compute

$$Q_r(r) = -D\frac{d}{dr}\nabla^2 w = -\frac{4AD}{r} \tag{5.73}$$

and with reference to (5.72b), we have

$$-\frac{P}{2\pi a} = -\frac{4AD}{a} = Q_r(a)$$

so that

$$A = \frac{P}{8\pi D} \tag{5.74a}$$

Additionally, from (5.72) and (5.71),

$$w(b) = 0 = Ab^2 \ln b + Bb^2 + C \ln b + H \tag{5.74b}$$

$$\left(\frac{dw}{dr}\right)_{r=b} = 0 = A(b + 2b \ln b) + 2Bb + \frac{C}{b} \tag{5.74c}$$

$$\left(\frac{dw}{dr}\right)_{r=a} = 0 = A(a + 2a \ln a) + 2Ba + \frac{C}{a} \tag{5.74d}$$

Solving the system of equations (5.74), we readily obtain

$$B = \frac{P}{8\pi D}\left[-\frac{1}{2} + \frac{(a/b)\ln a - (b/a)\ln b}{b/a - a/b}\right] \tag{5.75a}$$

$$C = \frac{P}{4\pi D}\frac{\ln r \ln(b/a)}{(1/a^2 - 1/b^2)} \tag{5.75b}$$

$$H = \frac{P}{8\pi D}\left[\frac{1}{2}\frac{b}{a} + \frac{(1 - 2\ln b)\ln(b/a)}{1/a^2 - 1/b^2}\right] \tag{5.75c}$$

Substitution of (5.74a) and (5.75) into (5.71) results in the solution

$$\frac{8\pi Dw}{abP} = \frac{1}{2}\left(\frac{b}{a} - \frac{r^2}{ab}\right) + \frac{r^2}{ab}\left[\frac{(b/a)\ln(r/b) - (a/b)\ln(r/a)}{b/a - a/b}\right] + \frac{[1 + 2\ln(r/b)]\ln(b/a)}{b/a - a/b} \tag{5.76}$$

Using (5.6), (5.7), and (5.76), we obtain

$$-\frac{4\pi M_{rr}}{P} = 1 + (1+\nu)\frac{(b/a)\ln(r/b) - (a/b)\ln(r/a)}{b/a - a/b} - (1-\nu)\frac{ab}{r^2}\frac{\ln(b/a)}{(b/a - a/b)} \tag{5.77a}$$

$$-\frac{4\pi M_{\theta\theta}}{P} = \nu + (1+\nu)\frac{(b/a)\ln(r/b) - (a/b)\ln(r/a)}{b/a - a/b} + (1-\nu)\frac{ab}{r^2}\frac{\ln(b/a)}{(b/a - a/b)} \tag{5.77b}$$

$$\nabla^2 w = 1 + 2\frac{(b/a)\ln(r/b) - (a/b)\ln(r/a)}{b/a - a/b} \tag{5.77c}$$

$$Q_r = -D\frac{d}{dr}(\nabla^2 w) = -\frac{P}{2\pi r} \tag{5.77d}$$

### 5.1.7 Concentrated Loads (Axial Symmetry)

We now consider the action of a concentrated load of magnitude $P$ at the center of a circular plate with radius $a$, as shown in Fig. 5.11($a$). In this case $p \equiv 0$, and with reference to (5.64), we have

$$Q_r = -D\frac{d}{dr}(\nabla^2 w) = -\frac{4AD}{r}$$

We now invoke force equilibrium for a circular portion of the plate with radius $r$ [see Fig. 5.11($b$)]. In this case we require

$$-2\pi r Q_r = P = 8\pi AD$$

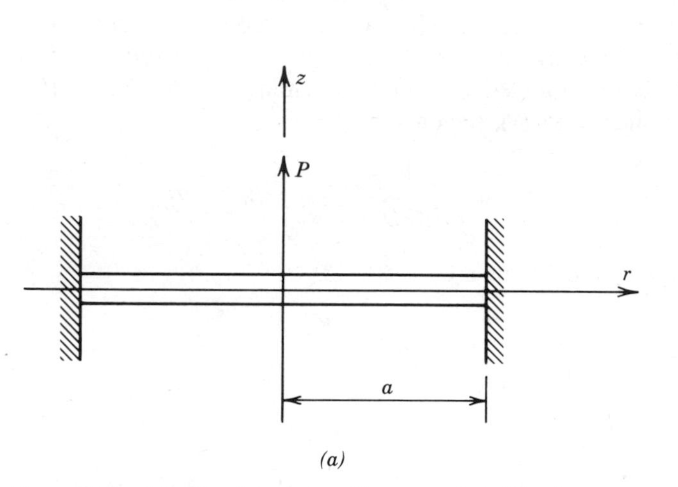


(a)

**Figure 5.11** ($a$) Circular plate with concentrated load at center.

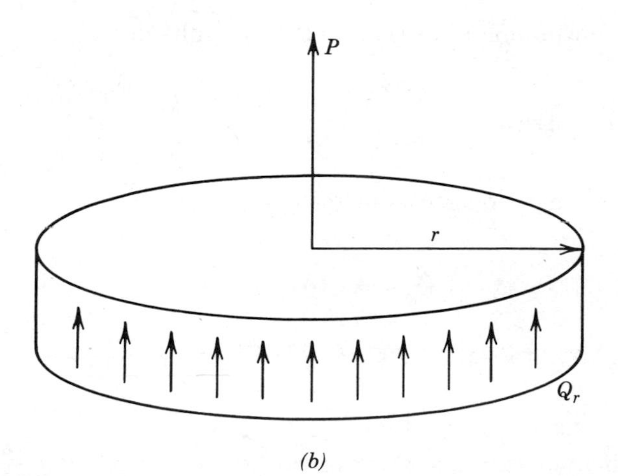


(b)

**Figure 5.11** ($b$) Circular plate element surrounding concentrated load.

so that

$$A = \frac{P}{8\pi D}$$

and therefore, the required singularity solution is

$$w = \frac{P}{8\pi D} r^2 \ln r \tag{5.78}$$

Equation (5.78) is the particular solution of (5.59) and corresponds to a concentrated load of magnitude $P$ acting at $r = 0$ and $p(r) \equiv 0$ for $0 < r \leq a$.

**Example:** We consider a circular plate clamped at the boundary $r = a$, as shown in Fig. 5.11($a$). The plate is acted upon by a concentrated load of magnitude $P$ acting at $r = 0$. Thus, we seek a solution of the equation $\nabla^4 w = 0$ that is bounded at $r = 0$ and displays the singularity (5.78) at $r = 0$. With reference to (5.71), such a solution is

$$w = \frac{P}{8\pi D} r^2 \ln r + Br^2 + H \tag{5.79}$$

The boundary conditions are

$$w(a) = \left(\frac{dw}{dr}\right)_{r=a} = 0 \tag{5.80}$$

Upon substitution of (5.80) into (5.79), we readily obtain

$$H = \frac{Pa^2}{16\pi D}, \qquad B = -(1 + 2\ln a)\frac{P}{16\pi D} \tag{5.81}$$

and further substitution of (5.81) into (5.79) results in

$$w(r) = \frac{Pa^2}{8\pi D}\frac{r^2}{a^2}\ln\frac{r}{a} + \frac{Pa^2}{16\pi D}\left(1 - \frac{r^2}{a^2}\right) \tag{5.82}$$

The maximum deflection occurs under the load:

$$w_{\max} = w(0) = \frac{Pa^2}{16\pi D} \tag{5.83}$$

With reference to the solution (5.82), it should be noted that

(a) deflections are bounded for $0 \leq r \leq a$ and
(b) bending moments and shear forces are unbounded at $r = 0$.

This "defect" of CPT is readily removed. For example, all field quantities $(M_{rr}, Q_r, \ldots)$ will remain bounded if the load is distributed over a circular region $r = \varepsilon > 0$, no matter how small (for details, see Exercise 5.10).

### 5.1.8 Clamped Circular Plate under Concentrated Load (Inversion)

In this section we shall use the method of inversion to generate the solution for the deflection surface of a clamped circular plate of radius $a$ under an arbitrarily placed concentrated load of magnitude $P$. The method is based on inversion transformation, a conformal transformation defined by

$$z'z = k^2 \tag{5.84a}$$

where

$$z' = x' + iy' = r'e^{i\theta'} \tag{5.84b}$$

$$z = x + iy = re^{i\theta} \tag{5.84c}$$

and where $k$, a real number, is the radius of inversion. A detailed study of (5.84) reveals that the transformation consists of an inversion of the modulus $r' = k^2/r$ and a reflection about the real axis $\theta' = -\theta$. The following properties of the transformation render it suitable for the present application: (a) If $w$ is a bi-harmonic function in the $r$-$\theta$ plane, then $r'^2 w$ is bi-harmonic in the $r'$-$\theta'$ plane. (b) The slopes in the $r'$-$\theta'$ plane are equal to

$$\frac{\partial}{\partial r'}(r'^2 w) = \frac{2a^2 w}{r} - a^2 \frac{\partial w}{\partial r} \tag{5.85a}$$

$$\frac{1}{r'}\frac{\partial}{\partial \theta'}(r'^2 w) = \frac{a^2}{r}\frac{\partial w}{\partial \theta} \tag{5.85b}$$

and therefore a clamped boundary in the $r$-$\theta$ plane along which $w = \partial w/\partial r = (1/r)(\partial w/\partial \theta) = 0$ transforms into a clamped boundary in the $r'$-$\theta'$ plane. (c) Circles transform into circles. In particular, a circle with radius $a$ and with center on the real axis a distance $h$ from the origin will transform into a similarly located circle for which

$$h' = \frac{k^2}{h^2 - a^2} h \tag{5.86a}$$

$$a' = \frac{k^2}{h^2 - a^2} h \tag{5.86b}$$

A circle that passes through the origin ($h = a$) transforms into a straight line. If $k^2 = h^2 - a^2$, the circle transforms into itself.

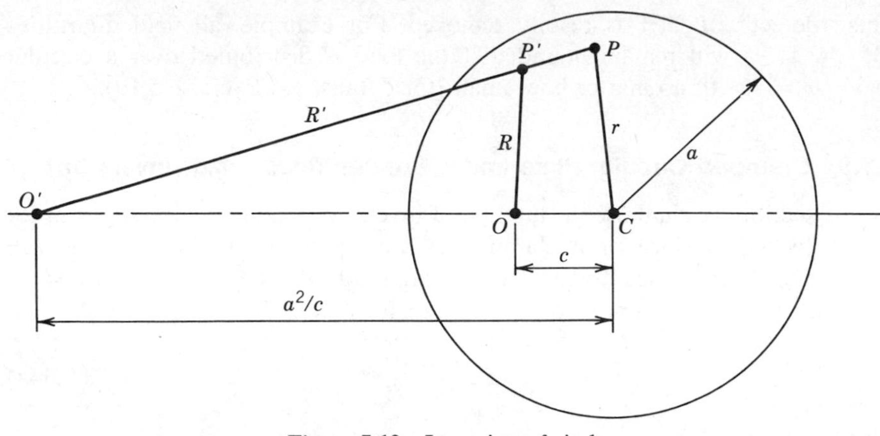


**Figure 5.12** Inversion of circle.

With reference to Fig. 5.12, we now use the method of inversion to find the deflected midsurface of a circular plate clamped at its edge $r = a$ and subjected to a concentrated load $F$ at the point $O$. We let $O'$ be the inverse point of $O$ with respect to the circle of radius $a$. We also let $c$ be the distance from $O$ to $C$. The solution for a clamped circular plate under a centrally located concentrated load $F$ is [see (5.82)]

$$w = \frac{Fa^2}{8\pi D}\left[\frac{r^2}{a^2}\ln\frac{r}{a} + \frac{1}{2}\left(1 - \frac{r^2}{a^2}\right)\right] \tag{5.87}$$

where $r$ denotes the distance of a generic point $P$ from $C$. We now invert from $O'$ with the constant of inversion equal to $k^2 = a^4/c^2 - a^2$. In this case the circle inverts into itself, $C$ inverts into $O$, and $P$ inverts into $P'$. Let $OP' = R$ and $O'P' = R'$. Then, from the geometry in Fig. 5.12, we have

$$\frac{R}{r} = \frac{R'}{a^2/c} \quad \text{or} \quad \frac{r}{a} = \frac{a}{c}\frac{R}{R'} \tag{5.88}$$

Upon substitution of (5.88) into (5.87), we obtain

$$R'^2 w = \frac{Fa^2}{8\pi D}\left[\frac{1}{2}\left(R'^2 - \frac{a^2}{c^2}R^2\right) + \frac{a^2}{c^2}R^2\ln\frac{aR}{cR'}\right] \tag{5.89}$$

However, $R'^2 w$ is bi-harmonic, and the boundary $a$ is again clamped. Hence,

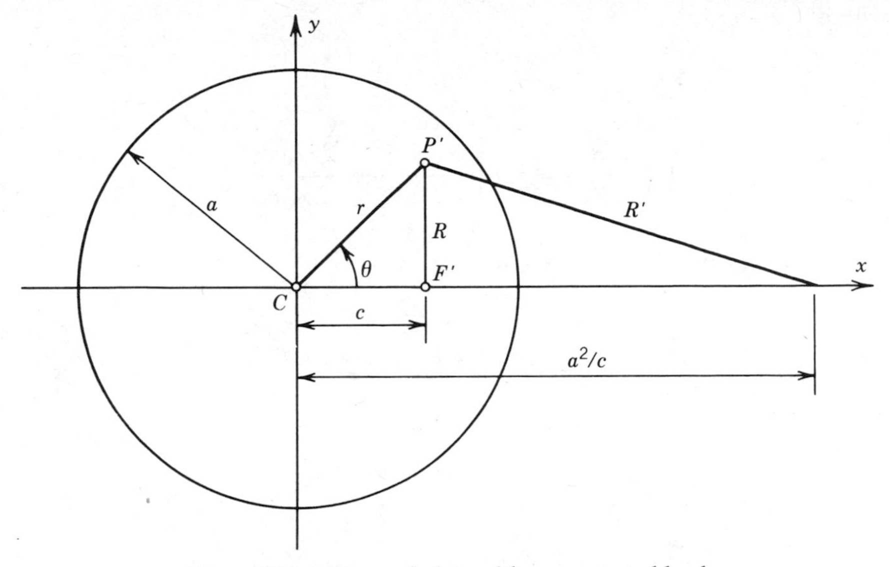


**Figure 5.13** Clamped plate with concentrated load.

if we set

$$w' = \frac{R'^2}{a^2} w \quad \text{and} \quad F' = \frac{F}{8\pi D} \frac{a^2}{c^2}$$

we obtain

$$w' = \frac{F'}{8\pi D} \left[ \frac{1}{2} \left( \frac{c^2}{a^2} R'^2 - R^2 \right) + R^2 \ln \left( \frac{a}{c} \frac{R}{R'} \right) \right] \tag{5.90}$$

for the midplane deflection of a clamped plate with radius $a$ under a transverse, concentrated load of magnitude $F'$ located at $r = c$. The representation (5.90) is not convenient, and we shall now transform it to Cartesian coordinates. With reference to Fig. 5.13, we have

$$R^2 = r^2 + c^2 - 2rc \cos \theta$$

$$R'^2 = \left( \frac{a^2}{c} \right)^2 + r^2 - 2 \frac{a^2}{c} r \cos \theta$$

Therefore,

$$\frac{c^2}{a^2} R'^2 = R^2 = a^2(1 - x^2)(1 - \xi^2)$$

where

$$x = \frac{r}{a}, \qquad \xi = \frac{c}{a}$$
$$\ln\left(\frac{a}{c}\frac{R}{R'}\right) = \frac{1}{2}\ln\left[\frac{x^2 + \xi^2 - 2x\xi\cos\theta}{1 + x^2\xi^2 - 2x\xi\cos\theta}\right] \tag{5.91}$$

and, upon substitution in (5.90), we obtain

$$w' = \frac{F'a^2}{16\pi D}\left[(1 - x^2)(1 - \xi^2) + (x^2 + \xi^2 - 2x\xi\cos\theta)\ln\frac{x^2 + \xi^2 - 2x\xi\cos\theta}{1 + x^2\xi^2 - 2x\xi\cos\theta}\right] \tag{5.92}$$

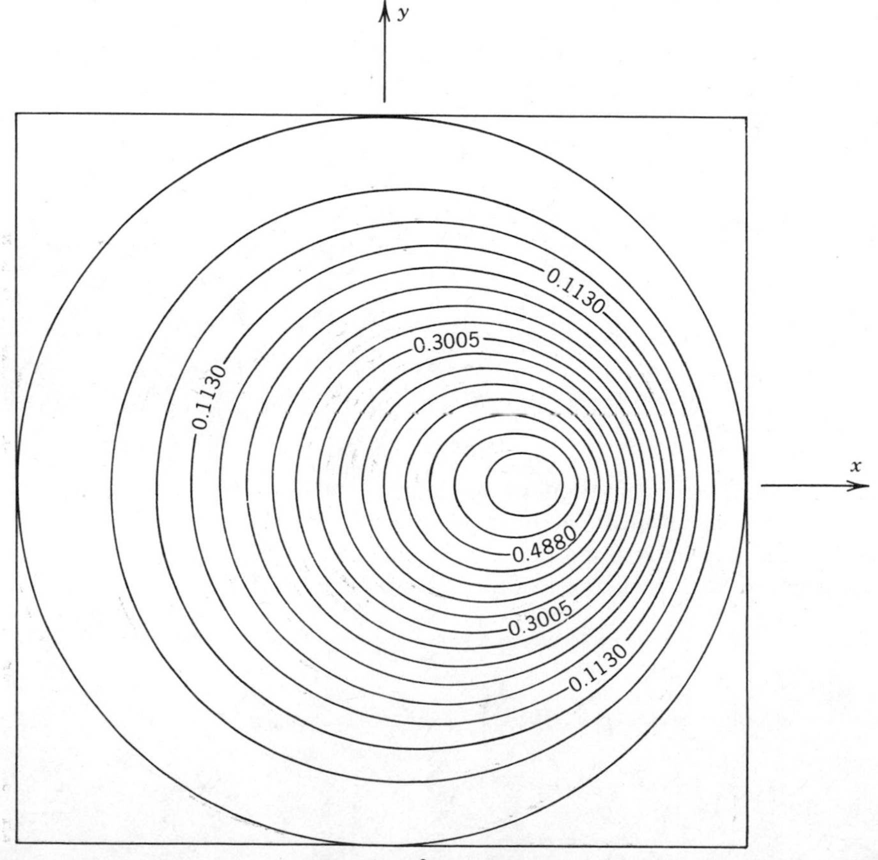


**Figure 5.14** Contour map $16\pi Dw'/a^2F' =$ constant where $\xi = \frac{1}{2}$. Circular plate under eccentrically placed concentrated load [see (5.92)].

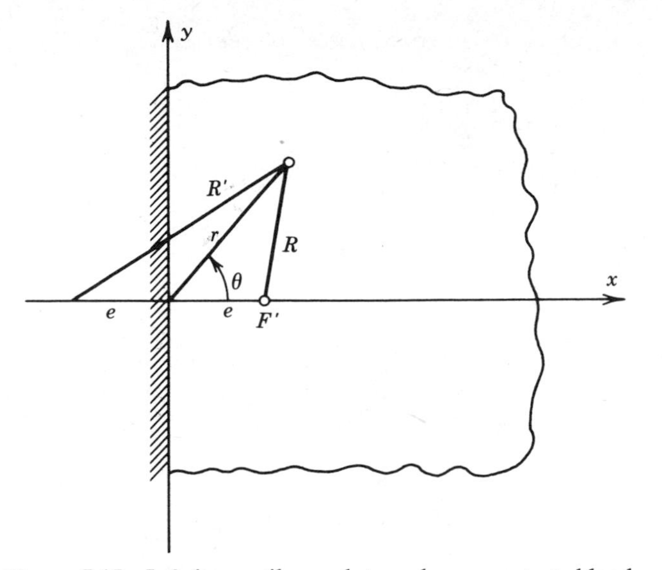


**Figure 5.15** Infinite cantilever plate under concentrated load.

Under the load we have $r = c$, $\theta = 0$, $x = \xi$, and

$$w' = \frac{F'a^2}{16\pi D}(1 - \xi^2)^2 = \frac{F'}{16\pi D}\frac{(a^2 - c^2)^2}{a^2}$$

The contour lines corresponding to (5.92) are shown in Fig. 5.14. For this figure the elevations are $16\pi Dw'/a^2F'$ and $\xi = \frac{1}{2}$.

Equation (5.90) can be used to obtain the solution of an infinite cantilever plate under a concentrated load, as shown in Fig. 5.15. Let $e = a - c =$ constant. Then we have

$$\frac{a^2}{c} = \frac{a^2}{a - e} = \frac{a}{1 - e/a}$$

and

$$\frac{a}{c} = \frac{1}{1 - e/a}$$

We now let the plate radius $a \to \infty$ and note that $a/c \to 1$. Thus, (5.90) reduces to

$$w' = \frac{F'}{8\pi D}\left[\frac{1}{2}(R'^2 - R^2) + R^2 \ln\frac{R}{R'}\right] \tag{5.93}$$

To write (5.93) in Cartesian coordinates, we set (see Fig. 5.15)

$$R'^2 = (e + x)^2 + y^2, \qquad R^2 = (e - x)^2 + y^2$$

and upon substitution into (5.85), we obtain

$$w' = \frac{F'}{8\pi D}\left\{2ex + \left[(e - x)^2 + y^2\right] \times \tfrac{1}{2}\ln\left[\frac{(e - x)^2 + y^2}{(e + x)^2 + y^2}\right]\right\} \tag{5.94}$$

Equation (5.94) characterizes the midplane deflection of an infinite cantilever plate clamped along $x = 0$ and acted upon by a concentrated load of magnitude $P$ at $(x, y) = (e, 0)$. A contour map of $8\pi Dw/F'e^2$ is shown in Fig. 5.16.

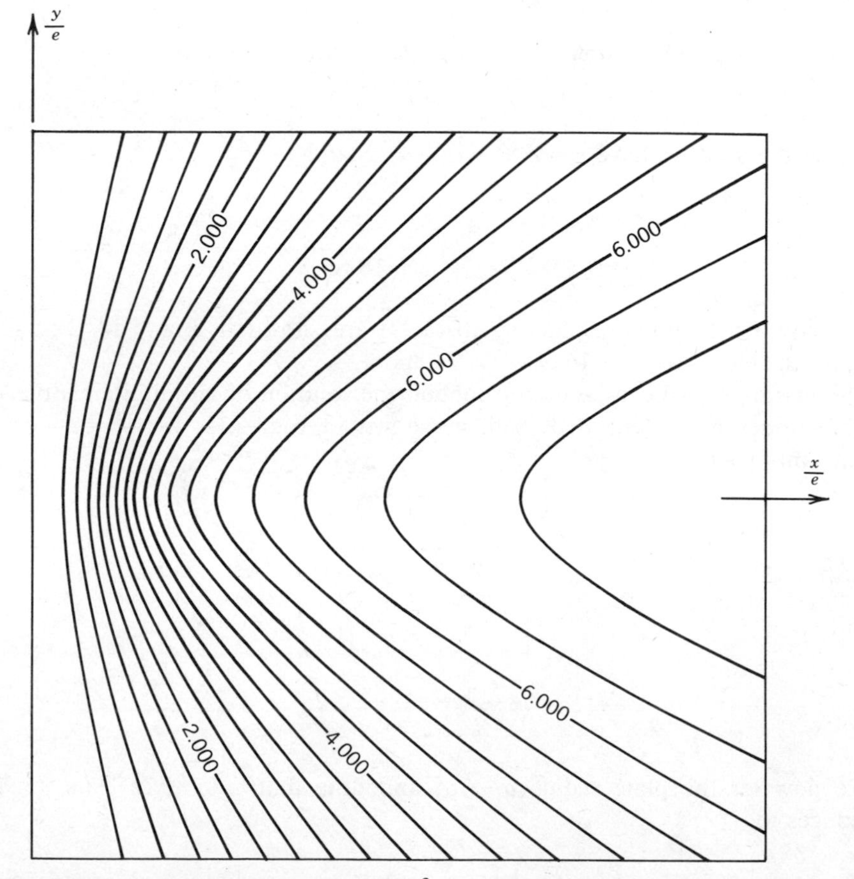


**Figure 5.16** Contour map $8\pi Dw'/F'e^2 = \text{constant}$. Cantilever plate under concentrated load [see (5.94)].

### 5.1.9 Elliptical Plate

In this section we shall consider the clamped, elliptical plate under a uniformly distributed load. When we considered the case of a clamped, circular plate under a uniformly distributed load in Section 5.1.6, we obtained the solution

$$w = \frac{p_0}{64D}(a^2 - r^2)^2 = \frac{p_0 a^4}{64D}\left(\frac{r^2}{a^2} - 1\right)^2 \tag{5.95}$$

In view of (5.95), for a similarly loaded and supported elliptical plate we attempt the generalization

$$w = C\left(\frac{x^2}{a^2} + \frac{y^2}{b^2} - 1\right)^2 = CF^2 \tag{5.96}$$

and we note that (5.96) satisfies the clamped boundary conditions

$$w = 0 \quad \text{and} \quad \frac{dw}{dn} = 0 \quad \text{on} \quad \frac{x^2}{a^2} + \frac{y^2}{b^2} = 1$$

We compute

$$C^{-1}\nabla^2 w = 4\left(\frac{1}{a^2} + \frac{1}{b^2}\right)F + 8\frac{x^2}{a^4} + 8\frac{y^2}{b^4}$$

$$D\nabla^4 w = 8\left[\frac{3}{a^4} + \frac{3}{b^4} + \frac{2}{a^2b^2}\right]CD = p_0$$

and thus conclude that

$$C = \frac{p_0}{8D}\frac{a^4b^4}{3a^4 + 3b^4 + 2a^2b^2}$$

Upon substitution into (5.96), we obtain the solution

$$w = \frac{p_0}{8D}\frac{a^4b^4\left(x^2/a^2 + y^2/b^2 - 1\right)^2}{3a^4 + 3b^4 + 2a^2b^2} \tag{5.97}$$

When $a = b$, (5.97) reverts to (5.95), providing a check on our solution. The maximum deflection is

$$w_{\max} = w(0,0) = \frac{p_0}{8D}\frac{a^4b^4}{3a^4 + 3b^4 + 2a^2b^2}$$

The bending moments are [see (5.2)]

$$M_{xx} = -\frac{p_0}{2}\frac{b^2x^2(3b^2+\nu a^2)+y^2a^2(b^2+3\nu a^2)-a^2b^2(b^2+\nu a^2)}{3a^4+3b^4+2a^2b^2} \tag{5.98a}$$

$$M_{yy} = -\frac{p_0}{2}\frac{x^2b^2(a^2+3\nu b^2)+y^2a^2(\nu b^2+3a^2)-a^2b^2(a^2+\nu b^2)}{3a^4+3b^4+2a^2b^2} \tag{5.98b}$$

The shear forces are [see (5.3)]

$$Q_x = -D\frac{\partial}{\partial x}\nabla^2 w = -p_0\frac{x(3b^4+a^2b^2)}{3a^4+3b^4+2a^2b^2} \tag{5.99a}$$

$$Q_y = -D\frac{\partial}{\partial y}\nabla^2 w = -p_0\frac{y(3a^4+a^2b^2)}{3a^4+3b^4+2a^2b^2} \tag{5.99b}$$

The total applied force is

$$W = \pi abp_0$$

The total reactive force is

$$R = \oint(V_x\,dy - V_y\,dx)$$

and we have $x = a\cos\phi$, $y = b\sin\phi$, and

$$\oint x\,dy = ab\int_0^{2\pi}\cos^2\phi\,d\phi = ab\pi = -\oint y\,dx$$

and we obtain (see Exercise 5.15)

$$R = -\pi abp_0$$

Hence, $W + R = 0$, and the plate is in a state of equilibrium.

## 5.2 SHEAR-DEFORMABLE PLATES (IMPROVED PLATE THEORY)

Sections 5.2.1–5.2.5 present solutions for specific, static plate problems formulated within the framework of a plate theory that accounts for shear as well as flexural deformations, as derived in Chapters 3 and 4. The basic assumptions of this theory are:

(a) The plate median surface does not stretch or contract.
(b) There is no thickness stretch.
(c) The plate deforms due to flexure as well as shear strain.

The basic equations of this theory, when referred to Cartesian coordinates, are:

*Equilibrium* [see (3.70)]

$$\frac{\partial Q_x}{\partial x} + \frac{\partial Q_y}{\partial y} + p = 0 \tag{5.100a}$$

$$\frac{\partial M_{xx}}{\partial x} + \frac{\partial M_{xy}}{\partial y} - Q_x = 0 \tag{5.100b}$$

$$\frac{\partial M_{yx}}{\partial x} + \frac{\partial M_{yy}}{\partial y} - Q_y = 0 \tag{5.100c}$$

*Stress-Displacement* [see (3.59)]

$$M_{xx} = D\left(\frac{\partial \psi_x}{\partial x} + \nu \frac{\partial \psi_y}{\partial y}\right) \tag{5.101a}$$

$$M_{yy} = D\left(\frac{\partial \psi_y}{\partial y} + \nu \frac{\partial \psi_x}{\partial x}\right) \tag{5.101b}$$

$$M_{xy} = \tfrac{1}{2}D(1 - \nu)\left(\frac{\partial \psi_x}{\partial y} + \frac{\partial \psi_y}{\partial x}\right) = M_{yx} \tag{5.101c}$$

$$Q_x = \kappa^2 Gh\left(\psi_x + \frac{\partial w}{\partial x}\right) \tag{5.102a}$$

$$Q_y = \kappa^2 Gh\left(\psi_y + \frac{\partial w}{\partial y}\right) \tag{5.102b}$$

When referred to polar coordinates, these equations become:

*Equilibrium* [see (3.73)]

$$\frac{\partial Q_r}{\partial r} + \frac{1}{r}\frac{\partial Q_\theta}{\partial \theta} + \frac{1}{r}Q_r + p = 0 \tag{5.103a}$$

$$\frac{\partial M_{rr}}{\partial r} + \frac{1}{r}\frac{\partial M_{r\theta}}{\partial \theta} + \frac{M_{rr} - M_{\theta\theta}}{r} - Q_r = 0 \tag{5.103b}$$

$$\frac{\partial M_{r\theta}}{\partial r} + \frac{1}{r}\frac{\partial M_{\theta\theta}}{\partial \theta} + \frac{2}{r}M_{r\theta} - Q_\theta = 0 \tag{5.103c}$$

*Stress-Displacement* [see (3.61)]

$$M_{rr} = D\left[\frac{\partial \psi_r}{\partial r} + \frac{\nu}{r}\left(\psi_r + \frac{\partial \psi_\theta}{\partial \theta}\right)\right] \tag{5.104a}$$

$$M_{\theta\theta} = D\left[\nu\frac{\partial \psi_r}{\partial r} + \frac{1}{r}\left(\psi_r + \frac{\partial \psi_\theta}{\partial \theta}\right)\right] \tag{5.104b}$$

$$M_{r\theta} = M_{\theta r} = \tfrac{1}{2}(1-\nu)D\left(\frac{\partial \psi_\theta}{\partial r} + \frac{1}{r}\frac{\partial \psi_r}{\partial \theta} - \frac{1}{r}\psi_\theta\right) \tag{5.104c}$$

$$Q_r = \kappa^2 Gh\left(\psi_r + \frac{\partial w}{\partial r}\right) \tag{5.105a}$$

$$Q_\theta = \kappa^2 Gh\left(\psi_\theta + \frac{1}{r}\frac{\partial w}{\partial \theta}\right) \tag{5.105b}$$

### 5.2.1 The Plate Strip

As a point of departure, we consider the plate strip $0 \le x \le a$, $-\infty < y < \infty$, as shown in Fig. 5.17. The plate is clamped along $x = a$, and it is acted upon by the uniformly distributed line load of magnitude $P$ along the edge $x = 0$. Consequently, we have a case of plane strain deformation with $\psi_y \equiv 0$ and $(\partial/\partial y)(\ \ ) \equiv 0$. The basic equations are (5.100)–(5.102). We have $Q_y \equiv M_{xy} \equiv p \equiv 0$, and

$$\frac{dQ_x}{dx} = 0, \quad \text{or } Q = C$$

But $Q_x(0) = -P$, so that $C_1 = -P$, and we have

$$Q_x = \frac{dM_{xx}}{dx} = -P, \quad \text{or} \quad M_{xx} = -Px + C_2$$

However, $M_{xx}(0) = 0$, and therefore, $C_2 = 0$. Application of (5.101a) results in

$$M_{xx} = D\frac{d\psi_x}{dx} = -Px \tag{5.106}$$

and therefore,

$$D\psi_x = -\tfrac{1}{2}Px^2 + C_3$$

But $\psi_x(a) = 0$, so that $C_3 = \frac{1}{2}Pa^2$, and

$$D\psi_x = \tfrac{1}{2}P(a^2 - x^2) \tag{5.107}$$

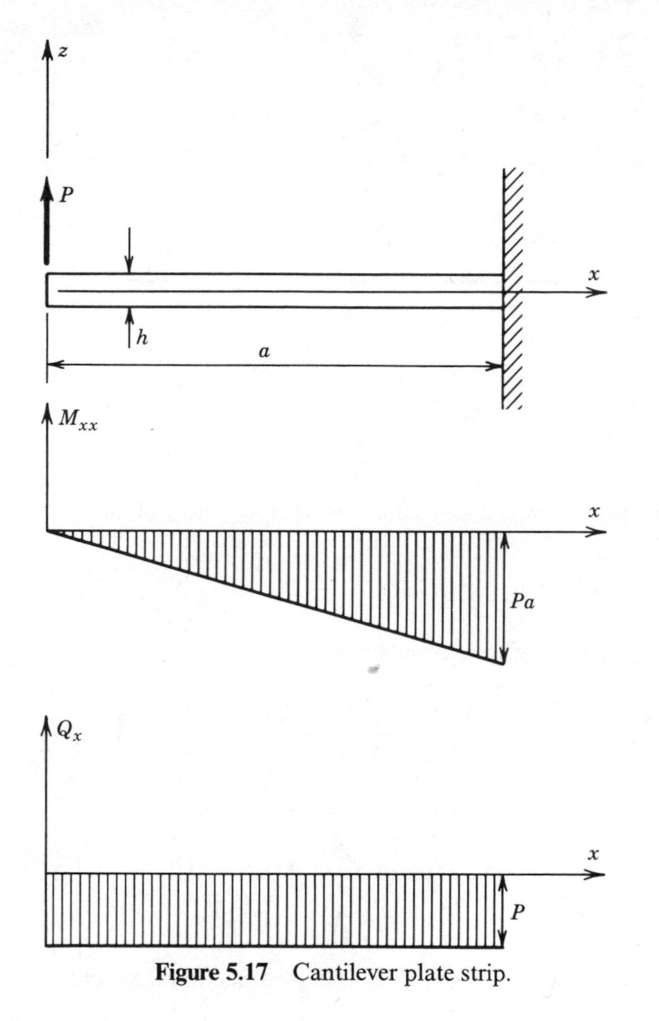


**Figure 5.17** Cantilever plate strip.

Furthermore,

$$\frac{dw}{dx} = \frac{Q_x}{\kappa^2 Fh} - \psi_x = -\frac{P}{\kappa^2 Gh} + \frac{P}{2D}(x^2 - a^2)$$

and upon integration, we obtain

$$\frac{Dw}{Pa^3} = -\frac{D}{\kappa^2 Gha^2}\left(\frac{x}{a}\right) - \frac{1}{2}\left(\frac{x}{a}\right) + \frac{1}{6}\left(\frac{x}{a}\right)^3 + C_4$$

When $x = a$, $w(a) = 0$, and therefore, $C_4 = \frac{1}{3} + D/(\kappa^2 Gha^2)$. Consequently,

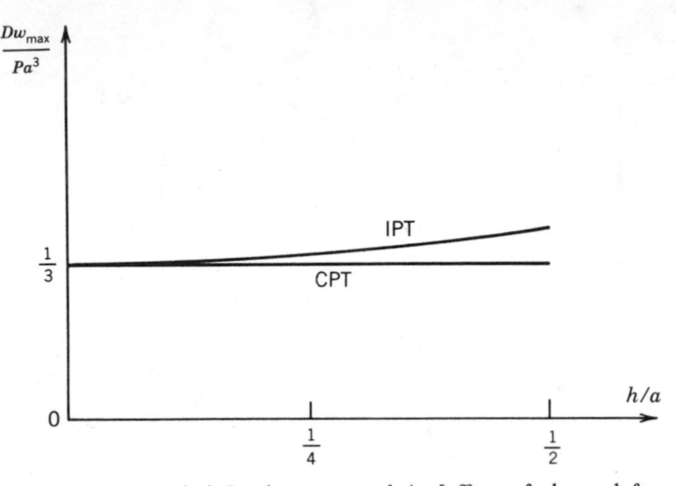


**Figure 5.18** Maximum end deflection versus $h/a$ [effect of shear deformation; see (5.109b)].

the equation of the deformed midsurface is

$$\frac{Dw}{Pa^3} = -\frac{1}{2}\left(\frac{x}{a}\right) + \frac{1}{6}\left(\frac{x}{a}\right)^3 + \frac{1}{3} + \frac{D}{\kappa^2 Gha^2}\left(1 - \frac{x}{a}\right) \tag{5.108}$$

The maximum deflection occurs at $x = 0$ and is given by

$$\frac{Dw_{\text{max}}}{Pa^3} = \frac{1}{3} + \frac{D}{\kappa^2 Gha^2} = \frac{1}{3} + \frac{1}{6(1-\nu)\kappa^2}\left(\frac{h}{a}\right)^2 \tag{5.109a}$$

For the special case $\nu = 0.3$, $\kappa = 0.9225$ (see Section 3.8) and

$$\frac{Dw_{\text{max}}}{Pa^3} = \frac{1}{3} + 0.280\left(\frac{h}{a}\right)^2 \tag{5.109b}$$

A graph of (5.109b) is shown in Fig. 5.18. We can neglect shear deformation by letting $\kappa^2 G \to \infty$ in (5.109a). In this case we obtain the solution within the framework of CPT, and $Dw_{\text{max}}/Pa^3 = \frac{1}{3}$. With reference to Fig. 5.18 it is seen that the maximum deflection of the plate is a function of the slenderness ratio $h/a$, and the contribution of shear strain to the total plate deformation increases with an increase in $h/L$ (thick plates).

Next we (briefly) consider a simply supported plate strip, as shown in Fig. 5.19. The load is characterized by

$$p(x, y) = \sum_{n=1}^{\infty} p_n \sin\frac{n\pi x}{a}$$

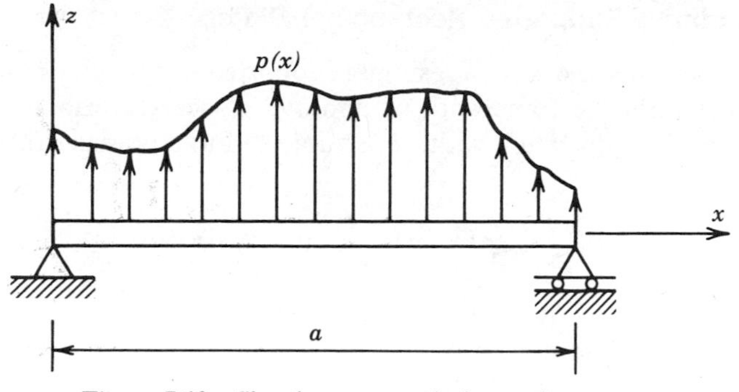


**Figure 5.19** Simply supported plate strip.

(see Exercise 5.1). Again we have a case of plane strain, and with reference to (5.100a)–(5.102b), we have

$$D\frac{d^3\psi_x^{(n)}}{dx^3} = -p_n \sin\frac{n\pi x}{a} \tag{5.110a}$$

$$\frac{d^2 w_n}{dx^2} = -\frac{p_n}{\kappa^2 Gh} - \frac{d\psi_x^{(n)}}{dx} \tag{5.110b}$$

where $w = \Sigma_{n=1}^{\infty} w_n$ and $\psi_x = \Sigma_{n=1}^{\infty} \psi_x^{(n)}$. The boundary conditions are

$$w(0) = M_{xx}(0) = w(a) = M_{xx}(a) = 0 \tag{5.111}$$

The solution of (5.110) subject to (5.111) is

$$\psi_x^{(n)} = -\frac{p_n a^3}{D\pi^3 n^3}\cos\frac{n\pi x}{a} \tag{5.112a}$$

$$\begin{aligned}\frac{w_n D}{p_n a^4} &= \frac{1}{n^4\pi^4}\left[1 + \frac{Dn^2\pi^2}{\kappa^2 Gha^2}\right]\sin\frac{n\pi x}{a} \\ &= \frac{1}{n^4\pi^4}\left[1 + \frac{n^2\pi^2}{6(1-\nu)\kappa^2}\left(\frac{h}{a}\right)^2\right]\sin\frac{n\pi x}{a}\end{aligned} \tag{5.112b}$$

With references to (5.112b), two observations are in order:

(a) The deflection $w$ reduces to the results of CPT when $\kappa^2 G \to \infty$.
(b) Shear deformation effects assume greater importance for higher harmonics of the load (due to $n^2$) and for thick plates [due to $(h/a)^2$].

### 5.2.2 Simply Supported Rectangular Plates

We now consider the case of a simply supported rectangular plate, as shown in Fig. 5.1, within the framework of improved (shear-deformable) plate theory. The plate of length $a$ and width $b$ is subjected to a pressure distribution given by

$$p(x, y) = F \sin Mx \sin Ny \tag{5.113}$$

where $F$, $M = m\pi/a$, and $N = n\pi/b$ are constants and $m, n = 1, 2, 3, \ldots$. The boundary conditions are

$$w(0, y) = M_{xx}(0, y) = \psi_y(0, y) = 0 \tag{5.114a}$$

$$w(x, 0) = M_{yy}(x, 0) = \psi_x(x, 0) = 0 \tag{5.114b}$$

$$w(a, y) = M_{xx}(a, y) = \psi_y(a, y) = 0 \tag{5.114c}$$

$$w(x, b) = M_{yy}(x, b) = \psi_x(x, b) = 0 \tag{5.114d}$$

Consequently, we need to find a solution of (5.100)–(5.102) subject to the loading (5.113) that satisfies the given boundary conditions (5.114):

$$w = A \sin Mx \sin Ny \tag{5.115a}$$

$$\psi_x = B \cos Mx \sin Ny \tag{5.115b}$$

$$\psi_y = C \sin Mx \cos Ny \tag{5.115c}$$

where $A$, $B$, and $C$ are constants.

Substitution of (5.113) and (5.115) into (5.101a, b) and (5.102a, b) and further substitution of the results into (5.100) results in three (algebraic) equations in the three unknowns $A$, $B$, and $C$. Upon solution, we obtain

$$\begin{aligned} A &= \frac{F}{D(M^2 + N^2)^2} + \frac{F}{\kappa^2 Gh(M^2 + N^2)} \\ B &= -\frac{FM}{D(M^2 + N^2)^2} \\ C &= -\frac{FN}{D(M^2 + N^2)^2} \end{aligned} \tag{5.116}$$

Equations (5.115) in conjunction with (5.116) provide a complete solution of the problem posed.

Because of the linearity of the present plate theory, the principle of superposition is applicable. The preceding solution in conjunction with the concept of a double Fourier series (see Exercise 5.3) can be used to generate solutions for simply supported rectangular plates subjected to arbitrarily distributed, transverse loads. If the load is given by

$$p(x, y) = \sum_{m=1}^{\infty} \sum_{n=1}^{\infty} p_{mn} \sin Mx \sin Ny \tag{5.117}$$

then the displacement field of the plate in this problem is characterized by

$$w = \sum_{m=1}^{\infty} \sum_{n=1}^{\infty} \left( \frac{1}{D(M^2 + N^2)} + \frac{1}{\kappa^2 Gh} \right) \frac{p_{mn} \sin Mx \sin Ny}{M^2 + N^2} \tag{5.118a}$$

$$\psi_x = - \sum_{m=1}^{\infty} \sum_{n=1}^{\infty} \frac{p_{mn} M \cos Mx \sin Ny}{D(M^2 + N^2)^2} \tag{5.118b}$$

$$\psi_y = - \sum_{m=1}^{\infty} \sum_{n=1}^{\infty} \frac{p_{mn} N \sin Mx \cos Ny}{D(M^2 + N^2)^2} \tag{5.118c}$$

where $M = m\pi/a$, $N = n\pi/b$, and $m, n = 1, 2, 3, \ldots$ . The values of the Fourier coefficients $p_{mn}$ depend upon the given load distribution $p(x, y)$. In this connection, see Exercises 5.3 and 5.4.

### 5.2.3 Twisting of Plate Strip

We now consider the deformation of a plate strip by opposing couples of magnitude $T$ applied to the ends, as shown in Fig. 5.20. This problem is similar to the one treated in Section 5.1.4, except that the effects of shear deformation will not be neglected in the present treatment. We shall utilize the stress-displacement relations and the equations of equilibrium as given by

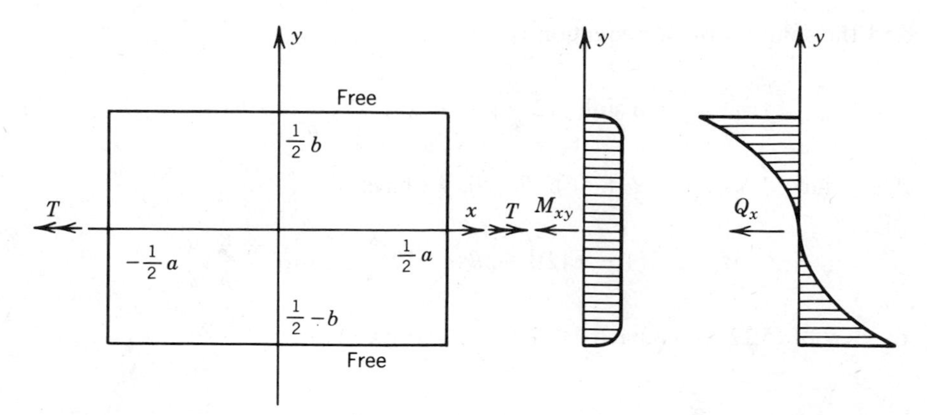


**Figure 5.20** Twisted plate.

(5.100)–(5.102). The boundary conditions are

$$\text{At } x = \pm\tfrac{1}{2}a: \quad w = \pm\tfrac{1}{2}\theta a y \tag{5.119a}$$

$$\psi_y = \mp\tfrac{1}{2}\theta a \tag{5.119b}$$

$$M_{xx} = 0 \tag{5.119c}$$

$$\text{At } y = \pm\tfrac{1}{2}b: \quad M_{yy} = 0 \tag{5.120a}$$

$$M_{xy} = 0 \tag{5.120b}$$

$$Q_y = 0 \tag{5.120c}$$

Based on the results of Section 5.1.4, and other physical consideration, we shall assume that $w = \theta xy$, $\psi_x = \psi_x(y)$, and $Q_y \equiv 0$, and since $Q_y = \kappa^2 Gh(\partial w/\partial y + \psi_y) = 0$, we obtain $\psi_y = -\theta x$. Also, since

$$M_{xx} = D\left(\frac{\partial \psi_x}{\partial x} + \nu\frac{\partial \psi_y}{\partial y}\right), \qquad M_{yy} = D\left(\frac{\partial \psi_y}{\partial y} + \nu\frac{\partial \psi_x}{\partial x}\right)$$

we can infer that $M_{xx} = M_{yy} = 0$. By direct calculation,

$$M_{xy} = \tfrac{1}{2}(1-\nu)D\left(-\theta + \frac{d\psi_x}{dy}\right)$$

$$Q_x = \tfrac{1}{2}(1-\nu)\frac{d^2\psi_x}{dy^2} = \kappa^2 Gh(\theta y + \psi_x)$$

or

$$\frac{d^2\psi_x}{dy^2} - \frac{12\kappa^2}{h^2}\psi_x = \frac{12\kappa^2}{h^2}\theta y$$

and the solution of this equation is

$$\psi_x = A\sinh\sqrt{12}\,\frac{\kappa}{h}y + B\cosh\sqrt{12}\,\frac{\kappa}{h}y - \theta y$$

Also, since $\psi_x(y) = -\psi_x(-y)$, $B = 0$, we have

$$M_{xy} = \tfrac{1}{2}(1-\nu)D\left(-2\theta + \sqrt{12}\,\frac{\kappa}{h}A\cosh\sqrt{12}\,\frac{\kappa}{h}y\right)$$

In view of (5.120e) we have $M_{xy}(\pm\tfrac{1}{2}b) = 0$, so that

$$A = \frac{2\theta}{\sqrt{12}\,(\kappa/h)\cosh\left[\sqrt{12}\,(\kappa/h)(b/2)\right]}$$

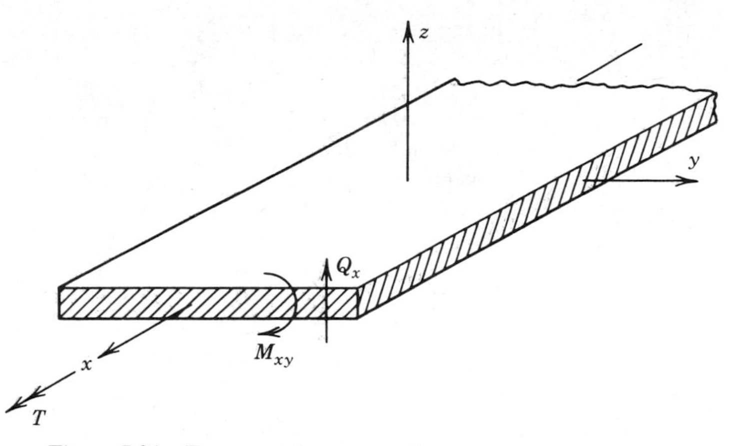


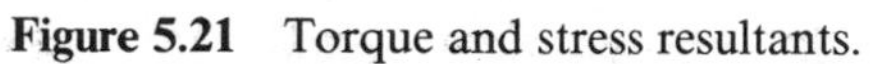
**Figure 5.21** Torque and stress resultants.

Thus, we conclude that

$$M_{xy} = (1-\nu) D\theta \left[ \frac{\cosh\left[\sqrt{12}\,\kappa(y/h)\right]}{\cosh\left[\sqrt{12}\,\kappa(b/2h)\right]} - 1 \right] \tag{5.121a}$$

$$Q_x = (1-\nu) D\theta \frac{\sqrt{12}\,\kappa}{h} \frac{\sinh\left[\sqrt{12}\,\kappa(y/h)\right]}{\cosh\left[\sqrt{12}\,\kappa(b/2h)\right]} \tag{5.121b}$$

With reference to Fig. 5.21, the torque acting on the plate is

$$T = \int_{-b/2}^{b/2} (yQ_x - M_{xy})\, dy \tag{5.122}$$

But

$$\int_{-b/2}^{b/2} yQ_x\, dy = \int_{-b/2}^{b/2} y \frac{dM_{xy}}{dy}\, dy$$
$$= (yM_{xy})_{-b/2}^{b/2} - \int_{-b/2}^{b/2} M_{xy}\, dy = -\int_{-b/2}^{b/2} M_{xy}\, dy$$

and in view of (5.122), we note that $Q_x$ and $M_{xy}$ contribute to the torque $T$ in equal measure. Hence,

$$T = kG\theta h^3 a$$

where

$$k = \frac{1}{3}\left[1 - \frac{2}{\sqrt{12}\,\kappa}\frac{h}{b}\tanh\left(\frac{\kappa\sqrt{12}}{2}\frac{b}{h}\right)\right] \tag{5.123}$$

**TABLE 5.3 Values of $k$ for Various Theories ($\kappa^2 = \frac{5}{6}$)**

| | $a/h = 1$ | $a/h = 2$ | $a/h = \infty$ |
|---|---|---|---|
| Improved plate theory | 0.139 | 0.228 | 0.333 |
| Classical plate theory | 0.333 | 0.333 | 0.333 |
| Elasticity theory | 0.141 | 0.229 | 0.333 |

If $\kappa^2 = \frac{5}{6}$ and $a/h > 3$, we have the useful approximation

$$k \cong \frac{1}{3}\left(1 - 0.632\frac{h}{b}\right)$$

Thus, the rate of twist (twist per unit of length) is

$$\theta = \frac{T}{kGah^3} \tag{5.124}$$

Table 5.3 presents a convenient comparison between the present, shear-deformable theory with CPT as well as three-dimensional elasticity theory (St. Venant). We note that the present theory compares favorably with the latter for $a/h > 2$. The present solution is adapted from E. Reissner, "The Effect of Transverse Shear Deformation on the Bending of Elastic Plates", Trans. ASME, Journal of Applied Mechanics, Vol. 12, No. 2, June 1945, pp. A-69 to A-77.

### 5.2.4 Circular Plates (Axial Symmetry)

We now consider the case of a circular plate loaded and supported in an axisymmetric manner so that $\partial(\ )/\partial\theta \equiv Q_\theta \equiv M_{r\theta} \equiv 0$. From (5.104) and (5.103b) we have

$$M_{rr} = D\left(\frac{d\psi_r}{dr} + \frac{\nu}{r}\psi_r\right) \tag{5.125a}$$

$$M_{\theta\theta} = D\left(\nu\frac{d\psi}{dr} + \frac{1}{r}\psi_r\right) \tag{5.125b}$$

$$rQ_r = r\frac{dM_{rr}}{dr} + M_{rr} - M_{\theta\theta} \tag{5.126a}$$

Upon substitution of (5.125) into (5.126), we obtain

$$\frac{d^2\psi_r}{dr^2} + \frac{1}{4}\frac{d\psi_r}{dr} - \frac{1}{r^2}\psi_r = \frac{Q_r}{D} = \frac{d}{dr}\left[\frac{1}{r}\frac{d}{dr}(r\psi_r)\right] = \frac{1}{r}\nabla^2(r\psi_r) \tag{5.126b}$$

and since [see (5.103a)]

$$\frac{d}{dr}(rQ_r) = -pr$$

we readily obtain

$$\frac{d}{dr}\left\{r\frac{d}{dr}\left[\frac{1}{r}\frac{d}{dr}(r\psi_r)\right]\right\} = -\frac{pr}{D} \tag{5.127}$$

In the following we shall assume that $p(r) = p_0 =$ constant. By successive integrations with respect to $r$, we obtain

$$r\frac{d}{dr}\left[\frac{1}{r}\frac{d}{dr}(r\psi_r)\right] = -\frac{p_0 r^2}{2D} + A_1$$

$$\frac{1}{r}\frac{d}{dr}(r\psi_r) = -\frac{p_0 r^2}{4D} + A_1 \ln r + B_1$$

$$r\psi_r = -\frac{p_0 r^4}{16D} + A_1\left(\frac{r^2}{2}\ln r - \frac{r^2}{4}\right) + B_1\frac{r^2}{2} + C_1$$

The constants of integration $A_1$, $B_1$, and $C_1$ are now redefined, resulting in

$$\psi_r = -\frac{p_0 r^3}{16D} + Ar\ln r + Br + \frac{C}{r} \tag{5.128}$$

If we substitute (5.128) into (5.126b), we readily obtain

$$Q_r = -\frac{p_0 r}{2} + \frac{2AD}{r} \tag{5.129}$$

and from (5.105a) we have

$$\begin{aligned}\frac{dw}{dr} &= \frac{Q_r}{\kappa^2 Gh} - \psi_r \\ &= -\frac{p_0 r}{2\kappa^2 Gh} + \frac{2AD}{\kappa^2 Ghr} + \frac{p_0 r^3}{16D} - Ar\ln r - Br - \frac{C}{r}\end{aligned}$$

Upon integration with respect to $r$, we obtain

$$\begin{aligned}w = &-\frac{p_0 r^2}{4\kappa^2 Gh} + \frac{2AD}{\kappa^2 Gh}\ln r + \frac{p_0 r^4}{64D} - A\left(\frac{r^2}{2}\ln r - \frac{r^2}{4}\right) \\ &- \frac{Br^2}{2} - C\ln r + H\end{aligned} \tag{5.130}$$

Equations (5.128) and (5.130) comprise the solution for circular plates under axisymmetric loading and boundary conditions.

**Example:** *Clamped Circular Plate under Uniformly Distributed Load*. The problem is identical to Example (a) in Section 5.1.6 (Fig. 5.8), except that we now use a shear-deformable plate theory. The applicable solution is provided by (5.128) and (5.130). We require $w$, $M_{rr}$, and $Q_r$ to be bounded at $r = 0$. Hence, $A = C = 0$, and we have

$$\psi_r = -\frac{p_0 r^3}{16D} + Br \tag{5.131a}$$

$$w = -\frac{p_0 r^2}{4\kappa^2 Gh} + \frac{p_0 r^4}{64D} - \frac{Br^2}{2} + H \tag{5.131b}$$

At the boundary we require

$$\psi_r(a) = w(a) = 0 \tag{5.132}$$

and upon substitution of (5.131) into (5.132), we obtain

$$B = \frac{p_0 a^2}{16D}, \qquad H = \frac{p_0 a^2}{4\kappa^2 Gh} + \frac{p_0 a^4}{64D}$$

so that

$$\psi_r = \frac{p_0 a^3}{16D}\frac{r}{a}\left(1 - \frac{r^2}{a^2}\right) \tag{5.133a}$$

$$w = \frac{p_0}{64D}(a^2 - r^2)^2 + \frac{p_0}{4\kappa^2 Gh}(a^2 - r^2) \tag{5.133b}$$

### 5.2.5 Concentrated Loads (Axial Symmetry, IPT)

In this section we repeat the problem that was formulated and solved in Section 5.1.7: the circular plate under a concentrically applied, concentrated load of magnitude $P$ applied at $r = 0$. However, we shall now postulate a plate that is shear deformable.

Since there is no distributed load, we shall set $p_0 = 0$ in (5.129). We consider equilibrium for a circular portion of the plate of radius $r$ (see Fig. 5.11). Then [see (5.129)]

$$Q_r = \kappa^2 Gh\left(\psi_r + \frac{dw}{dr}\right) = \frac{2AD}{r} = -\frac{P}{2\pi r}$$

or $A = -P/4\pi D$. Consequently, with reference to (5.128) and (5.130), the singular part of the solution for a concentrated load of magnitude $P$ at $r = 0$ for a shear-deformable plate is

$$w = \frac{Pr^2}{8\pi D}\left(\ln r - \tfrac{1}{2}\right) - \frac{P}{2\pi\kappa^2 Gh}\ln r \tag{5.134a}$$

$$\psi_r = -\frac{P}{4\pi D} r \ln r \tag{5.134b}$$

Equations (5.134) serve as a particular solution for (5.103)–(5.105) when $p \equiv 0$ and there is a concentrated load of magnitude $P$ at $r = 0$. It is also assumed that the boundary conditions will result in axisymmetric deformation of the circular plate.

**Example:** We now reconsider the example in Section 5.1.7 [see Fig. 5.11($a$)] except that we now use a shear-deformable model of the plate. Since there is a concentrated load at $r = 0$, the solution is [see (5.134), (5.128), and (5.130)]

$$w(r) = \frac{Pr^2}{8\pi D}\left(\ln r - \tfrac{1}{2}\right) - \frac{P}{2\pi\kappa^2 Gh}\ln r - \frac{Br^2}{r} - C\ln r + H \tag{5.135a}$$

$$\psi_r(r) = -\frac{P}{4\pi D} r \ln r + Br + \frac{C}{r} \tag{5.135b}$$

At the center we have $\psi_r(0) = 0$, and therefore $C = 0$. At the clamped boundary we require

$$w(a) = 0, \qquad \psi_r(a) = 0 \tag{5.136}$$

and upon substitution of (5.135) into (5.136), we obtain

$$B = \frac{P}{4\pi D}\ln a$$

$$H = \frac{Pa^2}{16\pi D} + \frac{P}{2\pi\kappa^2 Gh}\ln a$$

so that

$$w = \frac{Pa^2}{8\pi D}\left(\frac{r^2}{a^2} - \frac{2}{3\kappa^2(1-\nu)}\frac{h^2}{a^2}\right)\ln\frac{r}{a} + \frac{Pa^2}{16\pi D}\left(1 - \frac{r^2}{a^2}\right) \tag{5.137a}$$

$$\psi_r = -\frac{Pa}{4\pi D}\frac{r}{a}\ln\frac{r}{a} \tag{5.137b}$$

The following observations with reference to (5.137) are in order:

(a) The terms $r^2 \ln r$ and $r \ln r$ remain bounded as $r \to 0$.
(b) The term $\ln r$ becomes unbounded as $r \to 0$.
(c) In view of observations (a) and (b), we conclude that $w$ becomes unbounded as $r \to 0$ but $\psi_r$ remains bounded for $0 \le r \le a$.

This "defect" is readily removed: If we distribute the load $\pi\varepsilon^2 p_0 = P$ uniformly over the circular area with radius $\varepsilon > 0$, all quantities $w$, $\psi_r$, $M_{rr}$, $Q_r$, and so on, will remain bounded (see Exercise 5.21).

## EXERCISES

**5.1** Consider a sufficiently well-behaved function

$$p(x) = \sum_{m=1}^{\infty} p_m \sin(m\pi x/L), \qquad m = 1, 2, 3, \ldots,$$

where $p_m$ are constants. Multiply both sides of this equation by $\sin(i\pi x/L)$, $i = 1, 2, 3, \ldots$, and apply the operation $\int_0^L \ldots dx$ to each term. Use the orthonormality relation $\int_0^L \sin(m\pi x/L)\sin(i\pi x/L)\,dx = \frac{1}{2}L\delta_{mi}$, where $\delta_{mi}$ is the Kronecker delta symbol. Show (formally) that the Fourier coefficients are given by

$$p_m = \frac{2}{L}\int_0^L p(x)\sin\frac{m\pi x}{L}\,dx$$

**5.2** With reference to Exercise 5.1, find the Fourier coefficients $p_m$, $m = 1, 2, 3, \ldots$, for each of the following functions:

**(a)** $p(x) = p_0$, $0 < x < L$

**(b)** $p(x) = \left(\dfrac{x}{L}\right)p_0$, $0 < x < L$

**(c)** $p(x) = \begin{cases} 0, & 0 < x < c < L \\ p_0, & c < x < L \end{cases}$

where $p_0$ is a constant.

**5.3** Consider a sufficiently well-behaved function

$$p(x, y) = \sum_{m=1}^{\infty}\sum_{n=1}^{\infty} p_{mn} \sin\frac{m\pi x}{a}\sin\frac{n\pi y}{b}, \qquad m, n = 1, 2, 3, \ldots$$

Multiply both sides of this equation by $\sin(i\pi x/a)\sin(j\pi y/b)$, $i, j =$

$1, 2, 3, \ldots,$ and apply the operation $\int_0^a \int_0^b \ldots dx\,dy$ to each term. Use the orthonormality relation

$$\int_0^a \sin\frac{m\pi x}{a} \sin\frac{i\pi x}{b}\,dx = \frac{a}{2}\delta_{mi}$$

where $\delta_{mi}$ is the Kronecker delta symbol. Show (formally) that the (double) Fourier coefficients are given by

$$p_{mn} = \frac{4}{ab}\int_0^a \int_0^b p(x, y) \sin\frac{m\pi x}{a} \sin\frac{n\pi y}{b}\,dx\,dy$$

**5.4** With reference to Exercise 5.3, find the Fourier coefficients $p_{mn}$ for each of the following functions:

**(a)** $p(x, y) = p_0,\ 0 < x < a,\ 0 < y < b$

**(b)** $p(x, y) = \left(\frac{x}{a}\right) p_0,\ 0 < x < a,\ 0 < y < b$

**(c)** $p(x, y) = \begin{cases} 0, & 0 < x < c < a,\ 0 < y < b \\ p_0, & c < x < a,\ 0 < y < b \end{cases}$

where $p_0$ is a constant.

**5.5**

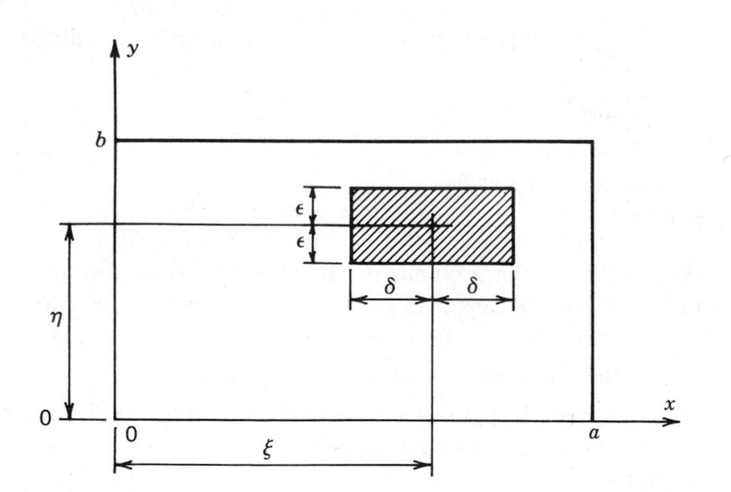


**(a)** Find the midplane deflection surface of a simply supported rectangular plate of dimensions $a : b : h$ under the distributed load

$$p(x, y) = \begin{cases} p_0 = \text{constant} & \text{for } \begin{Bmatrix} \xi - \delta \le x \le \xi + \delta, \\ \eta - \varepsilon \le y \le \eta + \varepsilon \end{Bmatrix} \\ = 0 & \text{elsewhere} \end{cases}$$

**(b)** Take the limit of the solution in (a) such that $\varepsilon \to 0$ and $\delta \to 0$ as $(p_0)(2\delta)(2\varepsilon) \to P$. This will be the solution for a rectangular plate under a concentrated load $P$ at $(x, y) = (\xi, \eta)$. Show that at the load point, the deflection remains bounded, but moment tensor components and shear force components become unbounded.

**5.6** Calculate the moment tensor components and the shear force vector components for the plate with deflection surface (5.12).

**5.7** Consider the problem of a simply supported rectangular plate. Assume the load distribution given in Exercise 5.4(b). Find the maximum deflection and the maximum bending moment. Calculate the corner forces. Use CPT.

**5.8** Consider the problem of a rectangular plate that is simply supported along the edges $y = 0$ and $y = b$. The edges $x = \pm\frac{1}{2}a$ are free. The plate is subjected to a uniformly distributed, transverse load of intensity $p_0$.

**(a)** Find the equation of the deformed plate median surface.

**(b)** Find the maximum deflection and the maximum bending moment. Use CPT.

**5.9** Consider a plate strip of unbounded length $-\infty < x < \infty$, $0 \le y \le b$. The edges $y = 0, b$ are simply supported. A line load characterized by $p(y) = \sum_{n=1}^{\infty} p_n \sin(n\pi y/b)$ force per unit length acts along the line $x = 0$. Find:

**(a)** Plate displacement field

**(b)** Moment tensor field

**(c)** Shear force vector field.

Use CPT.

**5.10** Consider the case of a clamped circular plate. A uniformly distributed load of intensity $p_0$ acts over the circular area $0 \le r \le \varepsilon$, and the rest of the plate is free from surface pressures.

**(a)** Find the midplane deflection of the plate.

**(b)** Take the limit of (a) as $\varepsilon \to 0$ such that $\pi\varepsilon^2 p_0 \to P$ and show that you obtain the expression (5.82).

Use CPT.

**5.11** A simply supported circular plate of radius $a$ is acted upon by a concentrated load of magnitude $P$ at the center. Find:

**(a)** Plate displacement field.

**(b)** Moment tensor field.

**(c)** Shear force vector field.

Use CPT.

**5.12** A simply supported, circular plate of radius $a$ is acted upon by a uniformly distributed load of intensity $p_0$. Find:

**(a)** Plate displacement field.

**(b)** Moment tensor field.

**(c)** Shear force vector field.

Use CPT.

**5.13** Consider a plate annulus $a \leq r \leq b$. The plate is clamped at $r = a$ and is free at $r = b$. It is acted upon by a uniformly distributed load of intensity $p_0$. Find:

**(a)** Plate displacement field.

**(b)** Moment tensor field.

**(c)** Shear force vector field.

Use CPT.

**5.14** Consider a semi-infinite plate strip $0 \leq x < \infty$, $0 \leq y \leq b$. The edges $y = 0, b$ and $x = 0$ are simply supported. A uniformly distributed load of magnitude $p_0$ acts upon the plate. Find:

**(a)** Plate displacement field.

**(b)** Moment tensor field.

**(c)** Shear force vector field.

**(d)** Corner forces at $(0, 0)$ and $(0, b)$.

Use CPT.

**5.15** In the case of a clamped circular plate under a uniformly distributed load (as in Section 5.1.9), show that the integrated shear force along the boundary is equal to $R = -\pi a^2 p_0$, where $a$ is the plate radius and $p_0$ is the load intensity.

**5.16** Consider the cantilever plate strip $0 \leq x \leq a$, $-\infty \leq y \leq \infty$ subjected to a uniformly distributed transverse load of intensity $p_0$. Find:

**(a)** Plate displacement field.

**(b)** Moment tensor field.

**(c)** Shear force vector field.

Assume a condition of plane strain. The plate is rigidly clamped at $x = a$ and is free at $x = 0$. Use (1) CPT and (2) improved (shear-deformable) plate theory and compare the results.

**5.17** Consider the simply supported plate strip $0 \leq x \leq a$, $-\infty \leq y \leq \infty$, as shown in Fig. 5.19. Assume that a line load of intensity $P$ (force per unit length) acts along $0 < x = b < a$. Find:

**(a)** Plate displacement field.

**(b)** Moment tensor field.

**(c)** Shear force vector field.

Assume a condition of plane strain. Use (1) CPT and (2) improved (shear-deformable) plate theory and compare the results.

**5.18** Solve the problem in Exercise 5.5 within the framework of improved (shear-deformable) plate theory. Show that in this case the deflection under the load becomes unbounded.

**5.19** Solve the problem in Exercise 5.7 within the framework of improved (shear-deformable) plate theory.

**5.20** Solve the problem in Exercise 5.9 within the framework of improved (shear-deformable) plate theory.

**5.21** Solve the problem in Exercise 5.10 within the framework of improved (shear-deformable) plate theory. Show that you obtain expression (5.137).

**5.22** Solve the problem in Exercise 5.11 within the framework of improved (shear-deformable) plate theory.

**5.23** Solve the problem in Exercise 5.12 within the framework of improved (shear-deformable) plate theory.

**5.24** Solve the problem in Exercise 5.13 within the framework of improved (shear-deformable) plate theory.

**5.25** *Thermo-elastic Plate Bending (CPT).* (*Note:* For the formulation of the present class of problems, we require some material from Chapter 7. However, the methods of solution are based on Sections 5.1–5.1.9 of the present chapter.)

Consider the problem of a thin plate with incremental temperature distribution $T(x, y, z)$. Assume that the plate is free to stretch by suitable boundary conditions on $C$. Then the plate membrane forces vanish, and $N_{\alpha\beta} \equiv 0$. Show that

$$Q_{\alpha,\alpha} = M_{\alpha\beta,\beta\alpha} = -D\nabla^4 w - \frac{1}{1-\nu}\nabla^2 M_T$$

[See (7.11) and (7.14) and Exercise 3.30.] Therefore,

$$D\nabla^4 w = p - \frac{1}{1-\nu}\nabla^2 M_T$$

where

$$M_T = \alpha E \int_{-h/2}^{h/2} zT\,dz$$

and

$$M_{\alpha\beta} = -D\left[(1-\nu)w_{,\alpha\beta} + \nu\nabla^2 w\,\delta_{\alpha\beta}\right] - \frac{M_T}{1-\nu}\delta_{\alpha\beta}$$

**5.26** Consider a rectangular plate simply supported along its boundary in the reference state. The plate is now heated such that $T(x, y, \frac{1}{2}h) = T_2 =$ constant, $T(x, y, -\frac{1}{2}h) = T_1 =$ constant. Assuming thermal equilibrium and $p \equiv 0$, find:

**(a)** Deformed median surface

**(b)** Components of moment tensor.

**(c)** Components of shear force vector.

Neglect membrane forces and refer to Exercise 5.25.

**5.27** Consider a circular plate that is clamped at its boundary $r = a$ in the reference state. The plate is now heated such that $T(x, y, \frac{1}{2}h) = T_2 =$ constant and $T(x, y, -\frac{1}{2}h) = T_1 =$ constant. Assuming thermal equilibrium and $p \equiv 0$, find:

**(a)** Deformed median surface.

**(b)** Components of moment tensor.

**(c)** Components of shear force vector.

Neglect membrane forces and refer to Exercise 5.25.

**5.28** Do Exercise 5.27, except that now the boundary is simply supported.

**5.29** Consider a plate (CPT) on an "elastic foundation." In this case the plate is surrounded by a medium exerting a force upon the plate that is proportional to the local deflection and which opposes it. Show that this behavior is characterized by

$$D\nabla^4 w + \kappa w = p(x, y)$$

where $\kappa$ is the foundation modulus (Winkler number).

**5.30** With reference to Exercise 5.29, find the equation of the deformed midplane of a rectangular plate that is simply supported and resting on an elastic foundation. The plate is subjected to a uniformly distributed load of intensity $p_0$.

**5.31** With reference to Exercise 5.29, find the equation of the deformed midplane of a circular plate that is clamped along its boundary $r = a$ and resting on an elastic foundation. The plate is subjected to a uniformly distributed load of intensity $p_0$.

**5.32** With reference to Exercise 5.29, find the equation of the deformed midplane of a circular plate that is clamped along its boundary $r = a$ and resting on an elastic foundation. The plate is subjected to a concentrated load of magnitude $P$ at its center $r = 0$.

**5.33** In the case of CPT, show that the equilibrium equation

$$D\nabla^4 w = p(x, y)$$

can be replaced by the equations

$$\nabla^2 M = -p(x, y), \qquad \nabla^2 w = -\frac{M}{D}$$

where

$$(1 + \nu) M = M_{xx} + M_{yy}$$

This observation is the starting point for a useful and effective numerical method of solution (due to Henri Marcus, 1885–1969).

**5.34** Consider a simply supported triangular plate under a uniformly distributed load with intensity $p_0$ (see Section 5.1.3 and Fig. 5.4). With reference to Exercise 5.33, show that

$$\nabla^2 M = -p_0$$

$$M = \frac{p_0}{4a}\left(x + \tfrac{1}{3}a\right)\left(x - \tfrac{2}{3}a + y\sqrt{3}\right)\left(x - \tfrac{2}{3}a - y\sqrt{3}\right)$$

and $M = 0$ along the plate edges. Show that

$$\nabla^2 w = -\frac{M}{D}$$

where $w(x, y)$ is given by (5.44) and $w = 0$ along the plate edges.

**5.35** Consider the case of a (circular) plate annulus loaded by boundary forces and moments. In this case $\nabla^4 w = \nabla^2(\nabla^2 w) = 0$, where

$$\nabla^2 w = \frac{\partial^2 w}{\partial r^2} + \frac{1}{r}\frac{\partial w}{\partial r} + \frac{1}{r^2}\frac{\partial^2 w}{\partial \theta^2}$$

Assume $w(r, \theta) = W_n \cos n\theta$. Show that

$$\left(\frac{d^2}{dr^2} + \frac{1}{r}\frac{d}{dr} - \frac{n^2}{r^2}\right)\left(\frac{d^2 W_n}{dr^2} + \frac{1}{r}\frac{dW_n}{dr} - \frac{n^2}{r^2} W_n\right) = 0$$

and

$$W_0 = A_0 r^0 + B_0 \ln r + C_0 r^2 + H_0 r^2 \ln r$$

$$W_1 = A_1 r^1 + B_1 r^{-1} + C_1 r^3 + H_1 r \ln r$$

$$W_n = A_n r^n + B_n r^{-n} + C_n r^{n+2} + H_n r^{2-n}, \qquad n \geq 2$$

**5.36** Use the results of Exercises 5.35 and 5.1 to find a solution of the problem of a clamped, circular plate subjected to an arbitrarily placed, concentrated load. (See H. Reismann, "Bending and Buckling of an Elastically Restrained Circular Plate," *Transactions ASME, Journal of Applied Mechanics*, Vol. 19, No. 2, 1952, pp. 167–172.)

# 6

# DYNAMICS OF PLATES

Many problems of modern technology require a detailed understanding of the motion (dynamics) of plates. The central problem of vibrations, free and forced motion of plates of bounded extent, and associated mathematical tools such as eigenfunctions and eigenvalues are discussed in Sections 6.2–6.7. The subject of wave motion in plates and associated physical concepts and phenomena are introduced in Sections 6.8–6.12.

## 6.1 EQUATIONS OF MOTION

To derive the pertinent equations of motion, we shall employ Hamilton's principle (see Section 2.5):

$$\int_{t_1}^{t_2} \delta L \, dt = -\int_{t_1}^{t_2} \delta \mathscr{W} \, dt \tag{6.1}$$

In (6.1), $L = T - U$ is the Lagrangian function and $T$ and $U$ are, respectively, the kinetic and potential energies of the plate. The symbol $\delta \mathscr{W}$ denotes the virtual work of all external forces acting upon the plate. In the present case we have

$$T = \frac{1}{2}\int_V \rho(\dot{\mathbf{u}} \cdot \dot{\mathbf{u}}) \, dV = \frac{1}{2}\int_A \int_{-h/2}^{h/2} \rho\left(z^2 \dot{\psi}_\alpha \dot{\psi}_\alpha + \dot{w}^2\right) dz \, dA$$

$$= \frac{1}{2}\int_A \left(\rho h \dot{w}^2 + \tfrac{1}{12}\rho h^3 \dot{\psi}_\alpha \dot{\psi}_\alpha\right) dA \tag{6.2}$$

where $\rho$ denotes the mass density of the plate material and where we have used (3.3). Upon taking the first variation of (6.2), we obtain

$$\delta T = \int_A \left(\rho h \dot{w} \, \delta \dot{w} + \tfrac{1}{12}\rho h^3 \dot{\psi}_\alpha \, \delta \dot{\psi}_\alpha\right) dA$$

and upon integration by parts and application of the conditions $\delta\psi_\alpha = \delta w = 0$

at $t = t_1$ and at $t = t_2$, we readily obtain

$$\int_{t_1}^{t_2} \delta T\, dt = -\int_{t_1}^{t_2} \int_A \left( \rho h \ddot{w}\, \delta w + \tfrac{1}{12} \rho h^3 \ddot{\psi}_\alpha\, \delta \psi_\alpha \right) dA\, dt \tag{6.3}$$

The strain energy in the plate is given by (4.5), and the virtual work of external and boundary forces is given by (4.1). Upon substitution of (6.3), (4.5), and (4.1) into (6.1), we obtain

$$\begin{aligned} &\int_{t_1}^{t_2} \int_A \left[ \left( -\rho h \ddot{w} + Q_{\alpha,\alpha} + p \right) \delta w + \left( -\tfrac{1}{12} \rho h^3 \ddot{\psi}_\alpha + M_{\alpha\beta,\beta} - Q_\alpha \right) \delta \psi_\alpha \right] dA\, dt \\ &\quad + \int_{t_1}^{t_2} \oint_C \left[ \left( M_{nn}^* - M_{nn} \right) \delta \psi_n + \left( M_{ns}^* - M_{ns} \right) \delta \psi_s + \left( Q_n^* - Q_n \right) \delta w \right] ds\, dt \\ &= 0 \end{aligned}$$

The time interval $t_2 – t_1$ as well as the area and shape of the plate are arbitrary. Therefore, the integrands of the triple and double integrals must vanish separately. Furthermore, the quantities $\delta w$ and $\delta \psi_\alpha$ are independent and arbitrary. Thus, we conclude that

$$\left. \begin{aligned} \rho h \ddot{w} &= Q_{\alpha,\alpha} + p \qquad &(6.4a) \\ \tfrac{1}{12} \rho h^3 \ddot{\psi}_\alpha &= M_{\alpha\beta,\beta} - Q_\alpha \qquad &(6.4b) \end{aligned} \right\} \quad \text{in } A$$

and the conditions (4.13) are to be specified on the boundary $C$. Equations (6.4) are the stress equations of motion of the plate, and (4.13b) are the associated, admissible boundary conditions. We note that the quantities $\rho h \ddot{w}$ and $\frac{1}{12} \rho h^3 \ddot{\psi}_\alpha$ characterize transverse and rotatory inertia, respectively, each per unit area of the plate. In the case of CPT, an analogous derivation will again result in the equations of motion (6.4). However, in this case the associated, admissible boundary conditions are now given by (4.19) (see Exercise 6.1), and it is customary to neglect (drop) the term $\frac{1}{12} \rho h^3 \ddot{\psi}_\alpha$ in (6.4b).

Upon substitution of (3.59) into (6.4), we obtain the displacement equations of motion of IPT:

$$\tfrac{1}{2} D \left[ (1 - \nu) \psi_{\alpha,\beta\beta} + (1 + \nu) \psi_{\beta,\beta\alpha} \right] - \kappa^2 G h \left( \psi_\alpha + w_{,\alpha} \right) = \tfrac{1}{12} \rho h^3 \ddot{\psi}_\alpha \tag{6.5a}$$

$$\kappa^2 G h \left( \psi_{\alpha,\alpha} + w_{,\alpha\alpha} \right) + p = \rho h \ddot{w} \tag{6.5b}$$

or, in vector notation,

$$\tfrac{1}{2} D \left[ (1 - \nu) \nabla^2 \boldsymbol{\psi} + (1 + \nu) \nabla (\nabla \cdot \boldsymbol{\psi}) \right] - \kappa^2 G h (\boldsymbol{\psi} + \nabla w) = \tfrac{1}{12} \rho h^3 \ddot{\boldsymbol{\psi}} \tag{6.6a}$$

$$\kappa^2 G h \left( \nabla \cdot \boldsymbol{\psi} + \nabla^2 w \right) + p = \rho h \ddot{w} \tag{6.6b}$$

In the case of CPT we drop the term $\frac{1}{12} \rho h^3 \ddot{\psi}_\alpha$ and combine (6.4a) with (6.4b) to obtain

$$\rho h \ddot{w} = M_{\alpha\beta,\beta\alpha} + p \tag{6.7}$$

Further substitution of (4.16) into (6.7) results in the displacement equation of motion

$$D \nabla^4 w + \rho h \ddot{w} = p(x, y, t) \tag{6.8}$$

## 6.2 FREE VIBRATIONS — GENERAL

We shall now consider the case of free harmonic vibrations for plates of bounded extent within the framework of CPT. The pertinent equation of motion is (6.8) with $p \equiv 0$, and it will be assumed that the boundary conditions are homogeneous, i.e., using (4.19), we prescribe

$$\left.\begin{aligned} &\text{Either } w = 0 \quad \text{or} \quad V_n = 0 \\ \text{and} \quad &\text{Either } \frac{\partial w}{\partial n} = 0 \quad \text{or} \quad M_{nn} = 0 \end{aligned}\right\} \text{ on } C \qquad \begin{aligned} &(6.9) \\ &(6.10) \end{aligned}$$

If free vibrations are characterized by

$$w_i(x, y, t) = W^{(i)}(x, y) \cos \Omega_i t, \qquad i = 1, 2, 3, \ldots \tag{6.11a}$$

$$w_j(x, y, t) = W^{(j)}(x, y) \cos \Omega_j t, \qquad j = 1, 2, 3, \ldots \tag{6.11b}$$

where $\Omega_i$ is the natural frequency and $W^{(i)}(x, y)$ is its associated mode shape, then upon substitution of (6.11) into (6.8), we obtain

$$D \nabla^4 W^{(i)} = \rho h \Omega_i^2 W^{(i)} \tag{6.12a}$$

$$D \nabla^4 W^{(j)} = \rho h \Omega_j^2 W^{(j)} \tag{6.12b}$$

Equations (6.12) have the appearance of the static equilibrium equation (4.17), and therefore, the reciprocal theorem (4.23b) is applicable if we set $p_i = \rho h \Omega_i^2 W^{(i)}$ and $p_j = \rho h \Omega_j^2 W^{(j)}$. In view of the fact that the present analysis is restricted to homogeneous boundary conditions, the application of (4.24) to (6.12) results in

$$\int_A \rho h \Omega_i^2 W^{(i)} W^{(j)} \, dA = \int_A \rho h \Omega_j^2 W^{(j)} W^{(i)} \, dA$$

or

$$\left(\Omega_i^2 - \Omega_j^2\right) \int_A \rho h W^{(i)} W^{(j)} \, dA = 0$$

If $\Omega_i^2 \neq \Omega_j^2$, we have the relation

$$\int_A \rho h W^{(i)} W^{(j)}\, dA = 0 \tag{6.13}$$

which is known as the orthogonality relation of the natural modes of oscillation.

The partial differential equation (6.12), which characterizes the eigenfunctions, is homogeneous. For this reason each eigenfunction $W^{(i)}$, $i = 1, 2, 3, \ldots,$ can be determined only to within an arbitrary, constant multiplier. This constant can be fixed by adjoining an arbitrary normalization condition. The normalization condition that will prove to be advantageous for our subsequent solution of the forced motion problem can be stated as

$$\int_A \rho h W^{(i)} W^{(j)}\, dA = 1 \tag{6.14}$$

Equations (6.13) and (6.14) can now be written in the form

$$\int_A \rho h W^{(i)} W^{(j)}\, dA = \delta_{ij} \tag{6.15}$$

where $\delta_{ij} = 1$ if $i = j$ and $\delta_{ij} = 0$ if $i \neq j$. The symbol $\delta_{ij}$ is the Kronecker delta symbol. Equation (6.15) is known as the orthonormality relation of the free vibration modes associated with (6.12).

## 6.3 FREE VIBRATIONS OF A SIMPLY SUPPORTED, RECTANGULAR PLATE

We now apply the theory developed in Section 6.2 to find the natural frequencies and associated mode shapes of a freely vibrating, simply supported, rectangular plate with dimensions $a : b : h$. We require a solution of (6.12) in the domain $0 \leq x \leq a$, $0 \leq y \leq b$ subject to the boundary conditions

$$W^{(i)}(x, 0) = M_{yy}^{(i)}(x, 0) = 0 \tag{6.16a}$$

$$W^{(i)}(x, b) = M_{yy}^{(i)}(x, b) = 0 \tag{6.16b}$$

$$W^{(i)}(0, y) = M_{xx}^{(i)}(0, y) = 0 \tag{6.16c}$$

$$W^{(i)}(a, y) = M_{xx}^{(i)}(a, y) = 0 \tag{6.16d}$$

where

$$M_{xx}(x, y, t) = M_{xx}^{(i)}(x, y) \cos \Omega_i t \tag{6.17a}$$

$$Q_x(x, y, t) = Q_x^{(i)}(x, y) \cos \Omega_i t \tag{6.17b}$$

and so on. The solution of (6.12), subject to the conditions (6.16), is given by

$$W^{(i)}(x, y) = A_{mn} \sin \frac{m\pi x}{a} \sin \frac{n\pi y}{b} \tag{6.18}$$

$m = 1, 2, 3, \ldots,\ n = 1, 2, 3, \ldots,$ where the $A_{mn}$ are constants, and

$$\Omega_i = \Omega_{mn} = \left(\frac{D}{\rho h}\right)^{1/2}\left[\left(\frac{m\pi}{a}\right)^2 + \left(\frac{n\pi}{b}\right)^2\right] \tag{6.19}$$

The correspondence between the integers $i$, $m$, and $n$ is established by ordering the natural frequencies so that $\Omega_1 \le \Omega_2 \le \Omega_3 \le \cdots$. For example, in the case of a plate with aspect ratio $a/b = 3/4$, we have

$$\Omega_i = \Omega_{mn} = \frac{1}{16}\left(\frac{D}{\rho h}\right)^{1/2}\left(\frac{\pi}{a}\right)^2 (16m^2 + 9n^2)$$

and we can construct a table of natural frequencies as shown in Table 6.1. The constants $A_{mn}$ in (6.18) are determined by subjecting the eigenfunctions $W^{(i)}$ to the normalization condition (6.14). The result is

$$A_{mn} = \left[\int_0^b \int_0^a \rho h \sin^2 \frac{m\pi x}{a} \sin^2 \frac{n\pi y}{b}\, dx\, dy\right]^{-1/2}$$

or

$$A_{mn} = \frac{2}{\sqrt{ab\rho h}}$$

**TABLE 6.1 Natural Frequencies of a Simply Supported, Rectangular Plate ($a/b = 3/4$)**

| $i$ | $m$ | $n$ | $16\left(\frac{a}{\pi}\right)^2\left(\frac{\rho h}{D}\right)^{1/2}\Omega_i = 16m^2 + 9n^2$ |
|---|---|---|---|
| 1 | 1 | 1 | 25 |
| 2 | 1 | 2 | 52 |
| 3 | 2 | 1 | 73 |
| 4 | 1 | 3 | 97 |
| 5 | 2 | 2 | 100 |
| 6 | 2 | 3 | 145 |
| 7 | 3 | 1 | 153 |
| 8 | 1 | 4 | 160 |
| 9 | 3 | 2 | 180 |
| 10 | 2 | 4 | 208 |
| 11 | 3 | 3 | 225 |
| 12 | 4 | 1 | 265 |

Consequently, the normalized eigenfunctions are given by

$$W^{(i)} = W^{(mn)} = \frac{2}{\sqrt{ab\rho h}} \sin \frac{m\pi x}{a} \sin \frac{n\pi y}{b} \tag{6.20}$$

## 6.4 FREE, AXISYMMETRIC VIBRATIONS OF A CIRCULAR PLATE

We shall now consider the axisymmetric, harmonic vibrations of a clamped, circular plate of radius $a$. The present problem is characterized by (6.12), where, however, the plate deflection surface is now referred to polar coordinates. Because of the assumed axial symmetry, $w = w(r)$, where $r$ is the radial coordinate. The mode shapes $W^{(i)}(r)$ must satisfy the boundary conditions

$$W^{(i)}(a) = \left( \frac{dW^{(i)}}{dr} \right)_{r=a} = 0 \tag{6.21}$$

For our present purpose, we write (6.12) in the form

$$(\nabla^2 + \lambda_i^2)(\nabla^2 - \lambda_i^2) W^{(i)} = 0 \tag{6.22}$$

where

$$\lambda_i^2 = \left( \frac{\rho h}{D} \right)^{1/2} \Omega_i \tag{6.23}$$

and

$$\nabla^2 = \frac{d^2}{dr^2} + \frac{1}{r}\frac{d}{dr}$$

Since $\lambda_i^2 > 0$, the solution of (6.22) can be written in the form

$$W^{(i)}(r) = A_i J_0(\lambda_i r) + C_i Y_0(\lambda_i r) + B_i I_0(\lambda_i r) + D_i K_0(\lambda_i r) \tag{6.24}$$

where $J_0$ and $Y_0$ are Bessel functions, and $I_0$ and $K_0$ are modified Bessel functions, all of order zero. If we require $W^{(i)}(0)$ to remain bounded for $0 \leq r \leq a$, then $C_i = D_i = 0$ because $Y_0(\lambda_i r) \to \infty$ and $K_0(\lambda_i r) \to \infty$ as $r \to 0$. Thus, we have

$$W^{(i)}(r) = A_i J_0(\lambda_i r) + B_i I_0(\lambda_i r) \tag{6.25}$$

Upon application of (6.21) to (6.25), we obtain

$$\begin{aligned} A_i J_0(\lambda_i a) + B_i I_0(\lambda_i a) &= 0 \\ A_i J_1(\lambda_i a) - B_i I_1(\lambda_i a) &= 0 \end{aligned} \tag{6.26a}$$

or

$$-\frac{B_i}{A_i} = \frac{J_0(\lambda_i a)}{I_0(\lambda_i a)} = -\frac{J_1(\lambda_i a)}{I_1(\lambda_i a)} \tag{6.26b}$$

In view of (6.23) and (6.26), the natural frequencies of vibration are obtained by finding the roots of the equation

$$J_0(\lambda_i a) I_1(\lambda_i a) + I_0(\lambda_i a) J_1(\lambda_i a) = 0 \tag{6.27}$$

Equation (6.27) has a denumerable infinity of positive, real roots, the lowest five of which are (approximately) given by

$$\begin{aligned} &\lambda_1 a = 3.1961, \quad \lambda_2 a = 6.3064, \quad \lambda_3 a = 9.4395 \\ &\lambda_4 a = 12.577, \quad \lambda_5 a = 15.716 \end{aligned} \tag{6.28}$$

The natural frequencies are related to the roots of (6.27) by the relation

$$\Omega_i = \frac{1}{a^2}\left(\frac{D}{\rho h}\right)^{1/2} (\lambda_i a)^2 \tag{6.29}$$

The eigenfunctions are now normalized by the substitution of (6.25) into (6.14). We obtain

$$\left[\int_0^a [A_i J_0(\lambda_i r) + B_i I_0(\lambda_i r)]^2 r\, dr\right]^{1/2} = \frac{1}{\sqrt{2\pi\rho h}}$$

However,

$$\int_0^a r J_0^2(\lambda_i r)\, dr = \frac{a^2}{2}[J_0^2(\lambda_i a) + J_1^2(\lambda_i a)] \tag{6.30a}$$

$$\int_0^a r I_0^2(\lambda_i r)\, dr = \frac{a^2}{2}[I_0^2(\lambda_i a) - I_1^2(\lambda_i a)] \tag{6.30b}$$

$$\int_0^a r J_0(\lambda_i r) I_0(\lambda_i r)\, dr = \frac{a}{2\lambda_i}[J_0(\lambda_i a) I_1(\lambda_i a) + I_0(\lambda_i a) J_1(\lambda_i a)] = 0 \tag{6.30c}$$

Consequently, the normalization condition becomes

$$A_i \left\{ \left[ J_0^2(\lambda_i a) + J_1^2(\lambda_i a) \right] + \frac{B_i^2}{A_i^2} \left[ I_0^2(\lambda_i a) - I_1^2(\lambda_i a) \right] \right\}^{1/2} = \frac{1}{\sqrt{\pi a^2 \rho h}} \tag{6.31}$$

Combining (6.26b) with (6.31), with the aid of (6.27), results in

$$A_i = \frac{1}{a\sqrt{2\pi\rho h}\, J_0(\lambda_i a)} \tag{6.32a}$$

$$B_i = \frac{-1}{a\sqrt{2\pi\rho h}\, I_0(\lambda_i a)} \tag{6.32b}$$

Upon substitution of (6.32) into (6.25), we obtain

$$W^{(i)}(r) = \frac{1}{a\sqrt{2\pi\rho h}} \left[ \frac{J_0(\lambda_i r)}{J_0(\lambda_i a)} - \frac{I_0(\lambda_i r)}{I_0(\lambda_i a)} \right] \tag{6.33}$$

Equation (6.33) characterizes the normalized mode shapes (eigenfunctions) for axisymmetric vibrations of a clamped, circular plate. The associated natural frequencies are the roots of (6.27). The five lowest frequencies are given by (6.29) with the aid of (6.28).

## 6.5 FORCED MOTION — GENERAL

A major problem in the application of the theory of elastic plates is concerned with the determination of the deformation and stresses in a plate subjected to time-dependent, transverse surface loads. A well-posed forced motion problem within the context of CPT can be stated in the following manner: Find the solution of the equation

$$D\nabla^4 w + \rho h \ddot{w} = p(x, y, t) \tag{6.34}$$

subject to the homogeneous boundary conditions

$$\left.\begin{array}{ll} & \text{Either } w = 0 \quad \text{or} \quad V_n = 0 \qquad (6.35\text{a}) \\ \text{and} & \\ & \text{Either } \dfrac{\partial w}{\partial n} = 0 \quad \text{or} \quad M_{nn} = 0 \qquad (6.35\text{b}) \end{array}\right\} \text{ on } C$$

For a unique solution, we impose the initial conditions

$$\left.\begin{aligned} w(x, y, 0) &= w_0(x, y) \\ \dot{w}(x, y, 0) &= \dot{w}_0(x, y) \end{aligned}\right\} \quad \text{in } A \qquad \begin{aligned} &(6.36a) \\ &(6.36b) \end{aligned}$$

In the preceding formulation of the forced motion problem, the quantities $\nu$, $D$, and $h$ are constants, $\rho = \rho(x, y)$, and $p = p(x, y, t)$. It can be shown (see Exercise 6.6) that the initial conditions (6.36) are sufficient to ensure a unique solution of (6.34) subject to (6.35).

It is now assumed that the solution of (6.34) can be characterized by the eigenfunction expansion

$$w(x, y, t) = \sum_{i=1}^{\infty} W^{(i)}(x, y) q_i(t) \tag{6.37}$$

where $W^{(i)}$ are the normalized eigenfunctions (mode shapes) that satisfy the homogeneous boundary conditions appropriate to the problem under consideration (see Section 6.2), and $q_i(t)$ are functions of time to be determined. To find $q_i(t)$, we substitute (6.37) into (6.34) and obtain

$$\sum_{i=1}^{\infty} \rho h W^{(i)}\left(\ddot{q}_i + \Omega_i^2 q_i\right) = p \tag{6.38}$$

where we have used (6.12). We now multiply both sides of (6.38) by $W^{(j)}$ and subsequently integrate both sides over the area $A$ of the plate. If, in addition, we interchange the order of the operations of integration and summation on the left-hand side of the resulting equation, we readily obtain

$$\sum_{i=1}^{\infty} \left(\ddot{q}_i + \Omega_i^2 q_i\right) \int_A \rho h W^{(i)} W^{(j)}\, dA = \int_A p W^{(j)}\, dA$$

Upon application of the orthonormality relation (6.15) to this equation, we obtain

$$\ddot{q}_i + \Omega_i^2 q_i = F_i(t) \tag{6.39}$$

where

$$F_i(t) = \int_A p W^{(i)}\, dA \tag{6.40}$$

The solution of (6.40) is

$$q_i(t) = q_i(0) \cos \Omega_i t + \frac{1}{\Omega_i} \dot{q}_i(0) \sin \Omega_i t + \frac{1}{\Omega_i} \int_0^t F_i(\tau) \sin \Omega_i (t - \tau)\, d\tau \tag{6.41}$$

To determine the value of $q_i(0)$ in (6.41), we proceed in the following manner: Set $t = 0$ in (6.37), write

$$w_0 = \sum_{i=1}^{\infty} W^{(i)} q_i(0) \tag{6.42}$$

and multiply both sides of (6.42) by $\rho h W^{(j)}$. Subsequently, both sides of the resulting equation are integrated over the area $A$, and the operations of integration and summation are interchanged. The result is

$$\int_A \rho h w_0 W^{(j)}\, dA = \sum_{i=1}^{\infty} q_i(0) \int_A \rho h W^{(i)} W^{(j)}\, dA = q_j(0)$$

where we have applied the relation (6.15). An entirely similar procedure can be applied to determine $\dot{q}_i(0)$. Thus, we obtain

$$q_i(0) = \int_A \rho h w_0 W^{(i)}\, dA \tag{6.43a}$$

$$q_i(0) = \int_A \rho h \dot{w}_0 W^{(i)}\, dA \tag{6.43b}$$

Equation (6.37), in conjunction with (6.41) and (6.43), provides a complete resolution of the forced motion problem as characterized by (6.34)–(6.36). The preceding development assumes that the pertinent mode shapes and associated frequencies are known.

In the case of homogeneous boundary conditions it is possible to utilize the eigenfunctions and associated eigenvalues of the pertinent free vibration problem to find a solution of the static deformation problem. Let us suppose that we wish to find a solution of (4.17) subject to admissible, homogeneous boundary conditions (6.9) and (6.10). The static solution is now characterized by

$$w(x, y) = \sum_{i=1}^{\infty} B_i W^{(i)}(x, y) \tag{6.44}$$

where $W^{(i)}$ are the eigenfunctions that satisfy (6.12) and the $B_i$ are constants to be determined. Upon substitution of (6.44) into (4.17) and subsequent utilization of (6.12), we readily obtain

$$\sum_{i=1}^{\infty} \rho h \Omega_i^2 B_i W^{(i)} = p \tag{6.45}$$

We now multiply (6.45) by $W^{(j)}$ and subsequently integrate over the area $A$.

Upon interchange of integration and summation, we obtain

$$\sum_{i=1}^{\infty} \Omega_i^2 B_i \int_A \rho h W^{(i)} W^{(j)}\, dA = \int_A p W^{(j)}\, dA$$

If (6.15) is now applied to this equation, we obtain the desired value of the constants:

$$B_i = \frac{1}{\Omega_i^2} \int_A p W^{(i)}\, dA \tag{6.46}$$

Thus, (6.44), in conjunction with (6.46), constitutes a formal solution of the static problem characterized by (4.17), (6.9), and (6.10) in terms of an eigenfunction expansion. In this resolution of the boundary value problem, it is assumed that the pertinent mode shapes and associated frequencies are known.

## 6.6 FORCED MOTION OF A SIMPLY SUPPORTED RECTANGULAR PLATE

In this section we consider the case of a simply supported, rectangular plate, as shown in Fig. 6.1. The plate is at rest in its equilibrium configuration for $t < 0$. At $t = 0$, a uniformly distributed load of intensity $p = p_0$ is suddenly applied over the rectangular area $2\varepsilon \times 2\delta$ with center at $(x, y) = (\xi, \eta)$. We wish to predict the motion for $t > 0$.

In the present case the pertinent normalized mode shape functions and associated frequencies are given by (6.20) and (6.19), respectively, and with the

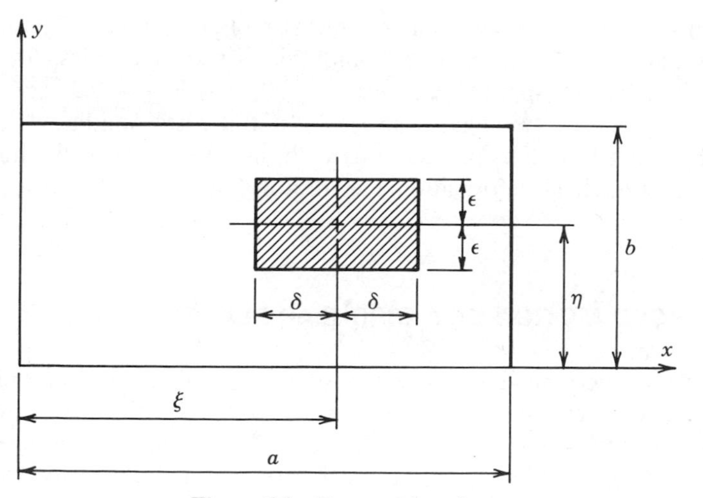


**Figure 6.1** Rectangular plate.

aid of (6.40), we obtain

$$F_{mn}(t) = \int_A p W_{mn}\, dA = p_0 \int_{\xi-\delta}^{\xi+\delta} \int_{\eta-\varepsilon}^{\eta+\varepsilon} \frac{2}{\sqrt{ab\rho h}} \sin\frac{m\pi x}{a} \sin\frac{n\pi y}{b}\, dx\, dy$$

$$= \frac{8p_0}{\pi^2 mn} \sqrt{\frac{ab}{\rho h}} \left(\sin\frac{m\pi\xi}{a} \sin\frac{n\pi\eta}{b}\right)\left(\sin\frac{m\pi\delta}{a} \sin\frac{n\pi\varepsilon}{b}\right) = F_{mn}^{(0)}$$

for $t > 0$. For the present case $w_0 = \dot{w}_0 = 0$, and therefore, with reference to (6.43), $q_i(0) = \dot{q}_i(0) = 0$. Thus, (6.41) becomes

$$q_{mn}(t) = \frac{1}{\Omega_{mn}} \int_0^t F_{mn}(\tau) \sin\Omega_{mn}(t-\tau)\, dt = \frac{F_{mn}^0}{\Omega_{mn}^2}(1 - \cos\Omega_{mn} t)$$

We can now assemble the desired forced motion solution (6.37):

$$\begin{aligned} w(x, y, t) &\\ = \frac{16p_0}{\pi^2 D} \sum_{m=1}^{\infty} \sum_{n=1}^{\infty} &\\ \times \frac{[\sin(m\pi\xi/a)\sin(n\pi\eta/b)][\sin(m\pi\delta/a)\sin(n\pi\varepsilon/b)][\sin(m\pi x/a)\sin(n\pi y/b)]}{mn[(m\pi/a)^2 + (n\pi/b)^2]^2} &\\ \times (1 - \cos\Omega_{mn} t), \qquad t \geq 0 & \end{aligned} \tag{6.47}$$

where

$$\Omega_{mn} = \left(\frac{D}{\rho h}\right)^{1/2} \left[\left(\frac{m\pi}{a}\right)^2 + \left(\frac{n\pi}{b}\right)^2\right]$$

Using (6.47), we can obtain (dimensionless) plots of center deflection $w$ and center moment $M_{xx}$ versus time $t$, and this is shown in Figs. 6.2 and 6.3, respectively, for the special case $h/a = 1/20$, $a = b$, $\nu = 0.3$, $\xi = \eta = a/2$, and $\delta = \varepsilon = h$. The two figures also exhibit the corresponding solution obtained by using IPT, thus providing us with an assessment of the importance of the effects of shear deformation and rotatory inertia for the present, special case.

## 6.7 FORCED MOTION OF A CIRCULAR PLATE

We now consider the dynamic response of a clamped, circular plate with radius $a$ and thickness $h$, as shown in Fig. 6.4. The plate is at rest in its equilibrium configuration for $t < 0$. A uniformly distributed transverse load of intensity $p = p_0$ is suddenly applied, at $t = 0$, over a circular area with radius $b$. We wish to predict the motion of the plate for $t \geq 0$.

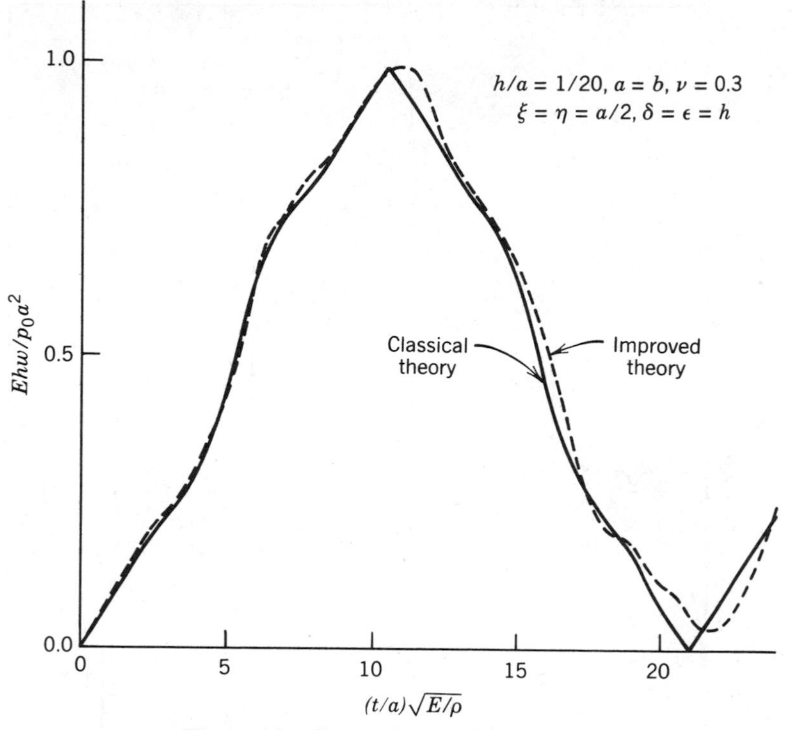


**Figure 6.2** Center deflection versus time.

The pertinent eigenfunctions and eigenvalues (or mode shapes and frequencies) have been determined in Section 6.4 and are listed in (6.33) and (6.29), respectively. Using (6.40), we have, for the present case,

$$F_i(t) = 2\pi \int_0^b p_0 W^{(i)} r\, dr$$

$$= \frac{2\pi p_0 b}{\sqrt{2\pi\rho h}\, a\lambda_i} \left[ \frac{J_1(\lambda_i b)}{J_0(\lambda_i a)} - \frac{I_1(\lambda_i b)}{I_0(\lambda_i a)} \right] = F_0^{(i)}$$

and in view of the fact that $w_0 = \dot{w}_0 = 0$, we infer from (6.43) that $q_i(0) = \dot{q}_i(0) = 0$. Hence, for the present problem (6.41) assumes the form

$$q_i(t) = \frac{1}{\Omega_i} \int_0^t F_i(\tau) \sin \Omega_i (t - \tau)\, d\tau$$

$$= \frac{F_0^{(i)}}{\Omega_i^2} (1 - \cos \Omega_i t)$$

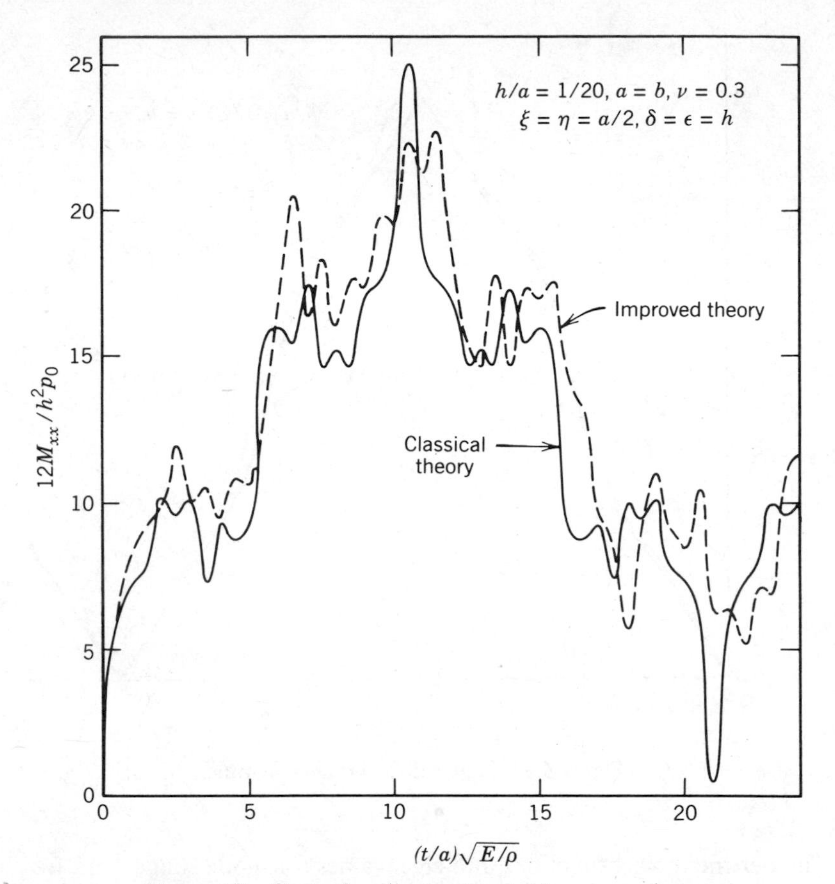


**Figure 6.3** Center bending moment $M_{xx}$ versus time.

We can now assemble the forced motion solution:

$$w(r, t) = \sum_{i=1}^{\infty} W^{(i)}(r) q_i(t)$$

or

$$w(r, t) = \frac{p_0 b a^3}{D} \sum_{i=1}^{\infty} \frac{1}{(\lambda_i a)^5} \left[ \frac{J_1(\lambda_i b)}{J_0(\lambda_i a)} - \frac{I_1(\lambda_i b)}{I_0(\lambda_i a)} \right] \left[ \frac{J_0(\lambda_i r)}{J_0(\lambda_i a)} - \frac{I_0(\lambda_i r)}{I_0(\lambda_i a)} \right] (1 - \cos \Omega_i t) \quad (6.48)$$

for $t \geq 0$. In this case $\lambda_i^4 = (\rho h/D)\Omega_i^2$, and $\lambda_i a$, $i = 1, 2, 3, \ldots$, are the roots of (6.27). Equation (6.48) can be used to obtain (dimensionless) plots of center deflection $w$ and bending moment $M_{rr}$ versus time $t$, and this is shown in

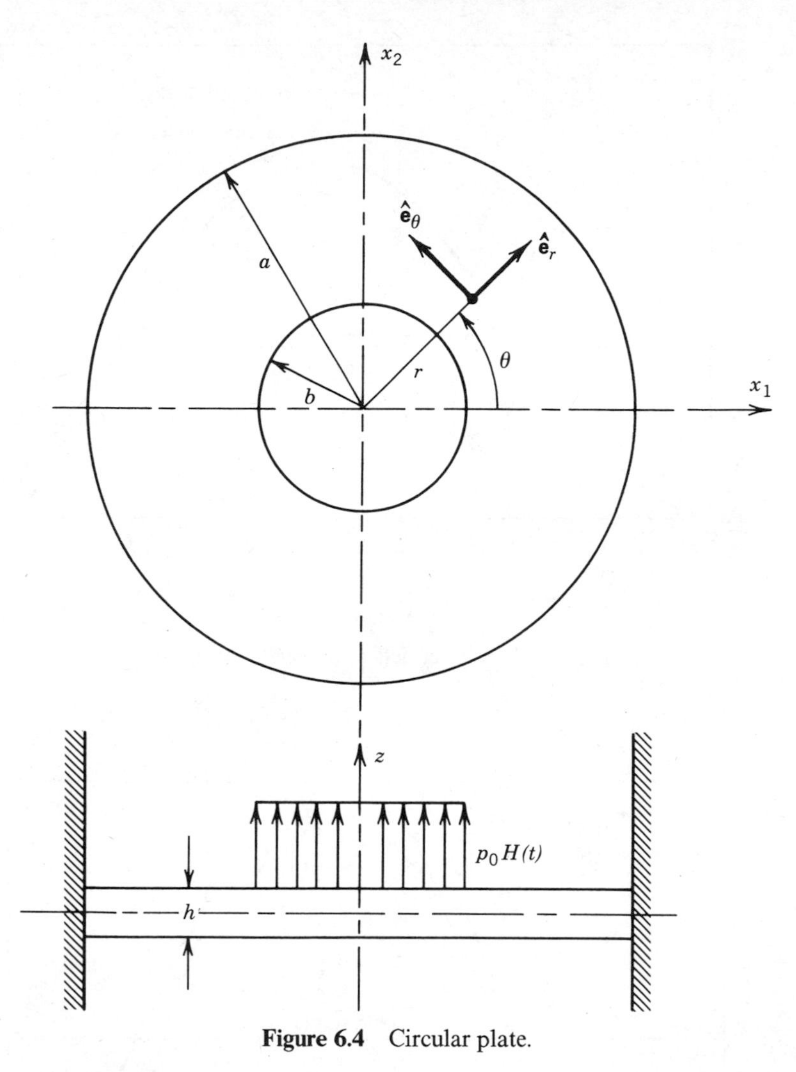


**Figure 6.4** Circular plate.

Figs. 6.5 and 6.6, respectively, for the special case $h/a = 1/10$, $b/a = 1/2$, and $\nu = 0.3$. The two figures also exhibit the corresponding solution obtained by using IPT, thus providing us with an assessment of the importance of the effects of shear deformation and rotatory inertia for the present, special case.

## 6.8 STANDING AND TRAVELING WAVES

In Section 6.2 we discussed standing waves as characterized by mode shapes and associated frequencies for plates of bounded extent. We shall now demonstrate that standing waves can be obtained by the judicious superposi-

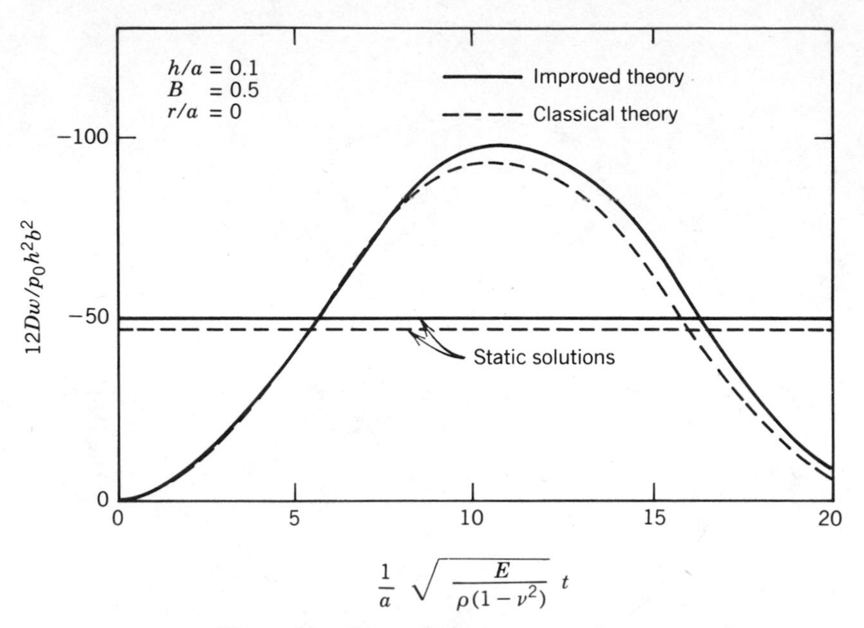


**Figure 6.5** Center deflection versus time.

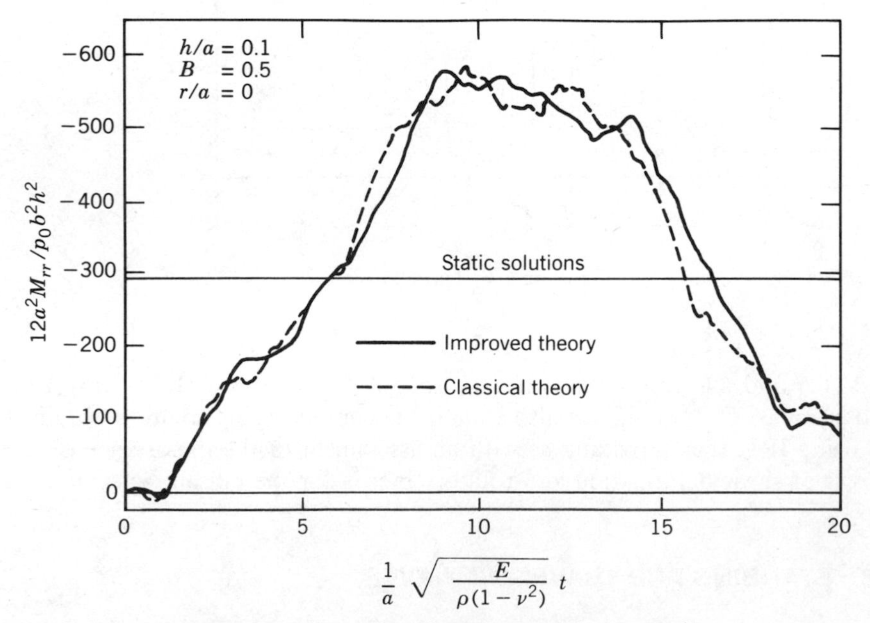


**Figure 6.6** Center bending moment $M_{rr}$ versus time.

tion of traveling waves. As a point of departure, we shall use the standing wave solution for a simply supported rectangular plate as characterized by (6.19) and (6.18). Upon substitution of (6.18) into (6.11) and with the aid of some trigonometric identities, it can be shown that

$$\begin{aligned} w(x, y, t) &= A_{mn} \sin k_1 x \sin k_2 y \cos \Omega_i t \\ &= -\tfrac{1}{4} A_{mn} \big[ \cos(\mathbf{k}_1 \cdot \mathbf{r} - \Omega_i t) + \cos(\mathbf{k}_1 \cdot \mathbf{r} + \Omega_i t) \\ &\qquad - \cos(\mathbf{k}_2 \cdot \mathbf{r} - \Omega_i t) - \cos(\mathbf{k}_2 \cdot \mathbf{r} + \Omega_i t) \big] \end{aligned} \tag{6.49}$$

where

$$k_1 = \frac{m\pi}{a}, \qquad k_2 = \frac{n\pi}{b}, \qquad \mathbf{r} = x\hat{\mathbf{e}}_x + y\hat{\mathbf{e}}_y$$

$$\mathbf{k}_1 = k_1\hat{\mathbf{e}}_x + k_2\hat{\mathbf{e}}_y = k\hat{\mathbf{k}}_1 = k(\hat{\mathbf{e}}_x \cos\alpha + \hat{\mathbf{e}}_y \sin\alpha)$$

$$\mathbf{k}_2 = k_1\hat{\mathbf{e}}_x - k_2\hat{\mathbf{e}}_y = k\hat{\mathbf{k}}_2 = k(\hat{\mathbf{e}}_x \cos\alpha - \hat{\mathbf{e}}_y \sin\alpha)$$

$$k^2 = |\mathbf{k}_1|^2 = |\mathbf{k}_2|^2 = k_1^2 + k_2^2 = \left(\frac{m\pi}{a}\right)^2 + \left(\frac{n\pi}{b}\right)^2 = \lambda_i^2 = \left(\frac{\rho h}{D}\right)^{1/2} \Omega_i$$

and

$$\tan\alpha = \frac{k_2}{k_1} = \frac{an}{bm}$$

With reference to (6.49), it can be inferred that the standing wave $w(x, y, t)$ is a superposition of four traveling waves.

Let us consider the expression $\cos(\mathbf{k}_1 \cdot \mathbf{r} - \Omega_i t)$. To an observer moving through space with velocity $\mathbf{v}_1 = (\Omega_i / k)\hat{\mathbf{k}}_1$, the phase $\psi_1 = \mathbf{k}_1 \cdot \mathbf{r} - \Omega_i t$ appears to be stationary, that is,

$$\frac{d\psi_1}{dt} = \mathbf{k}_1 \cdot \frac{d\mathbf{r}}{dt} - \Omega_i = \mathbf{k}_1 \cdot \mathbf{v}_1 - \Omega_i = \Omega_i - \Omega_i = 0$$

Thus, a stationary observer would see the wave form $\cos\psi_1$ advancing in the $\hat{\mathbf{k}}_1$ direction with speed $v$ (or phase velocity $\mathbf{v}_1$) such that

$$v = |\mathbf{v}_1| = \frac{\Omega_i}{k} = k\left(\frac{D}{\rho h}\right)^{1/2} = \left(\frac{D}{\rho h}\right)^{1/2} \left[\left(\frac{m\pi}{a}\right)^2 + \left(\frac{n\pi}{b}\right)^2\right]^{1/2} \tag{6.50}$$

For any fixed instant of time, the crests of this wave are found by setting $\cos\psi_1 = 1$ so that $\psi_1 = 0, \pm 2\pi, \pm 4\pi, \ldots$. In this manner the wave crests are defined by the equations $\mathbf{k}_1 \cdot \mathbf{r} - \Omega_i t = 2p\pi$, $p = 0, \pm 1, \pm 2, \ldots$, or by

$$x\cos\alpha + y\sin\alpha = \frac{\Omega_i t + 2p\pi}{k} \tag{6.51}$$

Equation (6.51) characterizes a family of equally spaced, parallel, straight lines

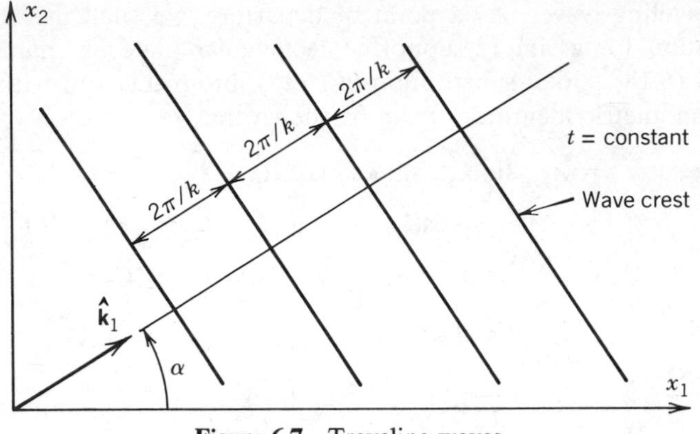


**Figure 6.7** Traveling waves.

progressing in the direction of the vector $\hat{\mathbf{k}}_1$, as shown in Fig. 6.7. We have shown that the expression $\cos(\mathbf{k}_1 \cdot \mathbf{r} - \Omega_i t)$ characterizes a straight-crested wave, and an entirely similar analysis can be carried out for the remaining three component waves in (6.49). Thus, we can say that the standing wave $w(x, y, t)$ given by (6.49) is the superposition of four straight-crested, traveling waves that advance with speed $v$ in the direction $\hat{\mathbf{k}}_1$, $-\hat{\mathbf{k}}_1$, $\hat{\mathbf{k}}_2$, and $-\hat{\mathbf{k}}_2$.

## 6.9 ENERGY FLUX

We shall now derive an expression for energy flux in the plate. We restrict the present discussion to flexural (or bending) deformations within the framework of CPT ($q_\alpha \equiv 0$). If the symbol $U$ denotes the strain energy in the plate, then with the aid of (3.60), (3.63a), (4.15), and (4.16b), we have

$$\dot{U} = \int_A \dot{W}_p \, dA = \int_A M_{\alpha\beta} \dot{m}_{\alpha\beta} \, dA$$

$$= -\int_A M_{\alpha\beta} \dot{w},_{\alpha\beta} \, dA = -\int_A \left[ M_{\alpha\beta,\beta} \dot{w},_\alpha - \left( M_{\alpha\beta} \dot{w},_\alpha \right),_\beta \right] dA$$

$$= \int_A \left[ Q_\alpha \dot{w},_\alpha - \left( M_{\beta\alpha} \dot{w},_\beta \right),_\alpha \right] dA$$

If $T$ denotes kinetic energy of the plate, then with the aid of (6.2) and (6.4a) and neglecting rotatory inertia, we obtain

$$\dot{T} = \int_A \rho h \ddot{w} \dot{w} \, dA = \int_A (Q_{\alpha,\alpha} + p) \dot{w} \, dA$$

Let the symbol $E^*$ denote the instantaneous energy density (energy per unit area) of the plate. Then

$$\int_A \dot{E}^* \, dA = \dot{T} + \dot{U} = \int_A (p\dot{w} - \nabla \cdot \mathbf{F}) \, dA$$

where $\nabla \cdot \mathbf{F} = F_{\alpha,\alpha}$ is the divergence of the energy flux vector $\mathbf{F}$, the components of which are

$$F_\alpha = -Q_\alpha \dot{w} + M_{\beta\alpha} \dot{w}_{,\beta} \tag{6.52}$$

Since the area $A$ of the plate is arbitrary, we obtain the energy continuity equation

$$\dot{E}^* = p\dot{w} - \nabla \cdot \mathbf{F} \quad \text{in } A \tag{6.53}$$

With the aid of Green's theorem, we find that

$$\int_A \nabla \cdot \mathbf{F} \, dA = \oint_C \mathbf{F} \cdot \hat{\mathbf{n}} \, ds = \oint_C F_n \, ds$$

where

$$\begin{aligned} F_n = F_\alpha n_\alpha &= -Q_\alpha n_\alpha \dot{w} + M_{\beta\alpha} n_\alpha \dot{w}_{,\beta} \\ &= -Q_n \dot{w} + M_\beta \dot{w}_{,\beta} = -Q_n \dot{w} + \mathbf{M} \cdot \nabla \dot{w} \end{aligned} \tag{6.54}$$

where we used (3.29) and (3.31). But in view of (3.33) and (4.14), we have

$$\mathbf{M} \cdot \nabla \dot{w} = M_{nn} \left( \frac{\partial \dot{w}}{\partial n} \right) + M_{ns} \left( \frac{\partial \dot{w}}{\partial s} \right)$$

Also, if the expression $(M_{ns} \dot{w})$ is single valued and continuous, we have $\oint_C (\partial / \partial s)(M_{ns} \dot{w}) \, ds = 0$, and it can be shown that

$$\oint_C M_{ns} \frac{\partial \dot{w}}{\partial s} \, ds = -\oint_C \dot{w} \frac{\partial M_{ns}}{\partial s} \, ds$$

Consequently,

$$-\int_A \nabla \cdot \mathbf{F} \, dA = -\oint_C F_\alpha n_\alpha \, ds = \oint_C \left[ V_n \dot{w} + M_{nn} \left( -\frac{\partial \dot{w}}{\partial n} \right) \right] ds$$

and

$$\dot{T} + \dot{U} = \int_A p\dot{w}\,dA - \oint_C F_n\,ds$$
$$= \int_A p\dot{w}\,dA + \oint_C \left[ V_n\dot{w} + M_{nn}\left( -\frac{\partial \dot{w}}{\partial n} \right) \right] ds \qquad (6.55)$$

In view of (6.55), the rate at which the total energy of the plate increases is equal to the rate at which energy flows into $A$. For example, if $p \equiv 0$ and $F_n > O(F_n < 0)$, there is energy leaving (entering) the plate. If $p \equiv 0$ and the boundary conditions are homogeneous, $\dot{T} + \dot{U} = 0$, and the total energy in the plate remains constant. It should be noted that the preceding arguments and results neglect any form of energy dissipation in the plate.

As an application of the previous formulation, we shall calculate the energy flux associated with the propagation of straight-crested waves

$$w = A\cos(\mathbf{k} \cdot \mathbf{r} - \Omega t) \qquad (6.56)$$

in a plate of unbounded extent, within the framework of CPT. The phase velocity of this wave is $\mathbf{v} = (\Omega/k)\hat{\mathbf{k}}$, and the propagation vector is $\mathbf{k} = k_\alpha \hat{\mathbf{e}}_\alpha = k\hat{\mathbf{k}}$, where $k = 2\pi/L = |\mathbf{k}|$ is the wave number and $L$ is the wavelength. In order that (6.56) satisfy (6.8) (with $p \equiv 0$), we must set

$$Dk^4 = \rho h \Omega^2, \quad \text{or } \mathbf{v} = \frac{\Omega}{k} = \sqrt{\frac{D}{\rho h}}\, k \qquad (6.57a)$$

Consequently, the phase velocity is

$$\mathbf{v} = \sqrt{\frac{D}{\rho h}}\, \mathbf{k} \qquad (6.57b)$$

and since it depends upon the wave number $k$, the medium through which the wave propagates (in the present case the plate) is said to be dispersive. The group velocity is

$$\mathbf{v}_G = \frac{d\Omega}{dk}\hat{\mathbf{k}} = 2\sqrt{\frac{D}{\rho h}}\, \mathbf{k} = 2\mathbf{v} \qquad (6.58)$$

The energy density in the plate is

$$E^* = \tfrac{1}{2}\rho h \dot{w}^2 + W_p = \tfrac{1}{2}Dk^4A^2 = \tfrac{1}{2}\rho h \Omega^2 A^2$$

and the energy flux vector has the components

$$F_\alpha = -Q_\alpha \dot{w} + M_{\beta\alpha}\dot{w},_\beta = Dk^4A^2 v_\alpha = \rho h \Omega^2 A^2 v_\alpha$$

or, in vector notation,

$$\mathbf{F} = Dk^4A^2\mathbf{v} = 2E^*\mathbf{v} = E^*\mathbf{v}_G \tag{6.59}$$

Equation (6.59) reveals that the flux vector is proportional to the group velocity, and the constant of proportionality is the local energy density $E^*$. The average speed at which energy is transported by a harmonic wave is defined to be the time average of the magnitude of energy flux $\langle F \rangle$ divided by the time average of the energy density $\langle E^* \rangle$:

$$\text{Speed of energy transport} = \frac{\langle F \rangle}{\langle E^* \rangle} \tag{6.60}$$

With reference to (6.59) and (6.60), we can interpret the group velocity $\mathbf{v}_G$ as the velocity with which energy is transported through the plate by the straight-crested, harmonic flexural wave (6.56).

As an example, we consider the case of a propagating wave of the type (6.56) with wavelength ratio $h/L = 1/20$ in a steel plate. In this case we have $E = 20.7 \times 10^{10}$ Pa, $\rho = 7.85 \times 10^3$ kg/m$^3$; and $\nu = 0.3$. From (6.57) we have

$$\frac{v^2}{c_p^2} = \frac{\pi^2}{3}\left(\frac{h}{L}\right)^2 \quad \text{where } c_p^2 = \frac{E}{\rho(1-\nu^2)}$$

and in the present case $c_p = 5383$ m/sec, and

$$v = \frac{\pi}{\sqrt{3}}\left(\frac{h}{L}\right)c_p = 488 \text{ m/sec}$$

Thus, the magnitudes of phase and group velocity are 488 and 976 m/sec, respectively, for this realistic example.

## 6.10 DISPERSION RELATIONS AND THE SHEAR COEFFICIENT

When deriving the basic equations of IPT in Section 3.8, we arbitrarily introduced a shear coefficient $\kappa^2$ into one of the plate stress-strain relations [see (3.58b)]. The value of this coefficient can now be determined in a rational manner. Toward this end, we consider straight-crested, harmonic waves propagating through the plate in the $x$-direction. The dispersion relations for this case within the framework of three-dimensional elasto-dynamics were first obtained by Sir Horace Lamb in 1917 (for a summary, see I. A. Viktorov, *Rayleigh and Lamb Waves*, Plenum, New York, 1967, Chapter 2).

We shall now determine the dispersion relations for IPT. Taking the divergence of (6.6a) results in

$$L(m) = \left( D\nabla^2 - \frac{\rho h^3}{12} \frac{\partial^2}{\partial t^2} - \kappa^2 G h \right) m = \kappa^2 G h \nabla^2 w$$

where $m = m_{\alpha\alpha} = \psi_{\alpha,\alpha} = \nabla \cdot \boldsymbol{\psi}$. Next, we set $p = 0$ and operate on (6.6b) with $L$ to obtain

$$\kappa^2 G h \nabla^2 w + L(\nabla^2 w) = \frac{\rho}{\kappa^2 G} L(\ddot{w})$$

or

$$\left( D\nabla^2 - \frac{\rho h^3}{12} \frac{\partial^2}{\partial t^2} \right) \left( \nabla^2 - \frac{\rho}{\kappa^2 G} \frac{\partial^2}{\partial t^2} \right) w = -\rho h \frac{\partial^2 w}{\partial t^2} \tag{6.61}$$

Assuming

$$w(x, t) = A \cos k(x - vt) \tag{6.62}$$

and substituting (6.62) into (6.61) results in

$$v^4 - \left( c_p^2 + \kappa^2 c_s^2 + \frac{12\kappa^2 c_s^2}{k^2 h^2} \right) v^2 + \kappa^2 c_s^2 c_p^2 = 0 \tag{6.63}$$

where

$$c_p^2 = \frac{E}{\rho(1 - \nu^2)}, \qquad c_s^2 = \frac{G}{\rho} \tag{6.64}$$

The roots of (6.63) are

$$\left.\begin{matrix} v_2^2 \\ v_1^2 \end{matrix}\right\} = \frac{c_p^2 + \kappa^2 c_s^2}{2} + \frac{6\kappa^2 c_s^2}{k^2 h^2}$$
$$\pm \left[ \left( \frac{c_p^2 - \kappa^2 c_s^2}{2} \right)^2 + \left( c_p^2 + \kappa^2 c_s^2 \right) \left( \frac{6\kappa^2 c_s^2}{k^2 h^2} \right) + \left( \frac{6\kappa^2 c_s^2}{k^2 h^2} \right)^2 \right]^{1/2} \tag{6.65}$$

In the limit, as $kh \to 0$, (6.65) yields

$$v_1^2 \to \frac{D}{\rho h} k^2 \quad \text{and} \quad v_2^2 \to \frac{12\kappa^2 c_s^2}{k^2 h^2} \tag{6.66a}$$

Conversely, as $kh \to \infty$, we obtain

$$v_1^2 \to \kappa^2 c_s^2 \quad \text{and} \quad v_2^2 \to c_p^2 \tag{6.66b}$$

The value of the shear coefficient $\kappa^2$ is now determined by matching the dispersion curve of the first mode of propagation $v_1(k)$ with the dispersion curve of the first antisymmetric mode of propagation obtained from the three-dimensional theory of elasticity. A comparison of (6.66a) with the corresponding expression of the three-dimensional theory reveals that both theories exhibit the same asymptotic behavior as $kh \to 0$. The results of Section 6.9 [i.e., (6.57a)] show that the dispersion relation of CPT is given by $v = (D/\rho h)^{1/2}k$. Consequently, the three theories under discussion exhibit excellent agreement for sufficiently long waves $kh = 2\pi h/L \to 0$. In the case of sufficiently short waves $kh = 2\pi h/L \to \infty$, three-dimensional elasticity theory predicts that wave speeds will approach the speed of Rayleigh surface waves; that is, $v$ will satisfy the equation

$$\left(2 - \frac{v^2}{c_s^2}\right)^2 = 4\left(1 - \frac{v^2}{c_s^2}\right)^{1/2}\left(1 - \gamma^2\frac{v^2}{c_s^2}\right)^{1/2} \tag{6.67}$$

where

$$\gamma^2 = \frac{c_s^2}{c_D^2} = \frac{1 - 2\nu}{2(1 - \nu)} \tag{6.68}$$

According to IPT, $v_1^2/c_s^2 = \kappa^2$ for sufficiently short waves [see (6.66b)]. The value of $\kappa^2$ that makes these two limits coincide is obtained by substituting $v^2/c_s^2 = \kappa^2$ into (6.67), resulting in

$$(2 - \kappa^2)^2 = 4(1 - \kappa^2)^{1/2}(1 - \gamma^2\kappa^2)^{1/2} \tag{6.69}$$

where $0 < \kappa^2 < 1$. A plot of (6.69) is shown in Fig. 6.8. With reference to this figure, we note that the variation of $\kappa$ with $\nu$ is nearly linear for $0.874 < \kappa < 0.955$ and $0 < \nu < \frac{1}{2}$. For example, if $\nu = 0.3$, then $\kappa = 0.9275$ or $\kappa^2 = 0.86$. The complete dispersion relation for the first antisymmetric mode of propagation as predicted by CPT, IPT, and the three-dimensional theory of elasticity are compared in Fig. 6.9 for $\nu = \frac{1}{2}$. With $\kappa = 0.9554$ the difference between the improved and elasticity theories is so small that it cannot be detected on the graph. The present method for the determination of $\kappa$ is due to R. D. Mindlin ("Influence of Rotatory Inertia and Shear on Flexural Motions of Isotropic Elastic Plates," *Journal of Applied Mechanics*, Vol. 18, No. 1, 1951, pp. 31–38).

## 6.11 EDGE WAVES

In this section we consider wave motion in a semi-infinite plate (CPT) with domain $-\infty < x < \infty, 0 \leq y < \infty$, as shown in Fig. 6.10. We shall consider

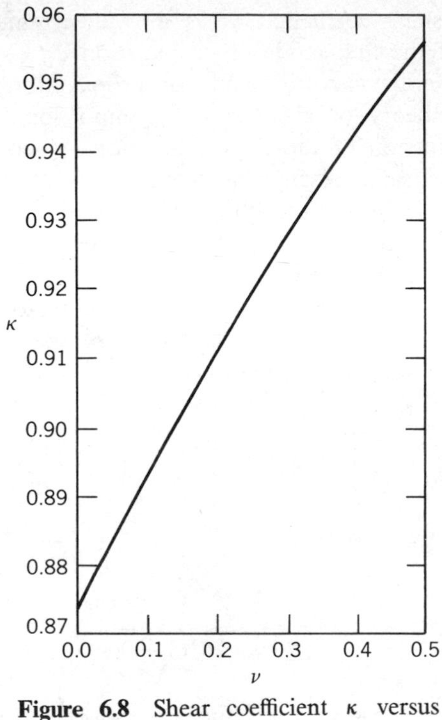


**Figure 6.8** Shear coefficient $\kappa$ versus Poisson's ratio $\nu$.

the possibility of a straight-crested flexural wave of the form

$$w(x, y, t) = f(y)\exp[ik(x - vt)] \tag{6.70}$$

Equation (6.70) characterizes a harmonic wave propagating in the positive $x$-direction with wave number $k = 2\pi/L$, where $L$ is the wavelength. Upon substitution of (6.70) into (6.8), we readily obtain the ordinary differential equation

$$\frac{d^4 f}{dy^4} - 2k^2\frac{d^2 f}{dy} + k^4(1 - \alpha^2)f = 0 \tag{6.71}$$

where $\alpha^2 = v^2/v^{*2}$ and where $v^* = (D/\rho h)^{1/2}k$ is the speed of straight-crested waves in a plate of unbounded extent (see Section 6.9). If we assume $f = e^{\lambda y}$ and substitute this expression into (6.71), we obtain $\lambda = \pm k(1 \pm \alpha)^{1/2}$, and the part of the solution that remains bounded in $0 \leq y < \infty$ is

$$f(y) = A\exp(-k\sqrt{1 - \alpha})y + B\exp(-k\sqrt{1 + \alpha})y \tag{6.72}$$

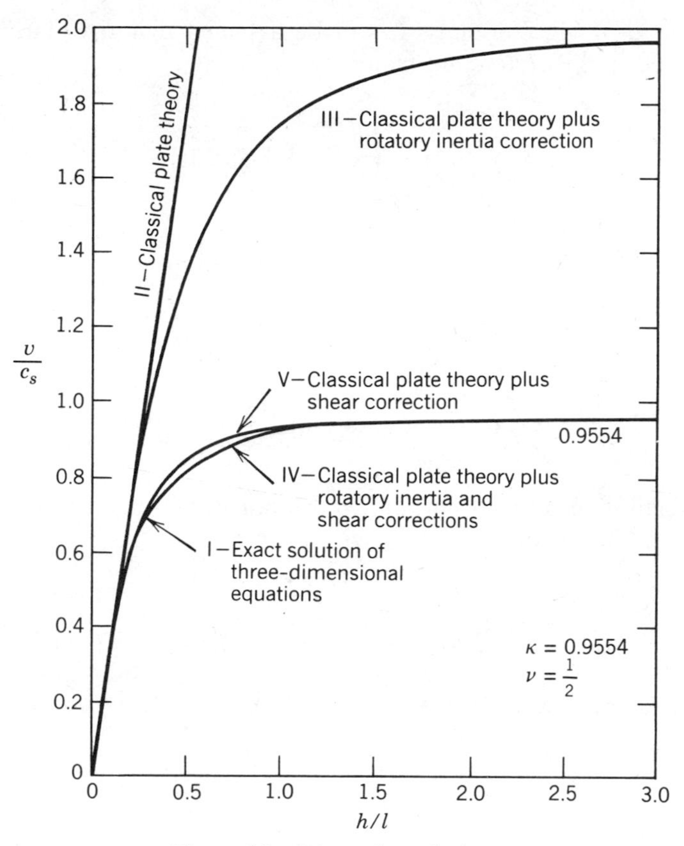

**Figure 6.9** Dispersion relations.

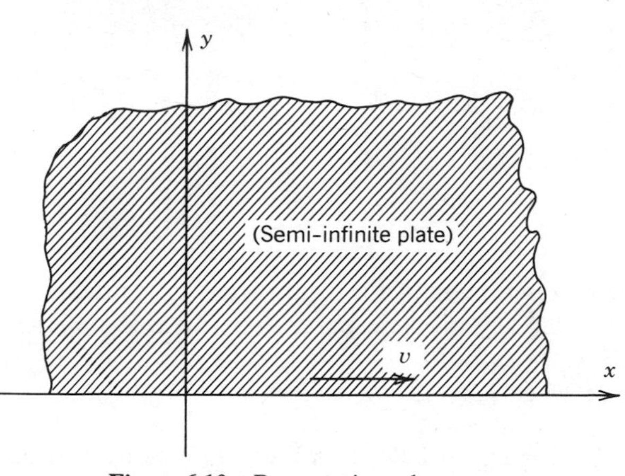


**Figure 6.10** Propagating edge wave.

where $A$ and $B$ are constants. It will be assumed that the plate boundary $y = 0$ is free, and therefore, we require

$$M_{yy}(x,0) = -D\left(\frac{\partial^2 w}{\partial y^2} + \nu\frac{\partial^2 w}{\partial x^2}\right)_{(x,0)} = 0 \tag{6.73a}$$

and

$$V_y(x,0) = \left(Q_y + \frac{\partial M_{yx}}{\partial x}\right)_{(x,0)} = 0 \tag{6.73b}$$

where

$$Q_y = -D\frac{\partial}{\partial y}(\nabla^2 w) \quad \text{and} \quad M_{yx} = -(1-\nu)D\frac{\partial^2 w}{\partial y \partial x}$$

Upon substitution of (6.70) into (6.73), we obtain two homogeneous simultaneous algebraic equations in $A$ and $B$:

$$A(-\alpha+\gamma) + B(\alpha+\gamma) = 0$$
$$A(1-\alpha)^{1/2}(-\alpha-\gamma) + B(1+\alpha)^{1/2}(\alpha-\gamma) = 0$$

where $\gamma = 1 - \nu$. Solving for $A/B$, we obtain

$$\frac{A}{B} = \frac{(1+\alpha)^{1/2}(\alpha-\gamma)}{(1-\alpha)^{1/2}(\alpha+\gamma)} = \frac{\alpha+\gamma}{\alpha-\gamma}$$

Consequently,

$$(1+\alpha)(\alpha-\gamma)^4 = (1-\alpha)(\alpha+\gamma)^4$$

and this can be reduced to

$$\alpha^4 + 2\alpha^2(3\gamma^2 - 2\gamma) + (\gamma^4 - 4\gamma^3) = 0$$

Solving

$$\alpha^2 = \gamma(2-3\gamma) \pm 2\gamma(1+2\gamma^2-2\gamma)^{1/2} \tag{6.74}$$

In view of the representation of the solution (6.72), we require that $0 < \alpha < 1$. Table 6.2 shows the key problem parameters for a useful range of values of Poisson's ratio $\nu$.

For example, when $\nu = 0.3$, we have

$$A^{-1}f(y) = \exp(-0.04359ky) + 0.1755\exp(-1.41354ky)$$

where $A$ is an arbitrary, nonzero constant. The amplitude function $f(y)$

**TABLE 6.2 Edge Wave Parameters**

| $\nu$ | $\gamma$ | $\alpha^2$ | $\alpha$ | $A/B$ | $B/A$ |
|---|---|---|---|---|---|
| 0.1 | 0.9 | 0.99997 | 0.9998 | 19.0361 | 0.0526 |
| 0.2 | 0.8 | 0.99939 | 0.9997 | 9.0120 | 0.1110 |
| 0.3 | 0.7 | 0.99621 | 0.9981 | 5.6964 | 0.1755 |
| 0.4 | 0.6 | 0.98533 | 0.9926 | 4.0565 | 0.2465 |
| 0.5 | 0.5 | 0.95711 | 0.9783 | 3.0907 | 0.3235 |

decreases monotonically from a maximum at $y = 0$. When $ky = 2\pi y/L = 23$, the value of $f$ is 0.312 times its value at $y = 0$. Consequently, when $y = 3.66$ times the wavelength, the wave amplitude is 31.2% of its maximum value at $y = 0$. Also, in this case $\alpha = 0.9981 = v/v^*$, and therefore, $v = 1.8104 c_p h/L$. In the case of a steel plate, we have (see Section 6.9) $c_p = 5383$ m/sec, and if we assume a wavelength ratio $L/h = 20$, we obtain the phase speed $v = 487$ m/sec.

## 6.12 THE IMPULSIVELY LOADED PLATE

Consider an unbounded plate subjected to a concentrated, pure impulse of magnitude $I$ at $r = 0$ and $t = 0$. We wish to find the dynamic response of the plate for $t > 0$ assuming that it is at rest and in its equilibrium configuration at $t = 0$. We shall investigate this problem within the framework of CPT using polar coordinates. Thus, we seek a solution of the equation

$$D\nabla^4 w + \rho h \ddot{w} = I\delta(r)\delta(t) \tag{6.75}$$

which satisfies the initial conditions

$$w(r,0) = \dot{w}(r,0) = 0 \tag{6.76}$$

The symbol $\delta$ denotes Dirac's delta function. The resolution of this problem is facilitated by the use of Laplace transforms and Hankel transforms (see, e.g., I. N. Sneddon, *Fourier Transforms*, McGraw-Hill, New York 1951, pp. 48–70). The Hankel transform of order zero of $w(r, t)$ is defined by

$$w^*(k, t) = \int_0^\infty r w(r, t) J_0(kr)\, dr \tag{6.77a}$$

and its inverse is

$$w(r, t) = \int_0^\infty k w^*(k, t) J_0(kr)\, dk \tag{6.77b}$$

where $J_0(kr)$ denotes the Bessel function of the first kind of order zero. The Hankel transform of a concentrated impulse is

$$[I\delta(r)]^* = \frac{I}{2\pi}\int_0^{2\pi}\int_0^{\infty}\delta(r)J_0(kr)r\,dr\,d\theta = \frac{I}{2\pi}J_0(0) = \frac{I}{2\pi} \quad (6.78)$$

It is assumed that $rw$, $r(\partial w/\partial r)$, $r(\partial^2 w/\partial r^2)$, and $r(\partial^3 w/\partial r^3)$ all vanish as $r$ approaches zero and as $r$ approaches infinity. By repeated integration by parts one can show that

$$(\nabla^4 w)^* = k^4 w^* \quad (6.79)$$

The equation of motion (6.75) is now transformed by multiplying both sides by $rJ_0(kr)$ and then integrating the result over $r$ between the limits zero and infinity. In view of (6.78) and (6.79), the result is

$$Dk^4 w^* + \rho h\ddot{w}^* = \frac{I}{2\pi}\delta(t) \quad (6.80)$$

We now take the Laplace transform of (6.80) to obtain, with the aid of (6.76),

$$Dk^4\overline{w}^* + \rho hp^2\overline{w}^* = \frac{I}{2\pi} \quad (6.81)$$

where

$$\overline{w}^* = \int_0^{\infty} w^*(k,t)\exp(-pt)\,dt$$

Thus,

$$w^* = \frac{I}{2\pi\sqrt{\rho hD}}\frac{1}{k^2}\frac{\lambda^2 k^2}{p^2 + \lambda^4 k^4} \quad (6.82)$$

where $\lambda^4 = D/\rho h$. The inverse Laplace transform of (6.82) is

$$w^*(k,t) = \frac{I}{2\pi\sqrt{\rho hD}}\frac{1}{k^2}\sin\lambda^2 k^2 t$$

and consequently,

$$w(r,t) = \frac{I}{2\pi\sqrt{\rho hD}}\int_0^{\infty}\sin\lambda^2 k^2 t J_0(kr)\frac{dk}{k} \quad (6.83)$$

By differentiating (6.83) with respect to $t$, the following expression is obtained

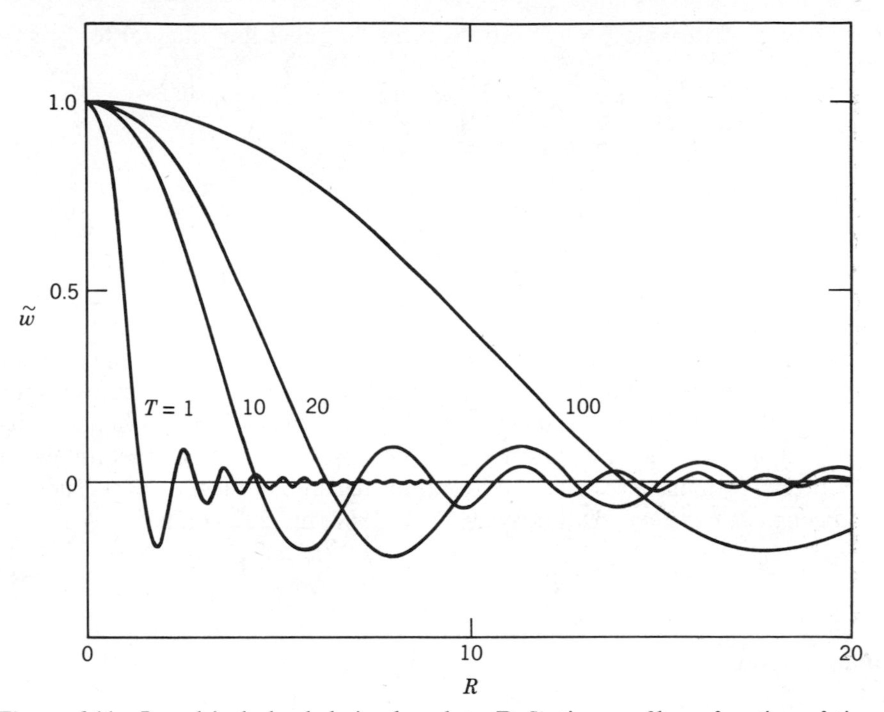


**Figure 6.11** Impulsively loaded circular plate. Deflection profile as function of time.

for the velocity:

$$\dot{w}(r,t) = \frac{I\lambda^2}{2\pi\sqrt{\rho h D}} \int_0^\infty k \cos \lambda^2 k^2 t J_0(kr)\, dk$$

$$= \frac{I}{4\pi\sqrt{\rho h D}} \sin \frac{r^2}{4\lambda^2 t} \tag{6.84}$$

(This integral is entry 4 of Section 6.728, p. 758, in I. S. Gradshteyn and I. M. Ryzhik, *Table of Integrals, Series, and Products*, 4th ed., Academic, New York, 1965.) If we set $\tau = 4\lambda^2 t$, then (6.84) can be written as

$$\frac{d\tilde{w}}{d\tau} = \frac{2}{\pi}\frac{1}{\tau} \sin \frac{r^2}{\tau}$$

where $\tilde{w} = (8/I)\sqrt{\rho h D}\, w$, and since $\tilde{w} = 0$ at $\tau = 0$,

$$\tilde{w} = \frac{2}{\pi} \int_0^\tau \sin \frac{r^2}{x} \frac{dx}{x}$$

The change of variable $\xi = r^2/x$ transforms the preceding integral to

$$\tilde{w} = \frac{2}{\pi}\int_{r^2/\tau}^{\infty}\frac{\sin\xi}{\xi}\,d\xi = \frac{2}{\pi}\int_{0}^{\infty}\frac{\sin\xi}{\xi}\,d\xi - \frac{2}{\pi}\int_{0}^{r^2/\tau}\frac{\sin\xi}{\xi}\,d\xi$$

or

$$\tilde{w} = 1 - \frac{2}{\pi}\,\mathrm{Si}\left(\frac{r^2}{\tau}\right) = 1 - \frac{2}{\pi}\,\mathrm{Si}\left(\frac{R^2}{T}\right) \tag{6.85}$$

where $R = r/h$ and $T = c_p t/h = \sqrt{(E/\rho)(1-\nu^2)}\,(t/h)$. The function, $\mathrm{Si}(x) = \int_0^x (\sin\xi/\xi)\,d\xi$ is called the sine integral. The dimensionless displacement $\tilde{w}(R, T)$ is shown in Fig. 6.11 plotted versus the dimensionless radius $R$ for various values of the dimensionless time $T$. For sufficiently large values of its argument $R^2/T$, the sine integral can be approximated by the following asymptotic formula (see, e.g., M. Abramowitz and I. Stegun, *Handbook of Mathematical Functions*, Dover, New York, 1965, pp. 232–233):

$$\mathrm{Si}\left(\frac{R^2}{T}\right) \cong \frac{\pi}{2} - \frac{T}{R^2}\cos\frac{R^2}{T}$$

and therefore,

$$\tilde{w} \cong \frac{2}{\pi}\frac{T}{R^2}\cos\frac{R^2}{T} \quad \text{for sufficiently large } \frac{R^2}{T} \tag{6.86}$$

## EXERCISES

**6.1** Derive the stress equations of motion (6.4) and the associated admissible boundary conditions (4.19) applicable to CPT. Use Hamilton's principle (6.1), and assume that the plate boundary $C$ is smooth (no corners).

**6.2** Repeat Exercise 6.1 for the case of a rectangular plate. In this case $C$ is not smooth at the four corners. What are the four "corner conditions"?

**6.3**

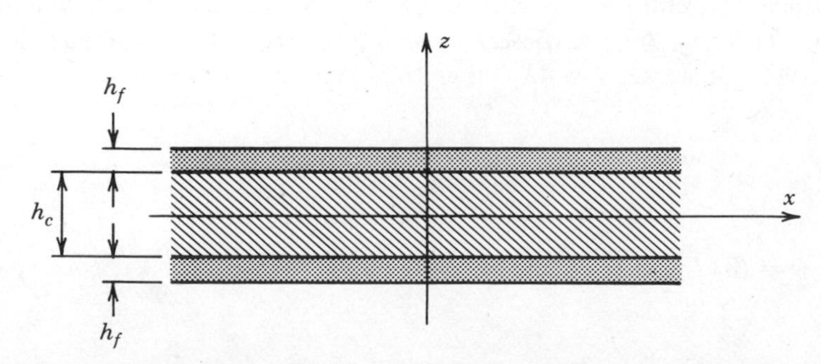


Using the method of Section 6.1, derive the equations of motion and associated, admissible boundary conditions for the sandwich plate shown in the figure. Assume that the facings are monolithically bonded to the core of the plate. Include the effects of shear deformation and rotatory inertia (see Exercise 3.36).

**6.4**

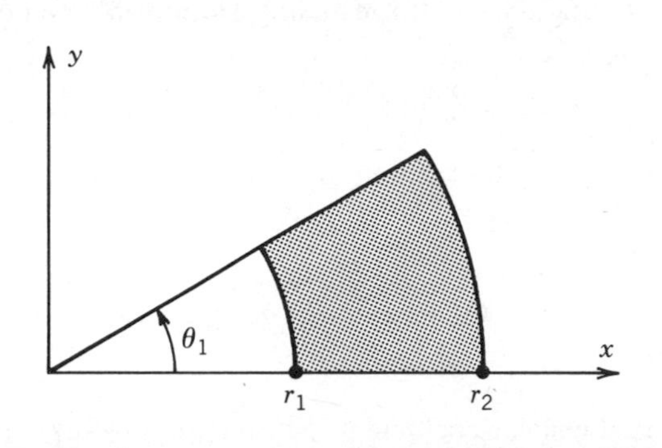


**(a)** Follow the procedure of Section 6.1 and apply Hamilton's principle to the plate element ($0 \leq r_1 < r < r_2, 0 < \theta < \theta_1 \leq 2\pi$) shown in the figure to obtain the stress equations of motion referred to polar coordinates $(r, \theta)$:

$$-\rho h \ddot{w} + \frac{\partial Q_r}{\partial r} + \frac{1}{r}\frac{\partial Q_\theta}{\partial \theta} + p + \frac{Q_r}{r} = 0$$

$$-\frac{\rho h^3}{12}\ddot{\psi}_r + \frac{\partial M_{rr}}{\partial r} + \frac{1}{r}\frac{\partial M_{r\theta}}{\partial \theta} + \frac{M_{rr} - M_{\theta\theta}}{r} - Q_r = 0$$

$$-\frac{\rho h^3}{12}\ddot{\psi}_\theta + \frac{\partial M_{r\theta}}{\partial r} + \frac{1}{r}\frac{\partial M_{\theta\theta}}{\partial \theta} + \frac{2M_{r\theta}}{r} - Q_\theta = 0$$

The admissible boundary conditions are: At $r = r_1$ and $r = r_2$ one member of each of the following pairs must be specified: $(w, Q_r)$, $(\psi_r, M_{rr})$, and $(\psi_\theta, M_{r\theta})$; at $\theta = 0$ and $\theta = \theta_1$ one member of each of the following pairs must be specified: $(w, Q_\theta)$, $(\psi_r, M_{r\theta})$, and $(\psi_\theta, M_{\theta\theta})$.

**(b)** Use the results of Exercise 6.4a to show that in the case of classical plate theory, the displacement equation of motion referred to polar coordinates is

$$D\nabla^4 w + \rho h \ddot{w} = p(r, \theta, t)$$

where

$$\nabla^2 = \frac{\partial^2}{\partial r^2} + \frac{1}{r}\frac{\partial}{\partial r} + \frac{1}{r^2}\frac{\partial^2}{\partial \theta^2}$$

What are the associated, admissible boundary conditions?

**6.5** Consider the mathematical model of a "shear plate," that is, a plate that deforms in shear only but remains rigid with respect to other types of deformation. Assume (and justify) the displacement field $u_1 = 0$, $u_2 = 0$, and $u_3 = w(x_1, x_2, t)$ and derive the stress equations of motion and admissible boundary conditions using Hamilton's principle. They are

$$-\rho h\ddot{w} + \nabla \cdot \mathbf{Q} + p = 0 \quad \text{in } A$$

and $w$ or $Q_n$ must be specified on $C$. Show that the stress-displacement relations are given by

$$Q_\alpha = \int_{-h/2}^{h/2} \tau_{z\alpha}\, dz = \kappa^2 Ghw_{,\alpha}$$

where $\kappa^2$ is the shear coefficient. Formulate a well-posed forced motion problem and discuss its solution. Derive an expression for energy flux in the plate.

**6.6** Show that the initial value problem characterized by (6.7), (6.35), and (6.36) has a unique solution.

**6.7** Show that the initial value problem characterized by (6.4) and (6.13) has a unique solution if, in addition, we specify

$$w(x, y, 0) = w_0(x, y) \quad \text{and} \quad \psi_\alpha(x, y, 0) = \psi_\alpha^{(0)}(x, y)$$

**6.8** Show that the natural frequencies associated with CPT are real.

**6.9** Show that the natural frequencies associated with IPT are real.

**6.10** Consider the static deformation of a simply supported, rectangular plate subjected to a uniformly distributed load of intensity $p_0$. Assume a solution in the form

$$w(x_1, x_2) = \sum_{m=1}^{\infty} \sum_{n=1}^{\infty} C_{mn} W_{mn}(x_1, x_2)$$

where $W_{mn}(x_1, x_2)$ are the normalized eigenfunctions given by (6.20) for CPT. Find the constants $C_{mn}$, $m = 1, 2, 3, \ldots,$ $n = 1, 2, 3, \ldots$ .

**6.11** Consider free harmonic vibrations of a rectangular plate. The edges $x = 0$ and $x = a$ are simply supported. Find the natural frequencies and associated mode shapes if **(a)** the edges $y = 0$ and $y = b$ are free, **(b)** the edges $y = 0$ and $y = b$ are clamped, **(c)** the edge $y = 0$ is simply supported and the edge $y = b$ is free, and **(d)** the edge $y = 0$ is simply supported and the edge $y = b$ is clamped.

Use CPT and assume that

$$w = Y(y) \sin \frac{m\pi x}{a} \cos \Omega t, \qquad m = 1, 2, 3, \ldots$$

**6.12** Consider free vibrations within the framework of IPT. Take

$$w = W^{(i)}(x, y) \cos \Omega_i t$$

$$\psi_\alpha = \Psi_\alpha^{(i)}(x, y) \cos \Omega_i t, \qquad i = 1, 2, 3, \ldots$$

where $W^{(i)}$ and $\Psi_\alpha^{(i)}$ are free vibration modes that satisfy homogeneous boundary conditions. Show that

$$\int_A \left( \rho h W^{(i)} W^{(j)} + \tfrac{1}{12} \rho h^3 \Psi_\alpha^{(i)} \Psi_\alpha^{(j)} \right) dA = \delta_{ij}$$

is the appropriate orthonormality relation if $\Omega_i \neq \Omega_j$.

**6.13** Determine the eigenfunctions for a simply supported, rectangular plate within the framework of IPT, that is, assume that

$$w(x_1, x_2, t) = W^{(m,n)}(x_1, x_2) \cos \Omega_{mn} t$$

$$\psi_1(x_1, x_2, t) = \Psi_1^{(m,n)}(x_1, x_2) \cos \Omega_{mn} t$$

$$\psi_2(x_1, x_2, t) = \Psi_2^{(m,n)}(x_1, x_2) \cos \Omega_{mn} t$$

and show that

$$W^{(m,n)} = A_{mn} \sin \frac{m\pi x_1}{a} \sin \frac{n\pi x_2}{b}$$

$$\Psi_1^{(m,n)} = B_{mn} \cos \frac{m\pi x_1}{a} \sin \frac{n\pi x_2}{b}$$

$$\Psi_2^{(m,n)} = C_{mn} \sin \frac{m\pi x_1}{a} \cos \frac{n\pi x_2}{b}$$

Normalize the eigenfunctions and find an expression for the natural frequencies $\Omega_{mn}$. (See the paper by H. Reismann and Y. C. Lee, "Forced Motion of Rectangular Plates," *Developments in Theoretical and Applied Mechanics*, Vol. 4, Pergamon, Oxford, 1970, pp. 3–18.)

**6.14** Find a resolution of the forced motion problem (6.5) and (4.13) with homogeneous boundary conditions within the framework of IPT. Assume that

$$w(x, y, t) = \sum_{i=1}^{\infty} W^{(i)}(x, y) q_i(t)$$

$$\psi_\alpha(x, y, t) = \sum_{i=1}^{\infty} \Psi_\alpha^{(i)}(x, y) q_i(t)$$

and use the initial conditions

$$w(x, y, 0) = w_0(x, y), \quad \dot{w}(x, y, 0) = \dot{w}_0(x, y)$$

$$\psi_\alpha(x, y, 0) = \psi_\alpha^{(0)}(x, y), \ \dot{\psi}_\alpha(x, y, 0) = \dot{\psi}_\alpha^{(0)}(x, y).$$

Show that

$$q_i(t) = q_i(0) \cos \Omega_i t + \frac{1}{\Omega_i} \dot{q}_i(0) \sin \Omega_i t + \frac{1}{\Omega_i} \int_0^t F_i(\tau) \sin \Omega_i(t - \tau)\, d\tau$$

where

$$q_i(0) = \int_A \left( \rho h w_0 W^{(i)} + \tfrac{1}{12} \rho h^3 \psi_\alpha^{(0)} \Psi_\alpha^{(i)} \right) dA$$

$$\dot{q}_i(0) = \int_A \left( \rho h \dot{w}_0 W^{(i)} + \tfrac{1}{12} \rho h^3 \dot{\psi}_\alpha^{(0)} \Psi_\alpha^{(i)} \right) dA$$

and $F_i(t) = \int_A p W^{(i)}\, dA$.

**6.15** Generalize the development at the end of Section 6.5. Consider the static problem (3.68) [or (4.9)] within the framework of IPT. Restrict the development to homogeneous boundary conditions (4.13b) on $C$. Assume

$$w = \sum_{i=1}^{\infty} A_i W^{(i)}(x, y), \qquad \psi_\alpha = \sum_{i=1}^{\infty} B_i \Psi_\alpha^{(i)}(x, y)$$

where $W^{(i)}$ and $\Psi_\alpha^{(i)}$ are free vibration modes. Devise a procedure to determine the constants $A_i$ and $B_i$.

**6.16** Consider free, axisymmetric harmonic vibrations of a circular plate with radius $a$. Calculate the three lowest natural frequencies using CPT if **(a)** the plate edge $r = a$ is simply supported and **(b)** the plate edge $r = a$ is free.

**6.17** Consider free harmonic vibrations of a complete circular or ring-shaped plate within the framework of CPT. Assume $w(r, \theta, t) = W(r, \theta) \cos \Omega t$

and show that $(\nabla^2 + \lambda^2)(\nabla^2 - \lambda^2)W = 0$, where $\lambda^4 = (\rho h/D)\Omega^2$ and

$$\nabla^2 = \frac{\partial^2}{\partial r^2} + \frac{1}{r}\frac{\partial}{\partial r} + \frac{1}{r^2}\frac{\partial^2}{\partial \theta^2}$$

Take $W(r, \theta) = R(r) \cdot T(\theta)$ and show that $R(r) = AJ_n(\lambda r) + BI_n(\lambda r) + CY_n(\lambda r) + FK_n(\lambda r)$, where $J_n$, $I_n$, $Y_n$, and $K_n$ are Bessel functions, $T(\theta) = H \cos n(\theta - \phi)$, $n = 0, 1, 2, 3, \ldots$, and $H$ and $\phi$ are constants.

**6.18** With reference to Exercise 6.17, consider the free, harmonic vibrations of a clamped, circular plate with radius $a$. Show that the natural frequencies are given by $\Omega_{mn} = [(\lambda a)^2_{mn}/a^2]\sqrt{D/\rho h}$, where $(\lambda a)_{mn}$ are the roots of the transcendental equation

$$J_n(\lambda a) I_{n+1}(\lambda a) + I_n(\lambda a) J_{n+1}(\lambda a) = 0$$

*Values of* $(\lambda a)_{mn}$

| $m$ | $n = 0$ | $n = 1$ | $n = 2$ | $n = 3$ |
|---|---|---|---|---|
| 0 | 3.196 | 4.611 | 5.906 | 7.143 |
| 1 | 6.306 | 7.799 | 9.197 | 10.537 |
| 2 | 9.440 | 10.958 | 12.402 | 13.795 |
| 3 | 12.577 | 14.108 | 15.579 | 17.005 |

Interpret the physical significance of the integers $m$ and $n$.

**6.19**

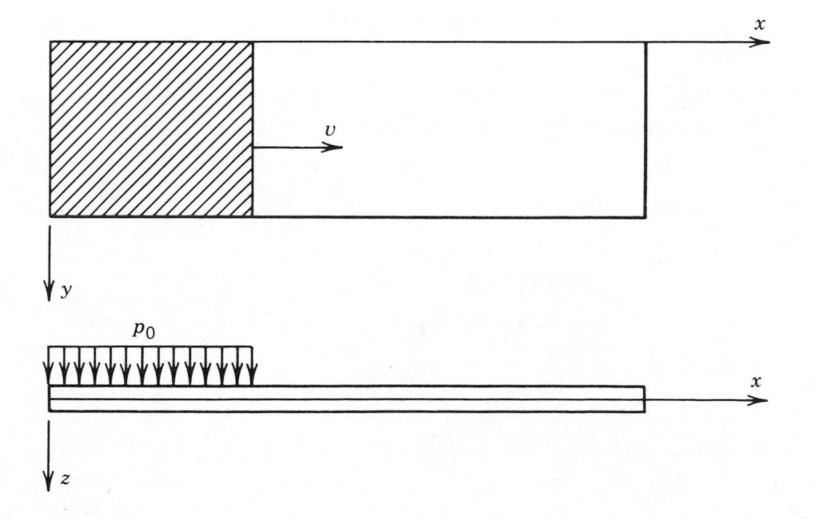


A uniformly distributed pressure of intensity $p_0$ propagates over the surface of a rectangular plate with constant speed $v$. The plate, which is simply supported, is at rest and undeformed at $t = 0$. The load front is at the edge $x = 0$ at $t = 0$. Derive a formula for the deflection of the plate for $0 \leq t \leq a/v$. Use CPT. Draw a graph of the deflection at $(x, y) = (\frac{1}{2}a, \frac{1}{2}b)$ for $0 \leq t \leq a/v$.

**6.20** Consider the initial value problem (6.34) and (6.36) with time-dependent (nonhomogeneous) boundary conditions

$$\left.\begin{aligned} w &= f_1(s,t) \quad \text{or} \quad V_n = f_1(s,t) \\ \frac{\partial w}{\partial n} &= f_2(s,t) \quad \text{or} \quad M_{nn} = f_2(s,t) \end{aligned}\right\} \quad \text{on } C \tag{$\alpha$}$$

Write the solution in the form

$$w = w_s(x, y, t) + \sum_{i=1}^{\infty} W^{(i)}(x, y) q_i(t)$$

where $W^{(i)}$ are the free vibration modes satisfying the following homogeneous boundary conditions:

$$\left.\begin{aligned} &\text{If } w = f_1(s,t), \quad \text{then } W^{(i)}(s,t) = 0 \\ &\text{If } V_n = f_2(s,t), \quad \text{then } V_n^{(i)}(s,t) = 0 \\ &\text{If } \frac{\partial w}{\partial n} = f_2(s,t), \quad \text{then } \left(\frac{\partial W^{(i)}}{\partial n}\right)_{(s,t)} = 0 \\ &\text{If } M_{nn} = f_2(s,t), \quad \text{then } M_{nn}^{(i)}(s,t) = 0 \end{aligned}\right\} \quad \text{on } C$$

The "quasi-static" solution $w_S(x, y, t)$ satisfies (6.34) and ($\alpha$), with inertia terms deleted in (6.34). Show that

$$\begin{aligned} q_i(t) = {} & [q_i(0) - F_i(0)] \cos \Omega_i t + \frac{1}{\Omega_i} [\dot{q}_i(0) - \dot{F}_i(0)] \sin \Omega_i t \\ & + F_i(t) - \Omega_i \int_0^t F_i(\tau) \sin \Omega_i (t - \tau)\, d\tau \end{aligned}$$

where

$$\Omega_i^2 F_i(t) = -\int_A p W^{(i)}\, dA + \oint_C \left[ \left(V_n^{(i)} - W^{(i)}\right) f_1 + \left(\frac{\partial W^{(i)}}{\partial n} - M_{nn}^{(i)}\right) f_2 \right] ds$$

and

$$q_i(0) - F_i(0) = \int_A \rho h w_0 W^{(i)}\, dA, \qquad \dot{q}_i(0) - \dot{F}_i(0) = \int_A \rho h \dot{w}_0 W^{(i)}\, dA.$$

**6.21** A rectangular plate is simply supported along $y = 0$, $y = b$, $x = 0$, and $x = a$. At the instant $t = 0$ a uniformly distributed bending moment of intensity $M_0$ is suddenly applied to the edge $y = 0$. Find the dynamic response of the plate assuming that it is at rest in its reference configuration at $t = 0$. Use CPT.

**6.22** Generalize Exercise 6.20 to facilitate the solution of initial value problems with admissible, nonhomogeneous boundary conditions within the framework of IPT. Let

$$w = w_S(x, y, t) + \sum_{i=1}^{\infty} W^{(i)}(x, y) q_i(t)$$

$$\psi_\alpha = \psi_\alpha^{(S)}(x, y, t) + \sum_{i=1}^{\infty} \Psi_\alpha^{(i)}(x, y) q_i(t)$$

where $w_S$ and $\psi_\alpha^{(S)}$ are solutions of the quasi-static problem and $W^{(i)}$ and $\Psi_\alpha^{(i)}$ are free vibration modes. Find an expression for $q_i(t)$.

**6.23** Obtain the response of a clamped, circular plate to a concentrated force that is suddenly applied at its center by letting $B = b/a$ tend to zero as $\pi b^2 p_0$ tends to $F$ in the solution of the example in Section 6.7.

**6.24** A simply supported, circular plate is subjected to a uniformly distributed, suddenly applied pressure of magnitude $p_0$. The plate is at rest in its reference configuration at $t = 0$ when the load is applied. Find the dynamic response within the framework of CPT.

**6.25** A simply supported, rectangular plate is subjected to a uniformly distributed, suddenly applied pressure of magnitude $p_0$. The plate is at rest in its reference configuration at $t = 0$ when the pressure is applied. Find the dynamic response within the framework of CPT.

**6.26** Find the steady-state response of a clamped, circular plate with radius $a$ under the influence of a concentrated, time harmonic load $F \cos \Omega t$ acting at $r = 0$. Use CPT and the results of Exercises 6.17 and 6.32.

**6.27** Consider a semi-infinite plate $-\infty < x < \infty, 0 \leq y < \infty$. A straight-crested, free, propagating harmonic wave (6.70) is advancing in the $x$-direction. Find the magnitudes of the phase and group velocity if (**a**) the plate is clamped along $y = 0$ and (**b**) the plate is simply supported along $y = 0$. Use CPT.

**6.28** Consider an infinite plate strip $-\infty < x < \infty, 0 \leq y \leq a$. A straight-crested, free, propagating harmonic wave (6.70) advances in the $x$-direction. Find the phase and group velocities and the shape of the deformed plate if (**a**) $y = 0$ is clamped and $y = a$ is clamped, (**b**) $y = 0$ and $y = a$ are simply supported and (**c**) $y = 0$ is clamped and $y = a$ is free. Use CPT.

**6.29** Consider wave motion in an infinite plate strip $-\infty < x < \infty, 0 \le y \le a$. Assume $w = f(y) \exp ik(x - vt)$ and assume that the edges $y = 0$ and $y = a$ are free. Find the dispersion relations $v$ and $v_G$ as a function of $h/L$ (see the example in Section 6.11). Use CPT.

**6.30** Consider a semi-infinite plate $0 \le x < \infty, -\infty < y < \infty$. A time harmonic incident, straight-crested wave with wave crests parallel to the $y$-axis is reflected from the edge $x = 0$. We wish to find the reflected wave (or waves) in the plate.

**(a)** Assume the edge $x = 0$ is clamped.

**(b)** Assume the edge $x = 0$ is simply supported.

**(c)** Assume the edge $x = 0$ is free.

Use CPT.

**6.31** Repeat Exercise 6.30 but use IPT.

**6.32** Find the steady-state response of an unbounded (infinite) plate under the influence of a concentrated, time harmonic load $F \cos \Omega t$ acting at $r = 0$. Use CPT. Show that

$$w(r, t) = -\frac{F}{8k^2 D}\left\{\left[Y_0(kr) + \frac{2}{\pi} K_0(kr)\right] \cos \Omega t - J_0(kr) \sin \Omega t\right\}$$

where $k^4 = (\rho h/D)\Omega^2$.

**6.33** An infinite elastic plate is subjected to a suddenly applied concentrated load of magnitude $F$ applied at $t = 0$. The plate is at rest in its reference configuration at $t = 0$. Using CPT, show that the solution is

$$w(r, t) = \frac{F\tau}{16\pi D}\left[\frac{\pi}{2} - \mathrm{Si}\left(\frac{r^2}{\tau}\right) + \frac{r^2}{\tau} \mathrm{Ci}\left(\frac{r^2}{\tau}\right) - \sin\left(\frac{r^2}{\tau}\right)\right]$$

where $\tau = 4\lambda^2 t$, $\lambda^4 = D/\rho h$, and

$$\mathrm{Si}(x) = \int_0^x \frac{\sin \xi}{\xi}\, d\xi \quad \text{and} \quad \mathrm{Ci}(x) = \int_\infty^x \frac{\cos \xi}{\xi}\, d\xi$$

are the sine and cosine integral functions, respectively. Show that for sufficiently large values of $r^2/\tau$, the solution can be approximated by the asymptotic formula

$$w(r, t) \cong -\frac{F\tau}{16\pi D}\left(\frac{\tau}{r^2}\right)^2 \sin\left(\frac{r^2}{\tau}\right)$$

**6.34** Demonstrate the existence of a plate stress potential using the first law of thermodynamics: $\dot{E}^* = \dot{T} + \dot{U} - \dot{Q}$, where $E^*$ is the work of external forces, $T$ is kinetic energy, $U$ is strain energy, and $Q$ is the heat absorbed by the plate. Show that

$$\dot{E}^* = \int_A p\dot{w}\,dA + \oint_C \left(\mathbf{M}^* \cdot \dot{\boldsymbol{\psi}} + Q_n^*\dot{w}\right) dS$$

$$= \dot{T} + \int_A \left(M_{\alpha\beta}\dot{m}_{\alpha\beta} + Q_\alpha \dot{q}_\alpha\right) dA$$

Assume an adiabatic process $\dot{Q} \equiv 0$ and show that $\dot{U} = \dot{E}^* - \dot{T} = \int_A (M_{\alpha\beta}\dot{m}_{\alpha\beta} + Q_\alpha \dot{q}_\alpha)\,dA$. But $U = \int_A W_p(m_{\alpha\beta}, q_\alpha)\,dA$, so that

$$\dot{W}_p = M_{\alpha\beta}\dot{m}_{\alpha\beta} + Q_\alpha \dot{q}_\alpha = \frac{\partial W_p}{\partial m_{\alpha\beta}}\dot{m}_{\alpha\beta} + \frac{\partial W_p}{\partial q_\alpha}\dot{q}_\alpha$$

and

$$M_{\alpha\beta} = \frac{\partial W_p}{\partial m_{\alpha\beta}}, \qquad Q_\alpha = \frac{\partial W_p}{\partial q_\alpha}$$

**6.35** Repeat Exercise 6.34 for the case of CPT.

**6.36** In order to account for energy dissipation (friction) in the plate, we write the equations of motion

$$\left.\begin{aligned} M_{\alpha\beta,\beta} - Q_\alpha - D_{\alpha\beta}\dot{\psi}_\beta + p_\alpha &= \tfrac{1}{12}\rho h^3 \ddot{\psi}_\alpha \\ Q_{\alpha,\alpha} - D_z\dot{w} + p_z &= \rho h \ddot{w} \end{aligned}\right\} \text{ in } A$$

and use the boundary conditions (4.13) with

$$M_\alpha^* = M_{\alpha\beta}n_\beta + d_{\alpha\beta}\dot{\psi}_\beta, \qquad Q_n^* = Q_\alpha n_\alpha + d_z\dot{w}$$

With reference to Exercise 6.34, it can be shown that $\dot{E}^* + \dot{Q} = \dot{T} + \dot{U}$, where

$$-\frac{1}{2}\dot{Q} = R(t) = \frac{1}{2}\int_A \left(D_{\alpha\beta}\dot{\psi}_\alpha\dot{\psi}_\beta + D_z\dot{w}^2\right) dA$$

$$+ \frac{1}{2}\oint_C \left(d_{\alpha\beta}\dot{\psi}_\alpha\dot{\psi}_\beta + d_z\dot{w}^2\right) ds$$

Show that energy dissipation imposes the following requirements upon the damping coefficients: $D_z > 0$ and $d_z > 0$ and the functions $D_{\alpha\beta}\dot{\psi}_\alpha\dot{\psi}_\beta$ and $d_{\alpha\beta}\dot{\psi}_\alpha\dot{\psi}_\beta$ must be positive definite. $R(t)$ is the Rayleigh dissipation function.

# 7

# INITIALLY STRESSED PLATES — ELASTIC INSTABILITY

In many practical applications, plates are in a state of initial membrane stress. When subsequently subjected to transverse pressure loads, their structural behavior and response can be entirely different from plates that are free from such internal stresses. The approach taken in this chapter, in a natural manner, leads to the formulation of problems dealing with the elastic stability (or instability) of plates.

## 7.1 BASIC EQUATIONS

In this section we consider the equilibrium of a plate in the deformed configuration, as shown in Fig. 7.1. The plate is acted upon by distributed transverse pressure forces $\hat{\mathbf{e}}_z \cdot p(x, y)\, dA$ and

(a) transverse boundary forces $\hat{\mathbf{e}}_z S_n\, ds$,
(b) membrane boundary forces $\hat{\mathbf{n}} N_{nn}\, ds$ and $\hat{\mathbf{s}} N_{ns}\, ds$, and
(c) boundary moments $\hat{\mathbf{n}} M_{ns}\, ds$ and $\hat{\mathbf{s}} M_{nn}\, ds$, as shown.

The unit vectors $\hat{\mathbf{n}}$ and $\hat{\mathbf{s}}$ are normal and tangential, respectively, to the boundary curve $C$ (see Fig. 3.4). The (vertical) boundary shear force $\hat{\mathbf{e}}_z S_n$ is parallel to the (vertical) $z$-axis. Initially, the plate is assumed to be in a state of plane stress characterized by the resultants

$$N_{\alpha\beta} = \int_{-h/2}^{h/2} \tau_{\alpha\beta}\, dz \tag{7.1}$$

We note that $N_{\alpha\beta}$ is a tensor of order two in a two-dimensional space. Associated with the stress tensor $N_{\alpha\beta}$ is the stress vector $N_\alpha$ such that

$$N_\alpha = N_{\alpha\beta} n_\beta \tag{7.2}$$

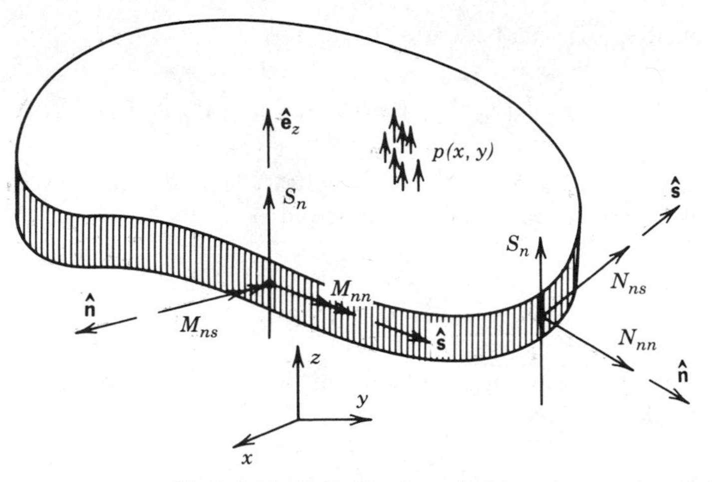


**Figure 7.1** Initially stressed plate.

[see Section 3.6 and (3.35)]. With reference to Fig. 7.1 and (3.5), we have (see Chapter 3)

$$\hat{\mathbf{n}} = n_\alpha \hat{\mathbf{e}}_\alpha, \qquad \hat{\mathbf{s}} = s_\alpha \hat{\mathbf{e}}_\alpha, \qquad \hat{\mathbf{n}} \cdot \hat{\mathbf{s}} = 0$$

$$n_\alpha n_\beta + s_\alpha s_\beta = \delta_{\alpha\beta}, \qquad n_\alpha s_\beta - n_\beta s_\alpha = e_{\alpha\beta} \tag{7.3}$$

$$e_{\alpha\beta}\hat{\mathbf{e}}_\beta = \hat{\mathbf{e}}_z \times \hat{\mathbf{e}}_\alpha, \qquad \mathbf{R} = \hat{\mathbf{e}}_\alpha x_\alpha$$

$$\begin{aligned} &S_n = S_\alpha n_\alpha, \qquad N_{nn} = N_\alpha n_\alpha, \qquad N_{ns} = N_\alpha s_\alpha \\ &N_\alpha = N_{\alpha\beta} n_\beta, \qquad M_{nn} = M_\alpha n_\alpha, \qquad M_{ns} = M_\alpha s_\alpha, \qquad M_\alpha = M_{\alpha\gamma} n_\gamma \end{aligned} \tag{7.4}$$

Considering all external forces acting upon the plate, static equilibrium requires that $\mathfrak{F} = \mathbf{0}$, or

$$\oint_C \hat{\mathbf{e}}_z S_n \, ds + \oint_C (\hat{\mathbf{n}} N_{nn} + \hat{\mathbf{s}} N_{ns}) \, ds + \int_A \hat{\mathbf{e}}_z p \, dA = \mathbf{0} \tag{7.5}$$

where $A$ denotes the area of the plate and $C$ denotes its boundary. With the aid of (7.3), (7.4), and the divergence theorem, we have

$$\oint_C \hat{\mathbf{e}}_z S_n \, ds = \oint_C \hat{\mathbf{e}}_z S_\alpha n_\alpha \, ds = \int_A (\hat{\mathbf{e}}_z S_\alpha)_{,\alpha} \, dA = \int_A \hat{\mathbf{e}}_z S_{\alpha,\alpha} \, dA$$

$$\begin{aligned} \oint_C (\hat{\mathbf{n}} N_{nn} + \hat{\mathbf{s}} N_{ns}) \, ds &= \oint_C \hat{\mathbf{e}}_\beta (n_\beta n_\alpha N_\alpha + s_\beta s_\alpha N_\alpha) \, ds \\ &= \oint_C \hat{\mathbf{e}}_\beta \delta_{\alpha\beta} N_\alpha \, ds = \oint_C \hat{\mathbf{e}}_\alpha N_\alpha \, ds = \oint_C \hat{\mathbf{e}}_\alpha n_\beta N_{\alpha\beta} \, ds \\ &= \int_A (\hat{\mathbf{e}}_\alpha N_{\alpha\beta})_{,\beta} \, dA = \int_A \hat{\mathbf{e}}_\alpha N_{\alpha\beta,\beta} \, dA \end{aligned}$$

Consequently, (7.5) can be written in the form

$$\int_A \left[\hat{\mathbf{e}}_z(S_{\alpha,\alpha} + p) + \hat{\mathbf{e}}_\alpha N_{\alpha\beta,\beta}\right] dA = \mathbf{0}$$

This equation is valid for plates with arbitrary area $A$, and the vectors $\hat{\mathbf{e}}_z$ and $\hat{\mathbf{e}}_\alpha$ are linearly independent. Thus, we require

$$S_{\alpha,\alpha} + p = 0 \tag{7.6a}$$

$$N_{\alpha\beta,\beta} = 0 \tag{7.6b}$$

For moment equilibrium, considering all external forces and moments acting upon the plate, we require $\mathfrak{M}_0 = \mathbf{0}$, or

$$\begin{aligned}
&\oint_C (\hat{\mathbf{s}} M_{nn} - \hat{\mathbf{n}} M_{ns})\, ds + \oint_C (\mathbf{R} + \hat{\mathbf{e}}_z w) \times (\hat{\mathbf{e}}_z S_n)\, ds \\
&\quad + \int_A \left[(\mathbf{R} + \hat{\mathbf{e}}_z w) \times (\hat{\mathbf{e}}_z p)\right] dA \\
&\quad + \oint_C \left(\vec{\mathbf{R}} + \hat{\mathbf{e}}_z w\right) \times (\hat{\mathbf{n}} N_{nn} + \hat{\mathbf{s}} N_{ns})\, ds \\
&\quad = \mathbf{0}
\end{aligned} \tag{7.7}$$

Hence, by (7.3), (7.4), and the divergence theorem, we have

$$\oint_C (\hat{\mathbf{s}} M_{nn} - \hat{\mathbf{n}} M_{ns})\, ds = \int_A \left[(\hat{\mathbf{e}}_\alpha \times \hat{\mathbf{e}}_z)(-M_{\alpha\gamma,\gamma})\right] dA$$

$$\begin{aligned}
\oint_C \left[(\mathbf{R} + \hat{\mathbf{e}}_z w) \times (\hat{\mathbf{e}}_z S_n)\right] ds &= \oint_C (\mathbf{R} \times \hat{\mathbf{e}}_z) S_n\, ds \\
&= \int_A \left[(\mathbf{R} \times \hat{\mathbf{e}}_z) S_{\alpha,\alpha} + (\hat{\mathbf{e}}_\alpha \times \hat{\mathbf{e}}_z) S_\alpha\right] dA
\end{aligned}$$

$$\int_A \left[(\mathbf{R} + w\hat{\mathbf{e}}_z) \times (\hat{\mathbf{e}}_z p)\right] dA = \int_A (\mathbf{R} \times \hat{\mathbf{e}}_z)\, p\, dA$$

$$\begin{aligned}
\oint_C \mathbf{R} \times (\hat{\mathbf{n}} N_{nn} + \hat{\mathbf{s}} N_{ns})\, ds &= \oint_C (\mathbf{R} \times \hat{\mathbf{e}}_\beta) N_\beta\, ds \\
&= \oint_C (\mathbf{R} \times \hat{\mathbf{e}}_\beta) n_\gamma N_{\beta\gamma}\, ds = \int_A \left[(\mathbf{R} \times \hat{\mathbf{e}}_\beta) N_{\beta\gamma}\right]_{,\gamma} dA \\
&= \int_A \left[(\mathbf{R} \times \hat{\mathbf{e}}_\beta) N_{\beta\gamma,\gamma} + (\hat{\mathbf{e}}_\gamma \times \hat{\mathbf{e}}_\beta) N_{\beta\gamma}\right] dA = \mathbf{0}
\end{aligned}$$

in view of (7.6b) and because

$$(\hat{\mathbf{e}}_\gamma \times \hat{\mathbf{e}}_\beta) N_{\beta\gamma} = \hat{\mathbf{e}}_z (N_{12} - N_{21}) = \mathbf{0}$$

Furthermore, we have

$$\oint_C [(\hat{\mathbf{e}}_z w) \times (\hat{\mathbf{n}} N_{nn} + \hat{\mathbf{s}} N_{ns})] \, ds$$

$$= \oint_C \left[ (\hat{\mathbf{e}}_z \times \hat{\mathbf{e}}_\beta)(\delta_{\alpha\beta} N_\alpha w) \right] ds$$

$$= \oint_C \left[ (\hat{\mathbf{e}}_z \times \hat{\mathbf{e}}_\beta)(w N_\beta) \right] ds = \oint_C \left[ (\hat{\mathbf{e}}_z \times \hat{\mathbf{e}}_\beta)(w n_{,\gamma} N_{\beta\gamma}) \right] ds$$

$$= \int_A \left[ (\hat{\mathbf{e}}_z \times \hat{\mathbf{e}}_\beta)(w N_{\beta\gamma}) \right]_{,\gamma} dA = \int_A (\hat{\mathbf{e}}_z \times \hat{\mathbf{e}}_\beta)(N_{\beta\gamma} w_{,\gamma}) \, dA$$

Consequently, (7.7) can be written in the form

$$\int_A \left[ (\hat{\mathbf{e}}_\alpha \times \hat{\mathbf{e}}_z)(-M_{\alpha\gamma,\gamma}) + (\mathbf{R} \times \hat{\mathbf{e}}_z)(S_{\alpha,\alpha} + p) + (\hat{\mathbf{e}}_\alpha \times \hat{\mathbf{e}}_z) S_\alpha \right.$$
$$\left. + (\hat{\mathbf{e}}_z \times \hat{\mathbf{e}}_\beta)(N_{\beta\gamma} w_{,\gamma}) \right] dA = \mathbf{0}$$

We now apply (7.6) and assume that the integrand is continuous. For arbitrary $A$, we obtain

$$(\hat{\mathbf{e}}_\alpha \times \hat{\mathbf{e}}_z)(Q_\alpha - M_{\alpha\gamma,\gamma}) = \mathbf{0}$$

where we set

$$Q_\alpha = S_\alpha - w_{,\gamma} N_{\alpha\gamma} \tag{7.8}$$

Thus, we conclude that

$$Q_\alpha = M_{\alpha\gamma,\gamma} \tag{7.6c}$$

Equations (7.6), in conjunction with the definition (7.8), constitute the stress equations of equilibrium of the initially stressed plate. We also note that equations (7.6) are identical to (9.34), and whereas these equations have been derived by the application of the laws of statics in this chapter, they are obtained by invoking variational principles in Section 9.2.

For a geometric interpretation of equations (7.8), we now refer to Fig. 7.2. In this figure $Q_x$ is the transverse shear force per unit length, normal to the *deformed* median surface of the plate. The symbol $S_x$ denotes the (vertical) shear force acting parallel to the $z$-axis upon the plate edge with normal $\hat{\mathbf{e}}_x$. For the present theory of initially stressed plates, the shear forces $Q_x$ and $S_x$, in general, are not parallel. Along the plate edge $x$ = constant, we have (see Fig. 7.2)

$$Q_x = N_{xx}(\hat{\boldsymbol{\varepsilon}} \cdot \hat{\mathbf{e}}_x) + N_{xy}(\hat{\boldsymbol{\varepsilon}} \cdot \hat{\mathbf{e}}_y) + S_x(\hat{\boldsymbol{\varepsilon}} \cdot \hat{\mathbf{e}}_z) \tag{7.9}$$

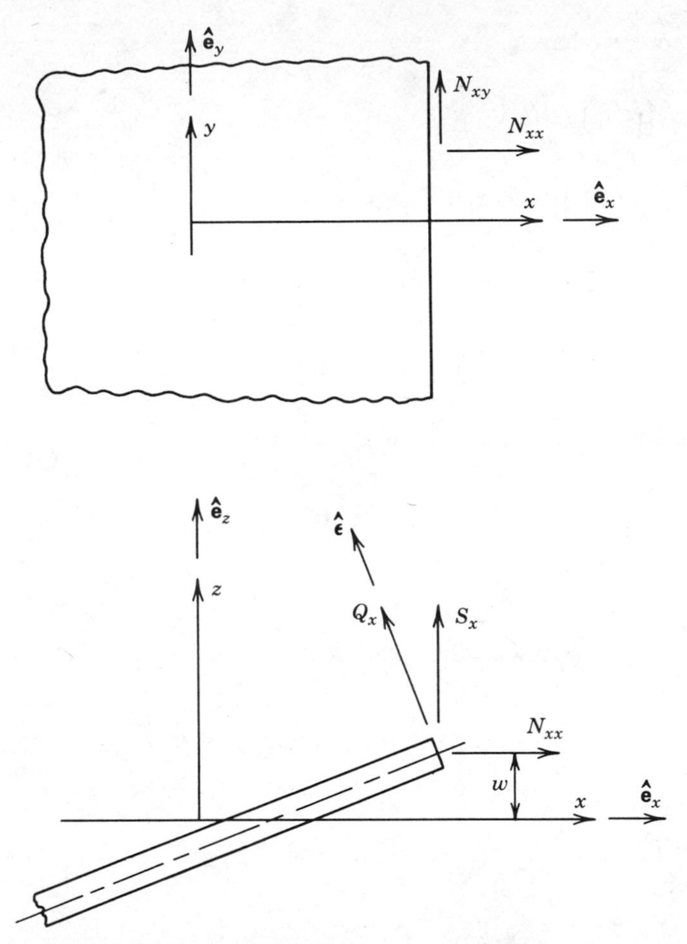


**Figure 7.2** Shear and membrane forces in deflected plate.

where $\hat{\boldsymbol{\varepsilon}}$ is a (local) unit vector that is normal to the deformed median surface of the plate. Let $\mathbf{r}$ be the position vector from the origin of coordinates to a generic point in the deformed median surface. Then

$$\mathbf{r} = \hat{\mathbf{e}}_x x + \hat{\mathbf{e}}_y y + \hat{\mathbf{e}}_z w(x, y)$$

$$\frac{\partial \mathbf{r}}{\partial x} = \hat{\mathbf{e}}_x + \hat{\mathbf{e}}_z \frac{\partial w}{\partial x}, \qquad \frac{\partial \mathbf{r}}{\partial y} = \hat{\mathbf{e}}_y + \hat{\mathbf{e}}_z \frac{\partial w}{\partial y}$$

$$\frac{\partial \mathbf{r}}{\partial x} \times \frac{\partial \mathbf{r}}{\partial y} = \hat{\mathbf{e}}_x \left( - \frac{\partial w}{\partial x} \right) + \hat{\mathbf{e}}_y \left( - \frac{\partial w}{\partial y} \right) + \hat{\mathbf{e}}_z$$

and we have

$$\left| \frac{\partial \mathbf{r}}{\partial x} \times \frac{\partial \mathbf{r}}{\partial y} \right|^2 = 1 + \left( \frac{\partial w}{\partial x} \right)^2 + \left( \frac{\partial w}{\partial y} \right)^2 \approx 1$$

in view of the small slope assumption (see Section 3.2). Using a well-known formula from differential geometry (see, e.g., E. Kreyszig, *Introduction to Differential Geometry and Riemannian Geometry*, University of Toronto Press, Toronto, 1975, p. 64) for the unit vector $\hat{\boldsymbol{\varepsilon}}$ normal to the deformed plate median surface, we obtain

$$\hat{\boldsymbol{\varepsilon}} = \frac{(\partial \mathbf{r}/\partial x) \times (\partial \mathbf{r}/\partial y)}{|(\partial \mathbf{r}/\partial x) \times (\partial \mathbf{r}/\partial y)|} = \hat{\mathbf{e}}_x \left( -\frac{\partial w}{\partial x} \right) + \hat{\mathbf{e}}_y \left( -\frac{\partial w}{\partial y} \right) + \hat{\mathbf{e}}_z$$

and therefore,

$$\hat{\boldsymbol{\varepsilon}} \cdot \hat{\mathbf{e}}_x = -\frac{\partial w}{\partial x}, \qquad \hat{\boldsymbol{\varepsilon}} \cdot \hat{\mathbf{e}}_y = -\frac{\partial w}{\partial y}, \qquad \hat{\boldsymbol{\varepsilon}} \cdot \hat{\mathbf{e}}_z = 1$$

Consequently, (7.9) assumes the form

$$Q_x = -N_{xx}\frac{\partial w}{\partial x} - N_{xy}\frac{\partial w}{\partial y} + S_x \tag{7.10a}$$

A similar analysis for the edge $y =$ constant results in

$$Q_y = -N_{yx}\frac{\partial w}{\partial x} - N_{yy}\frac{\partial w}{\partial y} + S_y \tag{7.10b}$$

We note that (7.10a) and (7.10b) are equivalent to the expanded form of (7.8) and therefore provide the motivation for this latter substitution.

The preceding analysis can be summarized as follows: The static equilibrium of a plate subjected to initial membrane prestress $N_{\alpha\beta}$ is characterized by

$$S_{\alpha,\alpha} + p = 0 \tag{7.11a}$$

$$N_{\alpha\beta,\beta} = 0 \tag{7.11b}$$

$$M_{\alpha\beta,\beta} = Q_\alpha \tag{7.11c}$$

where

$$S_\alpha = Q_\alpha + N_{\alpha\beta} w_{,\beta} \tag{7.11d}$$

relates vertical shear forces $S_\alpha$ to transverse shear forces $Q_\alpha$. Equations (7.11) have the invariant representation

$$\mathbf{div\,S} = \nabla \cdot \mathbf{S} = -p \tag{7.12a}$$

$$\mathbf{div}\,\mathcal{N} = \nabla \cdot \mathcal{N} = \mathbf{0} \tag{7.12b}$$

$$\mathbf{S} = \mathbf{Q} + \mathcal{N} \cdot \nabla w \tag{7.12c}$$

$$\mathbf{div}\,\mathcal{M} = \nabla \cdot \mathcal{M} = \mathbf{Q} \tag{7.12d}$$

where $\mathcal{N} = \hat{\mathbf{e}}_\alpha \hat{\mathbf{e}}_\beta N_{\alpha\beta}$ and $\mathcal{M} = \hat{\mathbf{e}}_\alpha \hat{\mathbf{e}}_\beta M_{\alpha\beta}$ are the membrane force tensor and moment tensor, respectively. Equations (7.11) and (7.12) revert to the corresponding equations (3.70) and (3.72) when the membrane force tensor components vanish (i.e., for $N_{\alpha\beta} \equiv 0$).

A variational derivation of (7.11) (or (7.12)) is provided in Section 9.2. There it is shown that admissible boundary conditions associated with (7.11) are given by the following: We need to specify

$$\left.\begin{array}{ll} & \text{Either } S_n \text{ or } w \\ \text{and} & \\ & \text{either } M_{nn} \text{ or } \dfrac{\partial w}{\partial n} \\ \text{and} & \\ & \text{either } N_{nn} \text{ or } u_n \\ \text{and} & \\ & \text{either } N_{ns} \text{ or } u_s \end{array}\right\} \text{ on } C \tag{7.13}$$

where $u_n$ and $u_s$ are boundary displacements parallel to the $x$-$y$ plane that are normal and tangential to $C$, respectively [see (9.35)].

If we neglect shear deformation (CPT), we can employ the moment curvature relation (4.16c)

$$Q_\alpha = M_{\alpha\beta,\beta} = -D(\nabla^2 w)_{,\alpha} \tag{7.14}$$

and upon combining (7.14) and (7.11), we obtain

$$D\nabla^4 w = p + \left(N_{\alpha\beta} w_{,\beta}\right)_{,\alpha} = p + N_{\alpha\beta} w_{,\alpha\beta} \tag{7.15a}$$

This equation can be written in the invariant form

$$D\nabla^4 w = p + \nabla \cdot (\mathcal{N} \cdot \nabla w) \tag{7.15b}$$

where $\mathcal{N} = \hat{\mathbf{e}}_\alpha \hat{\mathbf{e}}_\beta N_{\alpha\beta}$ is the membrane force tensor. Equation (7.15) is frequently used to solve for the deflection, moments, shear forces, and so on, in a transversely loaded plate that is in a known (or given) initial membrane stress state $N_{\alpha\beta} = N_{\alpha\beta}(x, y)$. If we set $p \equiv 0$, (7.15) can be used to calculate the critical (or buckling) load due to membrane forces in the plate.

## 7.2 STATICS

### 7.2.1 Rectangular Plates

We now consider the deflection of a rectangular plate of dimensions $a : b : h$ (see Fig. 7.3). The plate is in a state of initial membrane stress characterized by

$$\begin{bmatrix} N_{xx} & N_{xy} \\ N_{yx} & N_{yy} \end{bmatrix} = \begin{bmatrix} N & 0 \\ 0 & 0 \end{bmatrix} \tag{7.16}$$

The plate is simply supported, that is, we have

$$\begin{aligned} w(x,0) &= M_{yy}(x,0) = w(x,b) = M_{yy}(x,b) = 0 \\ w(0,y) &= M_{xx}(0,y) = w(a,y) = M_{yy}(a,y) = 0 \end{aligned} \tag{7.17}$$

In the spirit of Navier's solution (see Chapter 5), we assume that the transverse load $p(x, y)$, can be expanded into a double Fourier series

$$p(x,y) = \sum_{m=1}^{\infty} \sum_{n=1}^{\infty} p_{mn} \sin\frac{m\pi x}{a} \sin\frac{n\pi y}{b} \tag{7.18}$$

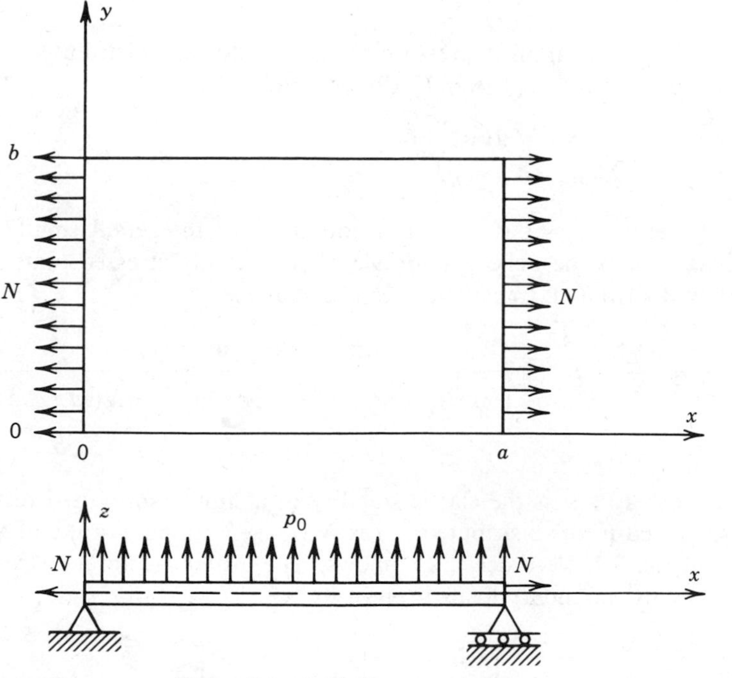


**Figure 7.3** Initially stressed plate.

where

$$p_{mn} = \frac{4}{ab} \int_0^a \int_0^b p(x, y) \sin \frac{m\pi x}{a} \sin \frac{n\pi y}{b} \, dx \, dy \tag{7.19}$$

In view of (7.16) and (7.18), we shall seek a solution of

$$D \nabla^4 w - N \frac{\partial^2 w}{\partial x^2} = p_{mn} \sin \frac{m\pi x}{a} \sin \frac{n\pi y}{b} \tag{7.20}$$

The function

$$w = w(x, y) = w_{mn} = a_{mn} \sin \frac{m\pi x}{a} \sin \frac{n\pi y}{b} \tag{7.21}$$

satisfies the boundary conditions (7.17). Upon substitution of (7.21) into (7.20), we obtain the relation

$$a_{mn} = \frac{p_{mn}}{D\left[(m\pi/a)^2 + (n\pi/b)^2\right]^2 + N(m\pi/a)^2} \tag{7.22}$$

Thus, a solution of (7.20) is given by

$$w_{mn} = \frac{p_{mn} \sin(m\pi x/a) \sin(n\pi y/b)}{D\left[(m\pi/a)^2 + (n\pi/b)^2\right]^2 + N(m\pi/a)^2} \tag{7.23}$$

If the transversely applied pressure load is uniformly distributed, we have $p(x, y) = p_0 = \text{const}$, and from (7.19) we obtain

$$p_{mn} = \frac{4}{ab} \frac{4abp_0}{\pi^2 mn} = \frac{16p_0}{\pi^2 mn}, \qquad m = 1, 3, 5, \ldots, \qquad n = 1, 3, 5, \ldots \tag{7.24}$$

The coefficients $a_{mn} = 0$ if $m$ or $n$ or both are even integers. Using (7.24) and (7.23), we obtain the deflected surface of an initially stressed plate under a uniformly distributed, transverse pressure load $p_0$:

$$w = \frac{16p_0}{\pi^2} \sum_{m=1}^{\infty} \sum_{n=1}^{\infty} \frac{\sin(m\pi x/a) \sin(n\pi y/b)}{mnD\left[(m\pi/a)^2 + (n\pi/b)^2\right]^2 + N(m\pi/a)^2},$$
$$m = 1, 3, 5, \ldots, \qquad n = 1, 3, 5, \ldots \tag{7.25}$$

If we wish to assess the elastic stability of a simply supported rectangular plate subjected to the distributed forces $N_{xx} = -P$ along $x = 0$ and $x = a$ as shown in Fig. 7.4, we proceed as follows: The initial membrane stress tensor induced by the boundary loads is given by

$$\begin{bmatrix} N_{xx} & N_{xy} \\ N_{yx} & N_{yy} \end{bmatrix} = \begin{bmatrix} -P & 0 \\ 0 & 0 \end{bmatrix} \tag{7.26}$$

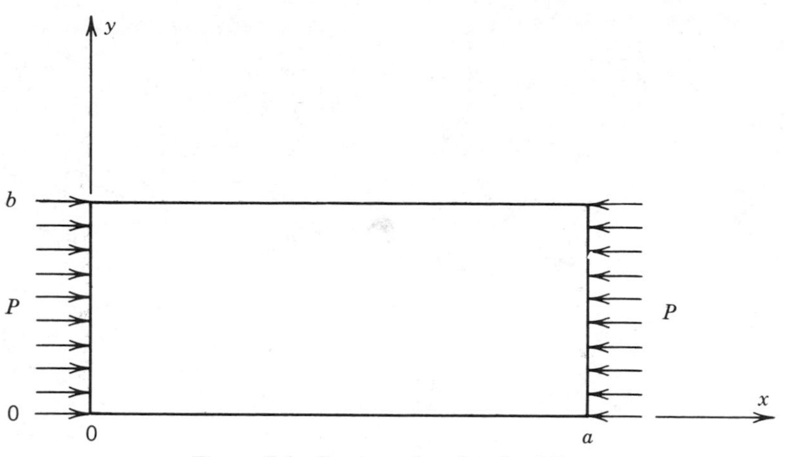


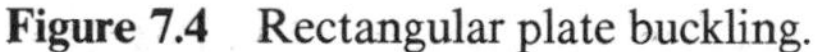

**Figure 7.4** Rectangular plate buckling.

and the boundary conditions are again characterized by (7.17). In the present case there are no transverse loads, so that $p \equiv 0$. Consequently, (7.15) reduces to

$$D \nabla^4 w + P \frac{\partial^2 w}{\partial x^2} = 0 \tag{7.27}$$

A solution of (7.27) that also satisfies the boundary conditions (7.17) is

$$w = A_{mn} \sin \frac{m\pi x}{a} \sin \frac{n\pi y}{b} \tag{7.28}$$

Upon substitution of (7.28) into (7.27), we readily obtain

$$P = \frac{\pi^2 D}{b^2} \left( \frac{mb}{a} + \frac{n^2 a}{mb} \right)^2 \tag{7.29}$$

We wish to find the lower bound of those values of $P$ for which the plate assumes an equilibrium configuration other than the trivial one characterized by $w \equiv 0$. With reference to (7.29), we note that $P$ will assume its smallest value when $n = 1$. In this case we can write

$$P_{\text{cr}} = k \frac{\pi^2 D}{b^2}, \quad \text{where } k = \left( \frac{mb}{a} + \frac{a}{mb} \right)^2 \tag{7.30}$$

A graph of $k$ versus $a/b$ is shown in Fig. 7.5 using $m$ as a parameter. The magnitude of the critical load and the number of half-waves for any given aspect ratio $a/b$ is determined by using the ordinate of the curve that yields

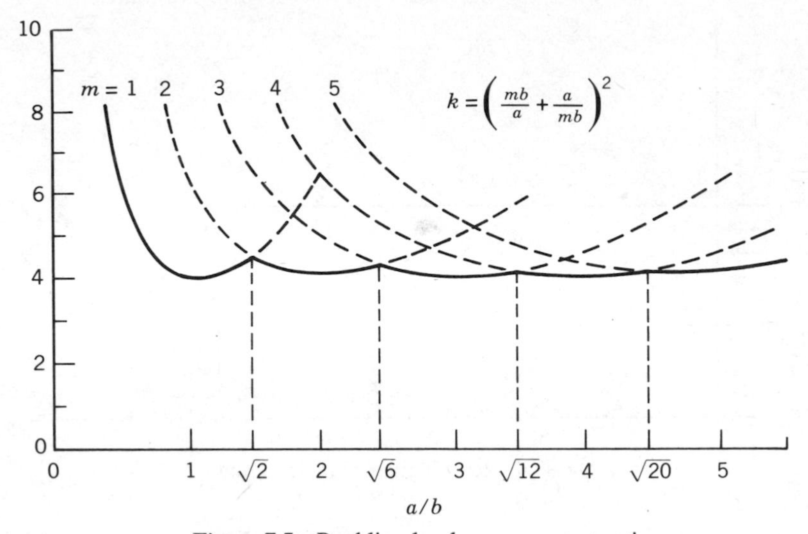


**Figure 7.5** Buckling load versus aspect ratio.

the smallest value of $k$. For example, if $a/b = 2.5$, we obtain $k = 4.133$ and $m = 3$ from Fig. 7.5. Thus, we conclude that the plate buckles into three half-waves in the $x$-direction when $P_{cr} = 4.133 \times (\pi^2 D/b^2)$. In view of the preceding analysis, it can be inferred that the plate can buckle into several half-waves in the direction of compression but only one half-wave in the direction that is at right angles to the direction of compression. By minimizing $k$ with respect to the parameter $mb/a$, it can be shown that a lower bound for $k$ is 4, and $k$ assumes this value for $a/b = 1, 2, 3, \ldots$ .

### 7.2.2 Circular Plates

We now consider a circular plate with radius $a$ and thickness $h$ clamped along its boundary, as shown in Fig. 7.6. The plate is subjected to a uniformly distributed radial membrane force $N_{rr} = N =$ constant along the circular boundary $r = a$. The plate is also subjected to a uniformly distributed transverse pressure of intensity $p_0$ acting over its surface. We wish to find the deflected surface of the deformed plate, neglecting shear deformation.

The plane stress field induced in the plate by the membrane force $N_{rr} = N$ at the boundary is readily shown to be (see Exercise 7.6)

$$\left.\begin{aligned} N_{rr} &= N_{\theta\theta} = N = \text{constant} \\ N_{r\theta} &= 0 \end{aligned}\right\} \quad \text{for } r < a \tag{7.31}$$

We now transform (7.15) to polar coordinates (see Exercise 7.7). For the

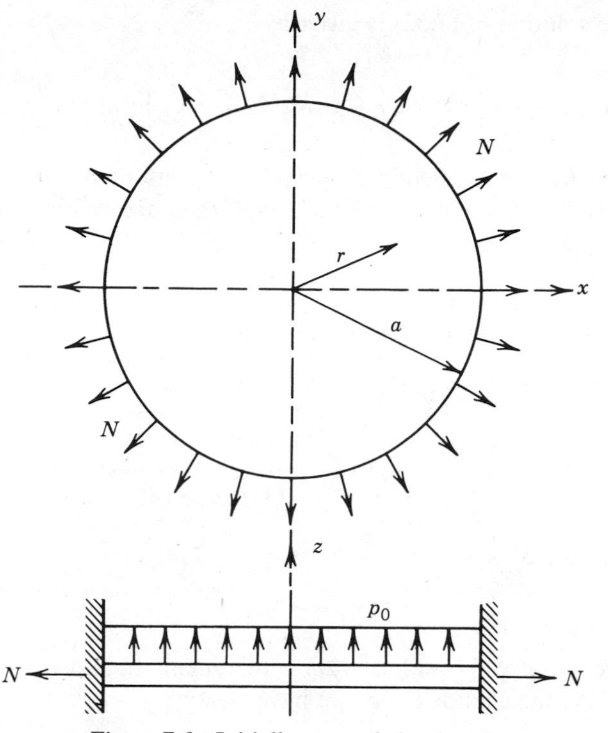


**Figure 7.6** Initially stressed circular plate.

present case of axial symmetry and in view of (7.31), we have

$$D \nabla^4 w = p_0 + N \nabla^2 w \tag{7.32}$$

where

$$\nabla^2 w = \frac{d^2 w}{dr^2} + \frac{1}{r}\frac{dw}{dr}$$

Equation (7.32) can be written in the form

$$\nabla^2(\nabla^2 - \lambda^2) w = \frac{p_0}{D}$$

where $\lambda^2 = N/D$, and we note that

$$\nabla^2 w - \lambda^2 w \equiv \frac{d^2 w}{dr^2} + \frac{1}{r}\frac{dw}{dr} - \lambda^2 w = 0$$

is a form of Bessel's equation of order zero. It is an elementary exercise to show that the solution of (7.32) is given by

$$w = C_1 I_0(\lambda r) + C_2 + C_3 K_0(\lambda r) + C_4 \ln\frac{r}{a} - \frac{p_0 r^2}{4N} \tag{7.33}$$

where $I_0$ and $K_0$ are the modified Bessel functions of the first and second kind, respectively, of order zero. Near $r = 0$, they are defined by the power series

$$I_0(x) = J_0(ix) = 1 + \left(\tfrac{1}{2}x\right)^2 + \frac{\left(\tfrac{1}{2}x\right)^4}{1^2 \cdot 2^2} + \cdots$$

$$I_0'(x) = I_1(x) = i^{-1}J_1(ix)$$

$$K_0(x) = -\left(\gamma + \ln\tfrac{1}{2}x\right)I_0(x) + \frac{\left(\tfrac{1}{2}x\right)^2}{(1!)^2} + \frac{\left(\tfrac{1}{2}x\right)^4}{(2!)^2}\left(1 + \tfrac{1}{2}\right) + \cdots$$

where $\gamma \approx 0.5772157$ and

$$K_0'(x) = -K_1(x)$$

If we require $w$, $dw/dr$, and $d^2w/dr^2$ to remain bounded at $r = 0$, then $C_3 = C_4 = 0$. At the clamped edge we have

$$w(a) = \left(\frac{dw}{dr}\right)_{r=a} = 0 \tag{7.34}$$

and upon substitution of (7.33) (with $C_3 = C_4 = 0$) into (7.34), the remaining constants are readily determined to be

$$C_1 = \frac{p_0 a^2}{2N(\lambda a)I_1(\lambda a)}$$

$$C_2 = \frac{p_0 a^2}{4N} - \frac{p_0 a^2 I_0(\lambda a)}{2N(\lambda a)I_1(\lambda a)}$$

These constants are now introduced into (7.33), yielding the complete, explicit solution to our problem:

$$\frac{2Dw(r)}{p_0 a^4} = \frac{1}{(\lambda a)^2}\left[\frac{1}{2}\left(1 - \frac{r^2}{a^2}\right) + \frac{I_0(\lambda r) - I_0(\lambda a)}{(\lambda a)I_1(\lambda a)}\right] \tag{7.35a}$$

This equation is useful for $N > 0$ but is awkward for the case $N < 0$. When $N < 0$, we have $\lambda^2 = -\beta^2 < 0$, where $\beta > 0$. Therefore,

$$\lambda = i\beta, \qquad a^2\lambda^2 = -a^2\beta^2, \qquad I_0(\lambda r) = I_0(i\beta r) = J_0(\beta r)$$

and

$$I_1(\lambda a) = I_1(i\beta a) = -i^{-1}J_1(\beta a)$$

Consequently, when $N < 0$, we can write (7.35a) in the form

$$\frac{2Dw(r)}{p_0 a^4} = \frac{1}{(\beta a)^2}\left[\frac{J_0(\beta r) - J_0(\beta a)}{(\beta a)J_1(\beta a)} - \frac{1}{2}\left(1 - \frac{r^2}{a^2}\right)\right] \qquad (7.35b)$$

where $\beta^2 = -\lambda^2 = -N/D > 0$. When $N = 0$, the solution of (7.32) is [see (5.68)]

$$\frac{2Dw(r)}{p_0 a^4} = \frac{1}{32}\left(1 - \frac{r^2}{a^2}\right)^2 \qquad (7.35c)$$

With reference to (7.35b), we note that the plate deflection $w(r)$ becomes unbounded if $J_1(\beta a) = 0$. The lowest positive root of this transcendental equation is $\beta_{cr} a \cong 3.832$ or $(\beta_{cr} a)^2 = -(\lambda_{cr} a)^2 = (3.832)^2$, so that

$$N_{cr} = \lambda_{cr}^2 D = -\frac{(3.832)^2 D}{a^2} \qquad (7.36)$$

is the largest value of $N$ for which (axisymmetric) plate buckling occurs. When $r = 0$, (7.35) reduce to

$$\frac{2Dw(0)}{p_0 a^4} = \frac{1}{(\alpha a)^2}\left[\frac{1}{2} + \frac{1 - I_0(\alpha a)}{(\alpha a) I_1(\alpha a)}\right], \qquad N > 0 \qquad (7.37a)$$

$$\frac{2Dw(0)}{p_0 a^4} = \frac{1}{(\alpha a)^2}\left[\frac{1 - J_0(\alpha a)}{(\alpha a) J_1(\alpha a)} - \frac{1}{2}\right], \qquad N < 0 \qquad (7.37b)$$

where $\alpha a = 3.832|N/N_{cr}|^{1/2}$, and

$$\frac{2Dw(0)}{p_0 a^4} = \frac{1}{32}, \qquad N = 0 \qquad (7.37c)$$

In view of (7.36), (7.35b) and (7.37b) are applicable only for $N_{cr} < N < 0$. With the aid of (7.37), we can construct a (dimensionless) graph of center deflection $w(0)$ versus membrane prestress $N$, as shown in Fig. 7.7. This graph clearly exhibits the strong nonlinear relation that exists between deflection and membrane prestress.

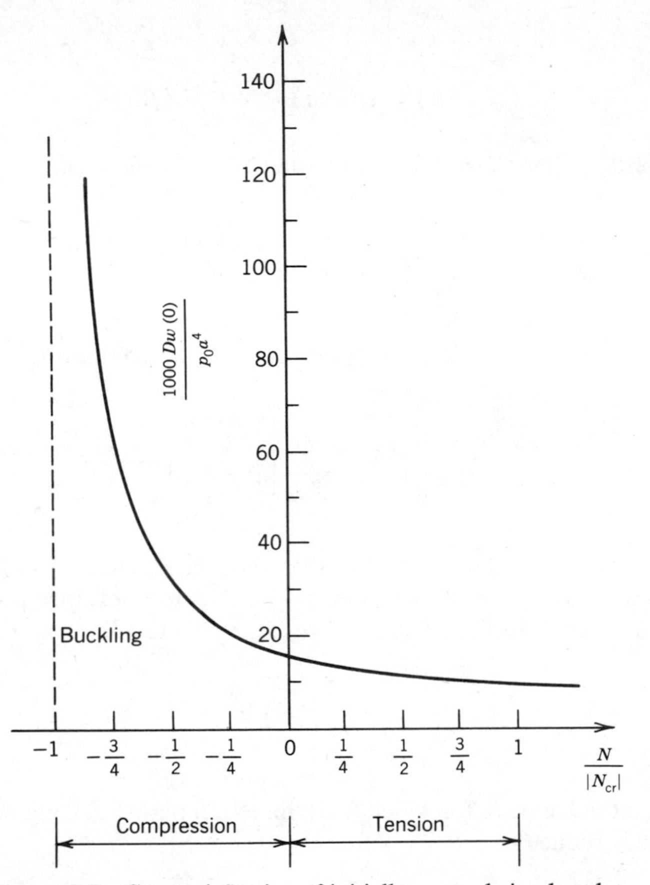


**Figure 7.7** Center deflection of initially stressed circular plate.

## 7.3 DYNAMICS

In this section we consider three different problems of vibration and wave motion in plates, which are in a state of initial stress. In each case the dynamic response as well as the stability of the motion are strongly affected by the magnitude of the prestress.

### 7.3.1 Rectangular Plates

We now consider free vibrations (in the sense of Section 6.3) of a simply supported rectangular plate with dimensions $a:b:h$. It is assumed that the plate is in a state of constant initial stress characterized by

$$\begin{bmatrix} N_{xx} & N_{xy} \\ N_{yx} & N_{yy} \end{bmatrix} = \begin{bmatrix} N & 0 \\ 0 & 0 \end{bmatrix} \quad \text{for} \begin{Bmatrix} 0 \le x \le a \\ 0 \le y \le b \end{Bmatrix} \tag{7.38}$$

If we wish to employ CPT, we use (7.15a), which, for the present case, reduces to

$$D\nabla^4 w - N\frac{\partial^2 w}{\partial x^2} = p \tag{7.39}$$

In the present, dynamical case, $p$ is the reversed, effective force per unit area

$$p = -\rho h \frac{\partial^2 w}{\partial t^2}$$

so that

$$D\nabla^4 w - N\frac{\partial^2 w}{\partial x^2} + \rho h \frac{\partial^2 w}{\partial t^2} = 0 \tag{7.40}$$

The boundary conditions corresponding to simple supports are given by (7.17). We now assume the solution

$$w = A_{mn} \sin\frac{m\pi x}{a} \sin\frac{n\pi y}{b} \begin{Bmatrix} \cos \Omega_{mn} t \\ \sin \Omega_{mn} t \end{Bmatrix} \tag{7.41}$$

Equation (7.41) satisfies the boundary conditions (7.17), and upon substitution of (7.41) into (7.40), we obtain

$$\frac{\rho h \Omega_{mn}^2}{D} = \left[\left(\frac{m\pi}{a}\right)^2 + \left(\frac{n\pi}{b}\right)^2\right]^2 + \frac{N}{D}\left(\frac{m\pi}{a}\right)^2 \tag{7.42}$$

or

$$\tilde{\Omega}_{mn}^2 = 1 + \tilde{N} \tag{7.43}$$

where

$$\tilde{\Omega}_{mn}^2 = \frac{\rho h \Omega_{mn}^2}{D\left[(m\pi/a)^2 + (n\pi/b)^2\right]^2} \tag{7.44}$$

is the square of the "dimensionless natural frequency" of the plate corresponding to the mode $(m, n)$, and

$$\tilde{N} = \frac{(N/D)(m\pi/a)^2}{\left[(m\pi/a)^2 + (n\pi/b)^2\right]^2} \tag{7.45}$$

is the "dimensionless preload" in the $x$-direction. A graph of (7.43) is shown in Fig. 7.8. We note that for any particular mode of vibration $(m, n)$ the square

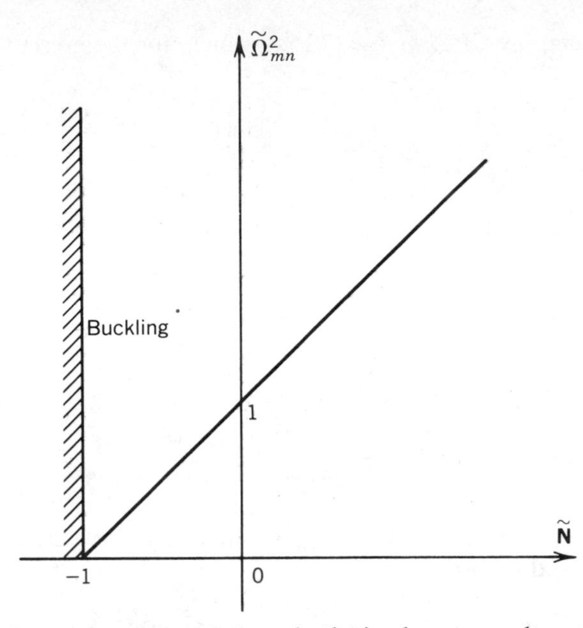


**Figure 7.8** Natural frequency versus preload (simply supported, rectangular plate).

of the frequency increases linearly with preload $\tilde{N}$. When $\tilde{N} = -1$ or, equivalently, when

$$\frac{N}{D} = \frac{N_{cr}}{D} = -\frac{\pi^2}{b^2}\left(\frac{mb}{a} + \frac{n^2 a}{mb}\right)^2 \tag{7.46}$$

the plate will buckle, and this result was obtained by static considerations in Section 7.2 (also see Fig. 7.5). The case $\tilde{N} = 0$ (no preload) reverts to the example treated in Section 6.3.

### 7.3.2 Circular Plates

In this section we consider free, axisymmetric vibrations of a clamped, circular plate, as shown in Fig. 7.6, with $p_0 \equiv 0$. The basic equation is given by (7.32), where

$$p_0 = -\rho h \frac{\partial^2 w}{\partial t^2} \tag{7.47}$$

is the inertia force per unit area. Thus, we have

$$D \nabla^4 w - N \nabla^2 w + \rho h \ddot{w} = 0 \tag{7.48}$$

where

$$\nabla^2 w = \frac{d^2 w}{dr^2} + \frac{1}{r}\frac{dw}{dr}$$

The boundary conditions are

$$w(a) = \left(\frac{dw}{dr}\right)_{r=a} = 0 \tag{7.49}$$

Assuming free harmonic vibrations, we have

$$w(x, t) = W_n(r)\begin{Bmatrix} \cos \Omega_n t \\ \sin \Omega_n t \end{Bmatrix} \tag{7.50}$$

and upon substitution of (7.50) into (7.48), we obtain

$$\nabla^4 W_n - \frac{N}{D}\nabla^2 W_n - \frac{\Omega_n^2 \rho h}{D} W_n = 0 \tag{7.51}$$

or

$$(\nabla^2 + \alpha_n^2)(\nabla^2 - \beta_n^2) W_n = 0 \tag{7.52}$$

where

$$\alpha_n^2 a^2 = \tfrac{1}{2}\sqrt{L^2 + \Gamma_n^2} - \tfrac{1}{2}L$$

$$\beta_n^2 a^2 = \tfrac{1}{2}\sqrt{L^2 + \Gamma_n^2} + \tfrac{1}{2}L$$

$$L = \frac{Na^2}{D}, \qquad \Gamma_n^2 = \frac{4\Omega_n^2 \rho h a^4}{D}$$

The expressions in parentheses in (7.52) are permutable, linear operators. By the theory of linear, differential equations, the solution of (7.52) is the sum of the solutions of

$$(\nabla^2 + \alpha_n^2) W_n = 0 \tag{7.53a}$$

and

$$(\nabla^2 - \beta_n^2) W_n = 0 \tag{7.53b}$$

Equations (7.53) are of the Bessel type. Consequently, the solution of (7.52) or (7.51) can be written as

$$W_n = A_n J_0(\alpha_n r) + B_n Y_0(\alpha_n r) + C_n I_0(\beta_n r) + D_n K_0(\beta_n r) \tag{7.54}$$

where $J_0$ and $Y_0$ are Bessel functions and $I_0$ and $K_0$ are modified Bessel functions, all four being of order zero. Deflection, slope, and so on, remain bounded at $r = 0$. Hence, $B_n = D_n = 0$. The remaining two constants are obtained with the aid of (7.49). Upon application of (7.49), we obtain

$$0 = A_n J_0(\alpha_n a) + C_n I_0(\beta_n a)$$
$$0 = A_n \alpha_n J_0'(\alpha_n a) + C_n \beta_n I_0'(\beta_n a)$$

or

$$-\frac{C_n}{A_n} = \frac{J_0(\alpha_n a)}{I_0(\beta_n a)} = \frac{\alpha_n J_0'(\alpha_n a)}{\beta_n I_0'(\beta_n a)}$$

But we have $J_0'(x) = -J_1(x)$ and $I_0'(x) = I_1(x)$, and therefore,

$$(\beta_n a) I_1(\beta_n a) J_0(\alpha_n a) + (\alpha_n a) J_1(\alpha_n a) I_0(\beta_n a) = 0 \qquad (7.55)$$

For a particular value of $N/D = \beta_n^2 - \alpha_n^2 > N_{\text{cr}}/D$, (7.55) has a denumerable infinity of real roots, which are readily related to the (dimensionless) frequency $\Gamma_n$, $n = 1, 2, 3, \ldots$. Figure 7.9 shows the first three natural frequencies as a function of radial preload $L$. When $\Omega_1 = \Gamma_1 = 0$, the plate will buckle. In this

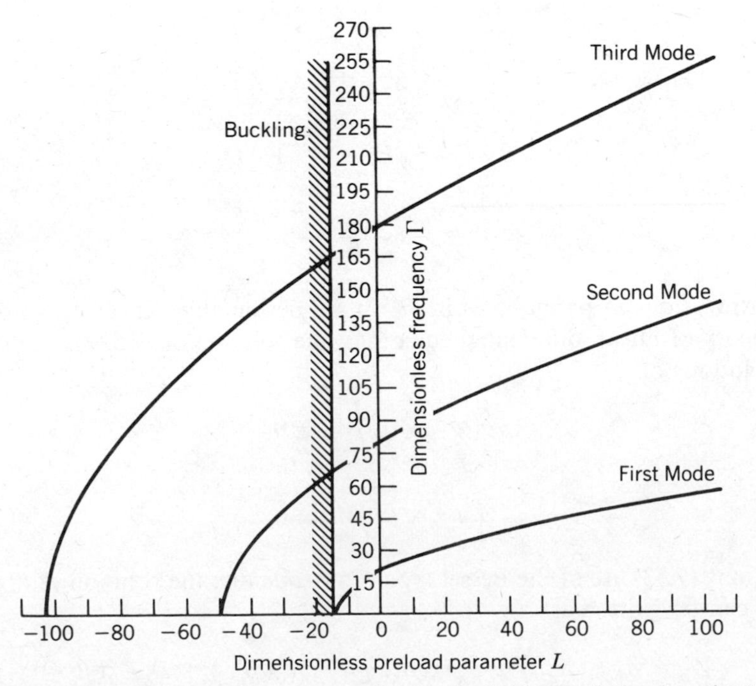


**Figure 7.9** Natural frequency versus radial preload (clamped circular plate); axial symmetry.

case $\alpha^2 a^2 = 0$, and $\beta^2 a^2 = L = -14.682 = N_{cr} a^2/D$. The special case $Na^2/D = L = 0$ results in the spectrum

$$\Gamma_1 = 20.429 \qquad (\alpha_1 a) = 3.196$$
$$\Gamma_2 = 79.531 \qquad (\alpha_2 a) = 6.306$$
$$\Gamma_3 = 178.189 \qquad (\alpha_3 a) = 9.439$$

and so on. Thus, the present solution reverts to the case treated in Section 6.4 [see (6.28)]. We note that the graphs in Fig. 7.9 are physically meaningful only if $L_{cr} = -14.682 < L$.

### 7.3.3 The Initially Stressed Plate Strip under Moving Pressure Load

In this section we shall determine the response of a thin, simply supported, infinite plate strip under the action of a transverse, moving pressure load. The distributed pressure load is assumed to propagate, with constant speed, in the direction that is parallel to the edges of the plate strip. In addition to the moving load, the plate strip is subjected to uniform, constant prestress parallel to its edges, acting in the plane of the plate strip (see Fig. 7.10).

Consider the thin, elastic plate strip $-\infty < x < \infty$, $0 \leq y \leq l$, under the action of the distributed load $q$ and under constant (static) prestress $N$ in the $x$-direction. The equation of motion for small deflections is

$$D\nabla^4 w - N\frac{\partial^2 w}{\partial x^2} + m\frac{\partial^2 w}{\partial t^2} = q \tag{7.56}$$

where $m = \rho h$ is the mass intensity per unit surface area of the plate. The

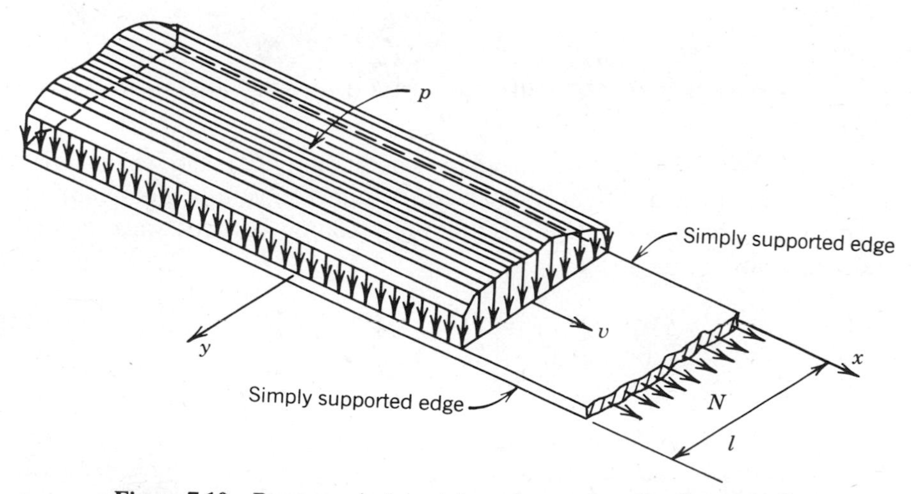


**Figure 7.10** Prestressed plate strip under moving, distributed load.

characterization (7.56) is subject to the limitations imposed by CPT [see (7.40)]. In addition, the effects of rotatory inertia and damping are neglected. The moments and shears are related to the deflections $w$ by the following equations:

$$M_{xx} = -D\left(\frac{\partial^2 w}{\partial x^2} + \nu\frac{\partial^2 w}{\partial y^2}\right) \tag{7.57}$$

$$M_{yy} = -D\left(\frac{\partial^2 w}{\partial y^2} + \nu\frac{\partial^2 w}{\partial x^2}\right) \tag{7.58}$$

$$M_{xy} = M_{yx} = -D(1-\nu)\frac{\partial^2 w}{\partial x\,\partial y} \tag{7.59}$$

$$Q_x = -D\frac{\partial}{\partial x}(\nabla^2 w) \tag{7.60}$$

$$Q_y = -D\frac{\partial}{\partial y}(\nabla^2 w) \tag{7.61}$$

Since the plate is simply supported at $y = 0$ and $y = l$, we require

$$\begin{aligned} w(x,0) &= (\nabla^2 w)_{(x,0)} = 0 \\ w(x,l) &= (\nabla^2 w)_{(x,l)} = 0 \end{aligned} \tag{7.62}$$

In the subsequent development, we shall require certain basic physical properties of the plate strip with respect to wave propagation in the $x$-direction. We consider the homogeneous (7.56) and assume the existence of waves of the type

$$w = w_n = \sin\frac{n\pi x}{l}\exp[i(k_n x - \Gamma_n t)], \qquad n = 1,2,\ldots \tag{7.63}$$

Equation (7.63) satisfies the boundary condition (7.62) and characterizes flexural waves traveling in the $x$-direction. By direct substitution of (7.63) into (7.56) (with $q = 0$), we obtain the following relation between wave number and frequency:

$$\Gamma_n(k_n) = \sqrt{\frac{D}{m}\left[k_n^2 + \left(\frac{n\pi}{l}\right)^2\right]^2 + \frac{N}{m}k_n^2} \tag{7.64}$$

In the case of an infinite plate strip, $k_n$ is restricted to real values in order that the deflection $w$ remain bounded in $-\infty < x < \infty$, $0 \le y \le l$. This will result in real values for $\Gamma_n$. Thus, (7.63) describes a traveling wave with wave number

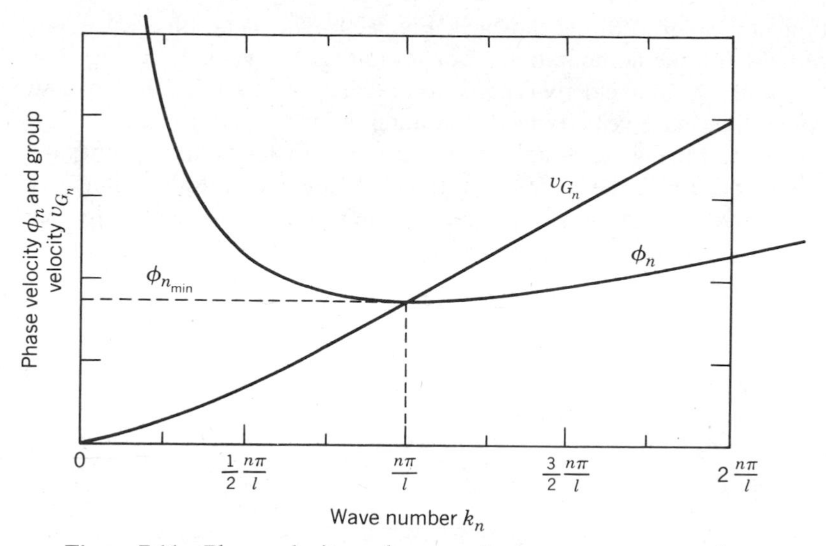


**Figure 7.11** Phase velocity and group velocity versus wave number.

$k_n$, frequency $\Gamma_n(k_n)$, and phase velocity

$$\phi_n(k_n) = \frac{\Gamma_n(k_n)}{k_n} = \sqrt{\frac{D}{mk_n^2}\left[k_n^2 + \left(\frac{n\pi}{l}\right)^2\right]^2 + \frac{N}{m}} \tag{7.65}$$

A typical plot of phase velocity versus wave number is shown in Fig. 7.11. We note that the phase velocity has a minimum

$$\phi_{n_{\min}} = \sqrt{\frac{N - N_{c_n}}{m}} \quad \text{when } k_n = \frac{n\pi}{l} \tag{7.66}$$

where $N_{c_n} = -(4n^2\pi^2 D)/l^2$ is that load $N$ acting in the plane of the plate in the $x$-direction that causes elastic buckling of the plate under static conditions with $n$ half-waves in the $y$-direction. When $k_n \to 0$ and $k_n \to \infty$, $\phi_n \to \infty$. For all $\phi_n > \phi_{n_{\min}}$ there exist two different wave numbers $k_n$ that correspond to the same phase velocity $\phi_n$. In subsequent developments, it will be necessary to determine the steady-state response for supercritical speeds of the load. This will be facilitated by consideration of the velocity of energy transport or its equivalent, group velocity. The group velocity for waves in the plate is obtained from (7.64) by [see (6.58)]

$$v_{G_n} = \frac{d\Gamma_n}{dk_n} = \frac{k_n\left\{2(D/m)\left[k_n^2 + (n\pi/l)^2\right] + N/m\right\}}{\sqrt{(D/m)\left[k_n^2 + (n\pi/l)^2\right]^2 + (N/m)k_n^2}} \tag{7.67}$$

A typical plot of (7.67) is shown in Fig. 7.11. The same physical parameters were used for the computation of phase and group velocity in Fig. 7.11. We note that the group velocity is equal to the phase velocity for the wave number at which the phase velocity has a minimum at $k_n = (n\pi)/l$.

In the following, we shall be concerned with loads that propagate with constant speed $v$ in the $x$-direction. It will be convenient to nondimensionalize the coordinates $x$ and $y$ and to describe the response of the plate in a moving coordinate system. This is accomplished by the change of variables

$$\xi = \frac{\pi(x - vt)}{l}, \qquad \eta = \frac{\pi y}{l} \tag{7.68}$$

which transforms (7.56) into

$$\frac{\partial^4 w}{\partial \xi^4} + 2\frac{\partial^4 w}{\partial \xi^2\, \partial \eta^2} + \frac{\partial^4 w}{\partial \eta^4} + \left(\frac{ml^2v^2}{\pi^2 D} - \frac{l^2 N}{\pi^2 D}\right)\frac{\partial^2 w}{\partial \xi^2} = \frac{l^4 q}{\pi^4 D} \tag{7.69}$$

The change of variables (7.68) may be given the following physical interpretation: An observer fixed with respect to the $x$-$y$ coordinate system will see the distributed load $q$ advance in the direction of the positive $x$-axis; to him the deflection of the plate will appear to depend on $x$, $y$, and $t$. However, an observer fixed with respect to the $\xi$-$\eta$ coordinate system will move with the advancing load distribution; to him the deflection surface will appear stationary, that is, independent of $t$, and a function of $\xi$ and $\eta$ alone. We note that by neglecting damped transients due to the starting of the motion, we have made the implicit assumption that the load has been moving for a sufficiently long period. Thus, we shall concentrate on the steady-state dynamical process as characterized by (7.69).

The moving distributed load is assumed to be a step function in the $\xi$-direction, but its $\eta$ dependence is arbitrary in the sense that it may be characterized by a Fourier sine series (see Fig. 7.10):

$$q(\xi, \eta) = \begin{cases} 0, & \xi > 0 \\ \sum_{n=1}^{\infty} pc_n \sin n\eta, & \xi < 0 \end{cases} \tag{7.70}$$

In the analysis that follows, we shall work with the $n$th harmonic component of the load and its corresponding deflection $w_n$, $n = 1, 2, \ldots$. Any particular case to be considered subsequently will be obtained by superimposing the appropriate number of component solutions.

To solve (7.69) with $q$ defined by (7.70), we assume a product solution that satisfies the boundary conditions (7.62):

$$\left.\begin{aligned} w_n^{(1)} &= \frac{pc_n l^4}{\pi^4 D} f_n^{(1)}(\xi) \sin n\eta, \qquad \xi > 0 \\ w_n^{(2)} &= \frac{pc_n l^4}{\pi^4 D} f_n^{(2)}(\xi) \sin n\eta, \qquad \xi < 0 \end{aligned}\right\} \tag{7.71}$$

where the superscripts (1) and (2) are attached to quantities pertaining to the region ahead and behind, respectively, of the moving load front. Upon substitution of (7.71) into (7.69), we obtain, in conjunction with (7.70),

$$\left.\begin{aligned} \frac{d^4 f_n^{(1)}}{d\xi^4} + 2\lambda_n n^2 \frac{d^2 f_n^{(1)}}{d\xi^2} + n^4 f_n^{(1)} &= 0, \quad \xi > 0 \\ \frac{d^4 f_n^{(2)}}{d\xi^4} + 2\lambda_n n^2 \frac{d^2 f_n^{(2)}}{d\xi^2} + n^4 f_n^{(2)} &= 1, \quad \xi < 0 \end{aligned}\right\} \tag{7.72}$$

where

$$\lambda_n = 2\Omega_n^2 - 2\beta_n - 1, \qquad \Omega_n^2 = -\frac{mv^2}{N_{c_n}}, \qquad \beta_n = -\frac{N}{N_{c_n}} \tag{7.73}$$

Assuming $f_n^{(i)} = e^{n\alpha_n\xi}$, $i = 1, 2$, $\alpha_n =$ constant, and substituting into the homogeneous (7.72), we obtain the characteristic equation

$$\alpha_n^4 + 2\lambda_n \alpha_n^2 + 1 = 0 \tag{7.74}$$

solving

$$\sqrt{2}\,\alpha_n = \pm\left(\sqrt{-\lambda_n - 1} \pm \sqrt{1 - \lambda_n}\right) \tag{7.75}$$

The roots of (7.75) are real for $\lambda_n < -1$, complex for $-1 < \lambda_n < 1$, and pure imaginary for $1 < \lambda_n$. The character of the roots as a function of preload and speed of propagation of the load is shown in Fig. 7.12.

Solutions of (7.72) must be bounded for $\xi \to \pm\infty$, and at $\xi = 0$, we require that the deflection, slope, moment, and shear be continuous. These continuity conditions may be expressed in terms of the functions $f_n^{(i)}(\xi)$, $i = 1, 2$, by the appropriate combination of (7.57), (7.60), (7.68), and (7.71):

$$\begin{aligned} f_n^{(1)}(0, \eta) &= f_n^{(2)}(0, \eta) \\ \left(\frac{df_n^{(1)}}{d\xi}\right)_{(0,\eta)} &= \left(\frac{df^{(2)}}{d\xi}\right)_{(0,\eta)} \\ \left(\frac{d^2 f_n^{(1)}}{d\xi^2}\right)_{(0,\eta)} &= \left(\frac{d^2 f_n^{(2)}}{d\xi^2}\right)_{(0,\eta)} \\ \left(\frac{d^3 f_n^{(1)}}{d\xi^3}\right)_{(0,\eta)} &= \left(\frac{d^3 f_n^{(2)}}{d\xi^3}\right)_{(0,\eta)} \end{aligned} \tag{7.76}$$

Solutions of (7.72) subject to the conditions prescribed by (7.76) and bounded at $\xi \to \pm\infty$ are found in the usual manner when the roots of the characteristic equation (7.74) are either real or complex. When these solutions

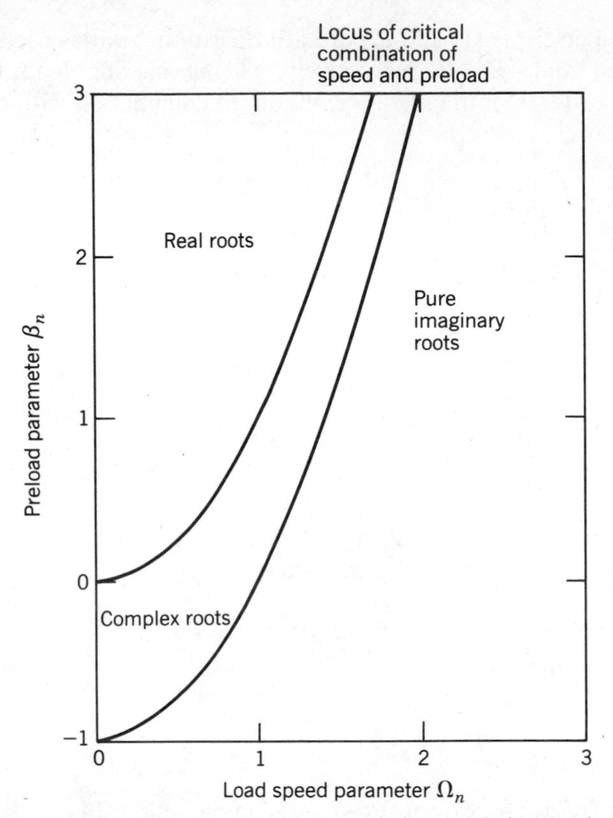


**Figure 7.12** Classification of roots of characteristic equation.

are combined with (7.71), they may be written as:

1. *Real Roots:* $\lambda_n \leq -1$

$$w_n^{(1)} = \frac{pc_n l^4 \sin n\eta}{n^4 \pi^4 D} \frac{\left[a_{n_1}^2 \exp(-nb_{n_1}\xi) - b_{n_1}^2 \exp(-na_{n_1}\xi)\right]}{2\left(a_{n_1}^2 - b_{n_1}^2\right)}, \qquad \xi \geq 0$$

$$w_n^{(2)} = \frac{pc_n l^4 \sin n\eta}{n^4 \pi^4 D} \left\{ \frac{\left[b_{n_1}^2 \exp(na_{n_1}\xi) - a_{n_1}^2 \exp(nb_{n_1}\xi)\right]}{2\left(a_{n_1}^2 - b_{n_1}^2\right)} + 1 \right\}, \qquad \xi \leq 0$$

(7.77)

where

$$\sqrt{2}\, a_{n_1} = \sqrt{1 - \lambda_n} + \sqrt{-1 - \lambda_n}$$
$$\sqrt{2}\, b_{n_1} = \sqrt{1 - \lambda_n} - \sqrt{-1 - \lambda_n}$$

2. Complex Roots: $-1 < \lambda_n < 1$,

$$w_n^{(1)} = \frac{pc_n l^4 \sin n\eta \exp(-na_{n_2}\xi)}{n^4\pi^4 D} \frac{1}{2}\left(\cos nb_{n_2}\xi + \frac{a_{n_2}^2 - b_{n_2}^2}{2a_{n_2}b_{n_2}} \sin nb_{n_2}\xi\right) \qquad \xi \geq 0,$$

$$w_n^{(2)} = \frac{pc_n l^4 \sin n\eta}{n^4\pi^4 D}\left[\left(-\cos nb_{n_2}\xi + \frac{a_{n_2}^2 - b_{n_2}^2}{2a_{n_2}b_{n_2}} \sin nb_{n_2}\xi\right)\frac{\exp(na_{n_2}\xi)}{2} + 1\right] \qquad \xi \leq 0, \tag{7.78}$$

where

$$\sqrt{2}\, a_{n_2} = \sqrt{1 - \lambda_n}, \qquad \sqrt{2}\, b_{n_2} = \sqrt{1 + \lambda_n}$$

In the case of pure imaginary roots ($1 < \lambda_n$), the solutions of Eq. (7.72) are

$$f_n^{(1)}(\xi) = C_1^{(1)} \cos na_{n_3}\xi + C_2^{(1)} \sin na_{n_3}\xi + C_3^{(1)} \cos nb_{n_3}\xi + C_4^{(1)} \sin nb_{n_3}\xi, \qquad \xi > 0$$

$$f_n^{(2)}(\xi) = C_1^{(2)} \cos nb_{n_3}\xi + C_2^{(2)} \sin nb_{n_3}\xi + C_3^{(2)} \cos na_{n_3}\xi + C_4^{(2)} \sin na_{n_3}\xi + \frac{1}{n^4}, \qquad \xi < 0, \tag{7.79}$$

where

$$\sqrt{2}\, a_{n_3} = \sqrt{1 + \lambda_n} + \sqrt{-1 + \lambda_n}$$

$$\sqrt{2}\, b_{n_3} = \sqrt{1 + \lambda_n} - \sqrt{-1 + \lambda_n}$$

and there are only four (7.76) to determine the eight constants of integration $C_i^{(j)}$, $i = 1, 2, 3, 4$, $j = 1, 2$. At this point we may use the concept of group velocity to determine the appropriate steady-state motion. The group velocity is the velocity of energy transport [see (6.58) and (6.60)], and the physically appropriate solution requires a flow of energy away from the load front. With reference to Fig. 7.11, we note that corresponding to a given phase velocity greater than $\phi_{n_{\min}}$, there are two wave numbers $k_{n_1}$ and $k_{n_2}$, one of them $k_{n_1} < n\pi/l$ and the other $n\pi/l < k_{n_2}$. For $k_{n_2}$, the group velocity is always greater than the phase velocity; therefore, this represents waves moving ahead of the load front at $\xi = 0$. For the smaller wave number $k_{n_1}$, the group velocity is always smaller than the phase velocity; therefore, $k_{n_1}$ corresponds to waves behind the load front. Since $a_{n_3} > b_{n_3}$, this argument enables us to set

$$C_3^{(1)} = C_4^{(1)} = C_3^{(2)} = C_4^{(2)} = 0$$

in (7.79). The remaining constants in (7.79) are now determined by applying continuity conditions expressed by (7.76), and the results are as follows:

3. *Pure Imaginary Roots:* $1 < \lambda_n$

$$w_n^{(1)} = -\frac{pc_n l^4 \sin n\eta}{n^4 \pi^4 D} \frac{b_{n_3}^2 \cos na_{n_3}\xi}{\left(a_{n_3}^2 - b_{n_3}^2\right)}, \qquad \xi \geq 0,$$
$$w_n^{(2)} = \frac{pc_n l^4 \sin n\eta}{n^4 \pi^4 D}\left(-\frac{a_{n_3}^2 \cos nb_{n_3}\xi}{\left(a_{n_3}^2 - b_{n_3}^2\right)} + 1\right), \qquad \xi \leq 0, \tag{7.80}$$

where

$$\sqrt{2}\, a_{n_3} = \sqrt{1 + \lambda_n} + \sqrt{-1 + \lambda_n}$$
$$\sqrt{2}\, b_{n_3} = \sqrt{1 + \lambda_n} - \sqrt{-1 + \lambda_n}$$

In the following discussion, the term *deflection profile* refers to the trace of the intersection of any plane $\eta$ = constant, $0 < \eta < \pi$, with the median plane of the plate as viewed in the plane $\eta$ = constant. A study of component solutions reveals that the character of the deflection profile is conveniently defined by a single parameter $\lambda_n = 2\Omega_n^2 - 2\beta_n - 1$, which involves all the physical parameters that enter the problem.

When $\lambda_n \leq -1$, we have an exponential, nonoscillatory deflection profile, as shown in Fig. 7.13($a$); and when $-1 < \lambda_n < 1$, the traveling wave is a damped sinusoid, as shown in Fig. 7.13($b$). For $-\infty < \lambda_n < 1$, $w_n \to 0$ as $\xi \to +\infty$, $w_n \to pc_n l^4 \sin n\eta/(n^4\pi^4 D)$ as $\xi \to -\infty$, the deflection profile is antisymmetric with respect to the point

$$(\xi, w_n) = \left(0, \frac{pc_n l^4 \sin n\eta}{2n^4\pi^4 D}\right) \quad \text{and} \quad w_n = \frac{pc_n l^4 \sin n\eta}{2n^4\pi^4 D}$$

at the load front $\xi = 0$. When $1 < \lambda_n$, we obtain a sinusoidal deflection profile as shown in Fig. 7.13($c$). In this case, the wavelength of waves ahead of the load front is smaller than that of waves behind the load front. The amplitude of waves ahead of the load front is smaller than that of waves behind the load front. Waves ahead of the load front oscillate about $w_n = 0$, while waves behind the load front oscillate about $w_n = pc_n l^4 \sin n\eta/(n^4\pi^4 D)$.

As $\lambda_n \to 1$, we obtain unbounded deflections (critical case). In this case, $\beta_n = \Omega_n^2 - 1$ (a plot of this equation is shown in Fig. 7.12). When the dimensionless load speed parameter $\Omega_n$ and the dimensionless preload parameter $\beta_n$ corresponding to the $n$th harmonic of the load lie on this curve, we obtain unbounded deflections. The case $\Omega_n = 0$, $\beta_n = -1$ corresponds to

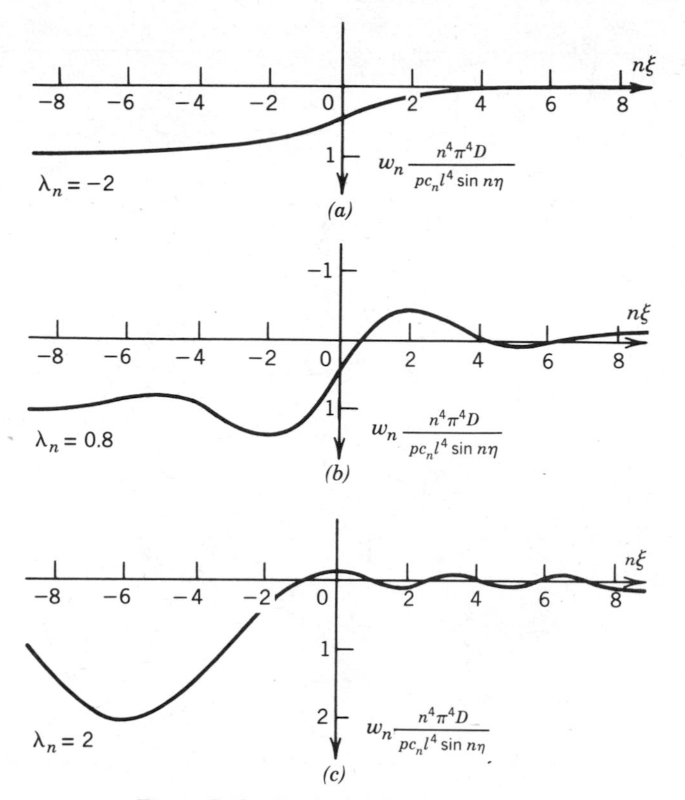


**Figure 7.13** Typical deflection profiles.

static, elastic buckling with $n$ half-waves in the $y$-direction. Thus, in the present analysis we have the restriction $\beta_n > -1$, and we require $N > -4\pi^2 D/l^2$.

Figure 7.12 indicates the pronounced effect of axial prestress upon response. The critical speed corresponding to a particular harmonic $n$ of the load distribution is reduced when the prestress is compressive ($-1 < \beta_n < 0$), and it is increased when the prestress is tensile ($0 < \beta_n$). In a similar manner, the preload of the plate strongly influences deflections and stresses that occur at load speeds other than critical (see Fig. 7.15).

The steady-state wavelength ratio of the deflection profile as a function of $\lambda_n$ is shown in Fig. 7.14. Wavelength ratio is defined as the ratio of actual wavelength to the (hypothetical) wavelength at $\lambda_n = 1$. We note that the wavelength ratio becomes unbounded for $\lambda_n \to -1$ and, as exhibited by case ($a$) in Fig. 7.13, remains unbounded for $\lambda_n < -1$.

A plot of dynamic amplification of deflection as a function of $\lambda_n$ is shown in Fig. 7.15. Dynamic amplification here is defined as the maximum negative deflection for $\xi > 0$. With this definition, the dynamic amplification vanishes for $-\infty < \lambda_n \leq -1$. For $-1 < \lambda_n < 1$ and $\xi < 0$, the maximum

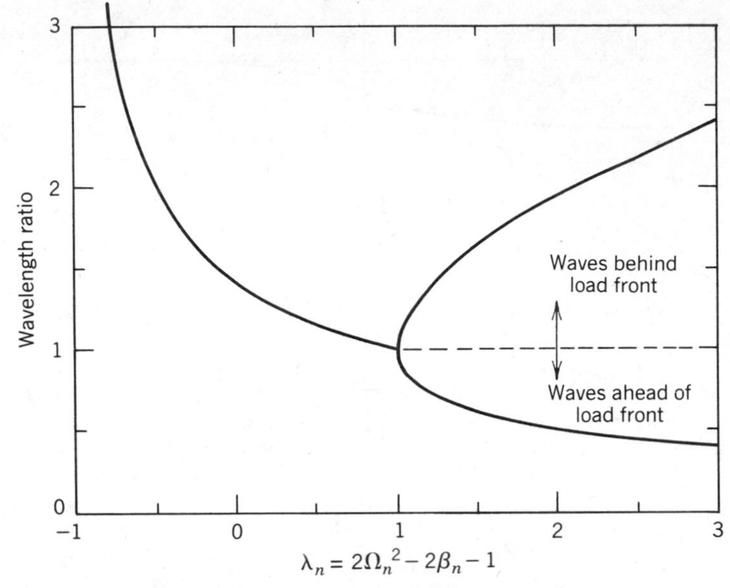


**Figure 7.14** Wavelength ration versus $\lambda_n$.

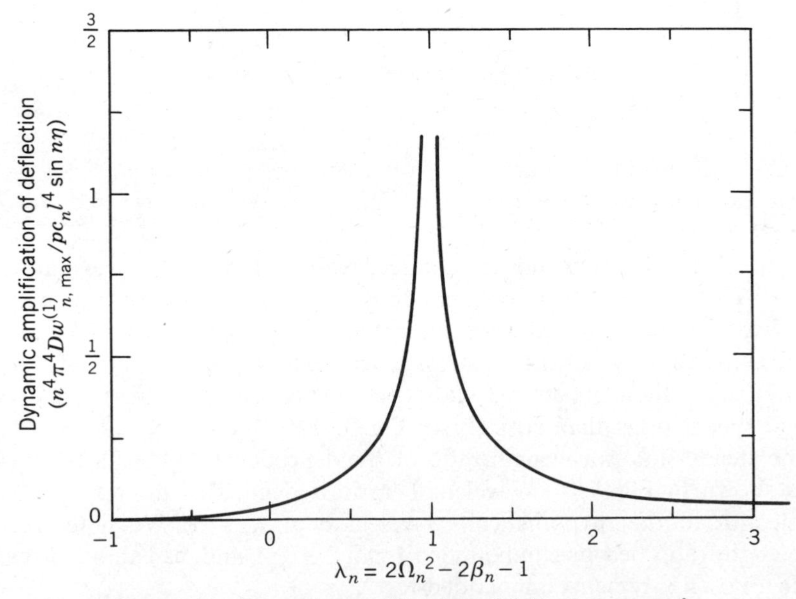


**Figure 7.15** Dynamic amplification of deflection versus $\lambda_n$.

deflection $w_n^{(2)} n^4 \pi^4 D/(p c_n l^4 \sin n\eta)$ is obtained by adding 1 to the value obtained from Fig. 7.15. For $1 < \lambda_n$ and $\xi < 0$, the maximum deflection $w_n^{(2)} n^4 \pi^4 D/(p c_n l^4 \sin n\eta)$ is obtained by adding 2 to the value obtained from Fig. 7.15. We also note that dynamic amplification approaches zero as $\lambda_n \to +\infty$.

In most practical cases, the load distribution in the $y$-direction is characterized either by a finite trigonometric series or a Fourier series, and to obtain a solution, a finite or an infinite number of component solutions must be superimposed. Since the basic partial differential equation of motion is linear and because each component solution satisfies the equation and the required boundary conditions, superposition is permissible and presents no special problems.

When considering superposition, we must distinguish between two distinct cases:

1. $mv^2 \leq N$ (or $\lambda_n \leq -1$ for all $n$), and for example, when $\xi \geq 0$, we have, from (7.77),

$$w^{(1)}(\xi, \eta) = \frac{pl^4}{2\pi^4 D} \sum_{n=1}^{\infty} \frac{c_n \left[a_{n_1}^2 \exp(-nb_{n_1}\xi) - b_{n_1} \exp(-na_{n_1}\xi)\right] \sin n\eta}{n^4 \left(a_{n_1}^2 - b_{n_1}^2\right)} \tag{7.81}$$

2. $mv^2 > N$ (or $-1 < \lambda_n < 1$, $1 < \lambda_n < \infty$ for all $n$), and for example, when $\xi \geq 0$, we have, from (7.78) and (7.80),

$$w^{(1)}(\xi, \eta) = -\frac{pl^4}{\pi^4 D} \sum_{n=1}^{\gamma} \frac{c_n b_{n_3}^2 \cos na_{n_3}\xi \sin n\eta}{n^4 \left(a_{n_3}^2 - b_{n_3}^2\right)}$$
$$+ \frac{pl^4}{2\pi^4 D} \sum_{\gamma+1}^{\infty} \frac{c_n}{n^4} \left( \cos nb_{n_2}\xi + \frac{a_{n_2}^2 - b_{n_2}^2}{2a_{n_2} b_{n_2}} \sin nb_{n_2}\xi \right)$$
$$\times \sin n\eta \exp(-na_{n_2}\xi) \tag{7.82}$$

where $\gamma$ is the greatest positive integer smaller than

$$(l/2\pi)\sqrt{(mv^2 - N)/D}.$$

We note that it is possible that $0 < (l/2\pi)\sqrt{(mv^2 - N)/D} < 1$. In this case, the finite series in (7.82) is deleted, and only the infinite series applies.

Throughout this analysis, we have neglected the effect of damping. This omission achieves a simplification in the analysis but results in an apparent indeterminacy in obtaining the steady-state solution for $1 < \lambda_n$. In the present investigation, we have used the concepts of phase and group velocity in conjunction with energy considerations to obtain the appropriate solution for this case. An identical result is obtained by solving the present problem under the added condition of viscous damping and then taking the limit of the solution as damping approaches zero. The equivalence of these two different approaches can be established. We also note (although this has not been shown) that in the presence of damping, deflections will remain bounded for $-\infty < \lambda_n < \infty$, and in all cases $w_n^{(1)} \to 0$ for $\xi \to +\infty$, and $w_n^{(2)} \to c_n p l^4 \sin n\eta/(n^4\pi^4 D)$ for $\xi \to -\infty$.

## EXERCISES

**7.1** A rectangular plate is simply supported at $x = 0, a$ so that $w(0, y) = w(a, y) = M_{xx}(0, y) = M_{xx}(a, y) = 0$. The plate is subjected to initial membrane forces $N_{xx}$ = constant parallel to the $x$-axis. A transverse, uniformly distributed load of intensity $p_0$ is applied. Find the deflection of the plate if:

**(a)** Edges $y = 0$ and $y = b$ are simply supported.

**(b)** Edges $y = 0$ and $y = b$ are free.

**(c)** Edges $y = 0$ and $y = b$ are clamped.

**(d)** Edge $y = 0$ is free and edge $y = b$ is clamped.

*Hint:* Assume $w(x, y) = \Sigma_{m=1}^{\infty} f_m(y) \sin(m\pi x/a)$, where $m = 1, 2, 3, \ldots$. Neglect shear deformations.

**7.2** A thin rectangular plate is simply supported at $x = 0, a$ so that $w(0, y) = w(a, y) = M_{xx}(0, y) = M_{xx}(a, y) = 0$. The plate is subjected to a compressive membrane force $N_{xx} \equiv -P$ = constant parallel to the $x$-axis. Find the buckling value $N_{\text{cr}}$ if:

**(a)** Edges $y = 0, b$ are free.

**(b)** Edges $y = 0, b$ are clamped.

**(c)** Edge $y = 0$ is free and edge $y = b$ is clamped.

Refer to the hint in Exercise 7.1.

**7.3** Normalize the eigenfunction characterizing free vibrations of a simply supported, initially stressed rectangular plate. The initial plane stress field is given by (7.16). Let

$$\int_0^b \int_0^a \rho h \left( A_{mn} \sin \frac{m\pi x}{a} \sin \frac{n\pi y}{b} \right)^2 dx\, dy = 1$$

and find the value of $A_{mn}$.

**7.4** A simply supported rectangular plate is at rest in its equilibrium position at $t = 0$. The plate is in a state of membrane prestress characterized by (7.16). At $t = 0^+$, a uniformly distributed, transverse load is suddenly applied and maintained for $t > 0$. Find the dynamic response of the plate using the method of Section 6.6 and the results of Exercise 7.3. Plot $w(\frac{1}{2}a, \frac{1}{2}b, t)$ and $M_{xx}(\frac{1}{2}a, \frac{1}{2}b, t)$ versus $t$. Find a quantitative relationship between preload $N$ and $w_{max}$ and $(M_{xx})_{max}$.

**7.5** Solve the problem of Section 7.2.1 using a plate theory that accounts for shear deformation as well as flexure. Assume that $\nu = 0.3$ (or $\kappa^2 = 0.86$), $b/a = \sqrt{2}$. Construct a nondimensional graph of $a^2 P_{cr}/h^3 E$ versus $h/a$. What is the effect of shear deformation upon $P_{cr}$?

**7.6** Consider a circular plate of radius $a$. The plate is subjected to radial forces $N_{rr} = \tau_{rr} h = N =$ constant along the boundary $r = a$. These applied forces act parallel to the plate median surface. Show that the stress field (7.3) is induced in the interior of the plate ($r < a$). *Hint:* For a plane, axisymmetric stress field we have $\nabla^4 \phi = 0$, and

$$\tau_{rr} = \frac{1}{r}\frac{\partial \phi}{\partial r} = hN_{rr}$$

$$\tau_{\theta\theta} = \frac{\partial^2 \phi}{\partial r^2} = hN_{\theta\theta}$$

$$\tau_{r\theta} = 0 = hN_{r\theta}$$

The stress function $\phi = A \ln r + Br^2 \ln r + Cr^2 + D$.

**7.7** Transform (7.15) to polar coordinates. Show that

$$D\nabla^4 w - p = \left(\frac{1}{r} - \frac{\partial}{\partial r}\right)\left(N_{rr}\frac{\partial w}{\partial r} + N_{r\theta}\frac{1}{r}\frac{\partial w}{\partial \theta}\right) + \frac{1}{r}\frac{\partial}{\partial \theta}\left(N_{\theta r}\frac{\partial w}{\partial r} + N_{\theta\theta}\frac{1}{r}\frac{\partial w}{\partial \theta}\right)$$

**7.8** Normalize the eigenfunctions $W_i(r) = A_i J_0(\alpha_i r) + C_i I_0(\beta_i r)$ characterizing the axisymmetric vibrations of a clamped, initially stressed plate (see Section 7.3.2). The initial stress field is given by $N_{rr} = N_{\theta\theta} = N =$ constant (see Exercise 7.6). Let

$$\int_0^a \rho h W_i^2 r\, dr = 1$$

Use (6.30a) and (6.30b). Note that

$$\int_0^a r J_0(\alpha_i r) I_0(\beta_i r)\, dr = \frac{a}{\alpha_i^2 + \beta_i^2}\left[\beta_i J_0(\alpha_i a) I_1(\beta_i a) + \alpha_i I_0(\beta_i a) J_1(\alpha_i a)\right] = 0$$

Show that

$$|A_i| a\sqrt{\frac{\rho h}{2}} = \left[2J_0^2(\alpha_i a) + \left(1 - \frac{\alpha_i^2}{\beta_i^2}\right) J_1^2(\alpha_i a)\right]^{-1/2}$$

$$|C_i| a\sqrt{\frac{\rho h}{2}} = \left[2I_0^2(\beta_i a) + \left(\frac{\beta_i^2}{\alpha_i^2} - 1\right) I_1^2(\beta_i a)\right]^{-1/2}$$

**7.9** Consider an initially stressed, clamped circular plate. The initial stress field is described in Section 7.3.2 and Exercise 7.6. At $t = 0^+$, a uniformly distributed, transverse pressure load of magnitude $p = p_0$ is suddenly applied and maintained for $t > 0$. Find the dynamic response of the plate. Assume $w(r, t) = \sum_{i=1}^{\infty} W_i(r) q_i(t)$ and use the method of Section 6.7 to determine $q_i(t)$. The normalized eigenfunctions can be obtained from Exercise 7.8. Plot $w(0, t)$ and $M_{rr}(0, t)$ versus $t$ for $t \geq 0$. Find a quantitative relationship between preload $N$ and $w_{\max}$ and $|M_{rr}(0, t)|_{\max}$.

**7.10** Find the natural frequencies of a simply supported circular plate. The plate is in a state of (initial) radial prestress characterized by (7.31). Assume axial symmetry, and use CPT.

**7.11** A simply supported circular plate is subjected to radial compression $N_{rr} = -P$ along its circular boundary $r = a$. Assume axial symmetry and find the buckling value $P_{\text{cr}}$. Use CPT.

**7.12** *Thermo-elastic Plate Bending (CPT).* Consider the problem of a thin plate with incremental temperature distribution $T(z)$ uniform with respect to $x$ and $y$. Assume that the plate is prevented from stretching by suitable boundary conditions on the boundary $C$. Then $u_\alpha \equiv 0$. Show that for this case

$$N_{\alpha\beta} = -\frac{N_T}{1 - \nu}\delta_{\alpha\beta}$$

$$(N_{\alpha\beta}, w_{,\beta})_{,\alpha} = -\frac{N_T}{1 - \nu}\nabla^2 w$$

$$Q_{\alpha,\alpha} = M_{\alpha\beta,\beta\alpha} = -D\nabla^2 w - \frac{1}{1 - \nu}\nabla^2 M_T$$

[see (7.11), (7.14), and Exercise 3.30], and therefore,

$$D\nabla^4 w + \frac{N_T}{1 - \nu}\nabla^2 w = p - \frac{1}{1 - \nu}\nabla^2 M_T$$

where

$$N_T = \alpha E \int_{-h/2}^{h/2} T\, dz$$

$$M_T = \alpha E \int_{-h/2}^{h/2} zT\, dz$$

and

$$M_{\alpha\beta} = -D\left[(1-\nu)w_{,\alpha\beta} + \nu\nabla^2 w\,\delta_{\alpha\beta}\right] - \frac{M_T}{1-\nu}\delta_{\alpha\beta}$$

**7.13** Consider a circular plate that is clamped at its boundary $r = a$ in the reference state. The plate is now heated such that $T(x, y, \frac{1}{2}h) = T_2 =$ constant and $T(x, y, -\frac{1}{2}h) = T_1 =$ constant. Assuming thermal equilibrium and $p \equiv 0$, find:

**(a)** Deformed median surface.

**(b)** Components of moment tensor.

**(c)** Components of shear force vector.

Refer to Exercise 7.12.

**7.14** Do Exercise 7.13, except that now the boundary is simply supported.

**7.15** Consider a rectangular plate that is simply supported along its boundary in the reference state. The plate is now heated such that $T(x, y, \frac{1}{2}h) = T_2 =$ constant and $T(x, y, -\frac{1}{2}h) = T_1 =$ constant. Assuming thermal equilibrium and $p \equiv 0$, find:

**(a)** Deformed median surface.

**(b)** Components of moment tensor.

**(c)** Components of shear force vector.

Refer to Exercise 7.12.

**7.16** The median plane of a plate of unbounded extent coincides with the $x$-$y$ plane. Assume that the plate is subjected to the constant prestress field

$$\begin{bmatrix} N_{xx} & N_{xy} \\ N_{yx} & N_{yy} \end{bmatrix} = \begin{bmatrix} \sigma_{xx} & 0 \\ 0 & \sigma_{yy} \end{bmatrix} (h)$$

where $\sigma_{xx} \geq 0$ and $\sigma_{yy} \geq 0$. A straight-crested harmonic wave, with wave crests parallel to the $y$-axis, propagates in the positive $x$-direction. Find expressions of phase and group velocity as a function of wave number (refer to Section 6.9).

**7.17** Repeat the problem in Section 7.3.3, except that now the plate is subjected to the moving line load $p(0, \eta) = \sum_{n=1}^{\infty} p_n \sin n\eta$ (force per unit length). (See H. Reismann, "Dynamic Response of an Elastic Plate Strip to a Moving Line Load," *Journal of the AIAA*, Vol. 1, No. 2, 1963, pp. 354–360.)

# 8

# NUMERICAL METHODS

The theory of elastic plates is characterized by partial differential equations, and the solution of these equations by classical methods is possible only for those cases where the plate geometry, applied loads, and boundary conditions are suitably restricted (see Chapters 5–7). Because of these limitations in the use of classical methods, a large number of approximate, numerical methods have been developed to calculate static deformations, natural frequencies and mode shapes, buckling loads, and so on, for many practical applications when classical analytical methods fail or become too cumbersome. These numerical methods have assumed increasing importance with the development and perfection of digital computers.

This chapter contains a (restricted) development of two of the principal numerical methods: (a) the method of finite differences and (b) the finite element method. In both methods the partial differential equations and associated boundary conditions of (classical) plate theory are converted to a system of (linear) algebraic equations. In the case of static deformation problems, these equations are nonhomogeneous, and many efficient numerical routines are available to generate a solution with the aid of a digital computer. In the case of plate problems concerned with free vibrations or buckling, we obtain a system of homogeneous equations leading to a (linear) algebraic eigenvalue-eigenvector problem. Here, again, we can use existing, efficient routines to obtain a solution for eigenvalues and eigenvectors with the aid of a digital computer. The refinements possible in the application of numerical methods seem to be without limit, and this chapter should be considered to be introductory in nature.

## 8.1 THE METHOD OF FINITE DIFFERENCES

The method of finite differences requires that all derivatives in the applicable partial differential field equations be replaced by their "central difference" approximations. With reference to the Cartesian grid shown in Fig. 8.1, we have the approximations

$$\left(\frac{\partial w}{\partial x}\right)_{x,y} = \frac{1}{k}[1 \quad -1]\begin{Bmatrix} w_{x} + \frac{1}{2}k,\ y \\ w_{x} - \frac{1}{2}k,\ y \end{Bmatrix} \tag{8.1a}$$

and

$$\left(\frac{\partial w}{\partial y}\right)_{x,y} = \frac{1}{l}[1 \quad -1]\begin{Bmatrix} w_{x,y} + \frac{1}{2}l \\ w_{x,y} - \frac{1}{2}l \end{Bmatrix} \tag{8.1b}$$

Consequently, we have

$$\left(\frac{\partial^2 w}{\partial x^2}\right)_{x,y} = \frac{1}{k}\left[\left(\frac{\partial w}{\partial x}\right)_{x+k/2,\,y} - \left(\frac{\partial w}{\partial x}\right)_{x-k/2,\,y}\right]$$

$$= \frac{1}{k^2}[1 \quad -2 \quad 1]\begin{Bmatrix} w_{x+k,\,y} \\ w_{x,\,y} \\ w_{x-k,\,y} \end{Bmatrix} \tag{8.2a}$$

$$\left(\frac{\partial^2 w}{\partial y^2}\right)_{x,y} = \frac{1}{l^2}[1 \quad -2 \quad 1]\begin{Bmatrix} w_{x,\,y+l} \\ w_{x,\,y} \\ w_{x,\,y-l} \end{Bmatrix} \tag{8.2b}$$

$$\left(\frac{\partial^2 w}{\partial x\,\partial y}\right)_{x,y} = \frac{\partial}{\partial y}\left(\frac{\partial w}{\partial x}\right) = \frac{1}{l}\left[\left(\frac{\partial w}{\partial x}\right)_{x,\,y+l/2} - \left(\frac{\partial w}{\partial x}\right)_{x,\,y-l/2}\right]$$

$$= \frac{1}{kl}[1 \quad -1 \quad 1 \quad -1]\begin{Bmatrix} w_{x+k/2,\,y+l/2} \\ w_{x-k/2,\,y+l/2} \\ w_{x-k/2,\,y-l/2} \\ w_{x+l/2,\,y-l/2} \end{Bmatrix} \tag{8.2c}$$

Equations (8.2) can be characterized by "computational modules," a mnemonic device shown in Fig. 8.2. This representation will prove useful for subsequent calculations.

We can utilize (8.2a) and (8.2b) to obtain the central difference approximation of the harmonic operator. By direct superposition of equations (8.2a) and (8.2b) or Figs. 8.2($a$) and 8.2($b$), we readily obtain the computational module in Fig. 8.3.

The preceding approach is now extended to higher order derivatives, and the reader is urged to verify the results shown in Fig. 8.4. By superposing the modules in Fig. 8.4, we obtain a central difference approximation for the bi-harmonic operator

$$\nabla^4 \equiv \frac{\partial^4}{\partial x^4} + 2\frac{\partial^4}{\partial x^2\,\partial y^2} + \frac{\partial^4}{\partial y^4} \tag{8.3}$$

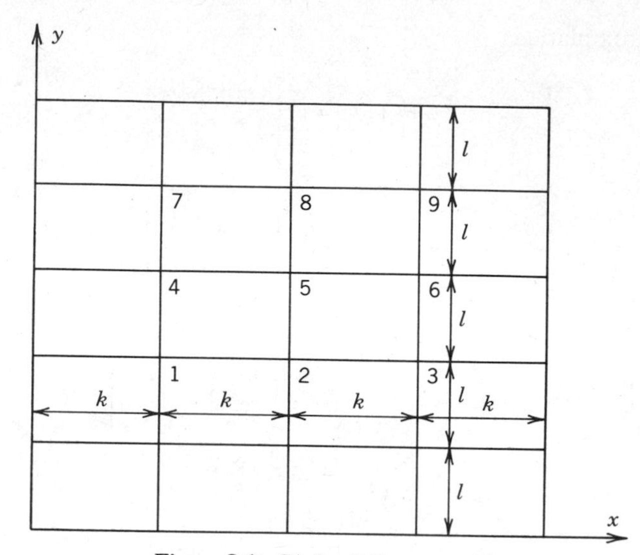


**Figure 8.1** Finite difference grid.

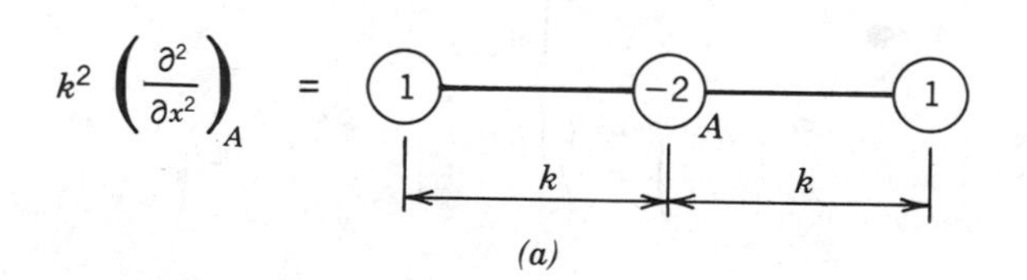


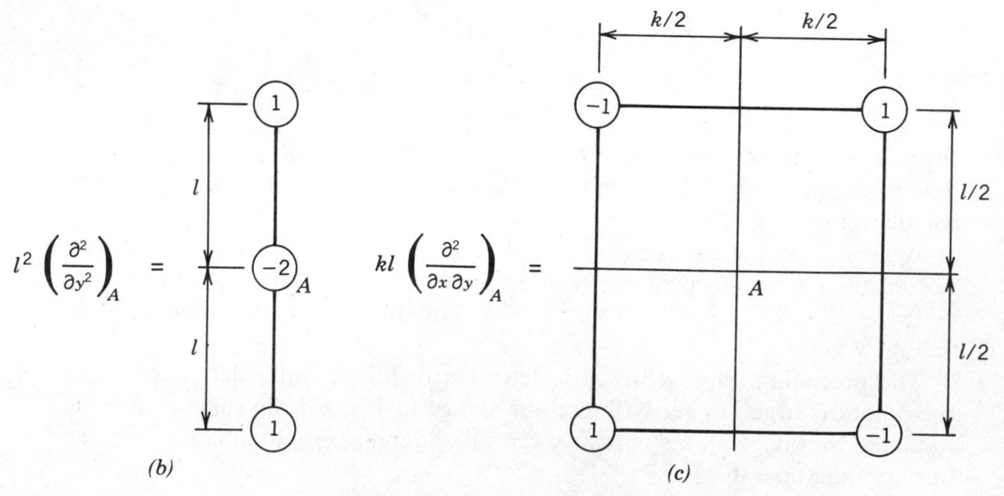


**Figure 8.2** Second derivatives.

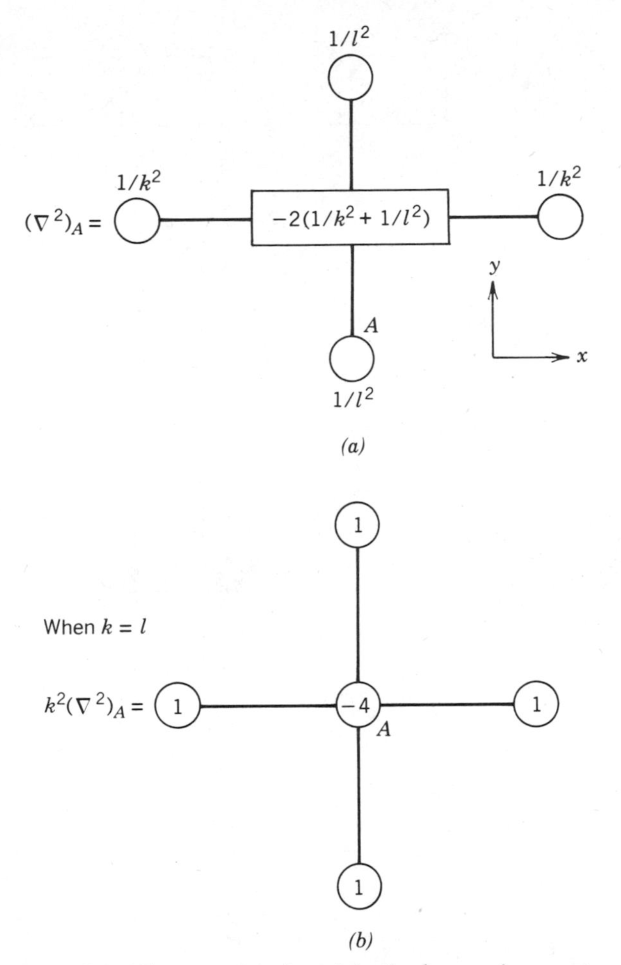


**Figure 8.3** Computational modules for harmonic operator.

and the corresponding computational module is shown in Fig. 8.5. In view of this result, we can write the displacement equation of equilibrium of CPT in the form

$$[\text{Coefficient}]\{w\} = \frac{p_i k^4}{D} \tag{8.4}$$

where the coefficients are the ones shown in Fig. 8.5, and where $\{w\}$ is a column vector of plate displacements. Equation (8.4), in conjunction with Fig. 8.5, relates the deflection $w_i$ at $i$ and at 12 other nodes in the vicinity of $i$ to the local load intensity $p_i$ and plate stiffness $D$.

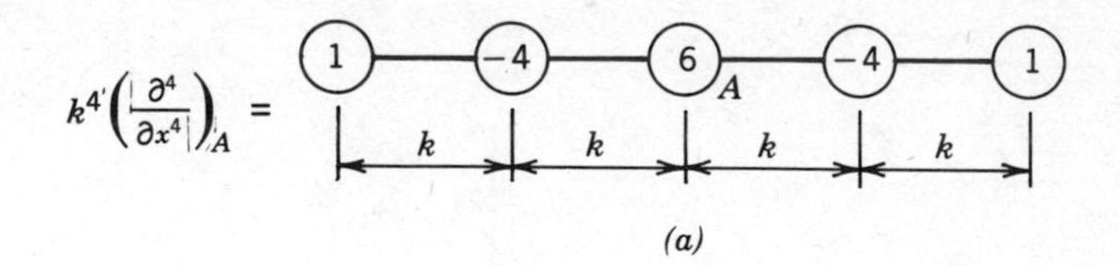


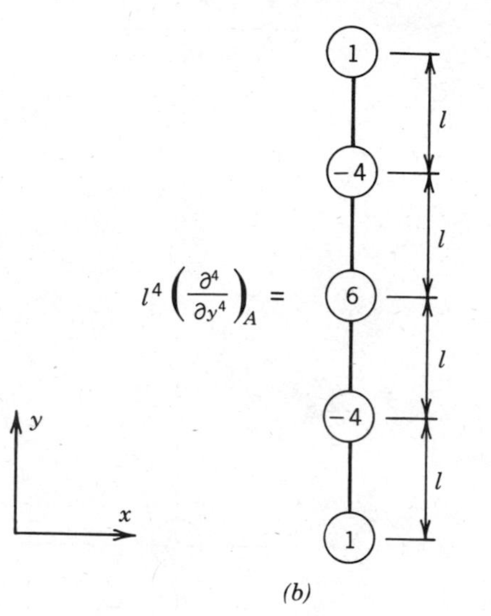


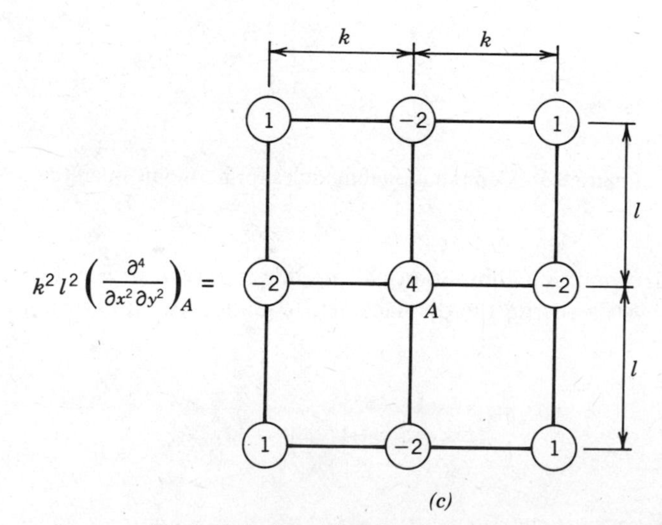


**Figure 8.4** Fourth-order derivatives.

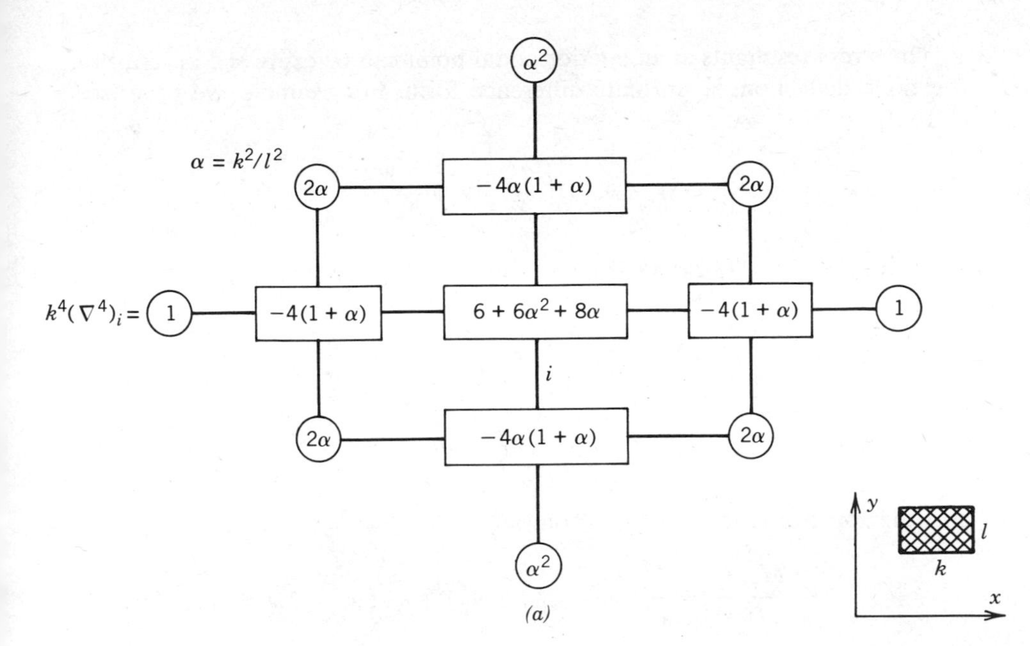


(a)

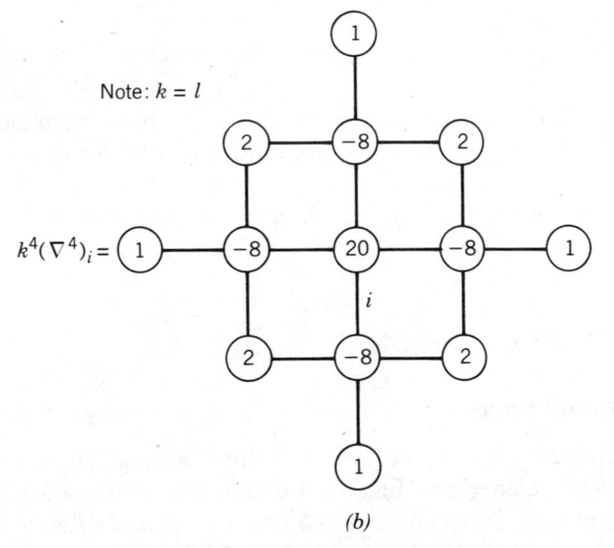


(b)

**Figure 8.5** Bi-harmonic operator.

The stress resultants at an interior nodal point can be expressed in terms of the node deflections $w_i$ in finite difference form. For example, we have [see (4.16a)]

$$M_{xx} = -D\left(\frac{\partial^2 w}{\partial x^2} + \nu\frac{\partial^2 w}{\partial y^2}\right) \tag{8.5}$$

and with the aid of (8.2a), (8.2b), and Fig. 8.1, we have

$$k^2\left(\frac{\partial^2 w}{\partial x^2}\right)_5 = w_4 - 2w_5 + w_6$$

$$l^2\left(\frac{\partial^2 w}{\partial y^2}\right)_5 = w_8 - 2w_5 + w_2$$

and the approximation of (8.5) becomes

$$-\frac{(M_{xx})_5}{D} = \frac{1}{k^2}(w_4 - 2w_5 + w_6) + \frac{\nu}{l^2}(w_8 - 2w_5 + w_2)$$

or

$$-\frac{k^2(M_{xx})_5}{D} = (w_4 + w_6) + \nu\alpha(w_8 + w_2) - 2(1 + \nu\alpha)w_5 \tag{8.6}$$

where $\alpha = (k/l)^2$. The computational module that corresponds to (8.6) is shown in Fig. 8.6 along with the computational modules that correspond to $M_{yy}$ and $M_{xy}$. In a similar manner, we can calculate the finite difference approximations of the shear forces $Q_x$ and $Q_y$, and these are shown in Fig. 8.7.

## 8.2 BOUNDARY CONDITIONS

### 8.2.1 Clamped Edge

Along a clamped edge we require vanishing normal slope and vanishing deflection. With reference to Fig. 8.8 we consider a clamped plate boundary along $x =$ constant. At node 2 we require $w_2 = 0$ and $(\partial w/\partial x)_2 = 0$. With the aid of (8.1a) we have (set $k = 2r$) $(\partial w/\partial x)_2 = (1/2r)(w_3 - w_1) = 0$ or $w_1 = w_3$ and $w_2 = 0$. Consequently, for a clamped edge the fictitious (or phantom) point 3 is assumed to have the same deflection as its image point 1. Similar considerations hold for other points along the clamped edge.

### 8.2.2 Simply Supported Edge

Along a simply supported edge we require the vanishing of deflection and normal bending moment. With reference to Fig. 8.8, we have, for the simply

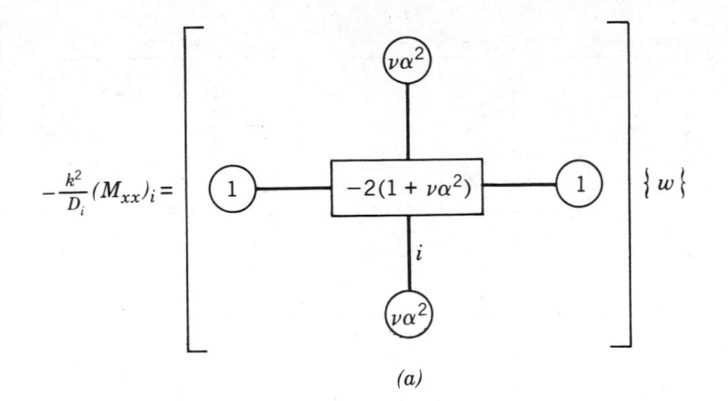


(a)

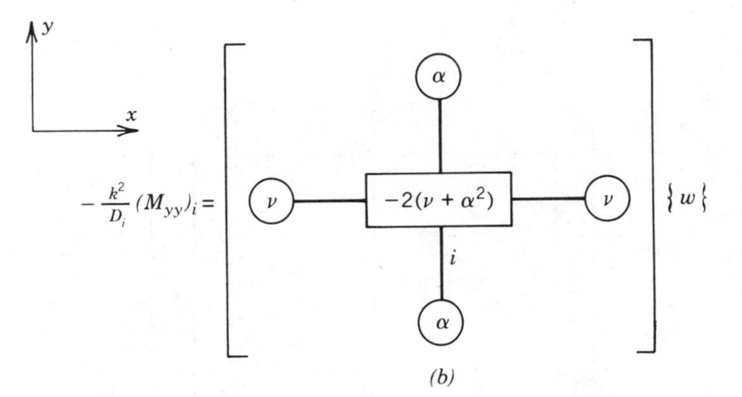


(b)

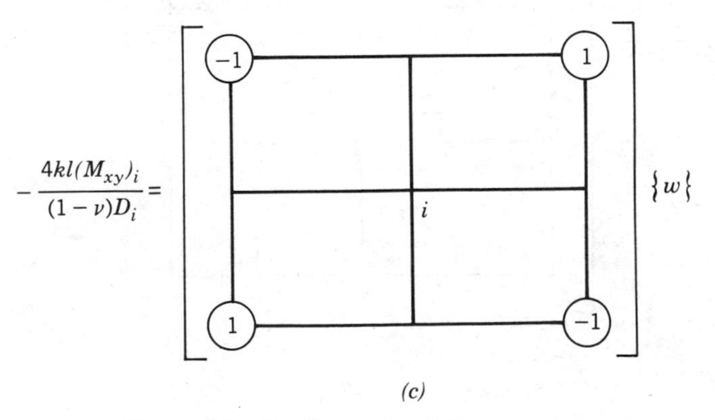


(c)

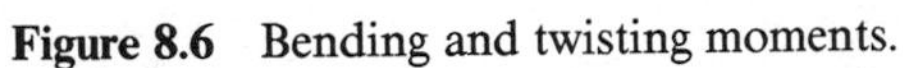

**Figure 8.6** Bending and twisting moments.

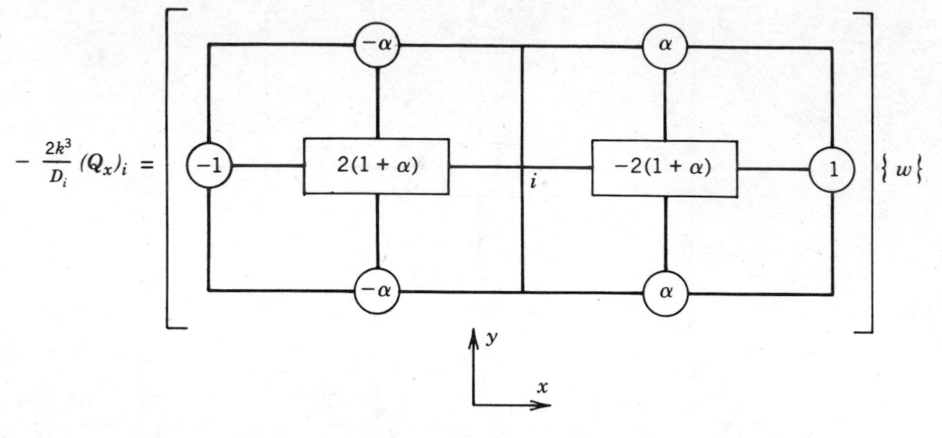


(a)

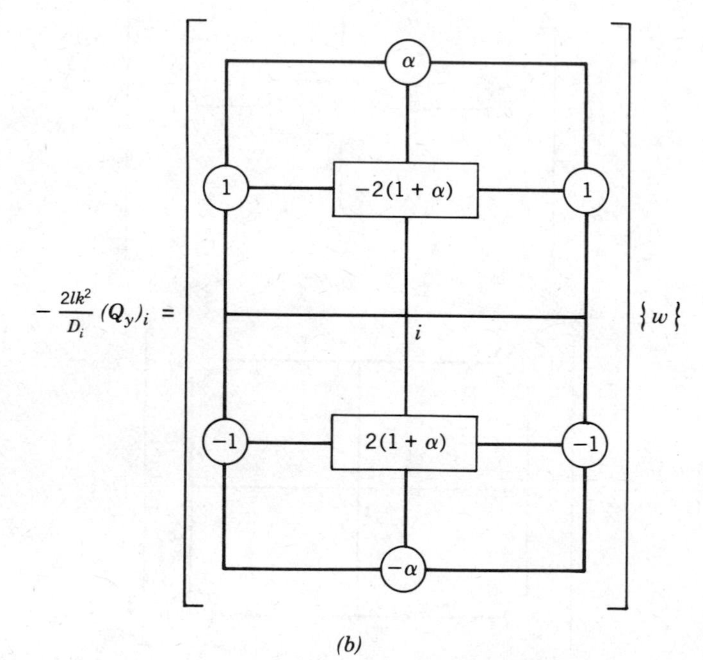


(b)

**Figure 8.7** Shear forces.

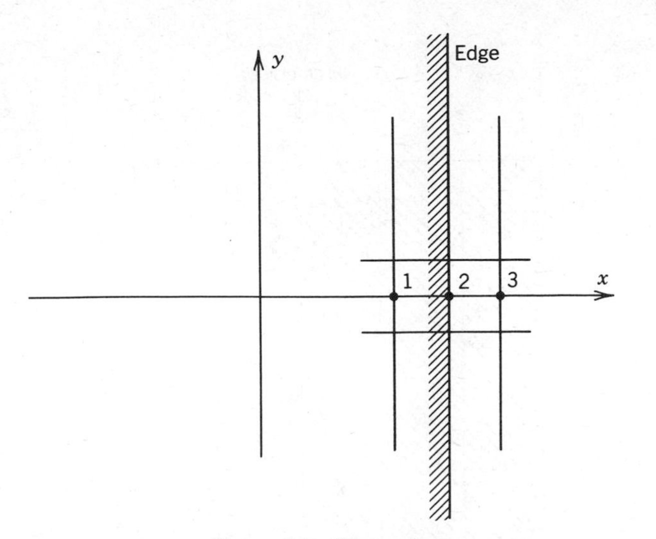


**Figure 8.8** Plate edge.

supported edge $x$ = constant;

$$w = 0 \quad \text{and} \quad M_{xx} = -D\left(\frac{\partial^2 w}{\partial x^2} + \nu \frac{\partial^2 w}{\partial y^2}\right) = 0$$

But along the edge $x$ = const we have $\partial^2 w/\partial y^2 = 0$, so that the condition along the simply supported edge becomes $w = 0$ and $\partial^2 w/\partial x^2 = 0$. At node 2 in Fig. 8.8, this condition assumes the form [see (8.2a) or Fig. 8.2($a$)]

$$w_2 = 0 \quad \text{and} \quad \frac{1}{k^2}(w_3 - 2w_2 + w_1) = 0$$

Consequently, we obtain $w_1 = -w_3$. Thus, nodes 3 and 1 deflect in opposite directions, but the magnitudes of their deflections are equal. Similar relations hold for all points along a simply supported edge.

### 8.2.3 Free Edge

Consider a plate with a free edge along $x$ = constant, as shown in Fig. 8.9. Along this edge we require [see (4.20a)]

$$-\frac{M_{xx}}{D} = \frac{\partial^2 w}{\partial x^2} + \nu \frac{\partial^2 w}{\partial y^2} = 0 \tag{8.7}$$

and

$$-\frac{V_x}{D} = \frac{\partial^3 w}{\partial x^3} + (2 - \nu)\frac{\partial^3 w}{\partial x \, \partial y^2} = 0 \tag{8.8}$$

If we write (8.7) in finite difference form and apply it at nodes $G$, $A$, and $J$ in

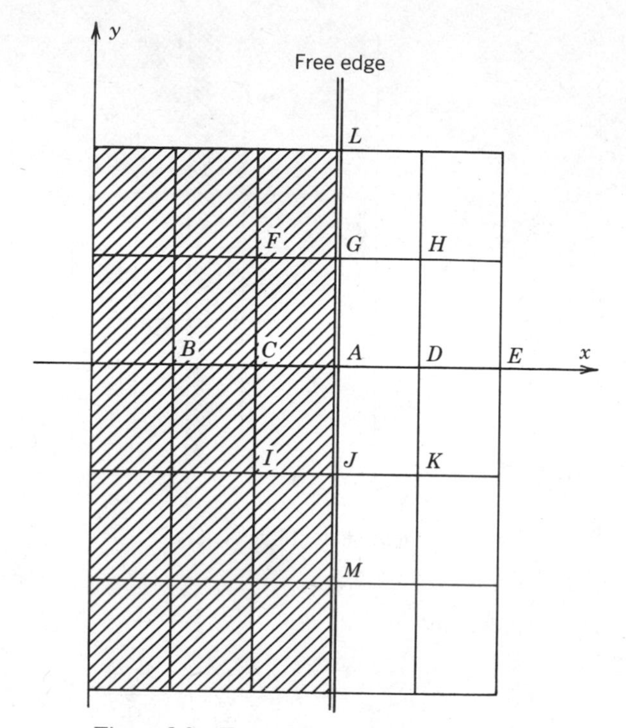


**Figure 8.9** Free edge and fictitious nodes.

Fig. 8.9, we obtain

$$\frac{1}{k^2}(w_H - 2w_G + w_F) + \frac{\nu}{l^2}(w_A - 2w_G + w_L) = 0 \tag{8.9a}$$

$$\frac{1}{k^2}(w_D - 2w_A + w_C) + \frac{\nu}{l^2}(w_J - 2w_A + w_G) = 0 \tag{8.9b}$$

$$\frac{1}{k^2}(w_K - 2w_J + w_I) + \frac{\nu}{l^2}(w_M - 2w_J + w_A) = 0 \tag{8.9c}$$

Conversion of (8.8) into finite difference form and application at node $A$ in Fig. 8.9 results in

$$\frac{1}{2k^3}(w_E - 2w_D + 2w_C - w_B) + \frac{2-\nu}{2k}\left(\frac{w_K - 2w_D + w_H}{l^2} - \frac{w_I - 2w_C + w_F}{l^2}\right) = 0 \tag{8.10}$$

Equation (8.4), in conjunction with Fig. 8.5, is now applied at $A$. There will be four fictitious nodes $H$, $D$, $K$, and $E$, and these can be eliminated with the aid of the four equations (8.9) and (8.10). The result is

$$Jw_A + k(w_J + w_G) + L(w_M + w_L) + H(w_I + w_F) + Fw_B = \frac{p_i k^4}{D} \tag{8.11}$$

where the coefficients $J$, $K$, $L$, $H$, and $F$ are defined in Fig. 8.10. We note that (8.11) no longer contains fictitious nodes and incorporates the requirements of equilibrium as well as the free edge boundary conditions.

Similar considerations apply to interior nodes and to the case of corners formed by two intersecting free edges. The results are summarized in Fig. 8.10, which is taken from A. Ghali and K.-J. Bathe, "Analysis of Plates in Bending using Large Finite Elements", *International Association for Bridge and Structural Engineering*, Vol. 30, No. 2, Zurich, 1970, p. 496. The computational modules in Fig. 8.10 simplify the formulation of plate problems with free edges and circumvent the need to use fictitious points (see the problem in Section 8.3.3). The equation of equilibrium associated with the coefficients in Fig. 8.10 is given by (8.4).

## 8.3 APPLICATION

The following examples are presented to demonstrate the application of finite difference techniques to plate problems. Extensive use is made of formulas and computational modules developed in Sections 8.1 and 8.2.

### 8.3.1 Bending of a Simply Supported Plate

We consider a simply supported, square plate with edges of length $a$. The plate is subjected to a transverse, uniformly distributed load of intensity $p_0$. In order to implement the finite difference procedure, we discretize the plate in the manner shown in Fig. 8.11. We have constructed a mesh such that $k = l = \frac{1}{6}a$. When labeling the nodes, we have taken advantage of the eightfold symmetry of the present problem. In addition, because the edges are simply supported, the fictitious nodes are labeled in accordance with the requirements derived in Section 8.2.2. We now apply (8.4) and utilize the computational module of Fig. 8.5($b$). Thus, we obtain six linear algebraic equations in the six unknowns $w_1, \ldots, w_6$, as shown in Fig. 8.12. These equations, in matrix notation, have the form $AX = B$, where the elements of the square matrix $A$ as well as the elements of the column vector $B$ are shown in Fig. 8.12. The solution is characterized by the column vector $X$, where $X^{\mathrm{T}} = \{w_1^*, w_2^*, \ldots, w_6^*\}$, where $w_i^* = w_i D/p_0 k^4$, $i = 1, 2, \ldots, 6$, are the dimensionless transverse deflections of the plate at the respective nodes. The solution, obtained with the aid of a digital computer, is also shown in Fig. 8.12. We note that the maximum deflection occurs at the center of the plate and is given by $w_1 = 5.2466 p_0 k^4/D = 0.00405 p_0 a^4/D$. The "exact" value can be obtained with the aid of (5.22). It is given by $w_1 = 0.00406 p_0 a^4/D$, and we find an error of 0.25% for the approximation obtained by finite difference methods. The maximum bending moment occurs at the center of the plate, and we use the computational module in Fig. 8.6($a$) or 8.6($b$). Upon application to node $i = 1$ in Fig. 8.11, we obtain

$$-\frac{k^2}{D}(M_{xx})_1 = -\frac{k^2}{D}(M_{yy})_1 = [-2(1+\nu)w_1 + 2w_2 + 2\nu w_2]\frac{p_0 k^4}{D}$$

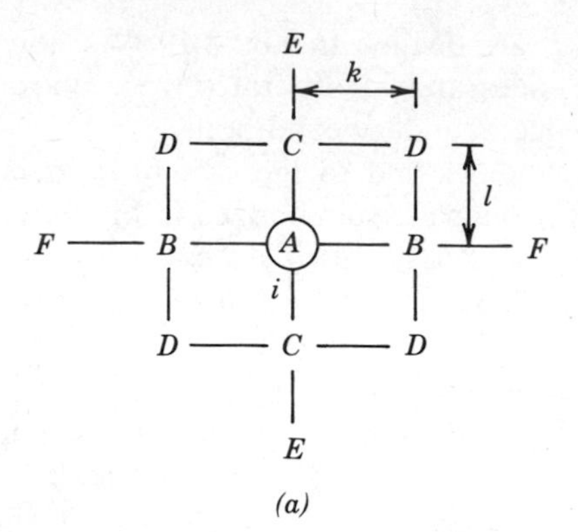


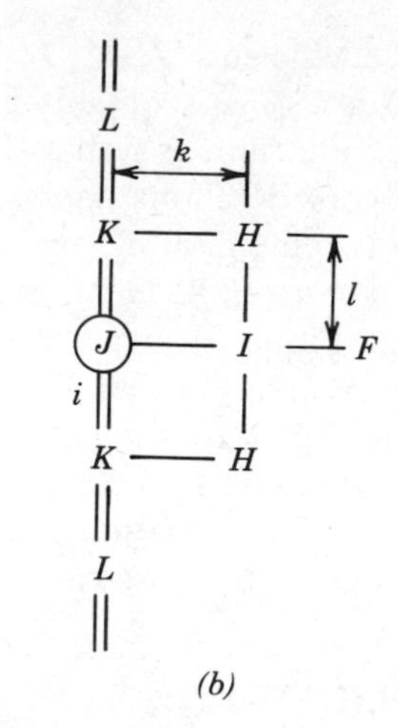


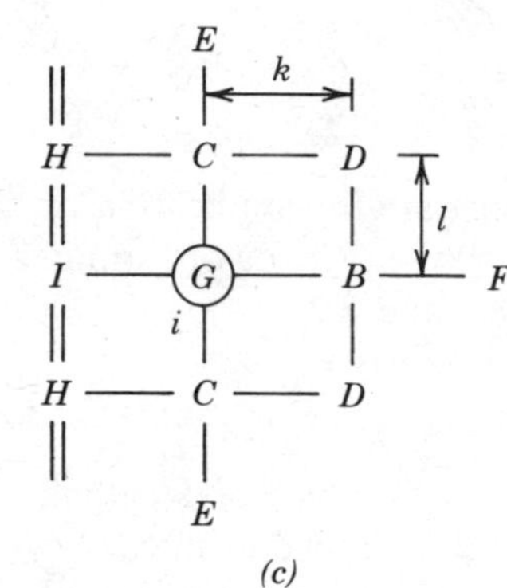


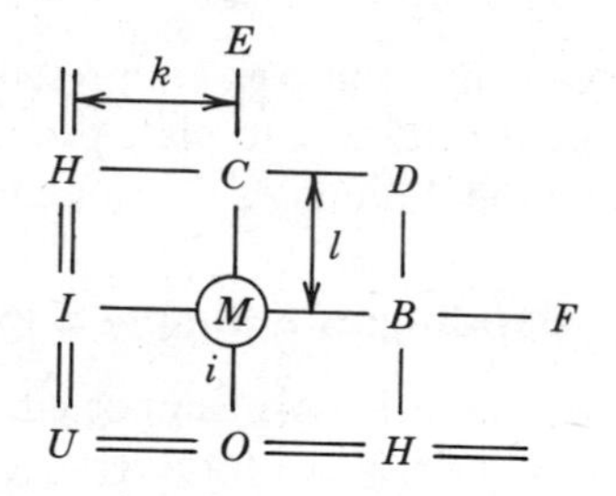


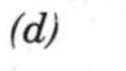


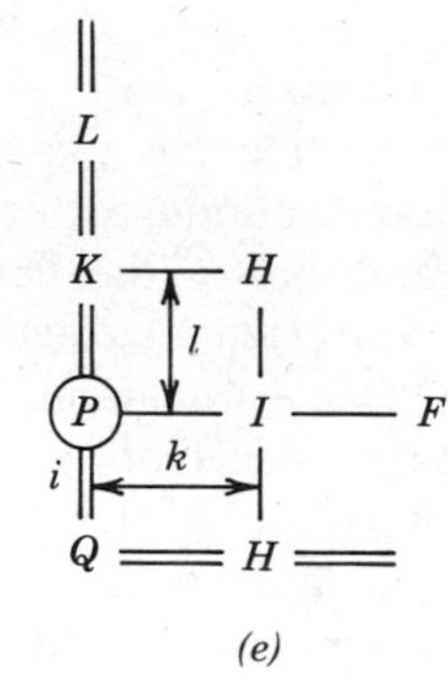


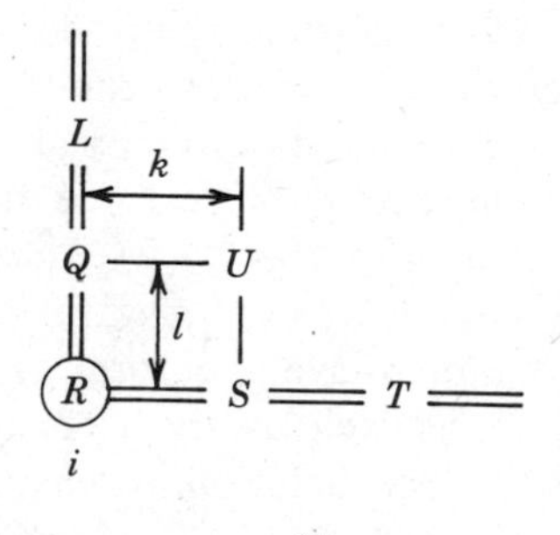


$A = 6 + 6\alpha^2 + 8\alpha$
$B = -4(1 + \alpha)$
$C = -4\alpha(1 + \alpha)$
$D = 2\alpha$
$E = \alpha^2$
$F = 1$
$G = 5 + 6\alpha^2 + 8\alpha$
$\alpha = (k/l)^2$

$H = \alpha(2 - \nu)$
$I = -2(2\alpha - \nu\alpha + 1)$
$J = 1 + 4\alpha(1 - \nu) + 3\alpha^2(1 - \nu^2)$
$K = -2\alpha[1 - \nu + \alpha(1 - \nu^2)]$
$L = \frac{1}{2}\alpha^2(1 - \nu^2)$
$M = 5 + 5\alpha^2 + 8\alpha$
$O = -2\alpha(2 - \nu + \alpha)$

$P = 1 + 4\alpha(1 - \nu) + \frac{5}{2}\alpha^2(1 - \nu^2)$
$Q = -2\alpha[1 - \nu + \frac{\alpha}{2}(1 - \nu^2)]$
$R = 2\alpha(1 - \nu) + \frac{1}{2}(1 + \alpha^2)(1 - \nu^2)$
$S = -2[\alpha(1 - \nu) + \frac{1}{2}(1 - \nu^2)]$
$T = \frac{1}{2}(1 - \nu^2)$
$U = 2\alpha(1 - \nu)$

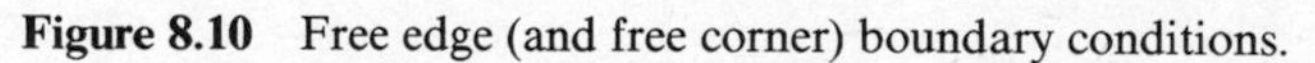
**Figure 8.10** Free edge (and free corner) boundary conditions.

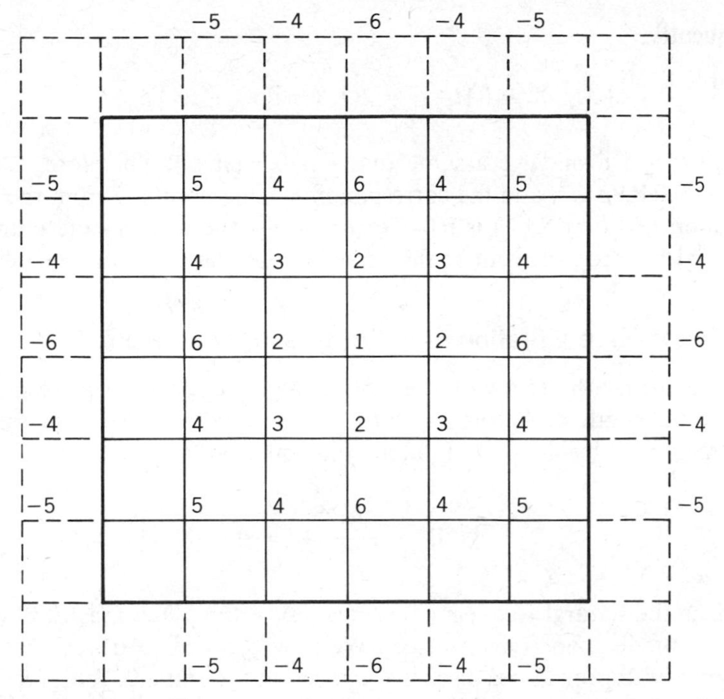


**Figure 8.11** Finite difference mesh (simply supported square plate, $k = l = \frac{1}{6}a$)

```
THE MATRIX A IS

    5.0000 -8.0000  2.0000    .0000    .0000  1.0000

   -8.0000 25.0000-16.0000  6.0000    .0000 -8.0000

    2.0000-16.0000 22.0000-16.0000  2.0000  4.0000

     .0000  6.0000-16.0000 44.0000-16.0000-16.0000

     .0000    .0000  2.0000-16.0000 18.0000  2.0000

    1.0000 -8.0000  4.0000-16.0000  2.0000 19.0000

THE VECTOR B IS

     .2500  1.0000  1.0000  2.0000  1.0000  1.0000

THE VECTOR X IS

    5.2466  4.5976  4.0312  2.4024  1.4392  2.7352
```

**Figure 8.12** The elements of the matrix equation $AX = B$ (simply supported square plate, $k = l = a/6$).

Consequently,

$$(M_{xx})_1 = (M_{yy})_1 = 2(1+\nu)(w_2 - w_1)p_0k^2$$

We have $k = \frac{1}{6}a$, and we assume that $\nu = 0.3$ (steel). Therefore, $(M_{xx})_1 = (M_{yy})_1 = 0.0469p_0a^2$, and the corresponding value obtained from the analytical solution (Section 5.1.1) is $0.0479p_0a^2$. Thus, the error in the numerically determined bending moment at the center of the plate is approximately 2%.

### 8.3.2 Transverse Vibration of a Simply Supported Plate

We now consider the free vibrations of a simply supported square plate with the edges of length $a$. According to (6.12), the mode shapes (eigenfunctions) are characterized by the partial differential equation

$$\nabla^4 W - \frac{\rho h \Omega^2}{D} W = 0 \tag{8.12}$$

where $\Omega$ is the natural frequency (eigenvalue) of the plate and $W(x, y)$ is the associated mode shape (eigenvector). We now proceed to discretize (8.12) in accordance with procedures discussed in Section 8.1 and 8.2. With reference to Fig. 8.11, we set $k = l = \frac{1}{6}a$. Assuming a mode shape with an eightfold symmetry, we arrive at the labeling of nodes shown in Fig. 8.11. The simple support boundary conditions are achieved by labeling the fictitious nodes in accordance with the rules obtained in Section 8.2.2. We now use the computational module in Fig. 8.5($b$) to convert the partial differential equation (8.12) to a system of linear algebraic equations. Using matrix notation, these equations can be written in the form

$$AW = \lambda BW \tag{8.13}$$

where the elements in the square matrices $A$ and $B$ are displayed in Fig. 8.13, and where $W$ is a column vector, the elements of which are proportional to the nodal displacements of the mode shape, so that

$$W^{\mathrm{T}} = [\beta W_1 \quad \beta W_2 \quad \dots \quad \beta W_6] \tag{8.14}$$

where $\beta \neq 0$ is an arbitrary real constant. In the present case, the eigenvalue is

$$\lambda = \frac{\rho h k^4 \Omega^2}{D} = \frac{\rho h a^4 \Omega^2}{(36)^2 D}$$

The (generalized) eigenvalue-eigenvector problem (8.13) is readily solved with the aid of a digital computer, and the lowest eigenvalue and its associated mode shape are, respectively, $\lambda = 0.2872$, $W^{\mathrm{T}} = [0.5963,\ 0.5164,\ 0.4472,$

```
THE MATRIX A IS

     5.0000 -8.0000  2.0000    .0000    .0000  1.0000

    -8.0000 25.0000-16.0000  6.0000    .0000 -8.0000

     2.0000-16.0000 22.0000-16.0000  2.0000  4.0000

      .0000  6.0000-16.0000 44.0000-16.0000-16.0000

      .0000    .0000  2.0000-16.0000 18.0000  2.0000

     1.0000 -8.0000  4.0000-16.0000  2.0000 19.0000

THE MATRIX B IS

      .2500    .0000    .0000    .0000    .0000    .0000

      .0000  1.0000    .0000    .0000    .0000    .0000

      .0000    .0000  1.0000    .0000    .0000    .0000

      .0000    .0000    .0000  2.0000    .0000    .0000

      .0000    .0000    .0000    .0000  1.0000    .0000

      .0000    .0000    .0000    .0000    .0000  1.0000
```

**Figure 8.13** The elements of the matrix equation $AW = \lambda BW$ (simply supported square plate, $k = l = a/6$, $\lambda = \rho\Omega^2 hk^4/D$).

0.2582, 0.1491, 0.2981]. Consequently, we have

$$\lambda = 0.2872 = \frac{\rho h a^4 \Omega^2}{(36)^2 D}$$

and

$$a^2\sqrt{\rho\frac{h}{D}}\,\Omega = 19.3$$

The exact value is [see (6.19)]

$$a^2\sqrt{\rho\frac{h}{D}}\,\Omega = 2\pi^2 = 19.7$$

and we infer that our approximation deviates from the exact solution by approximately 2.03%. Higher frequencies and associated mode shapes can be obtained in a similar manner, but in general, a relabeling of the nodes in Fig. 8.12 is required. In addition, in order to obtain comparable accuracy for higher frequencies, a successively finer mesh will be required.

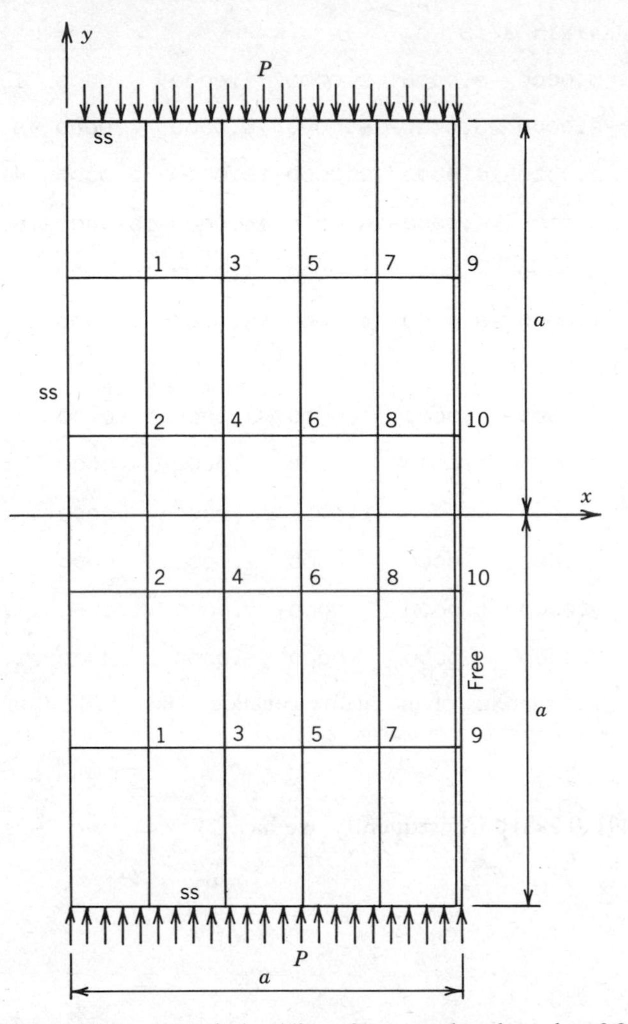


**Figure 8.14** Finite difference mesh; buckling of rectangular plate: $k = 0.2a$; $l = 0.4a$; $\alpha = (k/l)^2 = 0.25$.

### 8.3.3 Buckling of a Rectangular Plate

In this section we consider the elastic stability of a rectangular plate with aspect ratio $b/a = 2/1$, as shown in Fig. 8.14. The plate is simply supported along three edges, but the fourth edge is entirely unsupported (free). Parallel to its midplane, the plate is compressed by forces of intensity $P$ (force per unit length) along two simply supported, parallel edges of length $a$. We wish to find the value of $P$ that causes the plate to be in a state of incipient instability. With reference to (7.15), the partial differential equation that characterizes this phenomenon for the present situation is

$$D\nabla^4 w + P\frac{\partial^2 w}{\partial y^2} = 0 \tag{8.15}$$

We now discretize (8.15) in accordance with finite-difference methodology. With reference to Fig. 8.14, we take $k = 0.2a$, $l = 0.4a$, and $\nu = 0.25$. With the aid of Fig. 8.10, we calculate

$$\begin{array}{ll} A = 8.3750 & G = 7.375 \\ B = -5.00 & H = 0.4375 \\ C = -1.25 & I = -2.8750 \\ D = 0.5 & J = 1.9258 \\ E = 0.0625 & K = -0.4922 \\ F = 1.00 & L = 0.0293 \end{array}$$

We now use the computational modules in Figs. 8.10 and 8.2($b$) to obtain an equation for each node. For example, the equation at node 10 reads

$$Fw_6 + Hw_7 + (I+H)w_8 + (K+L)w_9 + (J+K)w_{10} = \lambda(0.5w_{10} - 0.5w_9)$$

where

$$\lambda = \frac{P}{D}\frac{k^4}{l^2} = \frac{Pa^2}{100D}$$

Similarly, we write an equation for each node in Fig. 8.14. The result, in matrix notation, is

$$AW = \lambda BW \tag{8.16}$$

where the elements of the square matrices $A$ and $B$ are tabulated in Fig. 8.15. Equation (8.16) characterizes a generalized algebraic eigenvalue-eigenvector problem, where $\lambda$ is an eigenvector (critical load) and $W$ is a column vector characterizing the eigenvector (buckled surface). We note that

$$W^{\mathrm{T}} = [\beta W_1 \quad \beta W_2 \quad \dots \quad \beta W_{10}]$$

where $\beta \neq 0$ is a real constant. A solution of the generalized eigenvector-eigenvalue problem (8.16) can be readily obtained with the aid of a digital computer. For the present case, the lowest eigenvalue is

$$\lambda = 0.067093 = \frac{P_{\mathrm{cr}}a^2}{100D}$$

so that

$$P_{\mathrm{cr}} = 6.7093\frac{D}{a^2}$$

```
THE MATRIX A IS

 7.3125 -1.1875 -5.0000   .5000  1.0000   .0000   .0000   .0000   .0000   .0000

-1.1875  6.1250   .5000 -4.5000   .0000  1.0000   .0000   .0000   .0000   .0000

-5.0000   .5000  8.3125 -1.1875 -5.0000   .5000  1.0000   .0000   .0000   .0000

  .5000 -4.5000 -1.1875  7.1250   .5000 -4.5000   .0000  1.0000   .0000   .0000

 1.0000   .0000 -5.0000   .5000  8.3125 -1.1875 -5.0000   .5000  1.0000   .0000

  .0000  1.0000   .5000 -4.5000 -1.1875  7.1250   .5000 -4.5000   .0000  1.0000

  .0000   .0000  1.0000   .0000 -5.0000   .5000  7.3125 -1.1875 -2.8750   .4375

  .0000   .0000   .0000  1.0000   .5000 -4.5000 -1.1875  6.1250   .4375 -2.4375

  .0000   .0000   .0000   .0000  1.0000   .0000 -2.8750   .4375  1.8965  -.4629

  .0000   .0000   .0000   .0000   .0000  1.0000   .4375 -2.4375  -.4629  1.4336

THE MATRIX B IS

 2.0000 -1.0000   .0000   .0000   .0000   .0000   .0000   .0000   .0000   .0000

-1.0000  1.0000   .0000   .0000   .0000   .0000   .0000   .0000   .0000   .0000

  .0000   .0000  2.0000 -1.0000   .0000   .0000   .0000   .0000   .0000   .0000

  .0000   .0000 -1.0000  1.0000   .0000   .0000   .0000   .0000   .0000   .0000

  .0000   .0000   .0000   .0000  2.0000 -1.0000   .0000   .0000   .0000   .0000

  .0000   .0000   .0000   .0000 -1.0000  1.0000   .0000   .0000   .0000   .0000

  .0000   .0000   .0000   .0000   .0000   .0000  2.0000 -1.0000   .0000   .0000

  .0000   .0000   .0000   .0000   .0000   .0000 -1.0000  1.0000   .0000   .0000

  .0000   .0000   .0000   .0000   .0000   .0000   .0000   .0000  1.0000  -.5000

  .0000   .0000   .0000   .0000   .0000   .0000   .0000   .0000  -.5000   .5000
```

**Figure 8.15** The elements of the matrix equation $AW = \lambda BW$ [rectangular plate buckling, $k = 0.2a$, $l = 0.4a$, $\lambda = Pa^2/(100D)$].

and the associated eigenvector is

$$\{W\} = [0.0749 \quad 0.1212 \quad 0.1476 \quad 0.2389 \quad 0.2168 \quad 0.3509 \quad 0.2829$$
$$\times 0.4577 \quad 0.3492 \quad 0.5650]^{\mathrm{T}}$$

The exact solution can be obtained by Lévy's method (see the hint in Exercise 7.1 and see Exercise 7.2). It results in

$$P_{\mathrm{cr}} = 6.8960 \frac{D}{a^2}$$

Consequently, the numerically determined buckling value is within 2.71% of the corresponding value obtained by analytical methods.

### 8.3.4 Shear Buckling of Rectangular Plates

In this section we briefly consider a simply supported plate with aspect ratio $b/a = 2/1$ subjected to a constant membrane stress (pure shear) of magnitude

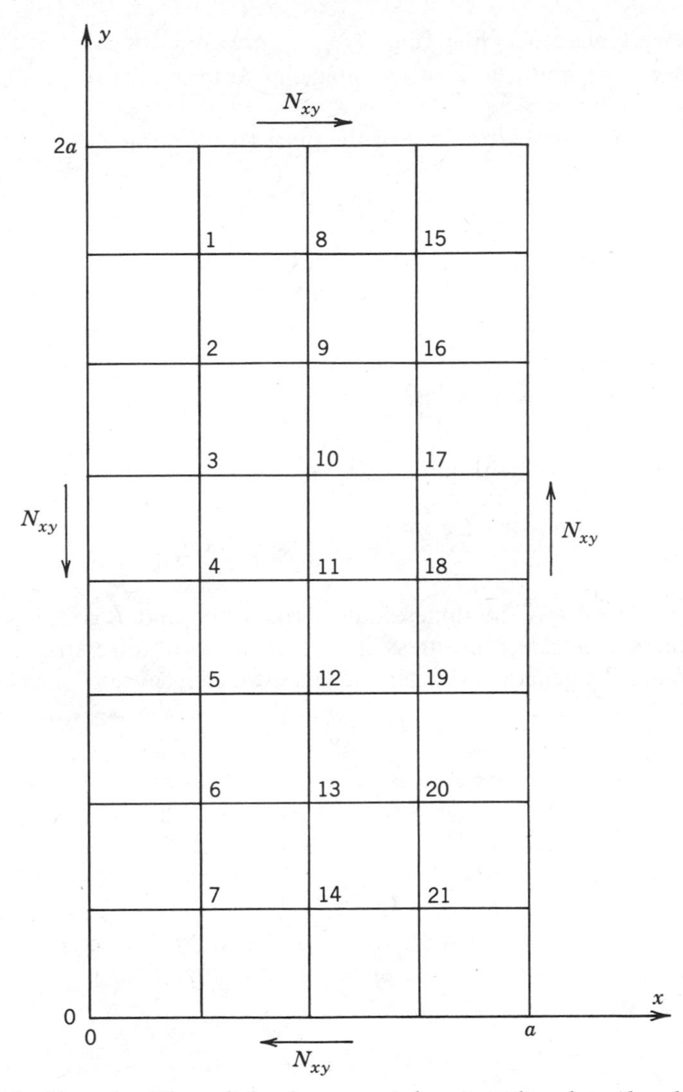


**Figure 8.16** Shear buckling of simply supported rectangular plate ($k = l = a/4$, 21 nodes).

$N_{xy}$, as shown in Fig. 8.16. The equation that characterizes the buckled surface of the plate [see (7.15)] is

$$D\nabla^4 w = 2N_{xy}\frac{\partial^2 w}{\partial x\,\partial y} \tag{8.17}$$

and we wish to determine $N_{xy}$ for incipient loss of stability. When (8.17) is discretized in accordance with the usual finite difference methodology, a system of algebraic equations is obtained. However, in this case the solution for the lowest plate buckling value $(N_{xy})_{cr}$ presents difficulties that are not easy to overcome, and it will be advantageous to formulate this problem from a dynamical point of view.

Let us consider free vibrations of the plate (see Section 6.2)

$$w = W(x, y)\cos \Omega t \tag{8.18}$$

where $\Omega$ is the natural frequency and $W(x, y)$ characterizes the associated mode shape. In this case (8.17) must be augmented by an inertia term, and we require

$$D\nabla^4 w = 2N_{xy}\frac{\partial^2 w}{\partial x\,\partial y} - \rho h\frac{\partial^2 w}{\partial t^2} \tag{8.19}$$

Upon substitution of (8.18) into (8.19), we obtain

$$\nabla^4 W - \beta\frac{\partial^2 W}{\partial x\,\partial y} = \alpha W \tag{8.20}$$

where $\alpha = \rho h\Omega^2/D$ is the dimensionless frequency and $\beta = 2N_{xy}/D$ is the dimensionless membrane prestress in shear. We now discretize (8.20) and readily obtain the generalized algebraic eigenvalue-eigenvector problem

$$AW = \lambda BW \tag{8.21}$$

where $A$ and $B$ are square matrices, $W$ is the column vector that characterizes the mode shape, and $\lambda = k^4\alpha = k^4\rho h\Omega^2/D$ is the eigenvalue of (8.21). The elements of the matrix $A$ are functions of $\beta$ (or $N_{xy}$). We can employ a digital computer to solve for $\lambda$ in (8.21) using a series of values of the parameter $N_{xy}$. The value $N_{xy}$ assumes when $\lambda = \Omega = 0$ is the buckling value $(N_{xy})_{cr}$ because in this case the period of the motion is unbounded. Such a series of computations have been performed for the plate shown in Fig. 8.16, and the result can be expressed by the equation

$$(N_{xy})_{cr} = \Gamma\frac{D}{a^2} \tag{8.22}$$

The constant $\Gamma$ depends upon the mesh size, as shown in Table 8.1 and Fig. 8.17. A previously obtained approximate solution of this problem using an energy method yields the result

$$(N_{xy})_{cr} = 65.14\frac{D}{a^2}$$

Consequently, with reference to Table 8.1, when using a mesh size resulting in 171 nodes, the buckling value obtained by application of the finite difference method deviates by approximately 1.32% from the value obtained by the energy method. The results in Table 8.1 and the curve in Fig. 8.17 were taken from an unpublished report by K. G. Solney and E. K. Marshman (State University of New York at Buffalo).

**TABLE 8.1 Convergence of Finite Difference Method**

| Number of Nodes | $\frac{k}{a} = \frac{l}{a}$ | $\Gamma$ |
|---|---|---|
| 21 | $\frac{1}{4}$ | 81.60 |
| 36 | $\frac{1}{5}$ | 74.00 |
| 55 | $\frac{1}{6}$ | 70.56 |
| 105 | $\frac{1}{8}$ | 67.84 |
| 171 | $\frac{1}{10}$ | 66.00 |

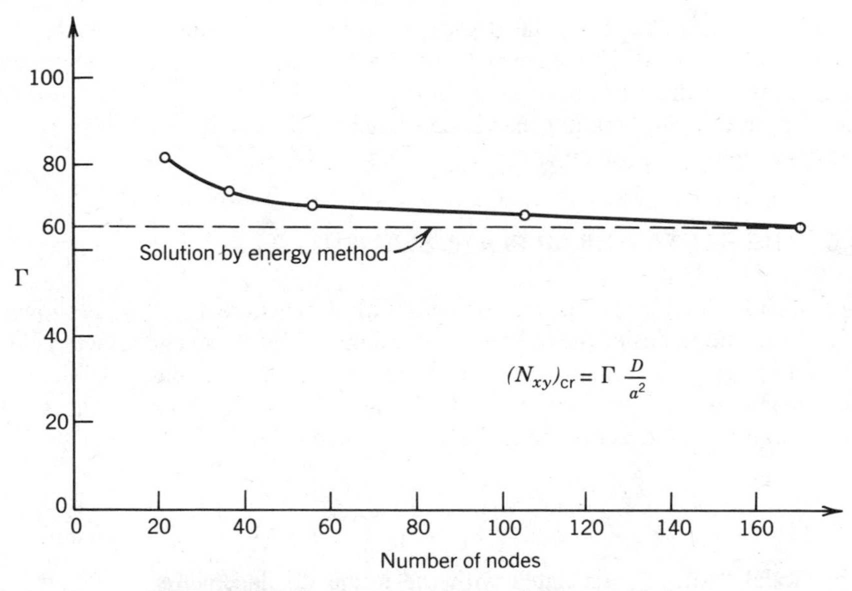


**Figure 8.17** Convergence of finite difference solution.

## 8.4 THE FINITE ELEMENT METHOD

The application and widespread use of the finite element method to solve problems in the theory of elastic plates is a relatively recent development. It is no coincidence that this occurred during the same period that the modern, high-speed digital computer was being developed. The finite element method is readily automated and depends upon the availability of a computer for its implementation.

The basic ideas of the finite element method are simple and physically appealing. The plate under consideration is viewed as an assembly of smaller plate (or finite) elements. These elements have a simple geometric shape. A possible displacement field in each element is assumed and expressed in terms of its values at certain points of the element called nodes. The nodal degrees of freedom, as they are called, determine the desired solution for each element. When the elements are pieced together to form the plate under consideration, there results a piecewise approximation for the displacements and stresses in the entire plate in terms of the (as yet) unknown nodal degrees of freedom. The total displacement field and its associated stress field is then required to satisfy either a variational principle, a conservation law, or an equilibrium condition. This requirement results in a set of simultaneous, algebraic equations that can be solved for the nodal degrees of freedom. In the subsequent development, we shall use the principle of minimum potential energy to obtain this set of equations. For a given element shape, many different finite element models of varying degrees of sophistication can be derived. These models differ in the number of nodal degrees of freedom and the element displacement field assumed. The elements selected for our subsequent treatment are elementary in the sense that the assumed element displacement field provides the simplest approximation that is consistent with certain continuity requirements at element boundaries.

## 8.5 THE RECTANGULAR PLATE ELEMENT

We consider a rectangular plate element with dimensions $c : b : h$, as shown in Fig. 8.18. The nodes of the element are designated by integers characterized by the letters $i$, $j$, $k$, and $l$. At each node $p$ we consider the possible nodal displacements $w_p$ (translation in the $z$-direction, $\theta_{xp}$ (rotation about the $x$-axis), and $\theta_{yp}$ (rotation about the $y$-axis), where

$$\theta_{xp} = \left(\frac{\partial w}{\partial y}\right)_p \quad \text{and} \quad \theta_{yp} = -\left(\frac{\partial w}{\partial x}\right)_p \tag{8.23}$$

The nodal "forces" associated with the nodal displacements $w_p, \theta_{xp}, \theta_{yp}$ are $F_{zp}, M_{xp}, M_{yp}$, respectively, and it will be convenient to define the element displacement vector

$$\delta(x, y) = \left[w(x, y) \quad \theta_x(x, y) \quad \theta_y(x, y)\right]^{\mathrm{T}} \tag{8.24}$$

If we wish to characterize the shape of the median surface of the rectangular plate element by its nodal displacements, then it is clear, with reference to (8.24) and Fig. 8.18, that this element will have $4 \times 3 = 12$ degrees of freedom. Consequently, we need to select a polynominal with 12 parameters (or constants), and the remainder of this analysis is directed toward the determination of these constants. The polynominal selected is

$$\begin{aligned} w = A_1 + A_2 x + A_3 y + A_4 x^2 + A_5 xy + A_6 y^2 + A_7 x^3 + A_8 x^2 y \\ + A_9 xy^2 + A_{10} y^3 + A_{11} x^3 y + A_{12} xy^3 \end{aligned} \tag{8.25a}$$

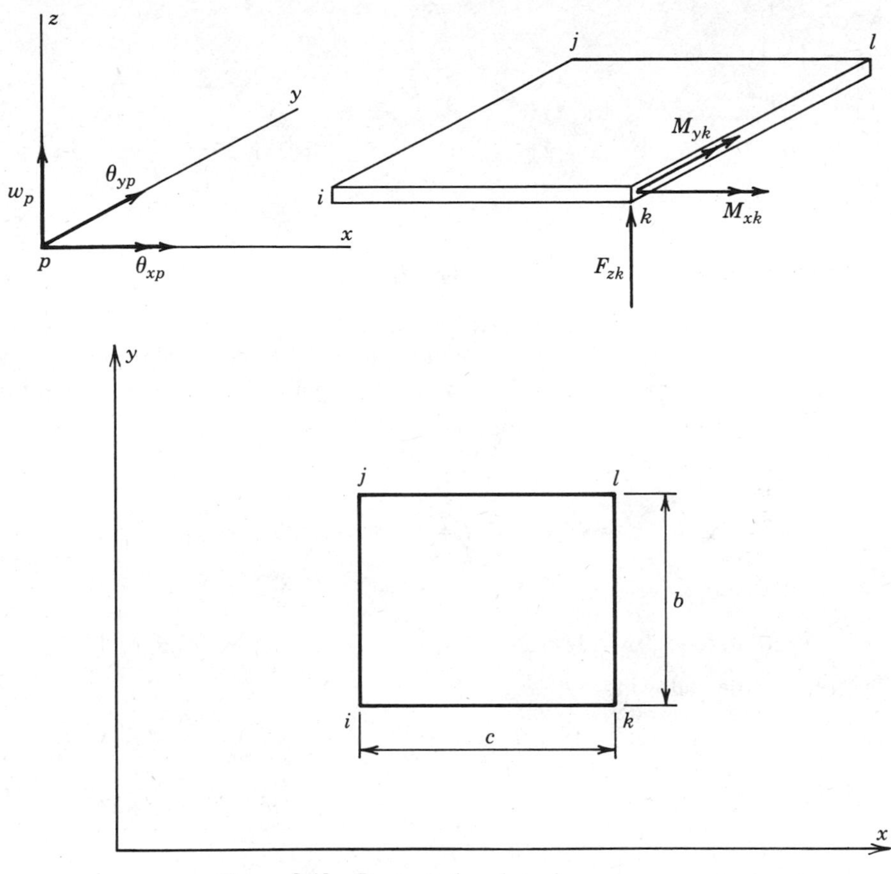


**Figure 8.18** Rectangular plate element.

so that

$$\theta_x = \frac{\partial w}{\partial y} = A_3 + A_5 x + 2A_6 y + A_8 x^2 + 2A_9 xy + 3A_{10} y^2$$

$$+A_{11} x^3 + 3A_{12} xy^2 \tag{8.25b}$$

$$\theta_y = -\frac{\partial w}{\partial x} = -A_2 - 2A_4 x - A_5 y - 3A_7 x^2 - 2A_8 xy$$

$$-A_9 y^2 - 3A_{11} x^2 y - A_{12} y^3 \tag{8.25c}$$

Thus, the displacement field $\{\delta(x, y)\}$ at any point can be written as

$$\{\delta(x, y)\} = [f(x, y)]\{A\} \tag{8.26}$$

where

$$[f(x,y)] = \begin{bmatrix} 1 & x & y & x^2 & xy & y^2 & x^3 & x^2y & xy^2 & y^3 & x^3y & xy^3 \\ 0 & 0 & 1 & 0 & x & 2y & 0 & x^2 & 2xy & 3y^2 & x^3 & 3xy^2 \\ 0 & -1 & 0 & -2x & -y & 0 & -3x^2 & -2xy & -y^2 & 0 & -3x^2y & -y^3 \end{bmatrix} \tag{8.27}$$

$$\{A\} = [A_1 \quad A_2 \quad A_3 \quad A_4 \quad A_5 \quad A_6 \quad A_7 \quad A_8 \quad A_9 \quad A_{10} \quad A_{11} \quad A_{12}]^{\mathrm{T}} \tag{8.28}$$

We shall now determine the displacement field $\{\delta(x, y)\}$ in terms of the nodal displacements $\{\delta^e\} = \{\delta_i \delta_j \delta_k \delta_l\}^{\mathrm{T}}$ relative to a local coordinate system the origin of which coincides with the center of the plate element. Upon substitution of the nodal coordinates into (8.26) such that

$$\begin{aligned} &\text{at Node } i\text{:} \quad x = -\tfrac{1}{2}c, \quad y = -\tfrac{1}{2}b, \quad \{\delta_i\} = \{w_i \theta_{xi} \theta_{yi}\}^{\mathrm{T}} \\ &\text{at Node } j\text{:} \quad x = -\tfrac{1}{2}c, \quad y = \tfrac{1}{2}b, \quad \{\delta_j\} = \{w_j \theta_{xj} \theta_{yj}\}^{\mathrm{T}} \\ &\text{at Node } k\text{:} \quad x = \tfrac{1}{2}c, \quad y = -\tfrac{1}{2}b, \quad \{\delta_k\} = \{w_k \theta_{xk} \theta_{yk}\}^{\mathrm{T}} \\ &\text{at Node } l\text{:} \quad x = \tfrac{1}{2}c, \quad y = \tfrac{1}{2}b, \quad \{\delta_l\} = \{w_l \theta_{xl} \theta_{yl}\}^{\mathrm{T}} \end{aligned} \tag{8.29}$$

we obtain the following:

$$\begin{Bmatrix} w_i \\ \theta_{xi} \\ \theta_{yi} \\ \hline w_j \\ \theta_{xj} \\ \theta_{yj} \\ \hline w_k \\ \theta_{xk} \\ \theta_{yk} \\ \hline w_l \\ \theta_{xl} \\ \theta_{yl} \end{Bmatrix} = \left[\begin{array}{cccccccccccc} 1 & -\frac{1}{2}c & -\frac{1}{2}b & \frac{1}{4}c^2 & \frac{1}{4}bc & \frac{1}{4}b^2 & -\frac{1}{8}c^3 & -\frac{1}{8}bc^2 & -\frac{1}{8}b^2c & -\frac{1}{8}b^3 & \frac{1}{16}bc^3 & \frac{1}{16}b^3c \\ 0 & 0 & 1 & 0 & -\frac{1}{2}c & -b & 0 & \frac{1}{4}c^2 & \frac{1}{2}bc & \frac{3}{4}b^2 & -\frac{1}{8}c^3 & -\frac{3}{8}b^2c \\ 0 & -1 & 0 & c & \frac{1}{2}b & 0 & -\frac{3}{4}c^2 & -\frac{1}{2}bc & -\frac{1}{4}b^2 & 0 & \frac{3}{8}bc^2 & \frac{1}{8}b^3 \\ \hline 1 & -\frac{1}{2}c & \frac{1}{2}b & \frac{1}{4}c^2 & -\frac{1}{4}bc & \frac{1}{4}b^2 & -\frac{1}{8}c^3 & \frac{1}{8}bc^2 & -\frac{1}{8}b^2c & \frac{1}{8}b^3 & -\frac{1}{16}bc^3 & -\frac{1}{16}b^3c \\ 0 & 0 & 1 & 0 & -\frac{1}{2}c & b & 0 & \frac{1}{4}c^2 & -\frac{1}{2}bc & \frac{3}{4}b^2 & -\frac{1}{8}c^3 & -\frac{3}{8}b^2c \\ 0 & -1 & 0 & c & -\frac{b}{2} & 0 & -\frac{3}{4}c^2 & \frac{1}{2}bc & -\frac{1}{4}b^2 & 0 & -\frac{3}{8}bc^2 & -\frac{1}{8}b^3 \\ \hline 1 & \frac{1}{2}c & -\frac{1}{2}b & \frac{1}{4}c^2 & -\frac{1}{4}bc & \frac{1}{4}b^2 & \frac{1}{8}c^3 & -\frac{1}{8}bc^2 & \frac{1}{8}b^2c & -\frac{1}{8}b^3 & -\frac{1}{16}bc^3 & -\frac{1}{16}b^3c \\ 0 & 0 & 1 & 0 & \frac{1}{2}c & -b & 0 & \frac{1}{4}c^2 & -\frac{1}{2}bc & \frac{3}{4}b^2 & \frac{1}{8}c^3 & \frac{3}{8}b^2c \\ 0 & -1 & 0 & -c & \frac{1}{2}b & 0 & -\frac{3}{4}c^2 & \frac{1}{2}bc & -\frac{1}{4}b^2 & 0 & \frac{3}{8}bc^2 & \frac{1}{8}b^3 \\ \hline 1 & \frac{1}{2}c & \frac{1}{2}b & \frac{1}{4}c^2 & \frac{1}{4}bc & \frac{1}{4}b^2 & \frac{1}{8}c^3 & \frac{1}{8}bc^2 & \frac{1}{8}b^2c & \frac{1}{8}b^3 & \frac{1}{16}bc^3 & \frac{1}{16}b^3c \\ 0 & 0 & 1 & 0 & \frac{1}{2}c & b & 0 & \frac{1}{4}c^2 & \frac{1}{2}bc & \frac{3}{4}b^2 & \frac{1}{8}c^3 & \frac{3}{8}b^2c \\ 0 & -1 & 0 & -c & -\frac{1}{2}b & 0 & -\frac{3}{4}c^2 & -\frac{1}{2}bc & -\frac{1}{4}b^2 & 0 & -\frac{3}{8}bc^2 & -\frac{1}{8}b^3 \end{array}\right] \begin{Bmatrix} A_1 \\ A_2 \\ A_3 \\ A_4 \\ A_5 \\ A_6 \\ A_7 \\ A_8 \\ A_9 \\ A_{10} \\ A_{11} \\ A_{12} \end{Bmatrix} \tag{8.30}$$

Solving these 12 equations for the coefficients $A_i$, $i = 1, 2, \ldots, 12$, in terms of the nodal displacements, we obtain

$$A_1 = \tfrac{1}{4}(w_i + w_j + w_k + w_l) + \tfrac{1}{16}b(\theta_{xi} - \theta_{xj} + \theta_{xk} - \theta_{xl}) + \tfrac{1}{16}c(-\theta_{yi} - \theta_{yj} + \theta_{yk} + \theta_{yl})$$

$$A_2 = \frac{3}{4c}(-w_i - w_j + w_k + w_l) + \frac{b}{8c}(-\theta_{xi} + \theta_{xj} + \theta_{xk} - \theta_{xl}) + \tfrac{1}{8}(\theta_{yi} + \theta_{yj} + \theta_{yk} + \theta_{yl})$$

$$A_3 = \frac{3}{4b}(-w_i + w_j - w_k + w_l) + \tfrac{1}{8}(-\theta_{xi} - \theta_{xj} - \theta_{xx} - \theta_{xl}) + \frac{c}{8b}(\theta_{yi} - \theta_{yj} - \theta_{yk} + \theta_{yl})$$

$$
\begin{aligned}
A_4 &= \frac{1}{4c}(\theta_{yi} + \theta_{yj} - \theta_{yk} - \theta_{yl}) \\
A_5 &= \frac{2}{bc}(w_i - w_j - w_k + w_l) + \frac{1}{4c}(\theta_{xi} + \theta_{xj} - \theta_{xk} - \theta_{xl}) + \frac{1}{4b}(-\theta_{yi} + \theta_{yj} - \theta_{yk} + \theta_{yl}) \\
A_6 &= \frac{1}{4b}(-\theta_{xi} + \theta_{xj} - \theta_{xk} + \theta_{xl}) \\
A_7 &= \frac{1}{c^3}(w_i + w_j - w_k - w_l) + \frac{1}{2c^2}(-\theta_{yi} - \theta_{yj} - \theta_{yk} - \theta_{yl}) \\
A_8 &= \frac{1}{2bc}(-\theta_{yi} + \theta_{yj} + \theta_{yk} - \theta_{yl}) \\
A_9 &= \frac{1}{2bc}(\theta_{xi} - \theta_{xj} - \theta_{xk} + \theta_{xl}) \\
A_{10} &= \frac{1}{b^3}(w_i - w_j + w_k - w_l) + \frac{1}{2b^2}(\theta_{xi} + \theta_{xj} + \theta_{xk} + \theta_{xl}) \\
A_{11} &= \frac{2}{bc^3}(-w_i + w_j + w_k - w_l) + \frac{1}{bc^2}(\theta_{yi} - \theta_{yj} + \theta_{yk} - \theta_{yl}) \\
A_{12} &= \frac{2}{b^3c}(-w_i + w_j + w_k - w_l) + \frac{1}{b^2c}(-\theta_{xi} - \theta_{xj} + \theta_{xk} + \theta_{xl})
\end{aligned}
\tag{8.31}
$$

Substitution of the constants (8.31) into (8.25a) and subsequent rearrangement results in the displacement function expressed in terms of nodal displacements:

$$
w = [L]\{\delta^e\} = \left[[L]_i \mid [L]_j \mid [L]_k \mid [L]_l\right]\begin{Bmatrix}\delta_i \\ \delta_j \\ \delta_k \\ \delta_l\end{Bmatrix} \tag{8.32}
$$

where $[L]$ are the shape functions at each node given as follows:

$$
\begin{aligned}
[L]_i = \Bigg[ & \frac{1}{4} - \frac{3x}{4c} - \frac{3y}{4b} + \frac{2xy}{bc} + \frac{x^3}{c^3} + \frac{y^3}{b^3} - \frac{2x^2y}{bc^3} - \frac{2xy^2}{b^3c} \\
& \Bigg| \; \frac{b}{16} - \frac{bx}{8c} - \frac{y}{8} + \frac{xy}{4c} - \frac{y^2}{4b} + \frac{xy^2}{2bc} + \frac{y^3}{2b^2} - \frac{xy^3}{b^2c} \; \Bigg| \\
& -\frac{c}{16} + \frac{x}{8} + \frac{cy}{8b} + \frac{x^2}{4c} - \frac{xy}{4b} - \frac{x^3}{2c^2} - \frac{x^2y}{2bc} + \frac{x^3y}{bc^2} \Bigg]
\end{aligned}
\tag{8.33a}
$$

$$
\begin{aligned}
[L]_j = \Bigg[ & \frac{1}{4} - \frac{3x}{4c} + \frac{3y}{4b} - \frac{2xy}{bc} + \frac{x^3}{c^3} - \frac{y^3}{b^3} + \frac{2x^3y}{bc^3} + \frac{2xy^3}{b^3c} \\
& \Bigg| \; -\frac{b}{16} + \frac{bx}{8c} - \frac{y}{8} + \frac{xy}{4c} + \frac{y^2}{4b} - \frac{xy^2}{2bc} + \frac{y^3}{2b^2} - \frac{xy^3}{b^2c} \; \Bigg| \\
& -\frac{c}{16} + \frac{x}{8} - \frac{cy}{8b} + \frac{x^2}{4c} + \frac{xy}{4b} - \frac{x^3}{2c^2} + \frac{x^2y}{2bc} - \frac{x^3y}{bc^2} \Bigg]
\end{aligned}
\tag{8.33b}
$$

$$[L]_k = \Bigg[\frac{1}{4} + \frac{3x}{4c} - \frac{3y}{4b} - \frac{2xy}{bc} - \frac{x^3}{c^3} + \frac{y^3}{b^3} + \frac{2x^3y}{bc^3} + \frac{2xy^3}{b^3c}$$
$$\Bigg| \; \frac{b}{16} + \frac{bx}{8c} - \frac{y}{8} - \frac{xy}{4c} - \frac{y^2}{4b} - \frac{xy^2}{2bc} + \frac{y^3}{2b^2} + \frac{xy^3}{b^2c} \; \Bigg|$$
$$\frac{c}{16} + \frac{x}{8} + \frac{cy}{8b} - \frac{x^2}{4c} - \frac{xy}{4b} - \frac{x^3}{2c^2} + \frac{x^2y}{2bc} + \frac{x^3y}{bc^2}\Bigg] \tag{8.33c}$$

$$[L]_l = \Bigg[\frac{1}{4} + \frac{3x}{4c} + \frac{3y}{4b} + \frac{2xy}{bc} - \frac{x^3}{c^3} - \frac{y^3}{b^3} - \frac{2x^3y}{bc^3} - \frac{2xy^3}{b^3c}$$
$$\Bigg| \; -\frac{b}{16} - \frac{bx}{8c} - \frac{y}{8} - \frac{xy}{4c} + \frac{y^2}{4b} + \frac{xy^2}{2bc} + \frac{y^3}{2b^2} + \frac{xy^3}{b^2c} \; \Bigg|$$
$$\frac{c}{16} + \frac{x}{8} - \frac{cy}{8b} - \frac{x^2}{4c} + \frac{xy}{4b} - \frac{x^3}{2c^2} - \frac{x^2y}{2bc} - \frac{x^3y}{bc^2}\Bigg] \tag{8.33d}$$

For our present purpose, we define the (flexural) plate strains by the vector

$$\{\varepsilon\} = \left[-\frac{\partial^2 w}{\partial x^2} \quad -\frac{\partial^2 w}{\partial y^2} \quad 2\frac{\partial^2 w}{\partial x\,\partial y}\right]^{\mathrm{T}} \tag{8.34}$$

Upon substitution of (8.32) into (8.34), we obtain

$$\{\varepsilon\} = \left[-\frac{\partial^2 [L]}{\partial x^2} \quad -\frac{\partial^2 [L]}{\partial y^2} \quad 2\frac{\partial^2 [L]}{\partial x\,\partial y}\right]^{\mathrm{T}} \{\delta^e\} \tag{8.35}$$

or

$$\{\varepsilon\} = [B]\{\delta^e\} = \big[[B]_i \;\big|\; [B]_j \;\big|\; [B]_k \;\big|\; [B]_l\big] \begin{Bmatrix} \delta_i \\ \delta_j \\ \delta_k \\ \delta_l \end{Bmatrix} \tag{8.36}$$

where

$$[B]_i = \begin{bmatrix} -\partial^2 [L]_i/\partial x^2 \\ -\partial^2 [L]_i/\partial y^2 \\ 2\partial^2 [L]_i/\partial x\,\partial y \end{bmatrix}$$
$$= \left[\begin{array}{c|c|c} -\dfrac{6x}{c^3} + \dfrac{12xy}{bc^3} & 0 & -\dfrac{1}{2c} + \dfrac{3x}{c^2} + \dfrac{y}{bc} - \dfrac{6xy}{bc^2} \\ -\dfrac{6y}{b^3} + \dfrac{12xy}{b^3c} & \dfrac{1}{2b} - \dfrac{x}{bc} - \dfrac{3y}{b^2} + \dfrac{6xy}{b^2c} & 0 \\ 2\left(\dfrac{2}{bc} - \dfrac{6x^2}{bc^3} - \dfrac{6y^2}{b^3c}\right) & 2\left(\dfrac{1}{4c} + \dfrac{y}{bc} - \dfrac{3y^2}{b^2c}\right) & 2\left(-\dfrac{1}{4b} - \dfrac{x}{bc} + \dfrac{3x^2}{bc^2}\right) \end{array}\right] \tag{8.37a}$$

$$[B]_j = \begin{bmatrix} -\partial^2[L]_j/\partial x^2 \\ -\partial^2[L]_j/\partial y^2 \\ 2\partial^2[L]_j/\partial x\,\partial y \end{bmatrix}$$

$$= \left[\begin{array}{c|c|c} -\dfrac{6x}{c^3} - \dfrac{12xy}{bc^3} & 0 & -\dfrac{1}{2c} + \dfrac{3x}{c^2} - \dfrac{y}{bc} + \dfrac{6xy}{bc^2} \\ \dfrac{6y}{3} - \dfrac{12xy}{b^3c} & -\dfrac{1}{2b} + \dfrac{x}{bc} - \dfrac{3y}{b^2} + \dfrac{6xy}{b^2c} & 0 \\ 2\left(-\dfrac{2}{bc} + \dfrac{6x^2}{bc^3} + \dfrac{6y^2}{b^3c}\right) & 2\left(\dfrac{1}{4c} - \dfrac{y}{bc} - \dfrac{3y^2}{b^2c}\right) & 2\left(\dfrac{1}{4b} + \dfrac{x}{bc} - \dfrac{3x^2}{bc^2}\right) \end{array}\right] \tag{8.37b}$$

$$[B]_k = \begin{bmatrix} -\partial^2[L]_k/\partial x^2 \\ -\partial^2[L]_k/\partial y^2 \\ 2\partial^2[L]_k/\partial x\,\partial y \end{bmatrix}$$

$$= \left[\begin{array}{c|c|c} \dfrac{6x}{c^3} - \dfrac{12xy}{bc^3} & 0 & \dfrac{1}{2c} + \dfrac{3x}{c^2} - \dfrac{y}{bc} - \dfrac{6xy}{bc^2} \\ -\dfrac{6y}{b^3} - \dfrac{12xy}{b^3c} & \dfrac{1}{2b} + \dfrac{x}{bc} - \dfrac{3y}{b^2} - \dfrac{6xy}{b^2c} & 0 \\ 2\left(-\dfrac{2}{bc} + \dfrac{6x^2}{bc^3} + \dfrac{6y^2}{b^3c}\right) & 2\left(-\dfrac{1}{4c} - \dfrac{y}{bc} + \dfrac{3y^2}{b^2c}\right) & 2\left(-\dfrac{1}{4b} + \dfrac{x}{bc} + \dfrac{3x^2}{bc^2}\right) \end{array}\right] \tag{8.37c}$$

$$[B]_l = \begin{bmatrix} -\partial^2[L]_l/\partial x^2 \\ -\partial^2[L]_l/\partial y^2 \\ 2\partial^2[L]_l/\partial x\,\partial y \end{bmatrix}$$

$$= \left[\begin{array}{c|c|c} \dfrac{6x}{c^3} + \dfrac{12xy}{bc^3} & 0 & \dfrac{1}{2c} + \dfrac{3x}{c^2} + \dfrac{y}{bc} + \dfrac{6xy}{bc^2} \\ \dfrac{6y}{b^3} + \dfrac{12xy}{b^3c} & -\dfrac{1}{2b} - \dfrac{x}{bc} - \dfrac{3y}{b^2} - \dfrac{6xy}{b^2c} & 0 \\ 2\left(\dfrac{2}{bc} - \dfrac{6x^2}{bc^3} - \dfrac{6y^2}{b^3c}\right) & 2\left(-\dfrac{1}{4c} + \dfrac{y}{bc} + \dfrac{3y^2}{b^2c}\right) & 2\left(\dfrac{1}{4b} - \dfrac{x}{bc} - \dfrac{3x^2}{bc^2}\right) \end{array}\right] \tag{8.37d}$$

We shall now relate internal plate stresses

$$\{\sigma(x, y)\} = \begin{bmatrix} M_{xx} & M_{yy} & M_{xy} \end{bmatrix}^{\mathrm{T}} \tag{8.38}$$

to the plate strain components $\{\varepsilon(x, y)\}$ and to the nodal displacements $\{\delta^e\}$: In view of (4.16a) and (8.38), we have

$$\{\sigma\} = D\begin{bmatrix} 1 & \nu & 0 \\ \nu & 1 & 0 \\ 0 & 0 & \frac{1}{2}(1-\nu) \end{bmatrix}\{\varepsilon\} = [D']\{\varepsilon\} \tag{8.39a}$$

where

$$[D'] = D\begin{bmatrix} 1 & \nu & 0 \\ \nu & 1 & 0 \\ 0 & 0 & \frac{1}{2}(1-\nu) \end{bmatrix} \tag{8.39b}$$

Also, since $\{\varepsilon\} = [B]\{\delta^e\}$, we have

$$\{\sigma\} = [D'][B]\{\delta^e\} = [H]\{\delta^e\}$$

where $[H] = [D'][B] = [H_{i,j}]$ and we have

$$H_{1,1} = -\frac{6x}{c^3} + \frac{12xy}{bc^3} - \frac{6\nu y}{b^3} + \frac{12\nu xy}{b^3c} \qquad H_{2,1} = -\frac{6\nu x}{c^3} + \frac{12\nu xy}{bc^3} - \frac{6y}{b^3} + \frac{12xy}{b^3c}$$

$$H_{1,2} = \frac{\nu}{2b} - \frac{\nu x}{bc} - \frac{3\nu y}{b^2} + \frac{6\nu xy}{b^2c} \qquad H_{2,2} = \frac{1}{2b} - \frac{x}{bc} - \frac{3y}{b^2} + \frac{6xy}{b^2c}$$

$$H_{1,3} = -\frac{1}{2c} + \frac{3x}{c^2} + \frac{y}{bc} - \frac{6xy}{bc^2} \qquad H_{2,3} = -\frac{\nu}{2c} - \frac{3\nu x}{c^2} + \frac{\nu y}{bc} - \frac{6\nu xy}{bc^2}$$

$$H_{1,4} = -\frac{6x}{c^3} - \frac{12xy}{bc^3} + \frac{6\nu y}{b^3} - \frac{12\nu xy}{b^3c} \qquad H_{2,4} = -\frac{6\nu x}{c^3} - \frac{12\nu xy}{bc^3} + \frac{6y}{b^3} - \frac{12xy}{b^3c}$$

$$H_{1,5} = -\frac{\nu}{2b} + \frac{\nu x}{bc} - \frac{3\nu y}{b^2} + \frac{6\nu xy}{b^2c} \qquad H_{2,5} = -\frac{1}{2b} + \frac{x}{bc} - \frac{3y}{b^2} + \frac{6xy}{b^2c}$$

$$H_{1,6} = -\frac{1}{2c} + \frac{3x}{c^2} - \frac{y}{bc} + \frac{6xy}{bc^2} \qquad H_{2,6} = -\frac{\nu}{2c} + \frac{3\nu x}{c^2} - \frac{\nu y}{bc} + \frac{6\nu xy}{bc^2}$$

$$H_{1,7} = \frac{6x}{c^3} - \frac{12xy}{bc^3} - \frac{6\nu y}{b^3} \qquad H_{2,7} = \frac{6\nu x}{c^3} - \frac{12\nu xy}{bc^3} - \frac{6y}{b^3} - \frac{12xy}{b^3c}$$

$$H_{1,8} = \frac{\nu}{2b} + \frac{\nu x}{bc} - \frac{3\nu y}{b^2} - \frac{6\nu xy}{b^2c} \qquad H_{2,8} = \frac{1}{2b} + \frac{x}{bc} - \frac{3y}{b^2} - \frac{6xy}{b^2c}$$

$$H_{1,9} = \frac{1}{2c} + \frac{3x}{c^2} - \frac{y}{bc} - \frac{6xy}{bc^2} \qquad H_{2,9} = \frac{\nu}{2c} + \frac{3\nu x}{c^2} - \frac{\nu y}{bc} - \frac{6\nu xy}{bc^2}$$

$$H_{1,10} = \frac{6x}{c^3} + \frac{12xy}{bc^3} + \frac{6\nu y}{b^3} + \frac{12\nu xy}{b^3c} \qquad H_{2,10} = \frac{6\nu x}{c^3} + \frac{12\nu xy}{bc^3} + \frac{6y}{b^3} + \frac{12xy}{b^3c}$$

$$H_{1,11} = -\frac{\nu}{2b} - \frac{\nu x}{bc} - \frac{3\nu y}{b^2} - \frac{6\nu xy}{b^2c} \qquad H_{2,11} = -\frac{1}{2b} - \frac{x}{bc} - \frac{3y}{b^2} - \frac{6xy}{b^3c}$$

$$H_{1,12} = \frac{1}{2c} + \frac{3x}{c^2} + \frac{y}{bc} + \frac{6xy}{bc^2} \qquad H_{2,12} = \frac{\nu}{2c} + \frac{3\nu x}{c^2} + \frac{\nu y}{bc} + \frac{6\nu xy}{bc^2}$$

$$H_{3,1} = (1 - \nu)\left(\frac{2}{bc} - \frac{6x^2}{bc^3} - \frac{6y^2}{b^3c}\right)$$

$$H_{3,2} = (1 - \nu)\left(\frac{1}{4c} + \frac{y}{bc} - \frac{3y^2}{b^2c}\right)$$

$$H_{3,3} = (1 - \nu)\left(-\frac{1}{4b} - \frac{x}{bc} + \frac{3x^2}{bc^2}\right)$$

$$H_{3,4} = (1 - \nu)\left(-\frac{2}{bc} + \frac{6x^2}{bc^3} + \frac{6y^2}{b^3c}\right)$$

$$H_{3,5} = (1 - \nu)\left(\frac{1}{4c} - \frac{y}{bc} - \frac{3y^2}{b^2c}\right)$$

$$H_{3,6} = (1 - \nu)\left(\frac{1}{4b} + \frac{x}{bc} - \frac{3x^2}{bc^2}\right)$$

$$H_{3,7} = (1 - \nu)\left(-\frac{2}{bc} + \frac{6x^2}{bc^3} + \frac{6y^2}{b^3c}\right)$$

$$H_{3,8} = (1 - \nu)\left(-\frac{1}{4c} - \frac{y}{bc} + \frac{3y^2}{b^2c}\right)$$

$$H_{3,9} = (1 - \nu)\left(-\frac{1}{4b} + \frac{x}{bc} + \frac{3x^2}{bc^2}\right)$$

$$H_{3,10} = (1 - \nu)\left(\frac{2}{bc} - \frac{6x^2}{bc^3} - \frac{6y^2}{b^3c}\right)$$

$$H_{3,11} = (1 - \nu)\left(-\frac{1}{4c} + \frac{y}{bc} + \frac{3y^2}{b^2c}\right)$$

$$H_{3,12} = (1 - \nu)\left(\frac{1}{4b} - \frac{x}{bc} - \frac{3x^2}{bc^2}\right) \tag{8.40}$$

Thus the "stress-strain" relationship

$$\{\sigma^e\} = \{H^e\}\{\delta^e\} \tag{8.41a}$$

is obtained by substituting the nodal coordinates into (8.40). The result is

$$\{\sigma^e\} = \left[M_{xi}M_{yi}M_{xyi} \;\vdots\; M_{xj}M_{yj}M_{xyj} \;\vdots\; M_{xk}M_{yk}M_{xyk} \;\vdots\; M_{xl}M_{yl}M_{xyl}\right] \tag{8.41b}$$

and

$$
[H^e] = D \left[ \begin{array}{ccc:ccc:ccc:ccc}
\left(\frac{6}{c^2} + \frac{6\nu}{b^2}\right) & 4b\nu & -4c & -\frac{6\nu}{b^2} & \frac{2\nu}{b} & 0 & -\frac{6}{c^2} & 0 & -\frac{2}{c} & 0 & 0 & 0 \\
\left(\frac{6\nu}{c^2} + \frac{6}{b^2}\right) & 4b & -4c\nu & -\frac{6}{b^2} & \frac{2}{b} & 0 & -\frac{b\nu}{c^2} & 0 & -\frac{2\nu}{c} & 0 & 0 & 0 \\
-\frac{(1-\nu)}{bc} & -\frac{(1-\nu)}{c} & \frac{(1-\nu)}{c} & \frac{(1-\nu)}{bc} & 0 & -\frac{(1-\nu)}{b} & \frac{(1-\nu)}{bc} & \frac{(1-\nu)}{c} & 0 & -\frac{(1-\nu)}{bc} & 0 & 0 \\
\hdashline
-\frac{6\nu}{b^2} & -\frac{2\nu}{b} & 0 & \left(\frac{6}{c^2} + \frac{6\nu}{b^2}\right) & -\frac{4\nu}{b} & -\frac{4}{c} & 0 & 0 & 0 & -\frac{6}{c^2} & 0 & -\frac{2}{c} \\
-\frac{6}{b^2} & -\frac{2}{b} & 0 & \left(\frac{6\nu}{c^2} + \frac{6}{b^2}\right) & -\frac{4}{b} & -\frac{4\nu}{c} & 0 & 0 & 0 & -\frac{6\nu}{c^2} & 0 & -\frac{2\nu}{c} \\
-\frac{(1-\nu)}{bc} & 0 & \frac{(1-\nu)}{b} & \frac{(1-\nu)}{bc} & -\frac{(1-\nu)}{c} & -\frac{(1-\nu)}{b} & \frac{(1-\nu)}{bc} & 0 & 0 & -\frac{(1-\nu)}{bc} & \frac{(1-\nu)}{c} & 0 \\
\hdashline
-\frac{6}{c^2} & 0 & \frac{2}{c} & 0 & 0 & 0 & \left(\frac{6}{c^2} + \frac{6\nu}{b^2}\right) & \frac{4\nu}{b} & \frac{4}{c} & -\frac{6\nu}{b^2} & \frac{2\nu}{b} & 0 \\
-\frac{6\nu}{c^2} & 0 & \frac{2\nu}{c} & 0 & 0 & 0 & \left(\frac{6\nu}{c^2} + \frac{6}{b^2}\right) & \frac{4}{b} & \frac{4\nu}{c} & -\frac{6}{b^2} & \frac{2}{b} & 0 \\
-\frac{(1-\nu)}{bc} & -\frac{(1-\nu)}{c} & 0 & \frac{(1-\nu)}{bc} & 0 & 0 & \frac{(1-\nu)}{bc} & \frac{(1-\nu)}{c} & \frac{(1-\nu)}{b} & -\frac{(1-\nu)}{bc} & 0 & -\frac{(1-\nu)}{b} \\
\hdashline
0 & 0 & 0 & -\frac{6}{c^2} & 0 & \frac{2}{c} & -\frac{6\nu}{b^2} & -\frac{2\nu}{b} & 0 & \left(\frac{6}{c^2} + \frac{6\nu}{b^2}\right) & -\frac{4\nu}{b} & \frac{4}{c} \\
0 & 0 & 0 & -\frac{6\nu}{c^2} & 0 & \frac{2\nu}{c} & -\frac{6}{b^2} & -\frac{2}{b} & 0 & \left(\frac{6\nu}{c^2} + \frac{6}{b^2}\right) & -\frac{4}{b} & \frac{4\nu}{c} \\
-\frac{(1-\nu)}{bc} & 0 & 0 & \frac{(1-\nu)}{bc} & -\frac{(1-\nu)}{c} & 0 & \frac{(1-\nu)}{bc} & 0 & \frac{(1-\nu)}{b} & -\frac{(1-\nu)}{bc} & \frac{(1-\nu)}{c} & -\frac{(1-\nu)}{b}
\end{array} \right] \tag{8.42}
$$

The vector (8.41) is the "element or nodal force vector," while (8.42) is referred to as the "element bending moment matrix."

We shall now employ the principle of potential energy to derive the element stiffness matrix $[K^e]$ and the element nodal force vector $\{F^e\}$. Toward this end, we write the strain energy in the element as [see (3.62)]

$$\begin{aligned} U &= \frac{1}{2}\int_A [\varepsilon]^{\mathrm{T}}\{\sigma\}\,dA \\ &= \frac{1}{2}\int_A \{[\delta^e]^{\mathrm{T}}[B]^{\mathrm{T}}[D'][B][\delta^e]\}\,dA \\ &= \{\delta^e\}^{\mathrm{T}}\left[\int_A \frac{1}{2}[B]^{\mathrm{T}}[D'][B]\,dA\right]\{\delta^e\} \end{aligned}$$

or

$$U = \tfrac{1}{2}\{\delta^e\}^{\mathrm{T}}[K^e]\{\delta^e\} \tag{8.43}$$

where $A$ is the rectangular area $c \times b$ (see Fig. 8.18) and

$$[K^e] = [K_{i,j}] \tag{8.44}$$

is the (symmetric) stiffness matrix. We note that (8.36) and (8.39a), in conjunction with (8.37) and (8.40), were used to obtain the result (8.43).

Expressions for the 144 elements $K_{i,j}$, $i = 1, 2, \ldots, 12$, $j = 1, 2, \ldots, 12$, are listed in the FORTRAN program RPLATE in Section 8.6 under the heading "Calculation of Element Stiffness Matrix." The following FORTRAN code symbols are used: BY $= b$, CX $= c$, BC $= bc$, BOC $= b/c$, COB $= c/b$, BOCC $= b/c^2$, COBB $= c/b^2$, BOCCC $= b/c^3$, COBBB $= c/b^3$, POI $= \nu$, and DE $= Eh^3/[12(1-\nu^2)]$. The reader is urged to show that

$$K_{1,1} = \mathrm{ESM}(1,1) = 4b/c^3 + 4c/b^3 - 4\nu/(5bc) + 14/(5bc);$$

and so forth, where ESM$(I, J)$ denotes the entry $(I, J)$ of the element stiffness matrix.

The potential $V$ of external "forces" acting upon the element is

$$V = -\{\delta^e\}^{\mathrm{T}}\{Q^e\} - p\{\delta^e\}^{\mathrm{T}}\int_A [L]^{\mathrm{T}}\,dA \tag{8.45}$$

where it is assumed that $p$ is constant over the element area, and the vectors $\{Q^e\}$ and $\{\delta^e\}$ are given by

$$\{Q^e\} = \left[F_i \quad M_{xi} \quad M_{yi} \mid F_j \quad M_{xj} \quad M_{yj} \mid F_k \quad M_{xk} \quad M_{yk} \mid F_l \quad M_{xl} \quad M_{yl}\right]^{\mathrm{T}} \tag{8.46}$$

$$\{\delta^e\} = \left[w_i \quad \theta_{xi} \quad \theta_{yi} \mid w_j \quad \theta_{xj} \quad \theta_{yj} \mid w_k \quad \theta_{xk} \quad \theta_{yk} \mid w_l \quad \theta_{xl} \quad \theta_{yl}\right]^{\mathrm{T}} \tag{8.47}$$

When writing expression (8.45), we used the relation [see (8.32)]

$$w = \{\delta^e\}^{\mathrm{T}}[L]^{\mathrm{T}}$$

It will be convenient to write (8.45) in the form

$$V = -\{\delta^e\}^{\mathrm{T}}\{F^e\}$$

where

$$\{F^e\} = \{Q^e\} + p\int_A \{L\}^{\mathrm{T}}\, dA \tag{8.48}$$

is the element force vector. If $\Pi = U + V$ is the total potential energy associated with an element, we have

$$\Pi = \tfrac{1}{2}\{\delta^e\}^{\mathrm{T}}[K^e]\{\delta^e\} - [\delta^e]^{\mathrm{T}}\{F^e\} \tag{8.49}$$

If we apply the principle of minimum potential energy to (8.49), we obtain

$$\begin{aligned}\frac{\partial \Pi}{\{\partial \delta^e\}} &= \left[\frac{\partial \Pi}{\partial w_i} \quad \frac{\partial \Pi}{\partial \theta_{xi}} \quad \frac{\partial \Pi}{\partial \theta_{yi}} \quad \frac{\partial \Pi}{\partial w_j} \quad \cdots \quad \frac{\partial \Pi}{\partial \theta_{yl}}\right]^{\mathrm{T}} \\ &= [K^e]\{\delta^e\} - \{F^e\} = \{0\}\end{aligned} \tag{8.50}$$

where

$$\{F^e\} = \{Q^e\} + \{P^e\}$$

and

$$\{P^e\} = qcb\left[\frac{1}{4} \quad \frac{b}{24} \quad -\frac{c}{24} \;\middle|\; \frac{1}{4} \quad -\frac{b}{24} \quad -\frac{c}{24} \;\middle|\; \frac{1}{4} \quad \frac{b}{24} \quad \frac{c}{24} \;\middle|\; \frac{1}{4} \quad -\frac{b}{24} \quad \frac{c}{24}\right]^{\mathrm{T}} \tag{8.51}$$

The $12 \times 12$ system of linear equations

$$[K^e]\{\delta^e\} = \{F^e\} \tag{8.52}$$

characterizes the deformation of the rectangular plate element subjected to surface as well as nodal forces.

When the elements are assembled, the total potential energy of the plate is characterized by [see (8.49)]

$$\Pi = \sum_{e=1}^{E} \langle \tfrac{1}{2}\{\delta^e\}^{\mathrm{T}}[K^e]\{\delta^e\} - \{\delta^e\}^{\mathrm{T}}\{F^e\}\rangle \tag{8.53}$$

This expression contains $12E$ unknowns, and this is excessive. The value of $\{\delta^e\}$ for two adjacent elements that share a common nodal point must be the same. When these compatibility relations are incorporated into the analysis, we are left with the reduced displacement vector

$$\{\Delta\} = [\Delta_1 \quad \Delta_2 \quad \Delta_3 \quad \ldots \quad \Delta_M]^{\mathrm{T}} \tag{8.54}$$

where $M = 3N$, and $N$ is the total number of nodal points in the mesh.

Consequently, (8.54) characterizes the deformed plate, and its potential energy is

$$\Pi = \tfrac{1}{2}\{\Delta\}^{\mathrm{T}}[K]\{\Delta\} - \{\Delta\}^{\mathrm{T}}F \tag{8.55}$$

where $K$ is the $M \times M$ global stiffness matrix and $F$ is the $M \times 1$ generalized global force vector.

The principle of minimum potential energy requires that the vector $\{\Delta\}$ be selected such that $\Pi$ is a minimum. The necessary condition is

$$\frac{\partial \Pi}{\partial \{\Delta_i\}} = 0, \qquad i = 1, 2, \ldots, M$$

so that

$$[K]\{\Delta\} = \{F\} \tag{8.56}$$

In (8.56), $[K]$ is the $M \times M$ global stiffness matrix, and $\{F\}$ is the $M \times 1$ global force vector. The global stiffness matrix and the global force vector can be assembled from the $12 \times 12$ element stiffness matrices $[K^e]$ and the element force vectors $\{F^e\}$, respectively, where $e = 1, 2, \ldots, E$. The algorithm for the assembly process is well known and is available from books on finite element methods. The assembly process is readily computerized once the individual element properties are known.

The matrix $[K]$ in (8.56) is singular. This is a consequence of possible rigid body motions that must be prevented to obtain a unique solution. When appropriate displacement boundary conditions are applied to (8.56), there results a partition: The rows and columns corresponding to vanishing displacements can be deleted. The remaining equilibrium equations are then solved for the unknown displacements. Once all displacements have been determined, we can utilize (8.41) and (8.42) to determine bending and twisting moments at all element nodes. The order of the final system of algebraic equations will be equal to the total number of unknown displacements of the discretized plate. The coefficient matrix $[K]$ can always be written in banded, symmetric form. The entire process of formulation of element stiffness matrices, assembly of the global stiffness matrix, application of boundary conditions, arrangement in banded symmetric form, and solution is readily computerized. The procedure becomes very efficient when a digital computer is used, and it will be discussed from this point of view in the next section.

## 8.6 A COMPUTER PROGRAM

To carry out the arithmetic operations discussed in Section 8.5 would be prohibitive even for a plate with a very coarse mesh, and the finite element method depends upon a computer program for its full realization. Such a program has been developed and is shown in what follows.

```
C
C
C
C
C ********************************************************************
C
      PROGRAM RPLATE
C
C ********************************************************************
C
      CHARACTER*20 TITLE
      DIMENSION ND(4),NS(12),ESM(12,12),A(6000)
      DIMENSION RR(12,12),EFI(12),EF(12),Q(200),C(12,12)
      DIMENSION SMI(12,12),BMI(12),BM(12),RK(12,12)
      COMMON/TLE/TITLE
      DATA NCL/1/,ID1/0/,KE/0/,NBW/0/
C
C      DEFINITION OF CONTROL PARAMETERS
C
C      NP - NUMBER OF NODAL DISPLACEMENTS
C      NE - NUMBER OF ELEMENTS
C      NBW- BANDWIDTH OF THE SYSTEM OF EQUATIONS
C      ND(1)-NODE NUMBER AT X(1),Y(1)
C      ND(2)-NODE NUMBER AT X(2),Y(2)
C      ND(3)-NODE NUMBER AT X(3),Y(3)
C      ND(4)-NODE NUMBER AT X(4),Y(4)
C      EM - ELASTIC MODULUS OF PLATE MATERIAL
C      POI- POISSON'S RATIO OF PLATE MATERIAL
C      TH - PLATE THICKNESS
C      CX - ELEMENT LENGTH IN X-DIRECTION
C      BY - ELEMENT LENGTH IN Y-DIRECTION
C      NEL-ELEMENT NUMBER
C      Q - LOAD INTENSITY
C      NOTE:X(1)<X(3),X(2)<X(4),Y(1)<Y(2),Y(3)<Y(4)
C
C
C     INPUT OF TITLE CARD AND PARAMETER CARD
C
      OPEN(5,FILE='RPLATIN.DAT',STATUS='OLD')
      READ(5,3)TITLE
    3 FORMAT(A20)
      READ(5,*)NE,NP,EM,TH,POI,CX,BY
C
C     CALCULATION OF SEMI BANDWIDTH
C
      DO4 IK=1,NE
      READ(5,*)NEL,ND,X1,Y1,X2,Y2,X3,Y3,X4,Y4,Q(IK)
      LB1=3*(IABS(ND(1)-ND(2))+1)
      LB2=3*(IABS(ND(1)-ND(3))+1)
      LB3=3*(IABS(ND(1)-ND(4))+1)
      LB4=3*(IABS(ND(2)-ND(3))+1)
      LB5=3*(IABS(ND(2)-ND(4))+1)
      LB6=3*(IABS(ND(3)-ND(4))+1)
      LB=MAX0(LB1,LB2,LB3,LB4,LB5,LB6)
      IF(LB.LE.NBW) GO TO 4
      NBW=LB
      NELBW=NEL
    4 CONTINUE
C
C     POINTERS AND INITIALIZATION OF THE COLUMN VECTOR A.
C
      JGF=NP*NCL
      JGSM=2*JGF
      JEND=JGSM+NP*NBW
      DO13 I=1,JEND
```

```
   13 A(I)=0.0
C
C     OUTPUT OF TITLE AND DATA
C
      OPEN(6,FILE='RPLATOUT.DAT',STATUS='NEW')
      WRITE(6,24)TITLE,CX,BY,TH,EM,POI
   24 FORMAT(1H1,//////1X,
     1A20//1X,
     2'CX=',F10.2//1X,
     3'BY=',F10.2//1X,
     4'TH=',F9.3//1X,
     5'EM=',F12.1//1X,
     6'POI=',F9.3//1X,
     7'ELEMENT DATA',//1X,
     8'   NEL        NODE NUMBERS        X(1)        Y(1)        X(2)',
     9'      Y(2)        X(3)        Y(3)        X(4)        Y(4)       Q')
C
C     ASSEMBLY OF THE GLOBAL STIFFNESS MATRIX
C       AND THE GLOBAL FORCE VECTOR
C
C     INPUT AND ECHO PRINT OF THE ELEMENT DATA
C
      DO7 KK=1,NE
      READ(5,*)NEL,ND,X1,Y1,X2,Y2,X3,Y3,X4,Y4,Q(KK)
      WRITE(6,23)NEL,ND,X1,Y1,X2,Y2,X3,Y3,X4,Y4,Q(KK)
   23 FORMAT(1X,5(2X,I3),8(2X,F8.4),2X,F8.4)
C
C     CALCULATION OF GLOBAL DOF FROM THE NODE NUMBERS
C
      DO18 L=1,4
      NS(3*L-2)=3*ND(L)-2
      NS(3*L-1)=3*ND(L)-1
   18 NS(3*L)=3*ND(L)
C
C     CALCULATION OF ELEMENT STIFFNESS MATRIX
C
      BOCCC=BY/(CX*CX*CX)
      COBBB=CX/(BY*BY*BY)
      BOCC =BY/(CX*CX)
      COBB =CX/(BY*BY)
      BOC  =BY/CX
      COB  =CX/BY
      BC   =BY*CX
      DE   =EM*TH*TH*TH/(12.*(1-POI*POI))
C
      ESM( 1, 1)=4.*BOCCC+4.*COBBB-4.*POI/(5.*BC)+14./(5.*BC)
      ESM( 2, 1)=2.*COBB+4.*POI/(5.*CX)+1/(5.*CX)
      ESM( 2, 2)=4.*COB/3.-4.*POI*BOC/15.+4.*BOC/15.
      ESM( 3, 1)=-2.*BOCC-4.*POI/(5.*BY)-1./(5.*BY)
      ESM( 3, 2)=-POI
      ESM( 3, 3)=4.*BOC/3.-4.*POI*COB/15.+4.*COB/15.
      ESM( 4, 1)=2.*BOCCC-4.*COBBB+4.*POI/(5.*BC)-14./(5.*BC)
      ESM( 4, 2)=-2.*COBB+POI/(5.*CX)-1./(5.*CX)
      ESM( 4, 3)=-BOCC+4.*POI/(5.*BY)+1./(5.*BY)
      ESM( 4, 4)=4.*BOCCC+4.*COBBB-4.*POI/(5.*BC)+14./(5.*BC)
      ESM( 5, 1)=2.*COBB-POI/(5.*CX)+1./(5.*CX)
      ESM( 5, 2)=2.*COB/3.+POI*BOC/15.-BOC/15.
      ESM( 5, 3)=0.
      ESM( 5, 4)=-2.*COBB-4.*POI/(5.*CX)-1./(5.*CX)
      ESM( 5, 5)=4.*COB/3.-4.*POI*BOC/15.+4.*BOC/15.
      ESM( 6, 1)=-BOCC+4.*POI/(5.*BY)+1./(5.*BY)
      ESM( 6, 2)=0.
      ESM( 6, 3)=2.*BOC/3.+4.*POI*COB/15.-4.*COB/15.
      ESM( 6, 4)=-2.*BOCC-4.*POI/(5.*BY)-1./(5.*BY)
      ESM( 6, 5)=POI
      ESM( 6, 6)=4.*BOC/3.-4.*POI*COB/15.+4.*COB/15.
```

```
      ESM( 7, 1)=-4.*BOCCC+2.*COBBB+4.*POI/(5.*BC)-14./(5.*BC)
      ESM( 7, 2)=COBB-4.*POI/(5.*CX)-1./(5.*CX)
      ESM( 7, 3)=2.*BOCC-POI/(5.*BY)+1./(5.*BY)
      ESM( 7, 4)=-2.*BOCCC-2.*COBBB-4.*POI/(5.*BC)+14./(5.*BC)
      ESM( 7, 5)=COBB+POI/(5.*CX)-1./(5.*CX)
      ESM( 7, 6)=BOCC+POI/(5.*BY)-1./(5.*BY)
      ESM( 7, 7)=4.*BOCCC+4.*COBBB-4.*POI/(5.*BC)+14./(5.*BC)
      ESM( 8, 1)=COBB-4.*POI/(5.*CX)-1./(5.*CX)
      ESM( 8, 2)=2.*COB/3.+4.*POI*BOC/15.-4.*BOC/15.
      ESM( 8, 3)=0.
      ESM( 8, 4)=-COBB-POI/(5.*CX)+1./(5.*CX)
      ESM( 8, 5)=COB/3.-POI*BOC/15.+BOC/15.
      ESM( 8, 6)=0.
      ESM( 8, 7)=2.*COBB+4.*POI/(5.*CX)+1/(5.*CX)
      ESM( 8, 8)=4.*COB/3.-4.*POI*BOC/15.+4.*BOC/15.
      ESM( 9, 1)=-2.*BOCC+POI/(5.*BY)-1./(5.*BY)
      ESM( 9, 2)=0.
      ESM( 9, 3)=2.*BOC/3.+POI*COB/15.-COB/15.
      ESM( 9, 4)=-BOCC-POI/(5.*BY)+1./(5.*BY)
      ESM( 9, 5)=0.
      ESM( 9, 6)=BOC/3.-POI*COB/15.+COB/15.
      ESM( 9, 7)=2.*BOCC+4.*POI/(5.*BY)+1./(5.*BY)
      ESM( 9, 8)=POI
      ESM( 9, 9)=4.*BOC/3.-4.*POI*COB/15.+4.*COB/15.
      ESM(10, 1)=-2.*BOCCC-2.*COBBB-4.*POI/(5.*BC)+14./(5.*BC)
      ESM(10, 2)=-COBB-POI/(5.*CX)+1./(5.*CX)
      ESM(10, 3)=BOCC+POI/(5.*BY)-1./(5.*BY)
      ESM(10, 4)=-4.*BOCCC+2.*COBBB+4.*POI/(5.*BC)-14./(5.*BC)
      ESM(10, 5)=-COBB+4.*POI/(5.*CX)+1./(5.*CX)
      ESM(10, 6)=2.*BOCC-POI/(5.*BY)+1./(5.*BY)
      ESM(10, 7)=2.*BOCCC-4.*COBBB+4.*POI/(5.*BC)-14./(5.*BC)
      ESM(10, 8)=-2.*COBB+POI/(5.*CX)-1./(5.*CX)
      ESM(10, 9)=BOCC-4.*POI/(5.*BY)-1./(5.*BY)
      ESM(10,10)=4.*BOCCC+4.*COBBB-4.*POI/(5.*BC)+14./(5.*BC)
      ESM(11, 1)=COBB+POI/(5.*CX)-1./(5.*CX)
      ESM(11, 2)=COB/3.-POI*BOC/15.+BOC/15.
      ESM(11, 3)=0.
      ESM(11, 4)=-COBB+4.*POI/(5.*CX)+1./(5.*CX)
      ESM(11, 5)=2.*COB/3.+4.*POI*BOC/15.-4.*BOC/15.
      ESM(11, 6)=0.
      ESM(11, 7)=2.*COBB-POI/(5.*CX)+1./(5.*CX)
      ESM(11, 8)=2.*COB/3.+POI*BOC/15.-BOC/15.
      ESM(11, 9)=0.
      ESM(11,10)=-2.*COBB-4.*POI/(5.*CX)-1/(5.*CX)
      ESM(11,11)=4.*COB/3.-4.*POI*BOC/15.+4.*BOC/15.
      ESM(12, 1)=-BOCC-POI/(5.*BY)+1./(5.*BY)
      ESM(12, 2)=0.
      ESM(12, 3)=BOC/3.-POI*COB/15.+COB/15.
      ESM(12, 4)=-2.*BOCC+POI/(5.*BY)-1./(5.*BY)
      ESM(12, 5)=0.
      ESM(12, 6)=2.*BOC/3.+POI*COB/15.-COB/15.
      ESM(12, 7)=BOCC-4.*POI/(5.*BY)-1./(5.*BY)
      ESM(12, 8)=0.
      ESM(12, 9)=2.*BOC/3.+4.*POI*COB/15.-4.*COB/15.
      ESM(12,10)=2.*BOCC+4.*POI/(5.*BY)+1./(5.*BY)
      ESM(12,11)=-POI
      ESM(12,12)=4.*BOC/3.-4.*POI*COB/15.+4.*COB/15.
      DO40 I=1,12
      DO35 J=1,12
      ESM(I,J)=ESM(J,I)
   35 CONTINUE
   40 CONTINUE
C
      DO 65 I=1,12
      DO 60 J=1,12
      ESM(I,J)=ESM(I,J)*DE
```

```
   60 CONTINUE
   65 CONTINUE
C
C     CALCULATION OF FORCE VECTOR
C
      EFI( 1)= 1.
      EFI( 2)= BY/6.
      EFI( 3)=-CX/6.
      EFI( 4)= 1.
      EFI( 5)=-BY/6.
      EFI( 6)=-CX/6.
      EFI( 7)= 1.
      EFI( 8)= BY/6.
      EFI( 9)= CX/6.
      EFI(10)= 1.
      EFI(11)=-BY/6.
      EFI(12)= CX/6.
      DO80 M=1,12
   80 EF(M)=CX*BY*Q(KK)*EFI(M)/4.
C
C     INSERTION OF ELEMENT PROPERTIES INTO GLOBAL STIFFNESS MATRIX
C
      DO7 I=1,12
      II=NS(I)
      J5=JGF+II
      A(J5)=A(J5)+EF(I)
      DO17 J=1,12
      JJ=NS(J)
      JJ=JJ-II+1
      IF(JJ)17,17,16
   16 J5=JGSM+(JJ-1)*NP+II
      A(J5)=A(J5)+ESM(I,J)
   17 CONTINUE
    7 CONTINUE
C
C  MODIFICATION AND SOLUTION OF THE SYSTEM OF EQUATIONS
C
      CALL BDYVAL(A(JGSM+1),A(JGF+1),NP,NBW,NCL)
      CALL DCMPBC(A(JGSM+1),NP,NBW)
      CALL SLVBD(A(JGSM+1),A(JGF+1),A(1),NP,NBW,NCL,ID1)
C
C     CALCULATION OF BENDING AND TWISTING MOMENTS
C
      WRITE(6,53)
   53 FORMAT(//1X,'BENDING AND TWISTING MOMENTS',//1X,
     1'   NEL   ND          MXX                    MYY                MXY '
     2'         ND         MXX                 MYY                MXY  ')
      DO8 JK=1,NE
      READ(5,*)NEL,ND,X1,Y1,X2,Y2,X3,Y3,X4,Y4,Q(JK)
      DO19 I=1,4
      NS(3*I-2)=3*ND(I)-2
      NS(3*I-1)=3*ND(I)-1
   19 NS(3*I)=3*ND(I)
C
      DO43 I=1,12
      DO42 J=1,12
      SMI(I,J)=0.
   42 CONTINUE
   43 CONTINUE
C
      COB=CX/BY
      BOC=BY/CX
      ZB =6.*BOC+6.*COB*POI
      BZ =6.*COB+6.*BOC*POI
      QER=1.-POI
C
```

```
      SMI( 1, 1)=ZB
      SMI( 1, 2)=4.*CX*POI
      SMI( 1, 3)=-4.*BY
      SMI( 1, 4)=-6.*COB*POI
      SMI( 1, 5)=2.*CX*POI
      SMI( 1, 7)=-6.*BOC
      SMI( 1, 9)=-2.*BY
C
      SMI( 2, 1)=BZ
      SMI( 2, 2)=4.*CX
      SMI( 2, 3)=-4.*BY*POI
      SMI( 2, 4)=-6.*COB
      SMI( 2, 5)=2.*CX
      SMI( 2, 7)=-6.*BOC*POI
      SMI( 2, 9)=-2.*BY*POI
C
      SMI( 3, 1)=-QER
      SMI( 3, 2)=-QER*BY
      SMI( 3, 3)=QER*CX
      SMI( 3, 4)=QER
      SMI( 3, 6)=-QER*CX
      SMI( 3, 7)=QER
      SMI( 3, 8)=QER*BY
      SMI( 3,10)=-QER
C
      SMI( 4, 1)=-6.*COB*POI
      SMI( 4, 2)=-2.*CX*POI
      SMI( 4, 4)=ZB
      SMI( 4, 5)=-4.*CX*POI
      SMI( 4, 6)=-4.*BY
      SMI( 4,10)=-6.*BOC
      SMI( 4,12)=-2.*BY
C
      SMI( 5, 1)=-6.*COB
      SMI( 5, 2)=-2.*CX
      SMI( 5, 4)=BZ
      SMI( 5, 5)=-4.*CX
      SMI( 5, 6)=-4.*BY*POI
      SMI( 5,10)=-6.*BOC*POI
      SMI( 5,12)=-2.*BY*POI
C
      SMI( 6, 1)=-QER
      SMI( 6, 3)=CX*QER
      SMI( 6, 4)=QER
      SMI( 6, 5)=-BY*QER
      SMI( 6, 6)=-CX*QER
      SMI( 6, 7)=QER
      SMI( 6,10)=-QER
      SMI( 6,11)=BY*QER
C
      SMI( 7, 1)=-6.*BOC
      SMI( 7, 3)=2.*BY
      SMI( 7, 7)=ZB
      SMI( 7, 8)=4.*CX*POI
      SMI( 7, 9)=4.*BY
      SMI( 7,10)=-6.*COB*POI
      SMI( 7,11)=2.*CX*POI
C
      SMI( 8, 1)=-6.*BOC*POI
      SMI( 8, 3)=2.*BY*POI
      SMI( 8, 7)=BZ
      SMI( 8, 8)=4.*CX
      SMI( 8, 9)=4.*BY*POI
      SMI( 8,10)=-6.*COB
      SMI( 8,11)=2.*CX
C
```

```
      SMI( 9, 1)=-QER
      SMI( 9, 2)=-BY*QER
      SMI( 9, 4)=QER
      SMI( 9, 7)=QER
      SMI( 9, 8)=BY*QER
      SMI( 9, 9)=CX*QER
      SMI( 9,10)=-QER
      SMI( 9,12)=-CX*QER
C
      SMI(10, 4)=-6.*BOC
      SMI(10, 6)=2.*BY
      SMI(10, 7)=-6.*COB*POI
      SMI(10, 8)=-2.*CX*POI
      SMI(10,10)=ZB
      SMI(10,11)=-4.*CX*POI
      SMI(10,12)=4.*BY
C
      SMI(11, 4)=-6.*BOC*POI
      SMI(11, 6)=2.*BY*POI
      SMI(11, 7)=-6.*COB
      SMI(11, 8)=-2.*CX
      SMI(11,10)=BZ
      SMI(11,11)=-4.*CX
      SMI(11,12)=4.*BY*POI
C
      SMI(12, 1)=-QER
      SMI(12, 4)=QER
      SMI(12, 5)=-BY*QER
      SMI(12, 7)=QER
      SMI(12, 9)=CX*QER
      SMI(12,10)=-QER
      SMI(12,11)=BY*QER
      SMI(12,12)=-CX*QER
C
      DO89I=1,12
      BMI(I)=0.
      DO89J=1,12
   89 BMI(I)=BMI(I)+SMI(I,J)*A(NS(J))
      DE=(EM*TH*TH*TH)/(12*(1-POI*POI))
      DO90I=1,12
   90 BM(I)=DE*BMI(I)/(CX*BY)
      WRITE(6,52)NEL,ND(1),BM(1),BM(2),BM(3),ND(2),BM(4),BM(5),BM(6)
      WRITE(6,52)NEL,ND(3),BM(7),BM(8),BM(9),ND(4),BM(10),BM(11),BM(12)
   52 FORMAT(1X,2(2X,I3),3(2X,E13.4),3X,I3,3(2X,E13.4))
    8 CONTINUE
      WRITE(6,51)NBW,NELBW
   51 FORMAT(///,1X,17HSEMI-BANDWIDTH IS,I3,22H CALCULATED IN ELEMENT
     +I3,///)
      CLOSE(5)
      CLOSE(6)
      STOP
      END
C
C**********************************************************************
C
      SUBROUTINE BDYVAL(GSM,GF,NP,NBW,NCL)
      CHARACTER*20 TITLE
      DIMENSION GSM(NP,NBW),GF(NP,NCL),IB(6),BV(6)
      COMMON/TLE/TITLE
      WRITE(6,200) TITLE
  200 FORMAT(1X,///////,1X,A20)
C
C  INPUT OF THE NODAL FORCE VALUES
C
      WRITE(6,201)
  201 FORMAT(/1X,'BOUNDARY VALUES'//1X,'NODAL FORCES')
```

```
      DO216JM=1,NCL
      ID1=0
      INK=0
  202 READ(5,*)IB,BV
      ID=0
      DO204L=1,6
      IF(IB(L).LE.0) GOTO205
      ID=ID+1
      I=IB(L)
  204 GF(I,JM)=BV(L) + GF(I,JM)
      GOTO206
  205 INK=1
      IF(ID.EQ.0) GOTO216
  206 IF(ID1.EQ.1) GOTO222
      WRITE(6,217) JM
  217 FORMAT(1X,'LOADING CASE',I2)
  222 WRITE(6,207) (IB(L),BV(L),L=1,ID)
  207 FORMAT(1X,6(I3,E14.5,2X))
      IF(INK.EQ.1) GOTO216
      ID1=1
      GOTO202
C
C  INPUT OF THE PRESCRIBED NODAL VALUES
C
  216 CONTINUE
      WRITE(6,208)
  208 FORMAT(////,1X,'PRESCRIBED NODAL VALUES')
      INK=0
  209 READ(5,*) IB,BV
      ID=0
      DO221L=1,6
      IF(IB(L).LE.0)GOTO215
      ID=ID+1
      I=IB(L)
      BC=BV(L)
C
C  MODIFICATION OF THE GLOBAL STIFFNESS MATRIX AND THE
C  GLOBAL FORCE MATRIX USING THE METHOD OF DELETION OF
C  ROWS AND COLUMNS
C
      K=I-1
      DO211J=2,NBW
      M=I+J-1
      IF(M.GT.NP) GOTO210
      DO218JM=1,NCL
  218 GF(M,JM)=GF(M,JM)-GSM(I,J)*BC
      GSM(I,J)=0.0
  210 IF(K.LE.0) GOTO 211
      DO219JM=1,NCL
  219 GF(K,JM)=GF(K,JM)-GSM(K,J)*BC
      GSM(K,J)=0.0
      K=K-1
  211 CONTINUE
  212 IF(GSM(I,1).LT.0.05) GSM(I,1)=500000.
      DO220JM=1,NCL
  220 GF(I,JM)=GSM(I,1)*BC
  221 CONTINUE
      GOTO214
C
C  OUTPUT OF THE BOUNDARY VALUES, (BV)
C
  215 INK=1
      IF(ID.EQ.0) RETURN
  214 WRITE(6,207) (IB(L),BV(L),L=1,ID)
      IF(INK.EQ.1) RETURN
      GOTO209
```

```
      END
C
C***********************************************************************
C
      SUBROUTINE DCMPBC(GSM,NP,NBW)
      DIMENSION GSM(NP,NBW)
      IO=61
      NP1=NP-1
      DO226I=1,NP1
      MJ=I+NBW-1
      IF(MJ.GT.NP) MJ=NP
      NJ=I+1
      MK=NBW
      IF((NP-I+1).LT.NBW) MK=NP-I+1
      ND=0
      DO225J=NJ,MJ
      MK=MK-1
      ND=ND+1
      NL=ND+1
      DO225K=1,MK
      NK=ND+K
  225 GSM(J,K)=GSM(J,K)-GSM(I,NL)*GSM(I,NK)/GSM(I,1)
  226 CONTINUE
      RETURN
      END
C
C***********************************************************************
C
      SUBROUTINE SLVBD(GSM,GF,X,NP,NBW,NCL,ID)
      CHARACTER*20 TITLE
      DIMENSION GSM(NP,NBW),GF(NP,NCL),X(NP,NCL)
      COMMON/TLE/TITLE
      IO=61
      NP1=NP-1
      DO265KK=1,NCL
      JM=KK
C
C  DECOMPOSITION OF THE COLUMN VECTOR GF( )
C
      DO250I=1,NP1
      MJ=I+NBW-1
      IF(MJ.GT.NP) MJ=NP
      NJ=I+1
      L=1
      DO250J=NJ,MJ
      L=L+1
  250 GF(J,KK)=GF(J,KK)-GSM(I,L)*GF(I,KK)/GSM(I,1)
C
C  BACKWARD SUBSTITUTION FOR DETERMINATION OF X( )
C
      X(NP,KK)=GF(NP,KK)/GSM(NP,1)
      DO252K=1,NP1
      I=NP-K
      MJ=NBW
      IF((I+NBW-1).GT.NP) MJ=NP-I+1
      SUM=0.0
      DO251J=2,MJ
      N=I+J-1
  251 SUM=SUM+GSM(I,J)*X(N,KK)
  252 X(I,KK)=(GF(I,KK)-SUM)/GSM(I,1)
C
C  OUTPUT OF THE CALCULATED NODAL VALUES
C
      IF(ID.EQ.1) GOTO265
      WRITE(6,260) TITLE,KK
  260 FORMAT(1H1////1X,A20//1X,'NODAL VALUES, LOADING CASE',I2)
      WRITE(6,264) (I,X(I,KK),I=1,NP)
  264 FORMAT(1X,I3,E14.5,3X,I3,E14.5,3X,I3,E14.5,3X,I3,E14.5,3X,I3,E14.5
     1)
  265 CONTINUE
      RETURN
      END
```

The FORTRAN program RPLATE is designed to implement the required steps outlined in Section 8.5. All the coefficients in the system of equations (8.56) are stored in a single column vector $\{A\}$. The relative location of $[K]$, $\{\Delta\}$, and $\{F\}$ in the column vector are given by

$$\{A\} = \begin{Bmatrix} \{\Delta\} \\ \{F\} \\ \{K\} \end{Bmatrix} \tag{8.57}$$

The main program reads the title and the basic data, calculates the bandwidth, and reads the 144 entries of the element stiffness matrices $[K^e]$ and the 12 entries of the element force vectors $\{F^e\}$, where $e = 1, 2, \ldots, E$. It then assembles the global stiffness matrix $[K]$ and the global force vector $\{F\}$. The program then calls the subroutines BDYVAL, DCMPBC, and SLVBD. The subroutine BDYVAL reads the specified values of $\{F\}$ and $\{\Delta\}$ and modifies $[K]$ using the procedure of deletion of rows and columns. The input for loads requires the specification of six integers (degrees of freedom) followed by the six corresponding forces (or moments). Similarly, the input for displacements requires the specification of six integers (degrees of freedom) followed by the magnitude of the corresponding displacements (rotation). The format for the specification of forces as well as displacements is repeated until an integer zero is found for a global degree of freedom.

The subroutine DCMPBD decomposes the modified symmetric band matrix $[K]$ into an upper triangular matrix using the Gaussian elimination procedure. The subroutine SLVBD implements the second part of the solution process and is used with DCMPBD to obtain a solution for (8.56). SLVBD first decomposes $\{F\}$ and then solves for $\{\Delta\}$ using the method of backward substitution. The subroutine SLVBD also performs an output function. The three subroutines BDYVAL, DCMPBD, and SLVBD were taken from L. J. Segerling, *Applied Finite Element Analysis*, Wiley, New York, 1976, with permission of the author. The FORTRAN program RPLATE on disk may be obtained from the present author by suitable arrangement.

## 8.7 EXAMPLE: CANTILEVER PLATE

In the present section we demonstrate the application of the FORTRAN program RPLATE in Section 8.6 to the solution of the following problem: A square cantilever plate, shown in Fig. 8.19, is subjected to a uniformly distributed transverse load $p = 2\ \text{N/cm}^2$. We wish to find the deflection and associated moment tensor field if $h = 1$ cm, $E = 20.7 \times 10^6\ \text{N/cm}^2$ and $\nu = 0.3$ (steel).

The plate is divided into 16 square elements, each with dimensions $10 \times 10$ cm. In view of the symmetry about the $y$-axis, we need to consider only 8

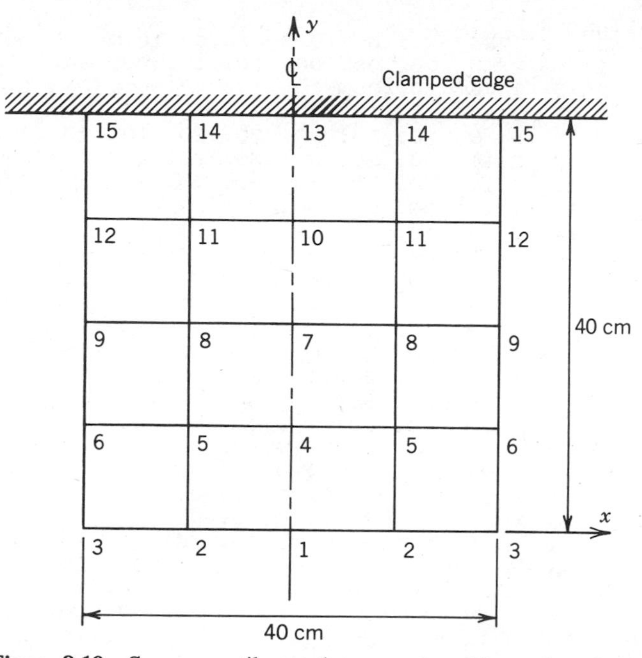


**Figure 8.19** Square cantilever plate: $c = b = 10$ cm; $h = 1$ cm.

elements, resulting in a total of 15 nodal points. As far as the boundary conditions are concerned, we need to specify vanishing displacements for $(w_i, \theta_{xi}, \theta_{yi})$ at nodes $i = 13, 14, 15$. To enforce symmetry along the $y$-axis, we must specify vanishing $\theta_{yi}$ at the nodes $i = 1, 4, 7, 10$. The complete data file is shown in Fig. 8.20, and it is the input to program RPLATE. The (incomplete) output file is shown in Fig. 8.21. The maximum deflection occurs at node 1, and it is equal to $w = 0.350$ cm. An experimental value $w = 0.353$ cm has been obtained by A. W. Leissa and N. F. Niedenfuhr, "A Study of the Cantilevered Square Plate Subjected to a Uniform Loading," *Journal of the Aerospace Sciences*, Vol. 29, No. 2, February 1962, pp. 162–169. Hence, there is an error of 0.78%. The bending moment distribution $M_{yy}$ along the clamped edge is shown in Fig. 8.22. With reference to this figure, the total (integrated) bending moment along the clamped edge is

$$-2[(1697)(5) + (1685)(10) + (1265)(5)] = -63{,}320 \text{ N-cm}$$

The distributed load causes an integrated bending moment of

$$-[(40)(40)(2)](20) = -64{,}000 \text{ N-cm}$$

along the clamped edge. Consequently, the summing of calculated nodal bending moments results in an error of 1.06%.

```
CANTILEVER PLATE
 8  45     20.7E6      1.0     0.3   10.0   10.0
 1    1   4   2   5   00.  00.  00.  10.  10.  00.  10.  10.   2.
 2    2   5   3   6   10.  00.  10.  10.  20.  00.  20.  10.   2.
 3    4   7   5   8    0.  10.  00.  20.  10.  10.  10.  20.   2.
 4    5   8   6   9   10.  10.  10.  20.  20.  10.  20.  20.   2.
 5    7  10   8  11   00.  20.  00.  30.  10.  20.  10.  30.   2.
 6    8  11   9  12   10.  20.  10.  30.  20.  20.  20.  30.   2.
 7   10  13  11  14   00.  30.  00.  40.  10.  30.  10.  40.   2.
 8   11  14  12  15   10.  30.  10.  40.  20.  30.  20.  40.   2.
 1    1   4   2   5   00.  00.  00.  10.  10.  00.  10.  10.   2.
 2    2   5   3   6   10.  00.  10.  10.  20.  00.  20.  10.   2.
 3    4   7   5   8    0.  10.  00.  20.  10.  10.  10.  20.   2.
 4    5   8   6   9   10.  10.  10.  20.  20.  10.  20.  20.   2.
 5    7  10   8  11   00.  20.  00.  30.  10.  20.  10.  30.   2.
 6    8  11   9  12   10.  20.  10.  30.  20.  20.  20.  30.   2.
 7   10  13  11  14   00.  30.  00.  40.  10.  30.  10.  40.   2.
 8   11  14  12  15   10.  30.  10.  40.  20.  30.  20.  40.   2.
  0   0   0   0   0   0      0.  0.  0.  0.  0.  0.
 37  38  39  40  41  42      0.  0.  0.  0.  0.  0.
 43  44  45  30  21  12      0.  0.  0.  0.  0.  0.
  3   0   0   0   0   0      0.  0.  0.  0.  0.  0.
 1    1   4   2   5   00.  00.  00.  10.  10.  00.  10.  10.   2.
 2    2   5   3   6   10.  00.  10.  10.  20.  00.  20.  10.   2.
 3    4   7   5   8    0.  10.  00.  20.  10.  10.  10.  20.   2.
 4    5   8   6   9   10.  10.  10.  20.  20.  10.  20.  20.   2.
 5    7  10   8  11   00.  20.  00.  30.  10.  20.  10.  30.   2.
 6    8  11   9  12   10.  20.  10.  30.  20.  20.  20.  30.   2.
 7   10  13  11  14   00.  30.  00.  40.  10.  30.  10.  40.   2.
 8   11  14  12  15   10.  30.  10.  40.  20.  30.  20.  40.   2.
```

**Figure 8.20** Data file (input to RPLATE).

## 8.8 A TRIANGULAR PLATE ELEMENT

It is easy to show that the equations

$$\nabla^2 M = -q \tag{8.58a}$$

$$\nabla^2 w = -\frac{M}{D} \tag{8.58b}$$

where

$$(1+\nu)M = -D(1+\nu)\nabla^2 w = M_{\alpha\alpha} \tag{8.59}$$

are, in every respect, equivalent to (5.1). Each equation in (8.58) can be associated with the deflection of a uniformly stretched and laterally loaded membrane. If we now consider a simply supported plate in the shape of a polygon, then along each rectilinear portion of the boundary we have $w = 0$ and $\partial^2 w/\partial s^2 = 0$. Since $M_{nn} = 0$ along this edge, we conclude that $\partial^2 w/\partial n^2 = 0$. Consequently, along a rectilinear, simply supported plate boundary we have

$$\nabla^2 w = \frac{\partial^2 w}{\partial s^2} + \frac{\partial^2 w}{\partial n^2} = \frac{\partial^2 w}{\partial x^2} + \frac{\partial^2 w}{\partial y^2} = -\frac{M}{D} = 0$$

```
CANTILEVER PLATE

NODAL VALUES, LOADING CASE 1
  1    .34992E+00    2  -.11723E-01    3   .00000E+00    4   .34806E+00    5  -.11726E-01
  6    .36192E-03    7   .34309E+00    8  -.11782E-01    9   .59078E-03   10   .23369E+00
 11   -.11438E-01   12   .00000E+00   13   .23176E+00   14  -.11439E-01   15   .39031E-03
 16    .22602E+00   17  -.11541E-01   18   .73622E-03   19   .12454E+00   20  -.10142E-01
 21    .00000E+00   22   .12287E+00   23  -.10083E-01   24   .36484E-03   25   .11619E+00
 26   -.10102E-01   27   .10085E-02   28   .37610E-01   29  -.68066E-02   30   .00000E+00
 31    .37070E-01   32  -.66753E-02   33   .15665E-03   34   .31880E-01   35  -.62267E-02
 36    .11973E-02   37   .00000E+00   38   .00000E+00   39   .00000E+00   40   .00000E+00
 41    .00000E+00   42   .00000E+00   43   .00000E+00   44   .00000E+00   45   .00000E+00

BENDING AND TWISTING MOMENTS

 NEL  ND      MXX          MYY          MXY       ND      MXX          MYY          MXY
  1    1   .7192E+02   .1588E+02   .5878E+00    4   .4143E+02  -.8025E+02   .9285E+00
  1    2   .6335E+02   .1879E+02  -.3180E+01    5   .4334E+02  -.8603E+02  -.2839E+01
  2    2   .6688E+02   .1985E+02  -.9516E+00    5   .4553E+02  -.8537E+02  -.7146E+01
  2    3   .2169E+02   .1215E+02  -.1648E+02    6   .2369E+02  -.8168E+02  -.2268E+02
  3    4   .4096E+02  -.8182E+02  -.3609E+01    7  -.6472E+02  -.3725E+03   .4241E+01
  3    5   .4242E+02  -.8912E+02  -.2293E+00    8  -.3396E+02  -.3760E+03   .7621E+01
  4    5   .4460E+02  -.8846E+02   .2222E+01    8  -.1940E+02  -.3716E+03   .1343E+02
  4    6   .2602E+02  -.7393E+02  -.3729E+02    9   .6018E+01  -.4132E+03  -.2608E+02
  5    7  -.6308E+02  -.3671E+03  -.7273E+01   10  -.2625E+03  -.8812E+03   .2416E+01
  5    8  -.3854E+02  -.3913E+03   .2035E+02   11  -.2050E+03  -.8576E+03   .3004E+02
  6    8  -.2398E+02  -.3869E+03   .5372E+01   11  -.2449E+03  -.8695E+03   .6733E+02
  6    9   .1355E+02  -.3881E+03  -.4730E+02   12   .6556E+02  -.9250E+03   .1466E+02
  7   10  -.2631E+03  -.8829E+03   .1026E+02   13  -.5091E+03  -.1697E+04  -.7160E+01
  7   11  -.1962E+03  -.8281E+03   .3105E+02   14  -.5056E+03  -.1685E+04   .1363E+02
  8   11  -.2360E+03  -.8400E+03   .1146E+02   14  -.5056E+03  -.1685E+04  -.4808E+02
  8   12   .4838E+02  -.9823E+03   .1495E+03   15  -.3796E+03  -.1265E+04   .9001E+02
```

**Figure 8.21** Data file (partial output from RPLATE).

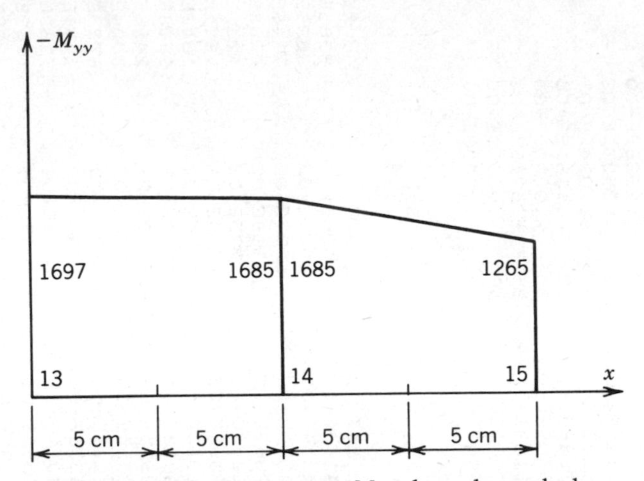


**Figure 8.22** Bending moment $M_{yy}$ along clamped edge.

Thus, in the case of a simply supported plate, we must first find a solution of (8.58a) with $M = 0$ on the boundary $C$. This solution is substituted into (8.58b), which is then solved for $w = w(x, y)$ with $w = 0$ on $C$.

If we wish to employ the finite element method to solve (8.58), we need to derive the appropriate stiffness matrix and force vector for a membrane element. This will now be done for a triangular element, as shown in Fig. 8.23.

Following the general procedure of Section 8.5, we assume a linear form for the deformed triangular element:

$$w = \alpha_1 + \alpha_2 x + \alpha_3 y = [1 \quad x \quad y]\{\alpha\} \tag{8.60}$$

where

$$\{\alpha\}^{\mathrm{T}} = [\alpha_1 \quad \alpha_2 \quad \alpha_3]$$

If, in (8.60), we successively substitute the nodal values at 1, 2, and 3 (see Fig.

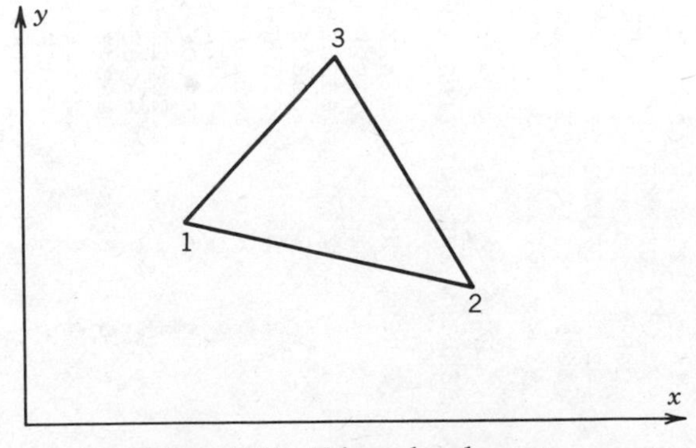


**Figure 8.23** Triangular element.

8.23), we obtain three equations in three unknowns:

$$\{\delta\} = \begin{Bmatrix} w_1 \\ w_2 \\ w_3 \end{Bmatrix} = [A]\{\alpha\} \tag{8.61}$$

where

$$[A] = \begin{bmatrix} 1 & x_1 & y_1 \\ 1 & x_2 & y_2 \\ 1 & x_3 & y_3 \end{bmatrix} \tag{8.62}$$

Consequently,

$$\{\alpha\} = [A]^{-1}\{\delta\} \tag{8.63}$$

where

$$[A]^{-1} = \frac{1}{2A}\begin{bmatrix} d_1 & d_2 & d_3 \\ b_1 & b_2 & b_3 \\ a_1 & a_2 & a_3 \end{bmatrix} \tag{8.64}$$

and

$$d_1 = x_2 y_3 - x_3 y_2, \qquad d_2 = x_3 y_1 - x_1 y_3, \qquad d_3 = x_1 y_2 - x_2 y_1 \tag{8.65a}$$
$$a_1 = x_3 - x_2, \qquad a_2 = x_1 - x_3, \qquad a_3 = x_2 - x_1 \tag{8.65b}$$
$$b_1 = y_2 - y_3, \qquad b_2 = y_3 - y_1, \qquad b_3 = y_1 - y_2 \tag{8.65c}$$

and $A$ is the area of the triangle.

With reference to (8.63) and (8.60) we have

$$w = [L]\{\delta\} = \{\delta\}^{\mathrm{T}}[L]^{\mathrm{T}} \tag{8.66}$$

where

$$[L] = [1 \quad x \quad y][A]^{-1} \tag{8.67}$$

The membrane "strain" field is defined by

$$\{\varepsilon\} = \begin{Bmatrix} \dfrac{\partial w}{\partial x} \\ \dfrac{\partial w}{\partial y} \end{Bmatrix} = \begin{bmatrix} 0 & 1 & 0 \\ 0 & 0 & 1 \end{bmatrix}\begin{Bmatrix} \alpha_1 \\ \alpha_2 \\ \alpha_3 \end{Bmatrix} = [B]\{\delta\} \tag{8.68}$$

where

$$[B] = \begin{bmatrix} 0 & 1 & 0 \\ 0 & 0 & 1 \end{bmatrix}[A]^{-1} \tag{8.69}$$

The potential energy of deformation is [see (8.43)]

$$U = \frac{1}{2}\int_A \{\varepsilon\}^{\mathrm{T}}\{\sigma\}\, dA = \{\delta\}^{\mathrm{T}}[K]\{\delta\} \tag{8.70}$$

where

$$[K] = \frac{N}{2}\int_A [B]^{\mathrm{T}}[B]\,dA \tag{8.71}$$

The integral in (8.71) is readily evaluated with the aid of area coordinates (see, e.g., L. T. Segerlind, *Applied Finite Element Analysis*, Wiley, New York, 1976). The result is

$$K_{ij} = \frac{N}{4A}(b_i b_j + a_i a_j), \qquad i, j = 1, 2, 3 \tag{8.72}$$

where $a_i$ and $b_i$ are given by (8.65b) and (8.65c), respectively.

The potential of external forces acting upon the element is

$$V = -w_1 F_1 - w_2 F_2 - w_3 F_3 - \int_A (p\,dA)w \tag{8.73}$$

where $F_1$, $F_2$, and $F_3$ are nodal forces and $p$ is the transverse pressure, assumed to be constant over the triangular element. With the aid of (8.66) we have

$$\int_A (p\,dA)w = p\{\delta\}^{\mathrm{T}}\int_A [L]^{\mathrm{T}}\,dA = \{\delta\}^{\mathrm{T}} pA \begin{Bmatrix} \frac{1}{3} \\ \frac{1}{3} \\ \frac{1}{3} \end{Bmatrix}$$

or

$$V = -\{\delta\}^{\mathrm{T}}\{Q\} - p\{\delta\}^{\mathrm{T}}\int_A [L]^{\mathrm{T}}\,dA \tag{8.74}$$

where

$$\{Q\}^{\mathrm{T}} = [F_1 \quad F_2 \quad F_3]$$

If we now define the force vector as

$$\{F\} = \begin{Bmatrix} F_1 \\ F_2 \\ F_3 \end{Bmatrix} + pA \begin{Bmatrix} \frac{1}{3} \\ \frac{1}{3} \\ \frac{1}{3} \end{Bmatrix} \tag{8.75}$$

then the total potential energy associated with the membrane element is

$$\Pi = U + V = \tfrac{1}{2}\{\delta\}^{\mathrm{T}}[K]\{\delta\} + \{\delta\}^{\mathrm{T}}\{F\} \tag{8.76}$$

Application of the principle of minimum potential energy to (8.76) results in

$$\frac{\partial \Pi}{\partial\{\delta\}} = [K]\{\delta\} + \{F\} = \{0\} \tag{8.77}$$

where the elements of $K$ are given by (8.72) and $\{F\}$ is characterized by

(8.75). With the aid of (8.68), (8.69), and (8.64), it is readily shown that for the present formulation, membrane "stresses" are obtained from

$$\{\sigma\} = \begin{Bmatrix} V_x \\ V_y \end{Bmatrix} = N \begin{Bmatrix} \dfrac{\partial w}{\partial x} \\ \dfrac{\partial w}{\partial y} \end{Bmatrix} = \frac{N}{2a} \begin{bmatrix} b_1 & b_2 & b_3 \\ a_1 & a_2 & a_3 \end{bmatrix} \begin{Bmatrix} w_1 \\ w_2 \\ w_3 \end{Bmatrix} \tag{8.78}$$

We now proceed in a manner completely analogous to Section 8.6. A FORTRAN computer program FISH2.FOR was written to implement the finite element solution of (8.58) when $M$ and $w$ vanish along the plate boundary (simply supported plate). Two input files are required: AIN.DAT provides the title, number of degrees of freedom, number of elements, bandwidth, membrane stress per unit length, intensity of transverse load, and specification of boundary conditions. BIN.DAT provides node numbers in

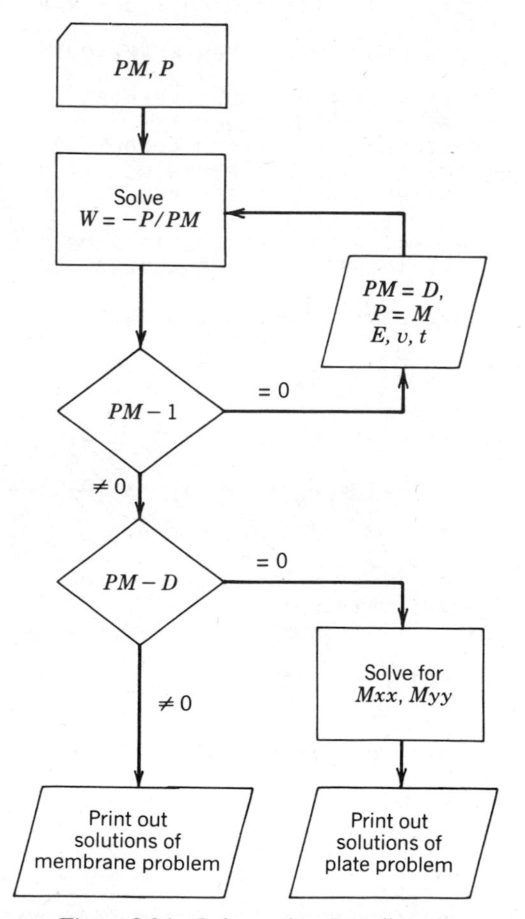


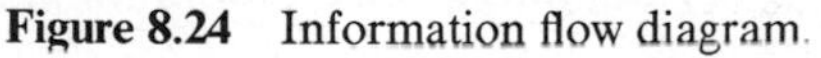
**Figure 8.24** Information flow diagram.

counterclockwise order and nodal coordinates for each element. The input variables are explained by comments at the beginning of the program. The calculation of plate bending moments is based on finite difference formulas that utilize the nodal deflections. An information flow diagram is shown in Fig. 8.24. The program was written and verified on the Digital Equipment VAX 8650 computer.

```
C *************************************************************
C *                       PROGRAM FISH2                       *
C *              COPYRIGHT,HERBERT REISMANN,1987               *
C *************************************************************

C *************************************************************
C          FOR USER'S REFERENCE (PROGRAM 'FISH2')
C          ------------------------------------
C      INPUT FILES:
C      1. FILENAME: AIN.DAT
C
C         THE 1st LINE: TITLE CARD ( JOB NAME )
C
C         THE 2nd LINE: NP,NE,NBW,PM,P,KOUT
C
C             NP: NUMBER OF NODAL POINTS
C             NE: NUMBER OF ELEMENTS
C            NBW: BANDWIDTH (FROM GR2OUT6.DAT)
C           KOUT: CONTROL FOR PUNCHING STRESSES
C             PM: MEMBRANE STRESS PER UNIT LENGTH
C                 (lbs/in.)
C               *GIVE PM=1 FOR PLATE PROBLEMS*
C              P: INTENSITY OF TRANSVERSE LOAD(lbs/in.sq)
C
C         THE 3rd LINE: 0 0 0 0 0 0 0.0 0.0 0.0 0.0 0.0 0.0
C
C         THE 4th "  ": 6 SUCCESSIVE BOUNDARY POINTS' # AND
C                       0.0 0.0 0.0 0.0 0.0 0.0
C
C         THE 5th "  ": . . . . .  .   .    .    .    .  .
C
C               ..... . .  .  .  .   .   .   .    .     .
C
C         END WITH    : 0 0 0 0 0 0 0.0 0.0 0.0 0.0 0.0 0.0
C
C
C      2. FILENAME: BIN.DAT
C                   IT'S OUTPUT OF PROGRAM GRID2(MECLIB)
C
C      3.OUTPUT FILE
C           FILENAME : FISH2OUT.DAT
C
C      4. COMPILATION & EXECUTION
C
C           THREE STATEMENTS
C
C            F FISH2
C            LINK FISH2
C            R FISH2
C *************************************************************
```

```
      DIMENSION NS(3),ESM(3,3),EF(3),B(3),C(3),PHI(3)
      DIMENSION A(16000),AC(400),CX(190),CY(190),W(190)
      REAL Mxx(190),Myy(190),M(190),NU
      COMMON/TLE/TITLE(20)
      OPEN(7,FILE='AIN.DAT',STATUS='OLD')
      OPEN(9,FILE='BIN.DAT',STATUS='OLD')
      OPEN(8,FILE='FISH2OUT.DAT',STATUS='NEW')
      DATA NCL/1/,ID1/0/,KE/0/

C     DEFINITION OF CONTROL PARAMETERS

C     NP     ---- NUMBER OF GLOBAL DISPLACEMENTS
C     NE     ---- NUMBER OF ELEMENTS
C     NBW    ---- BANDWIDTH
C     PM     ---- MEMBRANE FORCE PER UNIT LENGTH ,
C     P      ---- INTENSITTY OF TRANSVERSE LOAD (lb/insq)
C     KOUT   ---- CONTROL FOR PUNCHING STRESSES
C     NU     ---- POISSON'S RATIO

C     INPUT OF THE TITLE CARD AND PARAMETER CARD

      READ(7,5)TITLE
    5 FORMAT(20A4)
      READ(7,*)NP,NE,NBW,KOUT,PM,P
   15 FORMAT(4I3,6F10.1)

C     CALCULATION OF POINTERS AND INITIALIZATION OF
C     COLUMN VECTOR {A}

      JGF=NP*NCL
      JGSM=JGF*2
      JEND=JGSM+NP*NBW
      IF(PM.EQ.1)GOTO12
      IF(PM.NE.D)GOTO10
    1 DO I=1,JGF
      AC(I)=A(I)
      END DO
      GOTO 20

C     OUTPUT OF TITLE AND DATA HEADINGS

   10 WRITE(8,25) TITLE,PM,P
   25 FORMAT(1H1,//////1X,
     120A4//1X,
     2'MEMBRANE FORCE =',F7.2//1X,
     3'TRANSV.PRESSURE =',F7.2//1X,
     5'ELEMENT DATA',//1X,
     6'NEL    NODE NUMBERS          X(1)        Y(1)
     7X(2)          ','Y(2)        X(3)        Y(3)   ')

   12 WRITE(8,26)TITLE,P
   26 FORMAT(1H1,//////1X,
     120A4//1X,
     2'TRANSV.PRESSURE =',F7.2//1X,
     3'ELEMENT DATA',//1X,
     4'NEL    NODE NUMBERS          X(1)        Y(1)
     5X(2)          ','Y(2)        X(3)        Y(3)   ')
```

```
C*******************************************************
C     ASSEMBLING OF THE GLOBAL STIFFNESS
C     MATRIX AND THE GLOBAL FORCE MATRIX
C*******************************************************

C       INPUT AND ECHO PRINT OF THE ELEMENT DATA

   20 DO I=1,JEND
      A(I)=0.0
      END DO
      REWIND (9)
      DO 110 KK=1,NE
      READ(9,*)NEL,NS,X1,Y1,X2,Y2,X3,Y3
      IF(PM.EQ.D) GOTO 40
      WRITE(8,35)NEL,NS,X1,Y1,X2,Y2,X3,Y3
   35 FORMAT(1X,I3,2X,3I4,3X,6(2XF8.4))

C CALCULATION OF THE GLOBAL STIFFNESS MATRIX AND
C THE GLOBAL FORCE MATRIX

   40 B(1)=Y2-Y3
      B(2)=Y3-Y1
      B(3)=Y1-Y2
      C(1)=X3-X2
      C(2)=X1-X3
      C(3)=X2-X1
      AR4=(X2*Y3+X3*Y1+X1*Y2-X2*Y1-X3*Y2-X1*Y3)*2
      IF(PM.NE.D)GOTO 48
      SUM=0
      DO 43 I=1,3
      JJ=NS(I)
   43 SUM=SUM+AC(JJ)
      P=SUM/3
      DO 46 I=1,3
      IF(SUM.EQ.0)GOTO 45
      JJ=NS(I)
      EF(I)=P*(AR4/4)*AC(JJ)/SUM
   45 DO 46 J=1,3
   46 ESM(I,J)=(B(I)*B(J)+C(I)*C(J))*PM/AR4
      GOTO 60
   48 DO 50 I=1,3
      EF(I)=P*AR4/12
      DO 50 J=1,3
   50 ESM(I,J)=(B(I)*B(J)+C(I)*C(J))*PM/AR4

C  INSERTION OF ELEMENT PROPERTIES INTO THE
C  GLOBAL STIFFNESS MATRIX

   60 DO 110 I=1,3
      II=NS(I)
      DO 70 J=1,NCL
      J5=JGF+(J-1)*NP+II
      A(J5)=A(J5)+EF(I)
```

```
   70 CONTINUE
      DO 100 J=1,3
      JJ=NS(J)
      JJ=JJ-II+1
      IF(JJ)100,100,90
   90 J5=JGSM+(JJ-1)*NP+II
      A(J5)=A(J5)+ESM(I,J)
  100 CONTINUE
  110 CONTINUE

C  MODIFICATION AND SOLUTION OF THE SYSTEM OF EQUATIONS

      CALL BDYVAL(A(JGSM+1),A(JGF+1),NP,NBW,NCL,D,PM)
      CALL DCMPBC(A(JGSM+1),NP,NBW)
      CALL SLVBD(A(JGSM+1),A(JGF+1),A(1),NP,NBW,NCL,ID1,PM)

      IF(PM.NE.1) GOTO 130
      PRINT *,' '
      PRINT *,'MODULUS OF ELASTICITY E= ? Psi'
      PRINT *,' '
      READ(5,*) E
      PRINT *,' '
      PRINT *,'POISSON''S RATIO NU= ?'
      PRINT *,' '
      READ(5,*) NU
      PRINT *,' '
      PRINT *,'THICKNESS OF THE PLATE T= ? in.'
      PRINT *,' '
      READ(5,*) T
      D=E*T**3/(12.*(1.-NU**2))
      PM=D
      GOTO 1
  130 IF(PM.EQ.D) GOTO 180

C****************************************
C  CALCULATION OF THE ELEMENT RESULTANTS
C****************************************

C  INPUT OF THE ELEMENT DATA

      REWIND (9)
      DO 170 KK=1,NE
      READ(9,*)NEL,NS,X1,Y1,X2,Y2,X3,Y3
      IF(NEL.LT.0) STOP
      IF(KK.GT.1) GO TO 140
      IF(PM.EQ.D) GOTO 140
      WRITE(8,135) TITLE
  135 FORMAT(1X/////1X,20A4//1X,20HELEMENT SHEAR FORCES
     1////5X,58H ELEMENT          VX                   VY
     2VMAX        )
```

```
C  RETRIEVAL OF THE NODAL VALUES

  140 J1=JGSM+NEL
      A(J1)=0.0
      DO 150 I=1,3
      II=NS(I)
      PHI(I)=A(II)
  150 A(J1)=A(J1)+PHI(I)/3.

C  CALCULATION AND OUTPUT OF THE SHEAR FORCES VX,VY,
C  AND VMAX

      AR2=(X2*Y3+X3*Y1+X1*Y2-X2*Y1-X3*Y2-X1*Y3)
      B(1)=Y2-Y3
      B(2)=Y3-Y1
      B(3)=Y1-Y2
      C(1)=X3-X2
      C(2)=X1-X3
      C(3)=X2-X1
      J2=JGF+NEL
      J3=JGF+NE+NEL
      A(J2)=0.0
      A(J3)=0.0
      DO 160 I=1,3
      A(J2)=A(J2)+B(I)*PHI(I)/AR2
  160 A(J3)=A(J3)+C(I)*PHI(I)/AR2
      VX=A(J2)*PM
      VY=A(J3)*PM
      VMAX=SQRT(VX*VX+VY*VY)
  170 WRITE(8,175) NEL,VX,VY,VMAX
  175 FORMAT(7X,I3,3(5X,E12.5))
      GOTO 200


C ***************************************************
C     CALL SUBROUTINE TO CALCULATE THE BENDING
C     MOMENTS Mxx & Myy AT EACH NODAL POINT
C ***************************************************

  180 CALL BMC(AC(1),A(1),NP,NE,Mxx,Myy,NU,D)

      WRITE(8,185)E,NU,t
  185 FORMAT(1H1,//////1X,
     1'YOUNG''S MODULUS = ',F10.1,' psi',//1X,
     2'POISSON''S RATIO = ',F4.2//1X,
     3'PLATE THICKNESS = ',F4.2,' IN.')


  200 STOP
      END
```

```
C **********************************************************
      SUBROUTINE BDYVAL(GSM,GF,NP,NBW,NCL,D,PM)
      DIMENSION GSM(NP,NBW),GF(NP,NCL),IB(6),BV(6)
      COMMON/TLE/TITLE(20)
      IF(PM.EQ.D) GOTO 100
      WRITE(8,15) TITLE
   15 FORMAT(1X,///////,1X,20A4)

C  INPUT OF THE NODAL FORCE VALUES

      WRITE(8,25)
   25 FORMAT(/1X,'BOUNDARY VALUES'//1X,'NODAL FORCES')
      GOTO 200
  100 REWIND(7)
    5 FORMAT(20A4)
      READ(7,5)TITLE
      READ(7,*)K1,K2,K3,K4,A1,A2
  200 DO 270 JM=1,NCL
      ID1=0
      INK=0
  210 READ(7,*)IB,BV
      ID=0
      DO 220 L=1,6
      IF(IB(L).LE.0) GOTO230
      ID=ID+1
      I=IB(L)
  220 GF(I,JM)=BV(L) + GF(I,JM)
      GOTO240
  230 INK=1
      IF(ID.EQ.0) GOTO270
  240 IF(ID1.EQ.1) GOTO250
      IF(PM.EQ.D)GOTO 260
      WRITE(8,245) JM
  245 FORMAT(1X,'LOADING CASE',I2)
  250 IF(PM.EQ.D)GOTO260
      WRITE(8,255) (IB(L),BV(L),L=1,ID)
  255 FORMAT(1X,6(I3,E14.5,2X))
  260 IF(INK.EQ.1) GOTO270
      ID1=1
      GOTO210

C  INPUT OF THE PRESCRIBED NODAL VALUES

  270 CONTINUE
      IF(PM.EQ.D)GOTO 280
      WRITE(8,275)
  275 FORMAT(////,1X,'PRESCRIBED NODAL VALUES')
  280 INK=0
  290 READ(7,*) IB,BV
      ID=0
      DO 360 L=1,6
      IF(IB(L).LE.0)GOTO370
      ID=ID+1
      I=IB(L)
      BC=BV(L)
```

```
C  MODIFICATION OF THE GLOBAL STIFFNESS MATRIX AND THE
C  GLOBAL FORCE MATRIX USING THE METHOD OF DELETION OF
C  ROWS AND COLUMNS

      K=I-1
      DO 330 J=2,NBW
      M=I+J-1
      IF(M.GT.NP) GOTO310
      DO 300 JM=1,NCL
  300 GF(M,JM)=GF(M,JM)-GSM(I,J)*BC
      GSM(I,J)=0.0
  310 IF(K.LE.0) GOTO 330
      DO 320 JM=1,NCL
  320 GF(K,JM)=GF(K,JM)-GSM(K,J)*BC
      GSM(K,J)=0.0
      K=K-1
  330 CONTINUE
  340 IF(GSM(I,1).LT.0.05) GSM(I,1)=500000.
      DO 350 JM=1,NCL
  350 GF(I,JM)=GSM(I,1)*BC
  360 CONTINUE
      GOTO 380

C  OUTPUT OF THE BOUNDARY VALUES, (BV)

  370 INK=1
      IF(ID.EQ.0) RETURN
  380 IF(PM.EQ.D)GOTO 390
      WRITE(8,255) (IB(L),BV(L),L=1,ID)
  390 IF(INK.EQ.1) RETURN
      GOTO290
      END
C*****************************************************************
      SUBROUTINE DCMPBC(GSM,NP,NBW)
      DIMENSION GSM(NP,NBW)
      IO=61
      NP1=NP-1
      DO 20 I=1,NP1
      MJ=I+NBW-1
      IF(MJ.GT.NP) MJ=NP
      NJ=I+1
      MK=NBW
      IF((NP-I+1).LT.NBW) MK=NP-I+1
      ND=0
      DO 10 J=NJ,MJ
      MK=MK-1
      ND=ND+1
      NL=ND+1
      DO 10 K=1,MK
      NK=ND+K
   10 GSM(J,K)=GSM(J,K)-GSM(I,NL)*GSM(I,NK)/GSM(I,1)
   20 CONTINUE
      RETURN
      END
```

```
C*********************************************************************
      SUBROUTINE SLVBD(GSM,GF,X,NP,NBW,NCL,ID,PM)
      DIMENSION GSM(NP,NBW),GF(NP,NCL),X(NP,NCL)
      COMMON/TLE/TITLE(20)
      IO=61
      NP1=NP-1
      DO 60 KK=1,NCL
      JM=KK

C   DECOMPOSITION OF THE COLUMN VECTOR GF( )

      DO 10 I=1,NP1
      MJ=I+NBW-1
      IF(MJ.GT.NP) MJ=NP
      NJ=I+1
      L=1
      DO 10 J=NJ,MJ
      L=L+1
   10 GF(J,KK)=GF(J,KK)-GSM(I,L)*GF(I,KK)/GSM(I,1)

C   BACKWARD SUBSTITUTION FOR DETERMINATION OF X( )

      X(NP,KK)=GF(NP,KK)/GSM(NP,1)
      DO 30 K=1,NP1
      I=NP-K
      MJ=NBW
      IF((I+NBW-1).GT.NP) MJ=NP-I+1
      SUM=0.0
      DO 20 J=2,MJ
      N=I+J-1
   20 SUM=SUM+GSM(I,J)*X(N,KK)
      X(I,KK)=(GF(I,KK)-SUM)/GSM(I,1)
   30 CONTINUE

C   OUTPUT OF THE CALCULATED NODAL VALUES

      IF(ID.EQ.1) GOTO60
      IF(PM.NE.1) GOTO 40
      WRITE(8,35) TITLE,KK
   35 FORMAT(1H1////1X,20A4//1X,'NODAL MOMENT
     1INVARIANTS, LOADING CASE',I2)
      GOTO 50
   40 WRITE(8,45) TITLE,KK
   45 FORMAT(1H1////1X,20A4//1X,'NODAL DEFLECTIONS,
     1LOADING CASE',I2)
   50 WRITE(8,55) (I,X(I,KK),I=1,NP)
   55 FORMAT(/,4(I4,E13.4,3X))
   60 CONTINUE
      RETURN
      END
```

```
************************************************************
      SUBROUTINE BMC(M,W,NP,NE,Mxx,Myy,v,D)
      REAL M(NP),Mxx(NP),Myy(NP),MX1,MX2,MY1,MY2
      DIMENSION W(NP),NS(3),DX(3),DY(3),C(3)
      PI=3.1416
      DO 300 I=1,NP
      IF(M(I).EQ.0)GOTO280
      Mxx(I)=0
      Myy(I)=0
      REWIND (9)
      KOUNT=0
      DO 200 J=1,NE
      READ(9,*)NEL,NS,X1,Y1,X2,Y2,X3,Y3
      IF(NS(1).EQ.I)GOTO 10
      IF(NS(2).EQ.I)GOTO 20
      IF(NS(3).EQ.I)GOTO 30
      GOTO 200
   10 DY(1)=Y2-Y1
      DX(1)=X2-X1
      DY(2)=Y3-Y1
      DX(2)=X3-X1
      DY(3)=Y2-Y3
      DX(3)=X2-X3
      DW1=W(NS(2))-W(NS(1))
      DW2=W(NS(3))-W(NS(1))
      GOTO 40
   20 DY(1)=Y3-Y2
      DX(1)=X3-X2
      DY(2)=Y1-Y2
      DX(2)=X1-X2
      DY(3)=Y3-Y1
      DX(3)=X3-X1
      DW1=W(NS(3))-W(NS(2))
      DW2=W(NS(1))-W(NS(2))
      GOTO 40
   30 DY(1)=Y1-Y3
      DX(1)=X1-X3
      DY(2)=Y2-Y3
      DX(2)=X2-X3
      DY(3)=Y1-Y2
      DX(3)=X1-X2
      DW1=W(NS(1))-W(NS(3))
      DW2=W(NS(2))-W(NS(3))
   40 KOUNT=KOUNT+2
      H1=SQRT(DY(1)**2+DX(1)**2)
      H2=SQRT(DY(2)**2+DX(2)**2)
      DO 60 K=1,3
      IF(DX(K).EQ.0)GOTO 50
      C(K)=ATAN2(DY(K),DX(K))
      GOTO 60
   50 C(K)=PI/2
   60 CONTINUE
      R=-2*D/(SIN(C(3)-C(1))*SIN(C(3)+C(1))*SIN(2*C(2))+
     1SIN(C(1)-C(2))*SIN(C(1)+C(2))*SIN(2*C(3))+SIN(C(2)
     2-C(3))*SIN(C(2)+C(3))*SIN(2*C(1)))
```

```
      MX1=R*(SIN(C(2))*SIN(C(3))+v*COS(C(2))*COS(C(3))
     1)*SIN(C(3)-C(2))*DW1/H1**2
      MY1=R*(v*SIN(C(2))*SIN(C(3))+COS(C(2))*COS(C(3))
     1)*SIN(C(3)-C(2))*DW1/H1**2
      MX2=R*(SIN(C(3))*SIN(C(1))+v*COS(C(3))*COS(C(1))
     1)*SIN(C(1)-C(3))*DW2/H2**2
      MY2=R*(v*SIN(C(3))*SIN(C(1))+COS(C(3))*COS(C(1))
     1)*SIN(C(1)-C(3))*DW2/H2**2
      Mxx(I)=Mxx(I)+MX1+MX2
      Myy(I)=Myy(I)+MY1+MY2
      IF(KOUNT.EQ.12)GOTO 210
  200 CONTINUE
  210 Myy(I)=Myy(I)/2
      Mxx(I)=Mxx(I)/2
      GOTO 300
  280 Myy(I)=0
      Mxx(I)=0
  300 CONTINUE
      WRITE(8,315)
  315 FORMAT(////,' BENDING MOMENTS AT EACH NODAL
     1POINT',//,' NODAL POINT        Mxx
     2Myy       NODAL POINT     Mxx             Myy')
      WRITE(8,325)(I,Mxx(I),Myy(I),I=1,NP)
  325 FORMAT(/,2(5X,I3,5X,E11.4,2X,E11.4))
      RETURN
      END
```

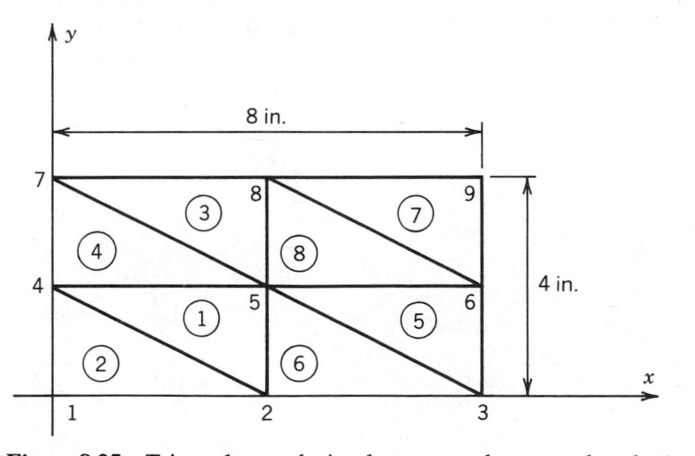


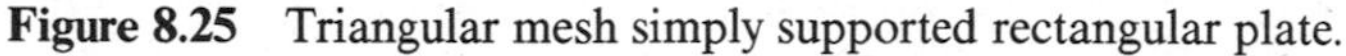

**Figure 8.25** Triangular mesh simply supported rectangular plate.

```
PROBLEM1-8ELEMENTS
9  8  8  1  1.  10.
0  0  0  0  0  0  0.0  0.0  0.0  0.0  0.0  0.0
1  2  3  6  9  8  0.0  0.0  0.0  0.0  0.0  0.0
7  4  0  0  0  0  0.0  0.0  0.0  0.0  0.0  0.0
0  0  0  0  0  0  0.0  0.0  0.0  0.0  0.0  0.0
```

**Figure 8.26** Input file AIN.DAT.

```
1     4     5     2     0.0    2.0    4.0    2.0    4.0    0.0
2     4     2     1     0.0    2.0    4.0    0.0    0.0    0.0
3     7     8     5     0.0    4.0    4.0    4.0    4.0    2.0
4     7     5     4     0.0    4.0    4.0    2.0    0.0    2.0
5     5     6     3     4.0    2.0    8.0    2.0    8.0    0.0
6     5     3     2     4.0    2.0    8.0    0.0    4.0    0.0
7     8     9     6     4.0    4.0    8.0    4.0    8.0    2.0
8     8     6     5     4.0    4.0    8.0    2.0    4.0    2.0
```

**Figure 8.26** Input file BIN.DAT (continued).

8.88

```
PROBLEM1-8ELEMENTS

TRANSV.PRESSURE =  10.00

ELEMENT DATA

NEL    NODE NUMBERS      X(1)     Y(1)     X(2)     Y(2)     X(3)     Y(3)
 1      4    5    2     0.0000   2.0000   4.0000   2.0000   4.0000   0.0000
 2      4    2    1     0.0000   2.0000   4.0000   0.0000   0.0000   0.0000
 3      7    8    5     0.0000   4.0000   4.0000   4.0000   4.0000   2.0000
 4      7    5    4     0.0000   4.0000   4.0000   2.0000   0.0000   2.0000
 5      5    6    3     4.0000   2.0000   8.0000   2.0000   8.0000   0.0000
 6      5    3    2     4.0000   2.0000   8.0000   0.0000   4.0000   0.0000
 7      8    9    6     4.0000   4.0000   8.0000   4.0000   8.0000   2.0000
 8      8    6    5     4.0000   4.0000   8.0000   2.0000   4.0000   2.0000



PROBLEM1-8ELEMENTS

BOUNDARY VALUES

NODAL FORCES


PRESCRIBED NODAL VALUES
 1    0.00000E+00    2    0.00000E+00    3    0.00000E+00    6    0.00000E+00
 9    0.00000E+00    8    0.00000E+00    7    0.00000E+00    4    0.00000E+00



PROBLEM1-8ELEMENTS

NODAL MOMENT INVARIANTS, LOADING CASE 1

 1    0.0000E+00    2    0.0000E+00    3    0.0000E+00    4    0.0000E+00
 5    0.1600E+02    6    0.0000E+00    7    0.0000E+00    8    0.0000E+00
 9    0.0000E+00
```

**Figure 8.27** Output file FISH2OUT.DAT.

We now consider the following example problem. A simply supported rectangular plate is acted upon by a uniformly distributed, transverse pressure load. We assume that $p = 10$ psi, $E = 30 \times 10^6$ psi, $\nu = 0.3$, $a = 8$ in., $b = 4$ in., and $h = 0.2$ in. The discretized plate with triangular mesh is shown in Fig. 8.25. The pertinent data files AIN.DAT and BIN.DAT are shown in Fig. 8.26. The output file FISH2OUT.DAT is shown in Fig. 8.27. This problem has been solved for discretization into 8, 32, 72, 128, and 200 elements, and the resulting convergence to the exact solution is exhibited in Tables 8.2 and 8.3.

```
PROBLEM1-8ELEMENTS

NODAL DEFLECTIONS, LOADING CASE 1

  1   0.0000E+00     2   0.0000E+00     3   0.0000E+00     4   0.0000E+00
  5   0.1165E-02     6   0.0000E+00     7   0.0000E+00     8   0.0000E+00
  9   0.0000E+00


BENDING MOMENTS AT EACH NODAL POINT

NODAL POINT       Mxx            Myy        NODAL POINT     Mxx            Myy

      1      0.0000E+00    0.0000E+00          2       0.0000E+00    0.0000E+00
      3      0.0000E+00    0.0000E+00          4       0.0000E+00    0.0000E+00
      5      0.7040E+01    0.1376E+02          6       0.0000E+00    0.0000E+00
      7      0.0000E+00    0.0000E+00          8       0.0000E+00    0.0000E+00
      9      0.0000E+00    0.0000E+00


   YOUNG'S MODULUS = 30000000.0 psi

   POISSON'S RATIO = 0.30

   PLATE THICKNESS = 0.20 IN.
```

**Figure 8.27** Output file FISH2OUT.DAT (continued).

**TABLE 8.2 Center Deflection (in.)**

| Total Number of Elements | Solution of FISH2 | Exact Solution | Percentage of Error |
|---|---|---|---|
| 8 | $0.1165 \times 10^{-2}$ | $0.1180 \times 10^{-2}$ | 1.27 |
| 32 | $0.1174 \times 10^{-2}$ | $0.1180 \times 10^{-2}$ | 0.51 |
| 72 | $0.1177 \times 10^{-2}$ | $0.1180 \times 10^{-2}$ | 0.25 |
| 128 | $0.1178 \times 10^{-2}$ | $0.1180 \times 10^{-2}$ | 0.17 |
| 200 | $0.1179 \times 10^{-2}$ | $0.1180 \times 10^{-2}$ | 0.08 |

**TABLE 8.3 Bending Moments at Center (in.-lb / in)**

| Total Number of Elements | Solution of FISH2 | | Exact Solution | | Percentage of Error | |
|---|---|---|---|---|---|---|
| | $M_{xx}$ | $M_{yy}$ | $M_{xx}$ | $M_{yy}$ | $M_{xx}$ | $M_{yy}$ |
| 8 | $0.7040 \times 10$ | $0.1376 \times 10^2$ | $0.7411 \times 10$ | $0.1627 \times 10^2$ | 5.01 | 15.43 |
| 32 | $0.7327 \times 10$ | $0.1550 \times 10^2$ | $0.7411 \times 10$ | $0.1627 \times 10^2$ | 1.13 | 4.73 |
| 72 | $0.7377 \times 10$ | $0.1591 \times 10^2$ | $0.7411 \times 10$ | $0.1627 \times 10^2$ | 0.46 | 2.21 |
| 128 | $0.7394 \times 10$ | $0.1606 \times 10^2$ | $0.7411 \times 10$ | $0.1627 \times 10^2$ | 0.23 | 1.3 |
| 200 | $0.7402 \times 10$ | $0.1614 \times 10^2$ | $0.7411 \times 10$ | $0.1627 \times 10^2$ | 0.12 | 0.8 |

## EXERCISES

**8.1** Consider the problem of a clamped, square plate. The plate is subjected to a uniformly distributed load of intensity $p_0$. Find the deflection field. Find the maximum deflection, and find the bending moment at the center of the plate. Use the method of finite differences.

**8.2** A square plate is clamped along one side and simply supported along the remaining three sides. If a uniformly distributed load with intensity $p_0$ is acting upon the plate, find the deflection field. Find and locate the extreme deflection and the extreme bending moment. Use the method of finite differences ($\nu = 0.3$).

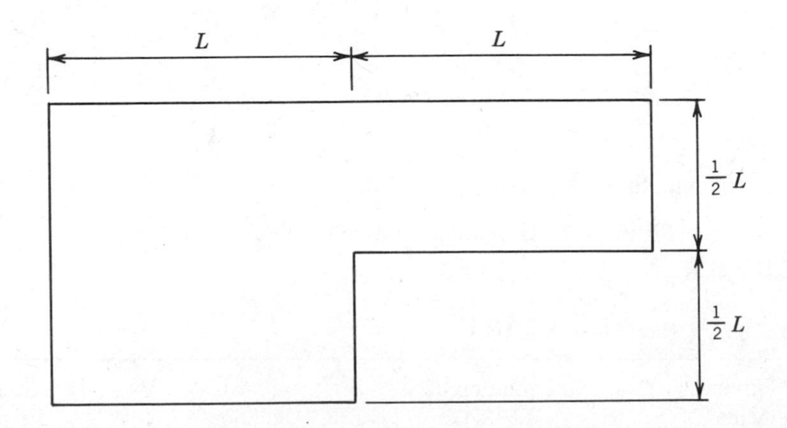


With reference to the figure, find the deflection field of the plate. Assume that all edges are clamped and that the plate is acted upon by a uniformly distributed load of intensity $p_0$. Estimate the location of the point of maximum deflection and find its magnitude. Use the method of finite differences.

**8.4**

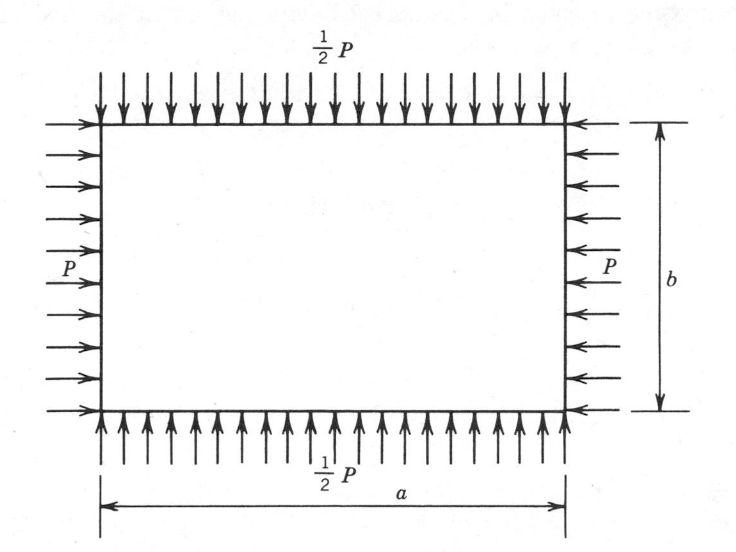


With reference to the figure, find the lowest value of $P$ for which the rectangular plate buckles. The plate is clamped along all four edges. Assume $b/a = 1, 1/2$. Use the method of finite differences.

**8.5** Consider a square plate supported at its four corners by point forces. The plate edges are free. Find the deflection of the plate if it is acted upon by a uniformly distributed load of intensity $p_0$. Use the method of finite differences. Assume $\nu = 0.3$.

**8.6** Consider the plate described in Exercise 8.5. Assume $p_0 = 0$. Find the lowest natural frequency of the plate and its associated mode shape. Use the method of finite differences.

**8.7** A square plate is subjected to a constant membrane shear force $N_{xy}$ applied along the simply supported plate edges. Determine the constant $\Gamma$ in the formula $(N_{xy})_{cr} = \Gamma D/a^2$, where $(N_{xy})_{cr}$ is the shear force at the point of incipient (elastic) instability and $a$ is the length of the plate edges. Use the method of finite differences.

**8.8** Consider the cantilever plate shown in Fig. 8.19 using the data in Section 8.7. Find the lowest natural frequency and associated mode shape. Use the method of finite differences.

**8.9** Solve the problem in Section 8.7 using the finite difference method.

**8.10** With reference to Exercise 8.3, assume that $p_0 \equiv 0$. Find the lowest natural frequency of the plate. Use the method of finite differences.

**8.11** Solve the problem in Exercise 8.1 with the aid of the finite element method. Plot a graph of maximum deflection versus grid size.

**8.12** Solve the problem in Section 8.3.1 with the aid of the finite element method.

**8.13** Solve the problem in Exercise 8.3 with the aid of the finite element method.

**8.14** Do Exercise 8.2 but use the method of finite elements. Plot a curve of maximum deflection versus mesh size.

**8.15** Solve the problem in Exercise 8.5 using the method of finite elements.

**8.16** Consider the decomposition proposed in Exercise 5.33. Use this idea to solve the problem in Section 8.3.1.

**8.17** Use the FORTRAN program in Section 8.8 to solve the problem in Section 5.1.3. Provide a detailed analysis of your results.

# 9

# NONLINEAR CONSIDERATIONS

In our treatment of plate theory in Chapters 3–6, the discussion was restricted to linear theories, both with respect to geometry of deformation as well as with respect to material behavior. This chapter is concerned with certain classes of geometrically nonlinear plate problems that occur in practical applications of plate theory. We shall continue to assume linearly elastic material behavior.

## 9.1 INEXTENSIONAL FINITE BENDING OF A PLATE STRIP

With reference to Fig. 9.1, we consider the plane strain elastic deformation of a plate strip $0 \leq s \leq L$, $-\infty < y < \infty$. The plate strip is acted upon by a line load of constant intensity $F$ along the edge $s = L$, where $s$ is the arc length measured along the plate median surface in the $x$-$y$ plane. The direction of the load $F$ remains parallel to the $z$-axis during deformation. We shall neglect deformations due to membrane forces, and therefore $L$ is a given constant in this problem. With reference to Exercise 3.25, we have $1/R_1$ and $1/R_2 = 0$ as curvatures in the planes $x$-$z$ and $y$-$z$, respectively, and therefore, according to Exercise 3.35, we have

$$M_{xx} = \frac{D}{R_1} = -D\frac{d\theta}{ds} \tag{9.1a}$$

$$M_{yy} = \frac{\nu D}{R_1} = \nu M_{xx} \tag{9.1b}$$

The bending moment at the generic point $s$ is (see Fig. 9.1)

$$M_{xx} = -F(x_0 - x) \tag{9.2}$$

and combining (9.1a) with (9.2) results in

$$D\frac{d\theta}{ds} = F(x_0 - x) \tag{9.3}$$

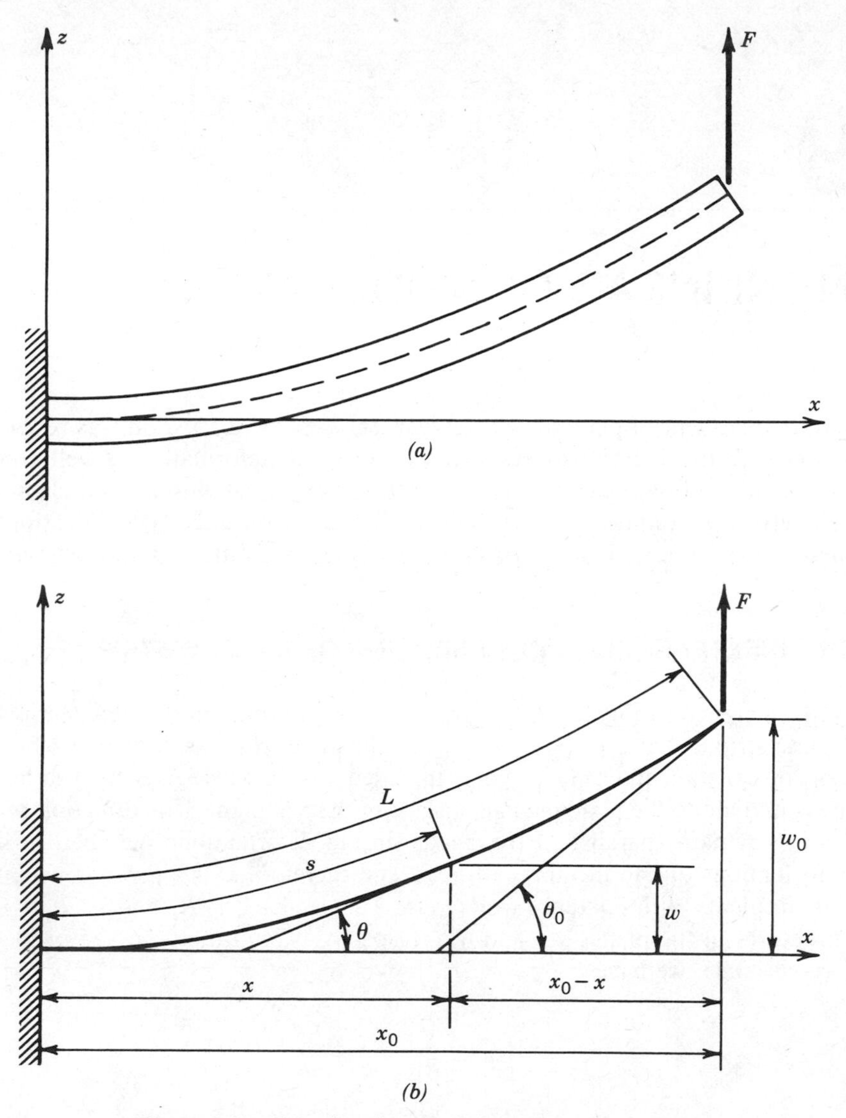


**Figure 9.1** Finite bending of plate strip (plane strain).

It will be convenient to continue the present analysis in nondimensional form. Toward this end, we set

$$\sigma = \frac{s}{L}, \qquad \eta = \frac{w}{L}, \qquad \xi = \frac{x}{L}, \qquad Q = \frac{FL^2}{D} \tag{9.4}$$

and $B = M_{xx}L/D$. With the aid of (9.4), we can write (9.3) in the nondimensional form

$$\frac{d\theta}{d\sigma} = Q(\xi_0 - \xi) \tag{9.5}$$

With the aid of Fig. 9.1($b$), we observe that

$$\frac{dx}{ds} = \frac{d\xi}{d\sigma} = \cos\theta \tag{9.6a}$$

$$\frac{dw}{ds} = \frac{d\eta}{d\sigma} = \sin\theta \tag{9.6b}$$

Differentiating (9.5) and using (9.6a), we obtain

$$\frac{d^2\theta}{d\sigma^2} = -Q\frac{d\xi}{d\sigma} = -Q\cos\theta$$

which can be integrated to read

$$\frac{1}{2}\left(\frac{d\theta}{d\sigma}\right)^2 = -Q\sin\theta + C$$

and since $B = d\theta/d\sigma$ vanishes for $\sigma = 1$, we have $C = Q\sin\theta_0$. Consequently,

$$\frac{d\theta}{d\sigma} = -B = \sqrt{2Q(\sin\theta_0 - \sin\theta)} \geq 0 \tag{9.7}$$

and

$$\sqrt{2a}\int_0^1 d\sigma = \int_0^{\theta_0} \frac{d\theta}{\sqrt{\sin\theta_0 - \sin\theta}} \tag{9.8}$$

The integral on the right-hand side of (9.8) can be transformed as follows: Let

$$2p^2\sin^2\phi = 1 + \sin\theta, \qquad 2p^2 = 1 + \sin\theta_0$$

Hence, when $\phi = \frac{1}{2}\pi$, $\theta = \theta_0$. Consequently,

$$\begin{aligned}
\sqrt{\sin\theta_0 - \sin\theta} &= \sqrt{2}\, p\cos\phi \\
\cos\theta\, d\theta &= 4p^2 \sin\phi\cos\phi\, d\phi \\
\cos\theta = \sqrt{1 - \sin^2\theta} &= 2p\sin\phi\sqrt{1 - p^2\sin^2\phi}
\end{aligned}$$

so that

$$d\theta = \frac{2p\cos\phi\, d\phi}{\sqrt{1 - p^2 \sin\phi}}$$

Consequently, (9.7) becomes

$$\sqrt{Q} = \int_{\phi_0}^{\pi/2} \frac{d\phi}{\sqrt{1 - p^2 \sin^2\phi}} \tag{9.9}$$

where $p^2 = \frac{1}{2}(1 + \sin\theta_0)$ and $\phi_0 = \sin^{-1}(1/\sqrt{2p^2})$. In view of (9.6b) and (9.7), we can write

$$d\eta = \frac{\sin\theta\, d\theta}{d\theta/d\sigma} = \frac{\sin\theta\, d\theta}{\sqrt{2Q}\sqrt{\sin\theta_0 - \sin\theta}}$$

so that

$$\sqrt{Q}\int_0^{\eta_0} d\eta = \sqrt{Q}\,\eta_0 = \int_{\phi_0}^{\pi/2} \frac{(2p^2\sin^2\phi - 1)\, d\phi}{\sqrt{1 - p^2\sin^2\phi}} \tag{9.10}$$

where $\eta_0$ is the (dimensionless) end deflection of the plate. We also note that

$$\xi_0 = -\frac{B}{Q} = \sqrt{\frac{2}{Q}\sin\theta_0} \tag{9.11}$$

and

$$B_0 = -Q\xi_0 = -\sqrt{2Q\sin\theta_0} \tag{9.12}$$

The definite integrals in (9.9) and (9.10) are readily evaluated by numerical methods. Thus, (9.9) provides a relationship between $\theta_0$ and $F$, and (9.10) is a formula for the end deflection $w_0$ if $\theta_0$ is known. Similarly, (9.12) gives the extreme bending moment (at the clamped end) when $\theta_0$ is known. Table 9.1 shows the results obtained in this manner with the aid of a digital computer.

It should be noted that a solution can be obtained with the aid of elliptic integrals for the present case. If

$$\mathfrak{F}(\phi_0, p) = \int_0^{\phi_0} \frac{dx}{\sqrt{1 - p^2\sin^2 x}} \tag{9.13a}$$

is the elliptic integral of the first kind and

$$\mathfrak{E}(\phi_0, p) = \int_0^{\phi_0} \sqrt{1 - p^2\sin^2 x}\, dx \tag{9.13b}$$

**TABLE 9.1 Problem Parameters: Finite Bending of Plate Strip**

| $Q$ | $\xi_0$ | $\eta_0$ | $\eta_{0_{CPT}}$ | $\theta_0$ | $B_0$ |
|---|---|---|---|---|---|
| .1001 | .9993 | .0333 | .0334 | .05 | −.1000 |
| .2007 | .9973 | .0666 | .0669 | .10 | −.2002 |
| .3025 | .9940 | .0998 | .1008 | .15 | −.3007 |
| .4059 | .9893 | .1329 | .1353 | .20 | −.4016 |
| .5117 | .9834 | .1657 | .1706 | .25 | −.5032 |
| .6204 | .9761 | .1984 | .2068 | .30 | −.6055 |
| .7327 | .9674 | .2308 | .2442 | .35 | −.7089 |
| .8495 | .9575 | .2628 | .2832 | .40 | −.8134 |
| .9715 | .9463 | .2946 | .3238 | .45 | −.9193 |
| 1.0997 | .9338 | .3259 | .3666 | .50 | −1.0269 |
| 1.2352 | .9200 | .3568 | .4117 | .55 | −1.1363 |
| 1.3792 | .9049 | .3873 | .4597 | .60 | −1.2480 |
| 1.5331 | .8885 | .4172 | .5110 | .65 | −1.3622 |
| 1.6987 | .8709 | .4466 | .5662 | .70 | −1.4794 |
| 1.8780 | .8520 | .4754 | .6260 | .75 | −1.6001 |
| 2.0733 | .8319 | .5037 | .6911 | .80 | −1.7247 |
| 2.2878 | .8104 | .5313 | .7626 | .85 | −1.8540 |
| 2.5249 | .7877 | .5583 | .8416 | .90 | −1.9889 |
| 2.7893 | .7637 | .5847 | .9298 | .95 | −2.1302 |
| 3.0870 | .7384 | .6104 | 1.0290 | 1.00 | −2.2793 |
| 3.4256 | .7116 | .6354 | 1.1419 | 1.05 | −2.4378 |
| 3.8156 | .6835 | .6598 | 1.2719 | 1.10 | −2.6079 |
| 4.2710 | .6538 | .6836 | 1.4237 | 1.15 | −2.7923 |
| 4.8121 | .6224 | .7068 | 1.6040 | 1.20 | −2.9950 |
| 5.4692 | .5891 | .7295 | 1.8231 | 1.25 | −3.2219 |
| 6.2893 | .5535 | .7518 | 2.0964 | 1.30 | −3.4814 |
| 7.3521 | .5152 | .7741 | 2.4507 | 1.35 | −3.7878 |
| 8.8064 | .4731 | .7966 | 2.9355 | 1.40 | −4.1661 |
| 10.9769 | .4253 | .8201 | 3.6590 | 1.45 | −4.6684 |
| 14.7969 | .3672 | .8466 | 4.9323 | 1.50 | −5.4332 |

is the elliptic integral of the second kind, then it can be shown that

$$\sqrt{Q} = \mathfrak{F}\left(\tfrac{1}{2}\pi, p\right) - \mathfrak{F}(\phi_0, p) \tag{9.14}$$

$$\sqrt{Q}\,\eta_0 = -2\mathfrak{E}\left(\tfrac{1}{2}\pi, p\right) + 2\mathfrak{E}(\phi_0, p) + \mathfrak{F}\left(\tfrac{1}{2}\pi, p\right) - \mathfrak{F}(\phi_0, p) \tag{9.15}$$

Extensive tables of the elliptic functions (9.13) are available, and these have been used to check the numerical results in Table 9.1. If CPT is used to solve the present problem, we readily obtain $d^2\eta/d\xi^2 = Q(1 - \xi)$, $\eta = \frac{1}{2}Q\xi^2(1 - \frac{1}{3}\xi)$, and $B = -Q(1 - \xi)$, so that $\eta_0 = \eta(1) = \frac{1}{3}Q$, $B_0 = B(0) = -Q$. Figures 9.2 and 9.3 show maximum deflection and extreme bending moment versus $F$, respectively. These graphs clearly show significant nonlinear elastic behavior when maximum deflections exceed $w_0/L = 0.25$.

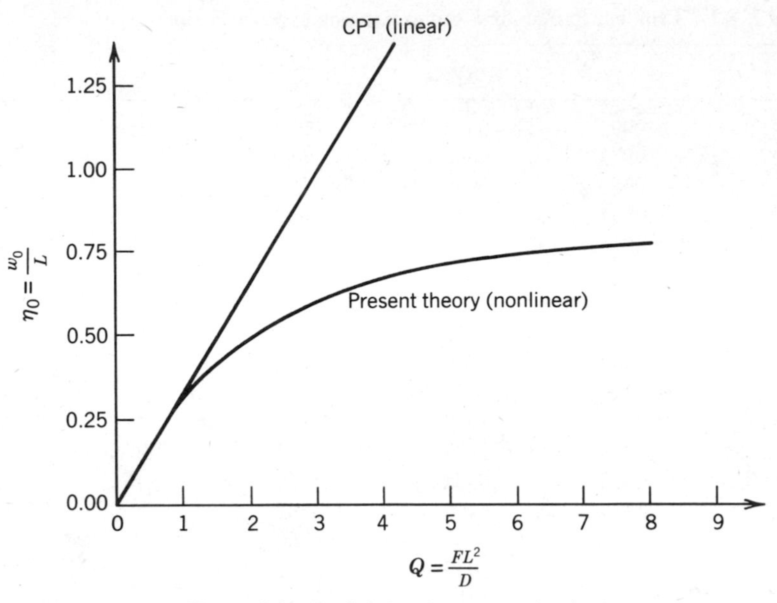


**Figure 9.2** End deflection versus load.

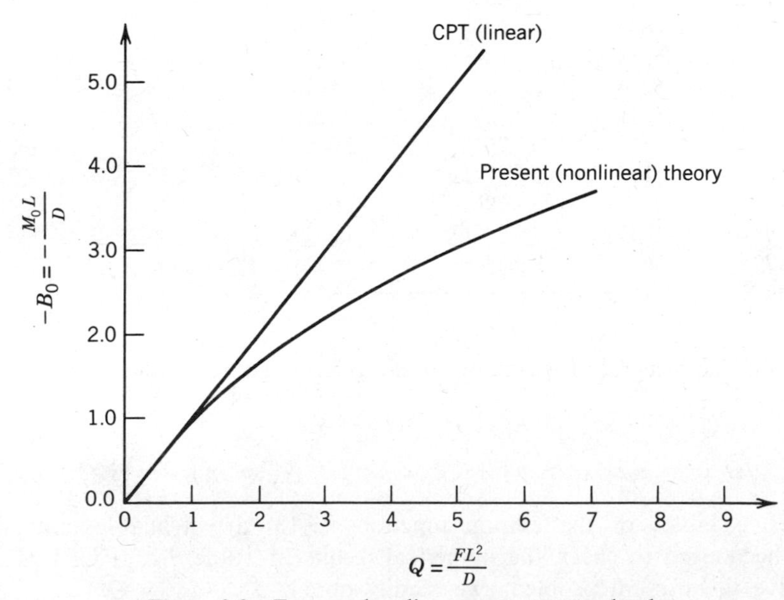


**Figure 9.3** Extreme bending moment versus load.

## 9.2 COUPLED BENDING AND STRETCHING (VON KÁRMÁN'S PLATE THEORY)

We shall now develop a nonlinear theory similar to CPT. The present theory also neglects shear deformations, but it accounts for the coexistence of flexure and membrane action in the plate. This effect will occur, for example, when the plate boundaries are restrained from moving freely in the plane of the plate. In such cases induced membrane forces will increase with an increase in plate deflections.

As a point of departure, we shall use the principle of virtual work:

$$\delta U = \delta \mathscr{W} \tag{9.16}$$

where $\delta U$ is the variation of strain energy and $\delta \mathscr{W}$ is the virtual work of all forces acting upon the plate. To account for the coupling of bending and stretching in the plate, it will be necessary to retain certain nonlinear terms in the strain tensor components. From three-dimensional theory we have (see Section 2.1)

$$2L_{ij}\,dx_i\,dx_j = 2d\mathbf{u} \cdot d\mathbf{r} + d\mathbf{u} \cdot d\mathbf{u} \tag{9.17}$$

where $L_{ij}$ is the Lagrangian strain tensor. The plate displacement field is (see 3.3)

$$\mathbf{u} = \hat{\mathbf{e}}_\alpha\left(u_\alpha^0 + z\psi_\alpha\right) + \hat{\mathbf{e}}_z w \tag{9.18}$$

where $u_\alpha^0$ is included to denote the components of the displacement vector relative to the axes $x_\alpha$ of points originally in the plate median surface. We have

$$\begin{aligned} d\mathbf{u} &= \hat{\mathbf{e}}_\alpha\left(du_\alpha^0 + z\,d\psi_\alpha\right) + \hat{\mathbf{e}}_z\,dw, \qquad d\mathbf{r} = \hat{\mathbf{e}}_\alpha\,dx_\alpha \\ du_\alpha^0 &= u_{\alpha,\beta}^0\,dx_\beta, \qquad d\psi_\alpha = \psi_{\alpha,\beta}\,dx_\beta, \qquad dw = w_{,\alpha}\,dx_\alpha \end{aligned} \tag{9.19}$$

and upon substitution of (9.19) into (9.17), we obtain $L_{13} = L_{31} = L_{23} = L_{32} = L_{33} = 0$ and

$$\begin{aligned} 2L_{ij}\,dx_i\,dx_j = 2L_{\alpha\beta}\,dx_\alpha\,dx_\beta &= 2\left(u_{\alpha,\beta}^0 + z\psi_{\alpha,\beta} + \tfrac{1}{2}w_{,\alpha}w_{,\beta}\right)dx_\alpha\,dx_\beta \\ &\quad + \left[\left(u_{\alpha,\beta}^0 + z\psi_{\alpha,\beta}\right)^2 dx_\alpha\,dx_\beta\right] \end{aligned}$$

and we shall drop (neglect) the second (or higher order) term on the right-hand side of this equation. Consequently,

$$L_{\alpha\beta} = \tfrac{1}{2}\left(u_{\alpha,\beta}^0 + u_{\beta,\alpha}^0\right) + \tfrac{1}{2}z\left(\psi_{\alpha,\beta} + \psi_{\beta,\alpha}\right) + \tfrac{1}{2}w_{,\alpha}w_{,\beta} \tag{9.20}$$

where we have imposed the requirement $L_{\alpha\beta} = L_{\beta\alpha}$. We note that the term

$\frac{1}{2}w_{,\alpha}w_{,\beta}$ is nonlinear, the remaining terms in (9.20) being linear in the plate displacements. For the present case of CPT we have the plate strain components

$$m_{\alpha\beta} = \frac{12}{h^3}\int_{-h/2}^{h/2} zL_{\alpha\beta}\,dz = \tfrac{1}{2}(\psi_{\alpha,\beta} + \psi_{\beta,\alpha}) = -w_{,\alpha\beta} \tag{9.21a}$$

$$n_{\alpha\beta} = \frac{1}{h}\int_{-h/2}^{h/2} L_{\alpha\beta}\,dz = \tfrac{1}{2}(u^0_{\alpha,\beta} + u^0_{\beta,\alpha}) + \tfrac{1}{2}w_{,\alpha}w_{,\beta} \tag{9.21b}$$

and therefore,

$$L_{\alpha\beta} = n_{\alpha\beta} + zm_{\alpha\beta} \tag{9.22}$$

An inspection of (9.21) reveals that coupling between stretching and bending arises because of the term $\frac{1}{2}w_{,\alpha}w_{,\beta}$. The strain energy density is

$$W = \tfrac{1}{2}\tau_{\alpha\beta}L_{\alpha\beta} = \tfrac{1}{2}\tau_{\alpha\beta}n_{\alpha\beta} + \tfrac{1}{2}z\tau_{\alpha\beta}m_{\alpha\beta}$$

and the plate strain energy density is

$$W_p = \int_{-h/2}^{h/2} W\,dz = \tfrac{1}{2}(M_{\alpha\beta}m_{\alpha\beta}) + \tfrac{1}{2}(N_{\alpha\beta}n_{\alpha\beta}) \tag{9.23}$$

where we have used (3.24b) and set

$$N_{\alpha\beta} = \int_{-h/2}^{h/2} \tau_{\alpha\beta}\,dz \tag{9.24}$$

The symbol $N_{\alpha\beta}$ denotes the membrane force intensity (force per unit of length). We can use Hooke's law to show that

$$M_{\alpha\beta} = D[(1-\nu)m_{\alpha\beta} + \nu m\delta_{\alpha\beta}] \tag{9.25a}$$

$$N_{\alpha\beta} = C[(1-\nu)n_{\alpha\beta} + \nu n\delta_{\alpha\beta}] \tag{9.25b}$$

where $m = m_{\alpha\alpha}$, $n = n_{\alpha\alpha}$, and

$$D = \frac{Eh^3}{12(1-\nu^2)}, \qquad C = \frac{Eh}{1-\nu^2}$$

[see (3.58a) and Exercise 9.6]. Upon substitution of (9.25) into (9.23), we obtain

$$W_p = \tfrac{1}{2}D[(1-\nu)m_{\alpha\beta}m_{\alpha\beta} + \nu m^2] + \tfrac{1}{2}C[(1-\nu)n_{\alpha\beta}n_{\alpha\beta} + \nu n^2] \tag{9.26}$$

and

$$\delta W_p = \frac{\partial W_p}{\partial m_{\alpha\beta}}\delta m_{\alpha\beta} + \frac{\partial W_p}{\partial n_{\alpha\beta}}\delta n_{\alpha\beta} = M_{\alpha\beta}\,\delta m_{\alpha\beta} + N_{\alpha\beta}\,\delta n_{\alpha\beta} \tag{9.27}$$

We now write

$$\delta U = \int_A \delta W_p \, dA = \delta U_1 + \delta U_2$$

where

$$\delta U_1 = \int_A \delta W_p^{(1)} \, dA, \qquad \delta U_2 = \int_A \delta W_p^{(2)} \, dA$$

$$\delta W_p^{(1)} = M_{\alpha\beta} \, \delta m_{\alpha\beta}, \qquad \delta W_p^{(2)} = N_{\alpha\beta} \, \delta n_{\alpha\beta}$$

With the aid of relations derived in Section 4.2, we have

$$\delta U_1 = \oint_C \left[ \left( Q_n + \frac{\partial M_{ns}}{\partial s} \right) \delta w + M_{nn} \delta \left( -\frac{\partial w}{\partial n} \right) \right] ds - \int_A M_{\alpha\beta,\beta\alpha} \, \delta w \, dA \quad (9.28)$$

To compute $\delta U_2$, we shall require the following auxiliary formulas, which are self-explanatory:

$$\mathbf{N} = N_{nn} \hat{\mathbf{n}} + N_{ns} \hat{\mathbf{s}}, \qquad \boldsymbol{\psi} = \psi_n \hat{\mathbf{n}} + \psi_s \hat{\mathbf{s}}, \qquad \delta \mathbf{u}^0 = \delta u_n^0 \, \hat{\mathbf{n}} + \delta u_s^0 \, \hat{\mathbf{s}}$$

$$N_{\alpha\beta} n_\beta \, \delta u_\alpha^0 = N_\alpha \delta u_\alpha^0 = \mathbf{N} \cdot \delta \mathbf{u}^0 = N_{nn} \delta u_n^0 + N_{ns} \delta u_s^0$$

$$\begin{aligned} N_{\alpha\beta} n_\beta \, \delta w,_\alpha &= -N_{\alpha\beta} n_\beta \, \delta \psi_\alpha = -N_\alpha \delta \psi_\alpha \\ &= -\mathbf{N} \cdot \delta \boldsymbol{\psi} = -(N_{nn} \delta \psi_n + N_{ns} \delta \psi_s) \\ &= N_{nn} \delta \left( \frac{\partial w}{\partial n} \right) + N_{ns} \delta \left( \frac{\partial w}{\partial s} \right) \end{aligned}$$

We have [see (9.21b)]

$$\begin{aligned} \delta U_2 &= \int_A N_{\alpha\beta} \, \delta n_{\alpha\beta} \, dA = \int_A \left[ N_{\alpha\beta} \, \delta u_{\alpha,\beta}^0 + (N_{\alpha\beta} w,_\alpha) \, \delta w,_\beta \right] dA \\ &= \oint_C (N_{nn} \delta u_n^0 + N_{ns} \delta u_s^0) \, ds - \int_A N_{\alpha\beta,\beta} \, \delta u_\alpha^0 \, dA \\ &\quad + \oint_C \left( N_{nn} \frac{\partial w}{\partial n} + N_{ns} \frac{\partial w}{\partial s} \right) \delta w \, ds - \int_A (N_{\alpha\beta} w,_\alpha),_\beta \, \delta w \, dA \end{aligned} \quad (9.29)$$

With the aid of (9.28) and (9.29), we obtain

$$\begin{aligned} \delta U &= \delta U_1 + \delta U_2 \\ &= \oint_C \left[ S_n \delta w + M_{nn} \delta \left( -\frac{\partial w}{\partial n} \right) + N_{nn} \delta u_n^0 + N_{ns} \delta u_s^0 \right] ds \\ &\quad - \int_A \left\{ \left[ M_{\alpha\beta,\beta\alpha} + (N_{\alpha\beta} w,_\alpha),_\beta \right] \delta w + N_{\alpha\beta,\beta} \, \delta u_\alpha^0 \right\} dA \end{aligned} \quad (9.30)$$

where, for the present case, on $C$ we have

$$S_n = Q_n + \frac{\partial M_{ns}}{\partial s} + N_{nn}\frac{\partial w}{\partial n} + N_{ns}\frac{\partial w}{\partial s} \tag{9.31}$$

The virtual work of external forces acting upon the plate is

$$\delta \mathscr{W} = \int_A p\,\delta w + \oint_C \left[ S_n^*\,\delta w + M_{nn}^*\,\delta\left(-\frac{\partial w}{\partial n}\right) + N_{nn}^*\,\delta u_n^* + N_{ns}^*\,\delta u_s^* \right] ds \tag{9.32}$$

where the starred symbols $S_n^*$, $N_{nn}^*$, $M_{nn}^*, \ldots$ denote specified boundary values on $C$. Upon substitution of (9.30) and (9.32) into (9.16), we obtain

$$\begin{aligned}
&\int_A \left\{ \left[ M_{\alpha\beta,\beta\alpha} + (N_{\alpha\beta} w_{,\alpha})_{,\beta} + p \right] \delta w + N_{\alpha\beta,\beta}\,\delta u_\alpha^0 \right\} dA \\
&\quad + \oint_C \left[ (S_n^* - S_n)\,\delta w + (M_{nn}^* - M_{nn})\,\delta\left(-\frac{\partial w}{\partial n}\right) \right. \\
&\quad \left. + (N_{nn}^* - N_{nn})\,\delta u_n^0 + (N_{ns}^* - N_{ns})\,\delta u_s^0 \right] ds = 0
\end{aligned} \tag{9.33}$$

The region $A$ and the boundary $C$ are arbitrary, and the variations $\delta w$, $\delta(-\partial w/\partial n)$, and $\delta u_n^0$ are independent and arbitrary. Therefore, (9.33) reduces to the system of partial differential equations

$$\left.\begin{aligned}
S_{\alpha,\alpha} + p &= 0 && (9.34\text{a})\\
N_{\alpha\beta,\beta} &= 0 && (9.34\text{b})\\
S_\alpha &= Q_\alpha + N_{\alpha\gamma} w_{,\gamma} && (9.34\text{c})\\
Q_\alpha &= M_{\alpha\beta,\beta} && (9.34\text{d})
\end{aligned}\right\} \text{ in } A$$

(see the results of Section 7.1). We also obtain the associated, admissible boundary conditions from (9.33): We require the specification of

$$\left.\begin{aligned}
&\text{Either } S_n \text{ or } w\\
\text{and}\quad &\text{either } M_{nn} \text{ or } \frac{\partial w}{\partial n}\\
\text{and}\quad &\text{either } N_{nn} \text{ or } u_n^0\\
\text{and}\quad &\text{either } N_{ns} \text{ or } u_s^0
\end{aligned}\right\} \text{ on } C \tag{9.35}$$

We also note the invariant form of (9.34):

$$\left.\begin{aligned}
\mathbf{div\,S} + p &= 0 && (9.36\text{a})\\
\mathbf{div}\,\boldsymbol{\mathscr{N}} &= \mathbf{0} && (9.36\text{b})\\
\mathbf{S} &= \mathbf{Q} + \boldsymbol{\mathscr{N}} \cdot \nabla w && (9.36\text{c})\\
\mathbf{Q} &= \mathbf{div}\,\boldsymbol{\mathscr{M}} && (9.36\text{d})
\end{aligned}\right\} \text{ in } A$$

where $\mathscr{M}$ and $\mathscr{N}$ are the moment and membrane stress tensors, respectively, and where $\mathbf{S}$ is the vertical shear force vector that acts normal to the undeformed median surface (see Section 7.1).

The stress equations of equilibrium can be converted to displacement equations of equilibrium by the relations

$$\begin{aligned} \operatorname{grad} \mathbf{Q} &= \operatorname{grad} \operatorname{div} \mathscr{M} = -D\nabla^4 w \\ Q_{\alpha,\alpha} &= M_{\alpha\beta,\beta\alpha} = -D\nabla^4 w \end{aligned} \tag{9.37}$$

(see Section 4.2) and

$$N_{\alpha\beta} = C\left[\tfrac{1}{2}(1-\nu)\left(u^0_{\alpha,\beta} + u^0_{\beta,\alpha} + w_{,\alpha}w_{,\beta}\right) + \nu\,\delta_{\alpha\beta}\left(u^0_{\gamma,\gamma} + \tfrac{1}{2}w_{,\gamma}w_{,\gamma}\right)\right] \tag{9.38}$$

[see (9.25) and (9.21)].

The present plate theory was first presented by Theodore von Kármán in *Encyklopädie der Mathematischen Wissenschaften*, Vol. $\mathrm{IV}_4$, p. 349, 1910. Note that in this formulation the slope of the median surface is still small in the sense that $w_{,\alpha}^2 \ll 1$, $\alpha = 1, 2$, and that shear deformations have been neglected. Equations (9.34) and (9.36) are nonlinear, and their solution, even in the case of simple problems, requires special techniques and usually considerable (numerical) computational effort. For an overview, see C. Y. Chia, *Nonlinear Analysis of Plates*, McGraw-Hill, New York, 1980.

## 9.3 UNIFORMLY LOADED MEMBRANE STRIP

We shall now consider the deformation of a "plate" strip of unbounded length and of width $L$, the domain of the plate being $-\tfrac{1}{2}L \le x \le \tfrac{1}{2}L$, $-\infty < y < \infty$, as shown in Fig. 9.4. The plate strip is supported along its immovable edges $x = \pm\tfrac{1}{2}L$. In the present case we assume that the plate flexural rigidity is negligible ($D \equiv 0$), and resistance to deformation is supplied entirely by the

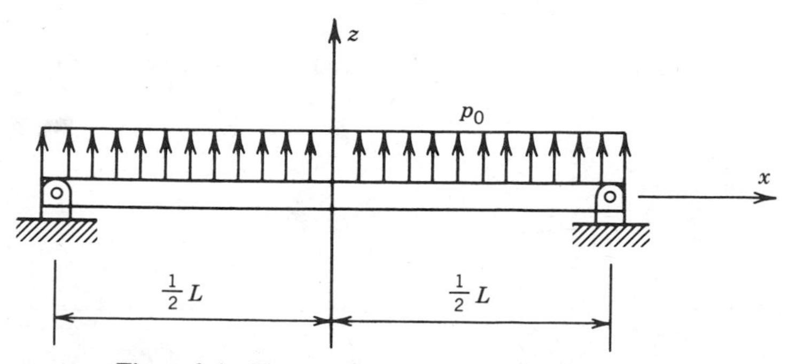


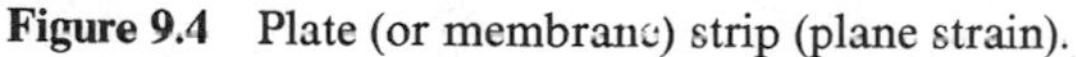

**Figure 9.4** Plate (or membrane) strip (plane strain).

membrane forces induced by the resulting deformation. The edges $x = \pm\frac{1}{2}L$ are fixed and are prevented from moving. The load is uniformly distributed with intensity $p_0$. Conditions of plane strain prevail, and we apply (9.34), (9.37), and (9.38) with $w = w(x)$, $u_x = u(x)$, $u_y \equiv 0$, $N_{xx} = N_{xx}(x)$, and $D \equiv 0$. Consequently, we have

$$\frac{dN_{xx}}{dx} = 0$$

so that $N_{xx}$ = constant, and

$$\frac{d^2w}{dx^2} = -\frac{p_0}{N_{xx}}$$

A solution of this equation that satisfies the conditions $w(\pm\frac{1}{2}L) = 0$ and $(dw/dx)_{x=0} = 0$ is

$$w = \frac{p_0 L^2}{2N_{xx}}\left(\frac{1}{4} - \frac{x^2}{L^2}\right) \tag{9.39}$$

From (9.38) we obtain the membrane forces:

$$N_{xx} = C\left[\frac{du}{dx} + \frac{1}{2}\left(\frac{dw}{dx}\right)^2\right] = \frac{N_{yy}}{\nu}$$

and therefore,

$$\frac{du}{dx} = \frac{N_{xx}}{C} - \frac{1}{2}\left(\frac{dw}{dx}\right)^2 = \frac{N_{xx}}{C} - \frac{1}{2}\frac{p_0^2 x^2}{N_{xx}^2}$$

A solution of this differential equation that satisfies the condition $u(0) = 0$ is

$$u(x) = x\left(\frac{N_{xx}}{C} - \frac{1}{6}\frac{p_0^2 x^2}{N_{xx}^2}\right) \tag{9.40}$$

If we now invoke the condition of a nonyielding support $u(\pm\frac{1}{2}L) = 0$, we obtain

$$\frac{N_{xx}}{C} = \frac{1}{24}\frac{p_0^2 L^2}{N_{xx}^2}$$

or

$$2\sqrt[3]{3}\,\frac{N_{xx}}{C} = \left(\frac{p_0 L}{C}\right)^{2/3} \tag{9.41}$$

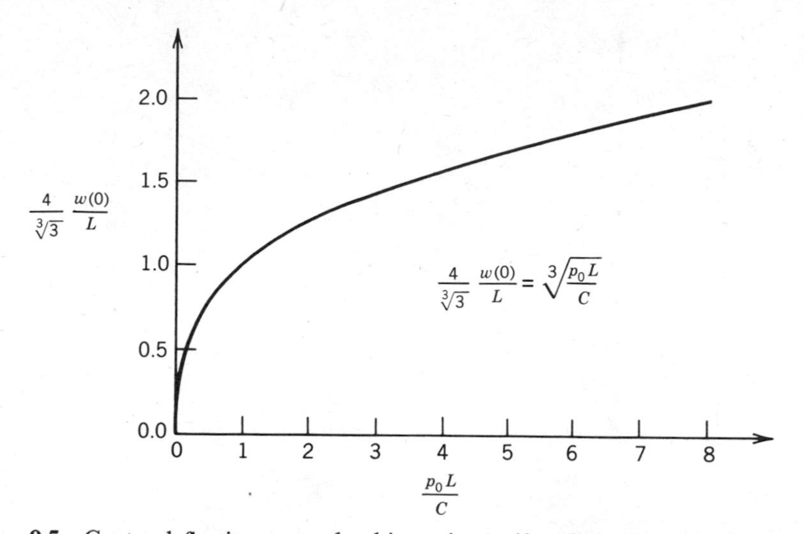


**Figure 9.5** Center deflection versus load intensity (uniformly loaded membrane strip).

With the aid of (9.41), we can eliminate $N_{xx}$ from (9.40). The result is

$$\frac{u(x)}{L} = \frac{1}{6}\left(\frac{24 p_0 L}{C}\right)^{2/3}\left(\frac{x}{L}\right)\left[\left(\frac{1}{2}\right)^2 - \left(\frac{x}{L}\right)^2\right] \tag{9.42}$$

Again, we use (9.41) to eliminate $N_{xx}$ from (9.39) and obtain

$$\frac{w(x)}{L} = \frac{1}{2}\left(\frac{24 p_0 L}{C}\right)^{1/3}\left[\left(\frac{1}{2}\right)^2 - \left(\frac{x}{L}\right)^2\right] \tag{9.43}$$

The maximum transverse displacement occurs at the center of the span, at $x = 0$, and it is given by

$$\frac{w(0)}{L} = \frac{\sqrt[3]{3}}{4}\sqrt[3]{\frac{p_0 L}{C}} \tag{9.44}$$

Equations (9.44) and (9.41) were used to construct the graphs in Figs. 9.5 and 9.6, respectively. These graphs clearly exhibit the nonlinear relations that exist between transverse deflection and load as well as between membrane force and load. We also note that by a combination of (9.41) and (9.44), we obtain

$$\frac{N_{xx}}{C} = \frac{8}{3}\left[\frac{w(0)}{L}\right]^2 \tag{9.45}$$

With reference to (9.45) it can be concluded that the membrane force intensity is proportional to the square of the center deflection.

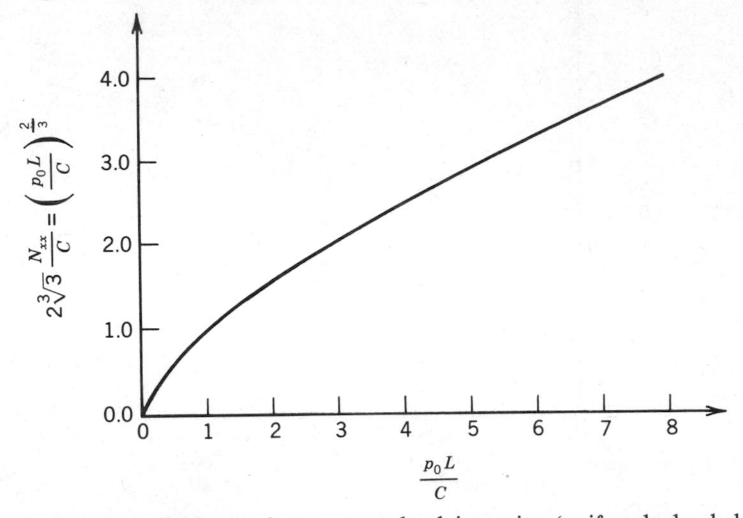


**Figure 9.6** Induced membrane force versus load intensity (uniformly loaded membrane strip).

## 9.4 THE UNIFORMLY LOADED PLATE STRIP

In this section we consider a deformed plate strip of unbounded length and width $L$, the domain of the plate being $0 \leq x \leq L$, $-\infty < y < \infty$, as shown in Fig. 9.4. The plate strip is simply supported along its immovable edges: $w(0, y) = w(L, y) = 0$, $M_{xx}(0, y) = M_{xx}(L, y) = 0$. In the present case we assume that the load is uniformly distributed with intensity $p_0$. Consequently, we have the condition of plane strain, and (9.34), (9.37), and (9.38) apply with $w = w(x)$, $u_x = u(x)$, $u_y \equiv 0$, $N_{xy} \equiv 0$, $M_{xy} \equiv 0$, $N_{xx} = N = \nu N_{yy}$, and $M_{xx} = M(x) = \nu M_{yy}$. Thus, we have [see (9.34), (9.37), and (9.38)]

$$D\frac{d^4w}{dx^4} - N\frac{d^2w}{dx^2} = p_0 = \frac{4p_0}{\pi}\sum_{m=1,3,5,\ldots}^{\infty}\frac{1}{m}\sin\frac{m\pi x}{L} \tag{9.46}$$

$$\frac{dN}{dx} = 0, \quad \text{or } N = \text{constant} \tag{9.47}$$

and using (9.38), we obtain

$$N = C\left[\frac{du}{dx} + \frac{1}{2}\left(\frac{dw}{dx}\right)^2\right] \tag{9.48}$$

or

$$\frac{du}{dx} = \frac{N}{C} - \frac{1}{2}\left(\frac{dw}{dx}\right)^2$$

To calculate the elongation of the plate in the $x$-direction, we write

$$\int_0^L \frac{du}{dx}\,dx = \frac{N}{C}\int_0^L dx - \frac{1}{2}\int_0^L \left(\frac{dw}{dx}\right)^2 dx$$

or

$$u(L) - u(0) = \frac{NL}{C} - \frac{1}{2}\int_0^L \left(\frac{dw}{dx}\right)^2 dx \tag{9.49}$$

In view of simple support conditions along $x = 0, L$, we can assume that the plate's deflected median surface can be characterized as

$$w = \sum_{m=1,3,5,\ldots}^{\infty} a_m \sin\frac{m\pi x}{L} \tag{9.50}$$

and upon substitution of (9.50) into (9.46), we obtain

$$a_m = \frac{4p_0L^4}{\pi^5 D}\,\frac{1}{m^3(m^2+Q)} \tag{9.51}$$

where

$$Q = \frac{NL^2}{\pi^2 D} \tag{9.52}$$

is the dimensionless membrane force per unit length in the $x$-direction. It can be readily shown that

$$\int_0^L \left(\frac{dw}{dx}\right)^2 dx = \sum_{m=1,3,5,\ldots}^{\infty} a_m^2 (m\pi)^2 \frac{1}{2L} \tag{9.53}$$

We now assume that $u(L) - u(0) = 0$ (rigid supports) and substitute (9.53) into (9.49). Thus, we obtain

$$Q = P^2 \sum_{m=1,3,5,\ldots}^{\infty} \frac{1}{(m^2+Q)^2 m^4} \tag{9.54}$$

where

$$P = \frac{2}{\pi^5}\,\frac{p_0L^4}{D}\sqrt{\frac{C}{D}} = \frac{4\sqrt{3}}{\pi^5}\,\frac{p_0L^4}{Dh} \tag{9.55}$$

We now define the dimensionless plate deflection as

$$W = \sqrt{3}\,\frac{w}{h} \tag{9.56}$$

If we substitute (9.51) into (9.50), utilize (9.55) and (9.56), and set $x = \frac{1}{2}L$, we obtain an expression for the "dimensionless" center deflection of the plate:

$$W = P \sum_{m=1,3,5,\ldots}^{\infty} \frac{\sin(m\pi/2)}{(m^2 + Q)m^3} \tag{9.57}$$

In the present case the bending moment

$$M_{xx} \equiv M = -D\frac{d^2w}{dx^2} \tag{9.58}$$

and we define the dimensionless bending moment by

$$B = \frac{ML^2}{2D\pi^2}\sqrt{\frac{C}{D}} \tag{9.59}$$

In view of (9.56)–(9.59), the dimensionless bending moment at the center of the plate $x = \frac{1}{2}L$ is

$$B = P \sum_{m=1,3,5,\ldots}^{\infty} \frac{\sin(m\pi/2)}{(m^2 + Q)m} \tag{9.60}$$

We have used (9.57), (9.60), and (9.54) to plot the graphs shown in Figs. 9.7, 9.8, and 9.9, respectively. The nonlinear behavior of deflection, bending moment, and membrane force is clearly exhibited in these graphs. In the case

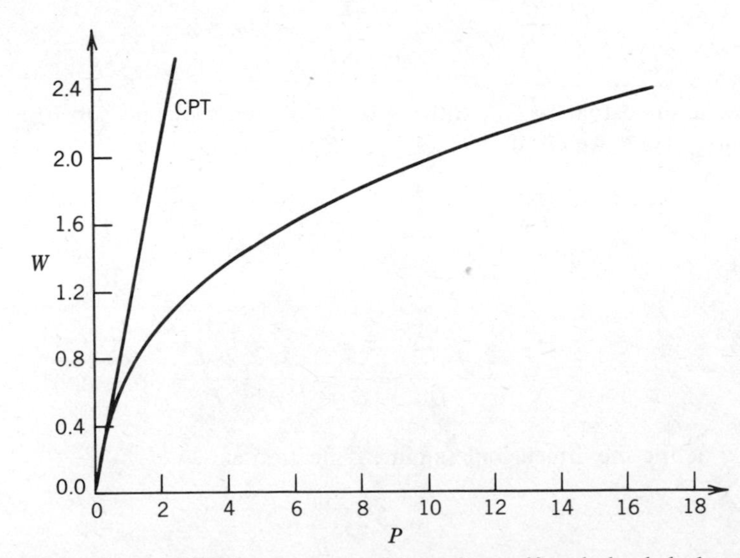


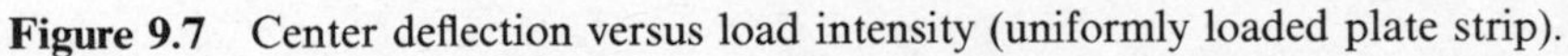
**Figure 9.7** Center deflection versus load intensity (uniformly loaded plate strip).

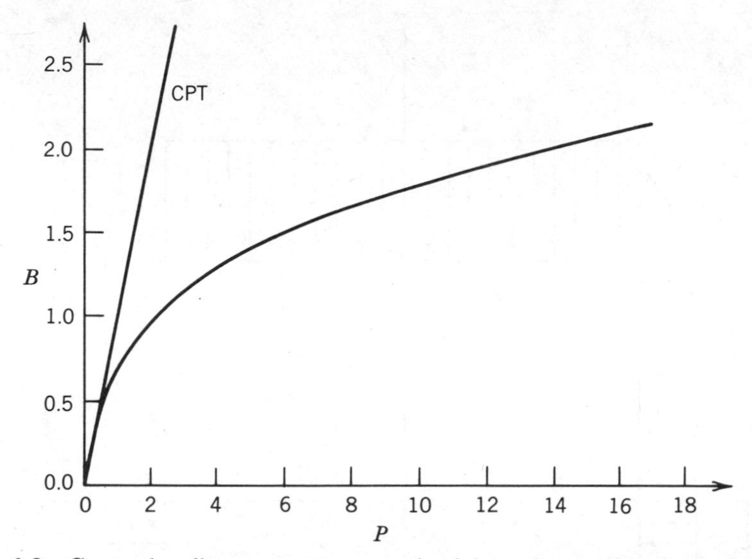


**Figure 9.8** Center bending moment versus load intensity (uniformly loaded plate strip).

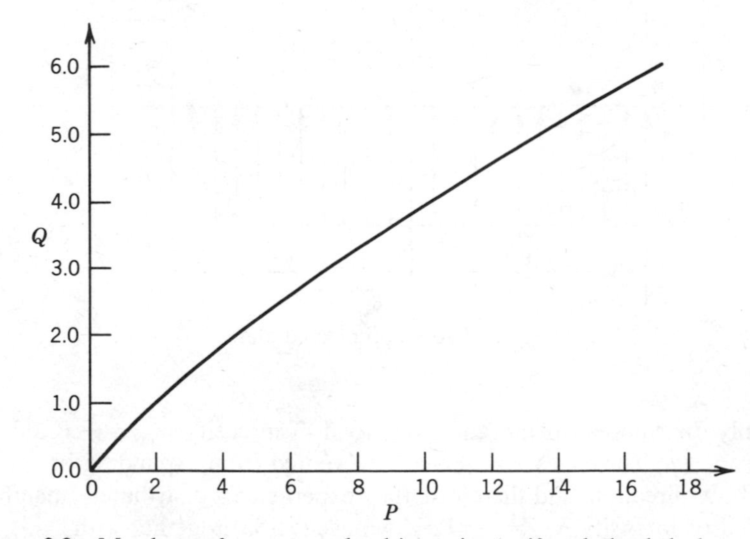


**Figure 9.9** Membrane force versus load intensity (uniformly loaded plate strip).

of center deflection and bending moment, a comparison with CPT is provided in Figs. 9.7 and 9.8. For this latter case, of course, $u(L) - u(0) \neq 0$.

## 9.5 POSTBUCKLING BEHAVIOR OF ELASTIC PLATES

Let us first consider the stability of a simply supported, elastic, square plate with dimensions $2a : 2a : h$, as shown in Fig. 9.10. The plate is subjected to a

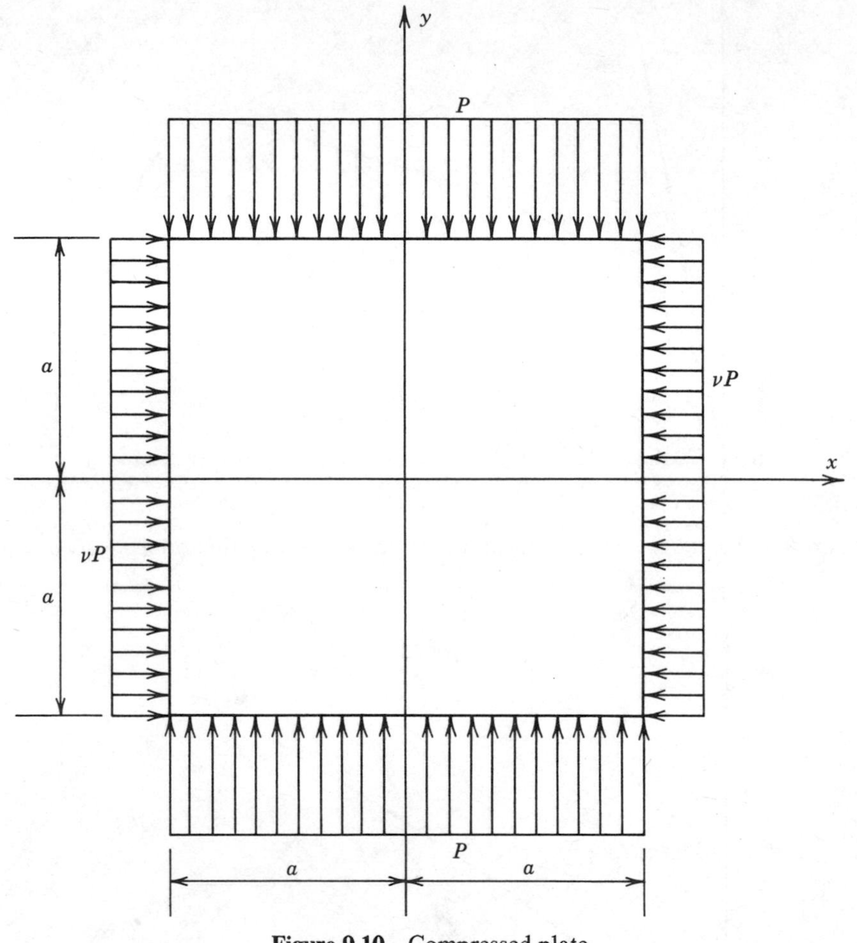


**Figure 9.10** Compressed plate.

uniformly distributed (membrane) edge load of intensity $N_{yy} = -P$ along the edges $y = \pm a$. The edges $x = \pm a$ are prevented from expanding or contracting in the $x$-direction, and therefore, they experience a distributed (membrane) edge load of intensity $N_{xx} = -\nu P$. If we use (7.15a) together with

$$\begin{bmatrix} N_{xx} & N_{xy} \\ N_{yx} & N_{yy} \end{bmatrix} = \begin{bmatrix} -\nu P & 0 \\ 0 & -P \end{bmatrix} \tag{9.61}$$

it can be readily established (for $\nu = 0.3$) that the plate will buckle when

$$P = P_{\text{cr}} = \frac{\pi^2 D}{(1+\nu)a^2} = 0.632 \frac{Eh^3}{(1-\nu^2)a^2} > 0 \tag{9.62}$$

When $P = P_{cr}$, the membrane strain $\varepsilon_{yy}$ assumes the value

$$(\varepsilon_{yy})_{cr} = -\frac{1-\nu^2}{E}\frac{P_{cr}}{h} = -0.632\frac{h^2}{a^2} \equiv e_{cr} \tag{9.63}$$

We now consider the case $e < e_{cr}$, that is, the plate has buckled but it remains elastic. For this case we assume the (approximate) displacement field

$$w = A\cos\frac{\pi x}{2a}\cos\frac{\pi y}{2a} \tag{9.64a}$$

$$u^0 = B\sin\frac{\pi x}{a}\cos\frac{\pi y}{2a} \tag{9.64b}$$

$$v^0 = B\sin\frac{\pi y}{a}\cos\frac{\pi x}{2a} + ey \tag{9.64c}$$

where $e < e_{cr}$ and where $A$ and $B$ are constants to be determined. We note that (9.64) satisfy the following boundary conditions:

$$w(x, \pm a) = M_{yy}(x, \pm a) = 0 \tag{9.65a}$$

$$w(\pm a, y) = M_{xx}(\pm a, y) = 0 \tag{9.65b}$$

$$u^0(\pm a, y) = 0 \tag{9.65c}$$

$$v^0(x, \pm a) = \pm ea \tag{9.65d}$$

It is assumed that the center of the plate remains stationary with respect to $x$ and $y$ and that the edges $y = \pm a$ remain straight and parallel to the $x$-axis and approach each other symmetrically during the process of deformation. The strain energy in the plate is

$$U = U_1 + U_2 = \int_{-a}^{a}\int_{-a}^{a}\left(W_p^{(1)} + W_p^{(2)}\right)dx\,dy \tag{9.66}$$

where [see (9.26)]

$$W_p^{(1)} = \tfrac{1}{2}D\left[(1-\nu)m_{\alpha\beta}m_{\alpha\beta} + \nu m^2\right] \tag{9.67a}$$

$$W_p^{(2)} = \tfrac{1}{2}C\left[(1-\nu)n_{\alpha\beta}n_{\alpha\beta} + \nu n^2\right] \tag{9.67b}$$

In view of (9.66), (9.67), and (9.21), we have

$$U_1 = \frac{\pi^4 A^2 D}{8a^2}$$

$$U_2 = \frac{Gh}{1-\nu}\left[4a^2e^2 + \frac{\pi^2}{4}A^2e(1+\nu) + \frac{\pi^2}{4}B^2(9-\nu) - \frac{\pi^2}{6}BA^2(5-3\nu) + \frac{16}{9}B^2(1+\nu) + \frac{5\pi^4}{256}\frac{A^4}{a^2}\right]$$

Hence, $U = U(A, B)$, and we can invoke the principle of minimum potential energy to obtain

$$\frac{\partial U}{\partial A} = 0 \tag{9.68a}$$

$$\frac{\partial U}{\partial B} = 0 \tag{9.68b}$$

Upon application of (9.68b), we obtain

$$-\frac{\pi^2}{12}\frac{A^2}{a}(5 - 3\nu) + \frac{\pi^2}{4}B(9 - \nu) + \frac{16}{9}(1 + \nu)B = 0$$

and for $\nu = 0.3$ this equation reduces to

$$B = 0.1418\frac{A^2}{a} \tag{9.69}$$

Upon application of (9.68a) and subsequent utilization of (9.69), we obtain

$$A(4.058h^2 + 6.42a^2e + 5.688A^2) = 0$$

For the buckled plate $A \neq 0$, and therefore, the expression in parentheses vanishes, resulting in

$$A = \sqrt{\frac{-6.42a^2e - 4.058h^2}{5.688}} \tag{9.70}$$

We obtain a real $A$ provided $-6.42a^2e \geq 4.058h^2$. For the limiting case $-6.42a^2e = 4.058h^2$, and

$$e = e_{cr} = -0.633\frac{h^2}{a^2}$$

which [see (9.63)] corresponds to

$$P_{cr} = \frac{e_{cr}E}{1 - \nu^2} = \frac{0.633h^3E}{(1 - \nu^2)a^2}$$

This result corresponds to the (lowest) elastic buckling load, as shown in (9.62) and (9.63). If we define the (dimensionless) strain parameter

$$\lambda = \frac{e}{e_{cr}} \tag{9.71}$$

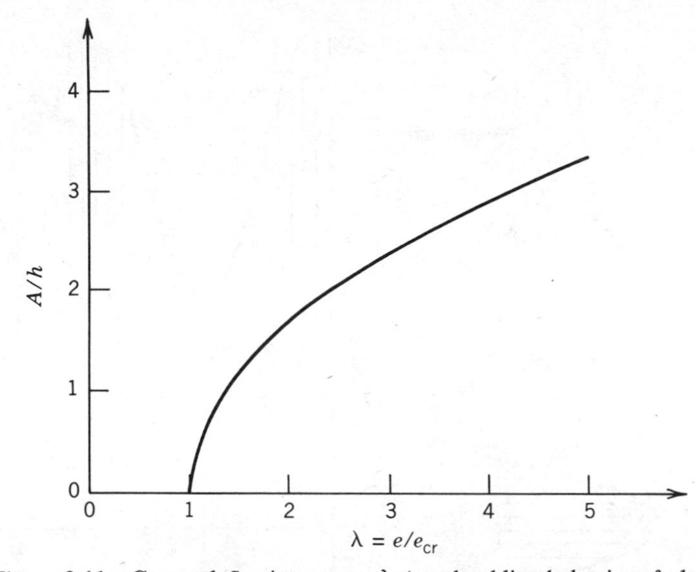


**Figure 9.11** Center deflection versus λ (postbuckling behavior of plate).

we have $e = \lambda e_{cr} = -0.632\lambda h^2/a^2$, and upon substitution in (9.70), we readily obtain

$$\frac{A}{h} = 0.845\sqrt{\lambda - 1}, \qquad \lambda \geq 1 \tag{9.72}$$

Note that $A/h = 0$ for $\lambda < 1$. A plot of $A/h$ versus $\lambda$ is shown in Fig. 9.11. We note the pronounced nonlinear relationship between plate center deflection $A$ and $\lambda$.

The membrane stress along the plate edge $y = a$, $z = 0$ is [see (9.20)]

$$\begin{aligned} \tau_{yy}(x, a) &= \frac{E}{1 - \nu^2}\left[L_{yy}(x, a) + \nu L_{xx}(x, a)\right] \\ &= 0.714\frac{E}{1 - \nu^2}\frac{h^2}{a^2}(\lambda - 1)\cos\frac{\pi x}{2a}\left(\frac{\pi^2}{8}\cos\frac{\pi x}{2a} - 0.142\pi\right) \\ &\quad -0.632\lambda\frac{h^2}{a^2}\frac{E}{1 - \nu^2} \end{aligned} \tag{9.73}$$

A plot of (9.73) and a similar plot for $\tau_{xx}(-a, y)$ is shown in Fig. 9.12 for the case $\lambda = e/e_{cr} = 10$. This graph reveals the nonuniform character of the distribution of compressive stresses that occurs in the postbuckling range. We also note that a major portion of the total load $F$ is now transmitted near the

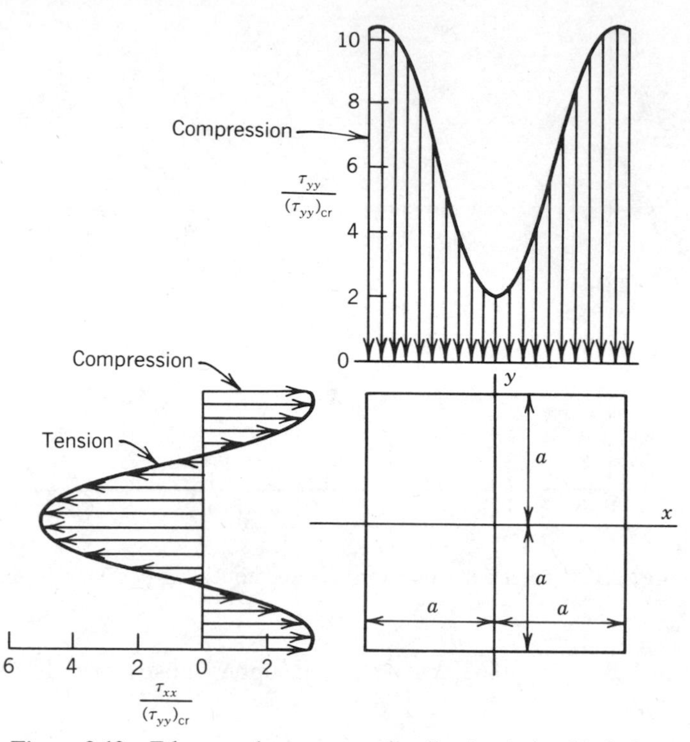


**Figure 9.12** Edge membrane stress distribution in buckled plate.

edges $x = \pm a$. The total compressive force after buckling is

$$F = -h\int_{-a}^{a} \tau_{yy}(x, a)\,dx$$
$$= 2a\left(0.623 + \frac{0.377}{\lambda}\right)(\lambda P_{cr}) = cP$$

where

$$c = 2a\left(0.623 + \frac{0.377}{\lambda}\right)$$

is the effective width of the plate and $P$ is now the *average* applied load per unit length. A (dimensionless) plot of $c/2a$ versus $\lambda$ is shown in Fig. 9.13. An inspection of Fig. 9.13 reveals that (a) the effective width decreases with an increase in $\lambda$ for $1 < \lambda$ and (b) there is a decrease in resistance to compression of the plate due to buckling. This resistance is equivalent to that of a plate having a width equal to $c$. The validity of the present approximate analysis is restricted to sufficiently small values of $A$, and improvements are possible by using more terms in the assumed solution (9.64).

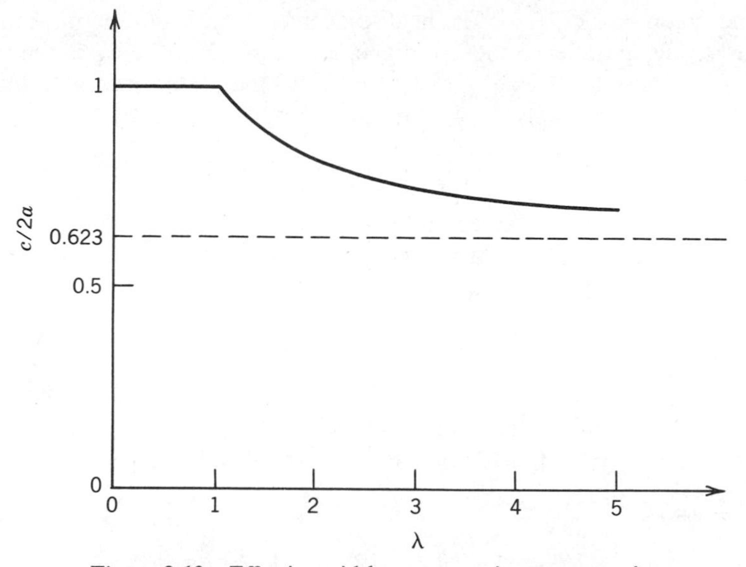


**Figure 9.13** Effective width versus strain parameter $\lambda$.

## 9.6 DYNAMICS OF SHEAR-DEFORMABLE PLATES

In the following development, we consider a plate that deforms by flexure and shear, as in Chapter 3, but is also free to stretch in its plane, such as the plate model in Section 9.2. The position of a material point in the plate is represented by

$$\mathbf{r} = \hat{\mathbf{e}}_x x + \hat{\mathbf{e}}_y y + \hat{\mathbf{e}}_z z$$

in the reference (or undeformed) configuration and

$$\mathbf{R} = \hat{\mathbf{e}}_x(x + u^0) + \hat{\mathbf{e}}_y(y + v^0) + \hat{\mathbf{e}}_z w + \hat{\boldsymbol{\eta}} z$$

in the deformed configuration. We note that the unit vector $\hat{\boldsymbol{\eta}} = \hat{\mathbf{e}}_x \psi_x + \hat{\mathbf{e}}_y \psi_y + \hat{\mathbf{e}}_z$ remains normal to the plate midsurface in the deformed configuration, and $\psi_x^2 + \psi_y^2 \ll 1$ ("small" rotations). Thus, the displacement vector of a generic point in the plate is

$$\mathbf{v} = \mathbf{R} - \mathbf{r} = \hat{\mathbf{e}}_x(u^0 + z\psi_x) + \hat{\mathbf{e}}_y(v^0 + z\psi_y) + \hat{\mathbf{e}}_z w$$

and consequently, the "displacement assumptions" are

$$v_1 = u^0(x, y, t) + z\psi_x(x, y, t) \tag{9.74a}$$

$$v_2 = v^0(x, y, t) + z\psi_y(x, y, t) \tag{9.74b}$$

$$v_3 = w(x, y, t) \tag{9.74c}$$

We note that $\mathbf{u} = \hat{\mathbf{e}}_x u^0 + \hat{\mathbf{e}}_y v^0$ is the displacement vector of a point in the plate median surface parallel to the $x$-$y$ plane, and $\boldsymbol{\psi} = \hat{\mathbf{e}}_x \psi_x + \hat{\mathbf{e}}_y \psi_y$ is the rotation (vector) of a line that was originally normal to the plate median surface. The displacement assumptions (9.74) can also be characterized by

$$\mathbf{v} = \mathbf{u}(x, y, t) + z\boldsymbol{\psi}(x, y, t) + \hat{\mathbf{e}}_z w(x, y, t) \tag{9.75}$$

We shall now derive the equations of motion of the plate as well as the associated, admissible boundary conditions with the aid of Hamilton's principle (see Sections 2.5 and 2.6 and Exercise 2.16):

$$\int_{t_1}^{t_2} (\delta T - \delta U + \delta \mathscr{W})\, dt = 0 \tag{9.76}$$

where $\delta T$, $\delta U$, and $\delta \mathscr{W}$ are the first variation of kinetic energy, first variation of potential energy, and virtual work of external forces, respectively. With the aid of (2.45), we have

$$\delta T = -\int_B \rho \ddot{v}_i\, \delta v_i\, dV = -\int_B \rho \ddot{\mathbf{v}} \cdot \delta \mathbf{v}\, dV \tag{9.77a}$$

and upon substitution of (9.75) into (9.77a), we obtain

$$\delta T = -\int_A \left[ \left( I_1 \ddot{\mathbf{u}} + I_2 \ddot{\boldsymbol{\psi}} \right) \cdot \delta \mathbf{u} + \left( I_2 \ddot{\mathbf{u}} + I_3 \ddot{\boldsymbol{\psi}} \right) \cdot \delta \boldsymbol{\psi} + I_1 \ddot{w}\, \delta w \right] dA \tag{9.77b}$$

where

$$I_1 = \int_{-h/2}^{h/2} \rho\, dz, \qquad I_2 = \int_{-h/2}^{h/2} \rho z\, dz, \qquad I_3 = \int_{-h/2}^{h/2} \rho z^2\, dz \tag{9.78}$$

are the inertia parameters of the plate. For the present, shear-deformable plate model, the plate strain energy density is [see (3.62) and (9.23)]

$$W_p = W_p(m_{\alpha\beta}, q_\alpha, n_{\alpha\beta})$$

and therefore,

$$\begin{aligned} \delta W_p &= \frac{\partial W_p}{\partial m_{\alpha\beta}} \delta m_{\alpha\beta} + \frac{\partial W_p}{\partial q_\alpha} \delta q_\alpha + \frac{\partial W_p}{\partial n_{\alpha\beta}} \delta n_{\alpha\beta} \\ &= M_{\alpha\beta}\, \delta m_{\alpha\beta} + Q_\alpha\, \delta q_\alpha + N_{\alpha\beta}\, \delta n_{\alpha\beta} \end{aligned} \tag{9.79}$$

According to (3.12) and (9.21b), the strain components associated with flexure, shear, and stretching are, respectively,

$$m_{\alpha\beta} = \tfrac{1}{2}(\psi_{\alpha,\beta} + \psi_{\beta,\alpha}) \tag{9.80a}$$

$$q_\alpha = \psi_\alpha + w_{,\alpha} \tag{9.80b}$$

$$n_{\alpha\beta} = \tfrac{1}{2}\left(u^0_{\alpha,\beta} + u^0_{\beta,\alpha}\right) + \tfrac{1}{2} w_{,\alpha} w_{,\beta} \tag{9.80c}$$

Upon substitution of (9.80) into (9.79), we obtain

$$\delta W_p = M_{\alpha\beta}\,\delta\psi_{\alpha,\beta} + N_{\alpha\beta}\,\delta u^0_{\alpha,\beta} + Q_\alpha\,\delta\psi_\alpha + S_\alpha\,\delta w,_\alpha \tag{9.81}$$

where

$$S_\alpha = Q_\alpha + N_{\alpha\beta} w,_\beta \tag{9.82}$$

With the aid of (9.81) and (9.80), we can now calculate the first variation of the (total) strain energy in the plate:

$$\delta U = \int_A \delta W_p\, dA = \oint_C (\mathbf{N}\cdot\delta u + \mathbf{M}\cdot\delta\psi + S_n\,\delta w)\, ds + \int_A [-\mathbf{div}\,\mathcal{N}\cdot\delta\mathbf{u} + (\mathbf{Q} - \mathbf{div}\,\mathcal{M})\cdot\delta\psi - \mathbf{div}\,\mathbf{S}\cdot\delta w]\, dA \tag{9.83}$$

where we have used the relations

$$\int_A M_{\alpha\beta}\,\delta\psi_{\alpha,\beta}\, dA = \oint_C \mathbf{M}\cdot\delta\psi\, ds - \int_A \mathbf{div}\,\mathcal{M}\cdot\delta\psi\, dA$$

$$\int_A N_{\alpha\beta}\,\delta u_{\alpha,\beta}\, dA = \oint_C \mathbf{N}\cdot\delta\mathbf{u}\, ds - \int_A \mathbf{div}\,\mathcal{N}\cdot\delta\mathbf{u}\, dA$$

$$\int_A S_\alpha\,\delta w,_\alpha\, dA = \oint_C S_n\,\delta w\, ds - \int_A \mathbf{div}\,\mathbf{S}\,\delta w\, dA$$

The virtual work of the applied forces is

$$\delta\mathcal{W} = \int_A p\,\delta w\, dA + \oint_{C_1} (\mathbf{M}^*\cdot\delta\psi + S_n^*\,\delta w + N^*\cdot\delta\mathbf{u})\, ds \tag{9.84}$$

We now substitute (9.77b), (9.83), and (9.84) into (9.76) and simplify. The result is

$$\int_{t_1}^{t_2}\int_A \Big[\big(\mathbf{div}\,\mathcal{N} - I_1\ddot{\mathbf{u}} - I_2\ddot{\psi}\big)\cdot\delta\mathbf{u} + \big(\mathbf{div}\,\mathcal{M} - \mathbf{Q} - I_2\ddot{\mathbf{u}} - I_3\ddot{\psi}\big)\cdot\delta\psi + (\mathbf{div}\,\mathbf{S} + p - I_1\ddot{w})\,\delta w\Big]\, dA\, dt$$
$$- \int_{t_1}^{t_2}\oint_C [(\mathbf{N} - \mathbf{N}^*)\cdot\delta\mathbf{u} + (\mathbf{M} - \mathbf{M}^*)\cdot\delta\psi + (S_n - S_n^*)\,\delta w]\, ds = 0$$

However, $t_2 - t_1$, $A$, and $C$ are arbitrary, and the virtual displacements $\delta\mathbf{u}$,

$\delta\psi$, and $w$ are arbitrary as well as independent. Consequently, in $A$ we require

$$\mathbf{div}\,\mathcal{N} = I_1\ddot{\mathbf{u}} + I_2\ddot{\psi} \tag{9.85a}$$

$$\mathbf{div}\,\mathcal{M} = \mathbf{Q} + I_2\ddot{\mathbf{u}} + I_3\ddot{\psi} \tag{9.85b}$$

$$\mathbf{div}\,\mathbf{S} + p = I_1\ddot{w} \tag{9.85c}$$

where

$$\mathbf{S} = \mathbf{Q} + \mathcal{N} \cdot \text{grad } w \tag{9.86}$$

On $C$ we need to specify

$$\text{Either } \mathbf{N} = \mathbf{N}^* \quad \text{or} \quad \mathbf{u} = \mathbf{u}^* \tag{9.87a}$$

and

$$\text{Either } \mathbf{M} = \mathbf{M}^* \quad \text{or} \quad \psi = \psi^* \tag{9.87b}$$

and

$$\text{Either } S_n \quad \text{or} \quad S_n^* \quad \text{or} \quad w = w^* \tag{9.87c}$$

Equations (9.85) are the equations of motion, and (9.87) are the associated, admissible boundary conditions. When referred to Cartesian coordinates, (9.85)–(9.87) become: In $A$ we require

$$\frac{\partial N_{xx}}{\partial x} + \frac{\partial N_{xy}}{\partial y} = I_1\ddot{u}_x + I_2\ddot{\psi}_x \tag{9.88a}$$

$$\frac{\partial N_{yx}}{\partial x} + \frac{\partial N_{yy}}{\partial x} = I_1\ddot{u}_y + I_2\ddot{\psi}_y \tag{9.88b}$$

$$\frac{\partial M_{xx}}{\partial x} + \frac{\partial M_{xy}}{\partial y} = Q_x + I_3\ddot{\psi}_x + I_2\ddot{u}_x \tag{9.89a}$$

$$\frac{\partial M_{yx}}{\partial x} + \frac{\partial M_{yy}}{\partial y} = Q_y + I_3\ddot{\psi}_y + I_2\ddot{u}_y \tag{9.89b}$$

$$\frac{\partial S_x}{\partial x} + \frac{\partial S_y}{\partial y} + p = I_1\ddot{w} \tag{9.90}$$

$$S_x = Q_x + N_{xx}\frac{\partial w}{\partial x} + N_{xy}\frac{\partial w}{\partial y} \tag{9.91a}$$

$$S_y = W_y + N_{yx}\frac{\partial w}{\partial x} + N_{yy}\frac{\partial w}{\partial y} \tag{9.91b}$$

On $C$ we need to specify

$$\text{Either } M_{nn} \quad \text{or} \quad \psi_n \tag{9.92a}$$

and

$$\text{Either } M_{ns} \quad \text{or} \quad \psi_s \tag{9.92b}$$

and

$$\text{Either } S_n \quad \text{or} \quad w \tag{9.93}$$

and

$$\text{Either } N_{nn} \quad \text{or} \quad u_n \tag{9.94a}$$

and

$$\text{Either } N_{ns} \quad \text{or} \quad u_s \tag{9.94b}$$

Applications of (9.88)–(9.94) can be found in Chapter 10.

## 9.7 DYNAMICS OF PLATES (COUPLING BETWEEN BENDING AND STRETCHING BUT RIGID WITH RESPECT TO SHEAR DEFORMATIONS)

As in Section 4.2, we now assume that material line elements that are originally normal to the plate median surface in the reference configuration remain normal after deformation. Consequently, we have $\boldsymbol{\psi} = -\operatorname{grad} w$, and the displacement assumption (9.75) now becomes

$$\mathbf{v} = \mathbf{u} - z \operatorname{grad} w + \hat{\mathbf{e}}_z w \tag{9.95}$$

and because independent variations $\delta\boldsymbol{\psi}$ are no longer allowed in $A$, the first variation of kinetic energy (9.77b) reduces to

$$\delta T = -\int_A [(I_1 \ddot{\mathbf{u}} - I_2 \operatorname{grad} \ddot{w}) \cdot \delta \mathbf{u} + I_1 \ddot{w}\, \delta w]\, dA \tag{9.96}$$

The virtual work of external forces is [see (9.32)]

$$\delta \mathscr{W} = \int_A p\, \delta w\, dA + \oint_{C_1} \left[ S_n^* \,\delta w + M_{nn}^* \,\delta\left(-\frac{\partial w}{\partial n}\right) + \mathbf{N}^* \cdot \delta \mathbf{u} \right] ds \tag{9.97}$$

where

$$\begin{aligned} S_n &= V_n + \mathbf{N} \cdot \operatorname{grad} w \\ &= Q_n + \frac{\partial M_{ns}}{\partial s} + N_{nn}\frac{\partial w}{\partial n} + N_{ns}\frac{\partial w}{\partial s} \end{aligned} \tag{9.98}$$

In the present case, the plate is rigid with respect to shearing deformations. Consequently, $q_\alpha \equiv 0$, and

$$W_p = W_p(m_{\alpha\beta}, n_{\alpha\beta})$$

so that

$$\begin{aligned}\delta W_p &= \frac{\partial W_p}{\partial m_{\alpha\beta}} \delta m_{\alpha\beta} + \frac{\partial W_p}{\partial n_{\alpha\beta}} \delta n_{\alpha\beta} \\ &= M_{\alpha\beta} \delta m_{\alpha\beta} + N_{\alpha\beta} \delta n_{\alpha\beta} \end{aligned} \tag{9.99}$$

We now set $\psi_{\alpha,\beta} = -w,_{\alpha\beta}$ and substitute (9.80) into (9.99). The result is

$$\delta W_p = -M_{\alpha\beta} \delta w,_{\alpha\beta} + N_{\alpha\beta} \delta u^0_{\alpha,\beta} + N_{\alpha\beta} w,_\alpha \delta w,_\beta \tag{9.100}$$

With the aid of (9.80) and (9.100), we can now calculate the first variation of the (total) strain energy of the plate. A straightforward calculation shows that

$$\int_A - M_{\alpha\beta} \delta w,_{\alpha\beta} \, dA = \int_A - \mathbf{div}\, \mathbf{Q} \, \delta w \, dA + \oint_C \left[ V_n \delta w + M_{nn} \delta\left( -\frac{\partial w}{\partial n} \right) \right] ds$$

$$\int_A N_{\alpha\beta} \delta u_{\alpha,\beta} \, dA = \oint_C \mathbf{N} \cdot \delta \mathbf{u} \, ds - \int_A \mathbf{div}\, \mathcal{N} \cdot \delta \mathbf{u} \, dA$$

$$\int_A N_{\alpha\beta} w,_\alpha \delta w,_\beta \, dA = \oint_C \mathbf{N} \cdot \operatorname{grad} w \, \delta w \, ds - \int_A \mathbf{div}(\mathcal{N} \cdot \operatorname{grad} w) \, \delta w \, dA$$

where we have used the following relations:

$$\begin{aligned} N_{\alpha\beta,\beta} \delta u_\alpha &= \mathbf{div}\, \mathcal{N} \cdot \delta \mathbf{u} \\ (N_{\alpha\beta} w,_\alpha),_\beta &= \mathbf{div}(\mathcal{N} \cdot \operatorname{grad} w) \\ M_{\alpha\beta,\beta\alpha} &= \mathbf{div}\, Q \\ \mathbf{Q} &= \mathbf{div}\, \mathcal{M} \end{aligned}$$

Consequently, we obtain

$$\begin{aligned} \delta U &= \int_A \delta W_p \, dA \\ &= \oint_C \left( S_n \delta w + M_{nn} \delta\left( -\frac{\partial w}{\partial n} \right) + \mathbf{N} \cdot \delta \mathbf{u} \right) ds \\ &\quad + \int_A (-\mathbf{div}\, \mathbf{S} \cdot \delta w - \mathbf{div}\, \mathcal{N} \cdot \delta \mathbf{u}) \, dA \end{aligned} \tag{9.101}$$

where

$$\mathbf{S} = \mathbf{Q} + \mathcal{N} \cdot \operatorname{grad} w \quad \text{in } A. \tag{9.102}$$

Upon substitution of (9.96), (9.97), and (9.101) into (9.76) and further simplification, we obtain

$$\int_{t_1}^{t_2} \int_A \left[ (\mathbf{div}\, \mathcal{N} - I_1 \ddot{\mathbf{u}} + I_2 \operatorname{grad} \ddot{w}) \cdot \delta\mathbf{u} + (\mathbf{div}\, \mathbf{S} + p - I_1 \ddot{w})\, \delta w \right] dA\, dt$$

$$+ \int_{t_1}^{t_2} \oint_C \left[ (S_n^* - S_n) w + (\mathbf{N}^* - \mathbf{N}) \cdot \delta\mathbf{u} \right.$$

$$\left. + (M_{nn}^* - M_{nn})\, \delta\left( -\frac{\partial w}{\partial n} \right) \right] ds\, dt = 0$$

However, $t_2 - t_1$, $A$, and $C$ are arbitrary, the virtual displacements $\delta\mathbf{u}$ and $\delta w$ are arbitrary and independent in $A$, and the virtual displacements $\delta w$, $\delta\mathbf{u}$, and $\delta(-\partial w/\partial n)$ are arbitrary and independent on $C_2$. Consequently, in $A$ we require

$$\mathbf{div}\, \mathcal{N} = I_1 \ddot{\mathbf{u}} - I_2 \operatorname{grad} \ddot{w} \tag{9.103a}$$

$$\mathbf{div}\, \mathbf{S} + p = I_1 \ddot{w} \tag{9.103b}$$

where

$$\mathbf{S} = \mathbf{Q} + \mathcal{N} \cdot \operatorname{grad} w \tag{9.104a}$$

$$\mathbf{Q} = \mathbf{div}\, \mathcal{M} \tag{9.104b}$$

On $C$ we need to specify

$$\text{Either } \mathbf{N} = \mathbf{N}^* \quad \text{or} \quad \mathbf{u} = \mathbf{u}^* \tag{9.105a}$$

and

$$\text{Either } M_{nn} = M_{nn}^* \quad \text{or} \quad \frac{\partial w}{\partial n} = \left( \frac{\partial w}{\partial n} \right)^* \tag{9.105b}$$

and

$$\text{Either } S_n = S_n^* \quad \text{or} \quad w = w^* \tag{9.105c}$$

Equations (9.103) are the equations of motion, and (9.105) are the associated, admissible boundary conditions. When referred to Cartesian coordinates,

(9.103)–(9.106) become

$$\frac{\partial N_{xx}}{\partial x} + \frac{\partial N_{xy}}{\partial y} = I_1 \ddot{u}_x - I_2 \frac{\partial \ddot{w}}{\partial x} \tag{9.106a}$$

$$\frac{\partial N_{yx}}{\partial x} + \frac{\partial N_{yy}}{\partial y} = I_1 \ddot{u}_y - I_2 \frac{\partial \ddot{w}}{\partial y} \tag{9.106b}$$

$$\frac{\partial S_x}{\partial x} + \frac{\partial S_y}{\partial y} + p = I_1 \ddot{w} \tag{9.106c}$$

$$Q_x = \frac{\partial M_{xx}}{\partial x} + \frac{\partial M_{xy}}{\partial y} \tag{9.107a}$$

$$Q_y = \frac{\partial M_{yx}}{\partial x} + \frac{\partial M_{yy}}{\partial y} \tag{9.107b}$$

$$S_x = Q_x + N_{xx} \frac{\partial w}{\partial x} + N_{xy} \frac{\partial w}{\partial y} \tag{9.108a}$$

$$S_y = Q_y + N_{yx} \frac{\partial w}{\partial x} + N_{yy} \frac{\partial w}{\partial y} \tag{9.108b}$$

On $C_1$ we require

$$\text{Either } S_n \quad \text{or} \quad w \tag{9.109a}$$

and

$$\text{Either } M_{nn} \quad \text{or} \quad \frac{\partial w}{\partial n} \tag{9.109b}$$

and

$$\text{Either } N_{nn} \quad \text{or} \quad u_n \tag{9.109c}$$

and

$$\text{Either } N_{ns} \quad \text{or} \quad u_s \tag{9.109d}$$

where $S_n$ is given by (9.98).

Applications of (9.106)–(9.109) can be found in Chapter 10.

## EXERCISES

**9.1** Derive a plate theory that accounts for coupled bending and stretching as well as for shear deformations. Use the method of Section 9.2.

**9.2** With reference to the von Kármán plate theory, show that

$$n_{\alpha\beta} = \frac{1}{Eh}\left[(1+\nu)N_{\alpha\beta} - \nu N\delta_{\alpha\beta}\right]$$

where $N = N_{\alpha\alpha}$. Also show that

$$\begin{aligned} 2n_{xy,xy} - n_{xx,yy} - n_{yy,xx} &= w,_{xx}w,_{yy} - w,_{xy}w,_{xy} \\ &= m_{xx}m_{yy} - m_{xy}^2 \\ &= \frac{1}{R_1 R_2} \end{aligned}$$

(see Exercise 3.26)

**9.3** Define a stress function $\phi(x, y)$ such that

$$N_{xx} = h\frac{\partial^2\phi}{\partial y^2}, \qquad N_{yy} = h\frac{\partial^2\phi}{\partial x^2}, \qquad N_{xy} = -h\frac{\partial^2\phi}{\partial x\,\partial y}$$

Show that in this case (9.34), in conjunction with (9.37) and Exercise 9.2, may be written in the form

$$\nabla^4 w = \frac{h}{D}L(w, \phi) + \frac{q}{D}, \qquad \nabla^4\phi = -\frac{E}{2}L(w, w)$$

where

$$L(w, \phi) = \phi,_{yy}w,_{xx} + \phi,_{xx}w,_{yy} - 2\phi,_{xy}w,_{xy}$$

**9.4** Transform the equations of Exercise 9.3 to plane polar coordinates $(r, \theta)$. Show that in this case

$$N_{rr} = h\left(\frac{1}{r}\frac{\partial\phi}{\partial r} + \frac{1}{r^2}\frac{\partial^2\phi}{\partial\theta^2}\right)$$

$$N_{\theta\theta} = h\frac{\partial^2\phi}{\partial r^2}$$

$$N_{r\theta} = -h\frac{\partial}{\partial r}\left(\frac{1}{r}\frac{\partial\phi}{\partial\theta}\right)$$

and

$$L(w, \phi) = \frac{\partial^2 w}{\partial r^2}\left(\frac{1}{r}\frac{\partial\phi}{\partial r} + \frac{1}{r^2}\frac{\partial^2\phi}{\partial\theta^2}\right) + \left(\frac{1}{r}\frac{\partial w}{\partial r} + \frac{1}{r^2}\frac{\partial^2 w}{\partial\theta^2}\right)\frac{\partial^2\phi}{\partial r^2}$$
$$-2\frac{\partial}{\partial r}\left(\frac{1}{r}\frac{\partial\phi}{\partial r}\right)\frac{\partial}{\partial r}\left(\frac{1}{r}\frac{\partial w}{\partial\theta}\right)$$

**9.5** Show that a form of (9.34) is given by

$$D\nabla^4 w = p + \nabla \cdot (\mathcal{N} \cdot \nabla w), \qquad \nabla \cdot \mathcal{N} = \mathbf{0}$$

where $\mathcal{N}$ is the membrane stress tensor $\mathcal{N} = \hat{\mathbf{e}}_\alpha \hat{\mathbf{e}}_\beta N_{\alpha\beta}$ and $\nabla$ is the gradient operator.

**9.6** Use Hooke's law to show that

$$N_{\alpha\beta} = C\left[(1-\nu) n_{\alpha\beta} + \nu n\, \delta_{\alpha\beta}\right] \qquad (9.25\text{b})$$

where $C = Eh/(1-\nu^2)$.

**9.7** Use the method of Section 9.4 (Fourier series) to solve the problems of Section 9.3. Show that the solution is identical to the one obtained in Section 9.3.

**9.8** Use the method of Section 9.4 (Fourier series) to obtain the solution for the deflection and membrane force of a partially loaded membrane strip: $p = p_0$ for $0 < x < \frac{1}{2}L$, $p = 0$ for $\frac{1}{2}L < x < L$. Assume that $w = 0$ along $x = 0, L$ and that $u(0) = u(L) = 0$.

**9.9** Use the method of Section 9.4 to solve for the deflection, bending moment, and membrane force in a partially loaded, simply supported plate strip: $p = p_0$ for $0 < x < \frac{1}{2}L$, $p = 0$ for $\frac{1}{2}L < x < L$. Assume $u(0) = u(L) = 0$.

**9.10** Show that the solution of the homogeneous equation (9.46) is $w(x) = C_1 + C_2 x + C_3 \cosh \lambda x + C_4 \sinh \lambda x$, where $\lambda = N/D > 0$. Find a solution of the problem treated in Section 9.4 in "closed form."

**9.11** Find a solution for the uniformly loaded plate strip with clamped edges: $w(-\frac{1}{2}L) = w(\frac{1}{2}L) = 0$, $(dw/dx)_{x=-L/2} = (dw/dx)_{x=L/2} = 0$, $u(-\frac{1}{2}L) = u(\frac{1}{2}L) = 0$. Provide a plot of center deflection, center bending moment, and membrane force versus load intensity.

**9.12** *Von Kármán Plate Theory*. Consider a material point in the plate. In the (unstrained) reference configuration, its position vector is $\mathbf{r} = \hat{\mathbf{e}}_x x + \hat{\mathbf{e}}_y y + \hat{\mathbf{e}}_z z$. When the plate is deformed, its position vector is characterized by

$$\mathbf{R} = \hat{\mathbf{e}}_x(x + u^0) + \hat{\mathbf{e}}_y(y + v^0) + \hat{\mathbf{e}}_z w(x, y) + \hat{\boldsymbol{\varepsilon}} z$$

where the unit vector $\hat{\boldsymbol{\varepsilon}}$ is normal to the plate median surface in the reference and deformed configuration.

**(a)** Show that the displacement vector of the material point is

$$\mathbf{u} = \mathbf{R} - \mathbf{r} = \hat{\mathbf{e}}_x u^0 + \hat{\mathbf{e}}_y v^0 + \hat{\mathbf{e}}_z (w - z) + \hat{\boldsymbol{\varepsilon}} z$$

**(b)** Points on the deformed median surface are characterized by $z = 0$, and their position vector is given by

$$\mathbf{R}_0 = \hat{\mathbf{e}}_x(x + u_0) + \hat{\mathbf{e}}_y(y + v^0) + \hat{\mathbf{e}}_z w(x, y)$$

Show that

$$\hat{\boldsymbol{\varepsilon}} = \frac{(\partial \mathbf{R}_0/\partial x) \times (\partial \mathbf{R}_0/\partial y)}{|(\partial \mathbf{R}_0/\partial x) \times (\partial \mathbf{R}_0/\partial y)|} \cong \hat{\mathbf{e}}_x\left(-\frac{\partial w}{\partial x}\right) + \hat{\mathbf{e}}_y\left(-\frac{\partial w}{\partial y}\right) + \hat{\mathbf{e}}_z$$

provided $|\text{grad } w| \ll 1$ (see Exercise 4.11).

**(c)** Using (a) and (b), show that the components of the displacement vector $\mathbf{u} = \mathbf{R} - \mathbf{r}$ are given by

$$u_x = u^0 - z\frac{\partial w}{\partial x}$$
$$u_y = v^0 - z\frac{\partial w}{\partial y}$$
$$u_z = w(x, y)$$

**9.13** *Von Kármán Plate Theory*. Show that (9.26) can be written in the form $W_p = \frac{1}{2}M_{\alpha\beta}m_{\alpha\beta} + \frac{1}{2}C[n^2 - 2(1 - \nu)\eta]$, where $n = n_{11} + n_{22}$ and $\eta = n_{11}n_{22} - n_{12}n_{21}$ are the first and second invariants, respectively, of the membrane strain tensor. If $2(1 - \nu)\eta \ll n^2$, show that $W_p = \frac{1}{2}M_{\alpha\beta}m_{\alpha\beta} + \frac{1}{2}Cn^2$ and use the principle of virtual work to show that in this case

$$\left.\begin{aligned} &D\nabla^4 w - Cn\nabla^2 w = p(x, y) \\ &\frac{N}{1 + \nu} = Cn = C\left(u^0_{\alpha,\alpha} + \tfrac{1}{2}w_{,\alpha}w_{,\alpha}\right) = \text{constant} \end{aligned}\right\} \text{ in } A$$

and we need to specify

$$\left.\begin{aligned} &\text{Either } S_n^* = Q_n = \frac{\partial M_{ns}}{\partial s} + Cn\frac{\partial w}{\partial n} \quad \text{or} \quad w \\ \text{and}\quad &\text{Either } N_{nn}^* = Cn \quad \text{or} \quad u_n^0 \\ \text{and}\quad &\text{Either } M_{nn}^* = M_{nn} \quad \text{or} \quad \frac{\partial w}{\partial n} \end{aligned}\right\} \text{ on } C$$

This useful approximation results in a considerable simplification of the von Kármán plate theory, and several solutions are provided in H. M. Berger, "A New Approach to the Analysis of Large Deflection of Plates," *Journal of Applied Mechanics*, Vol. 22, 1955, pp. 465–472.

**9.14**

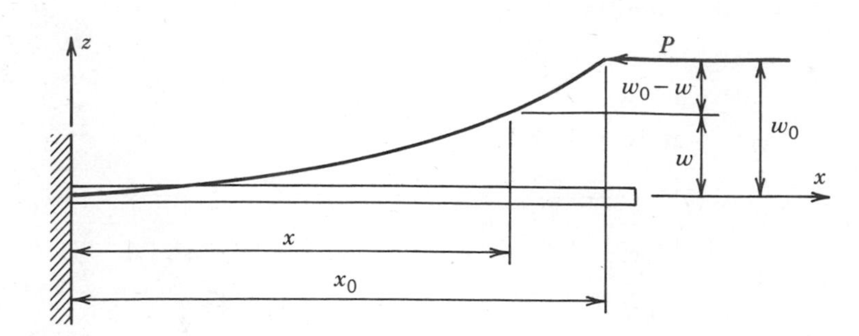


Use the method of Section 9.1 to study the stability and deformation of an elastic plate strip in plane strain subjected to a force per unit length $P$ parallel to the $x$-axis. The plate is clamped along $x = s = 0$ and free along $s = L$. Plot a dimensionless curve of $w_0$ versus $P$. Assume $EI$ = constant. How is this problem related to its linearized version? Provide a full discussion.

**9.15** Repeat the analysis in Section 9.5 but assume that the plate edges $x = \pm a$ remain straight and are allowed to move freely in the $x$-direction when the plate is compressed in the $y$-direction.

**9.16** What is the relationship between the topics discussed in Sections 9.7 and 9.2?

**9.17** Repeat the analysis of Section 9.7, except that now the boundary of the plate is a rectangle with corners at (0, 0), (a, 0), (a, b), and (0, b). Show that in addition to the boundary conditions 9.109, there will be four "corner conditions" to be specified at the corners of the plate (see Section 4.2).

# 10

# LAMINATED PLATES

**Robert C. Wetherhold**
Department of Mechanical and Aerospace Engineering
State University of New York,
Buffalo, New York 14260

A variety of structural plate forms are constructed as laminates in which a series of laminae or plies are stacked and joined to form a single structural unit. The individual laminae may be the traditional isotropic materials, or may be anisotropic materials such as polycrystalline polymers, grainy metals, or composite materials. Composite materials combine a reinforcement, often in the form of fibers, together with a surrounding matrix to produce a new material with a blend of properties superior to the properties of the constituents. There is a great deal of interest in laminates composed of composite materials for aerospace and transportation applications, where a judicious choice of matrix, reinforcement, and reinforcement geometry will achieve significant weight reductions at equivalent strength or stiffness. The emphasis in this chapter, then, will be on laminated plates composed of layers of composite materials. Although the notation to be introduced will be for the general case of anisotropic laminae, there will be direct reduction to the case of laminates of isotropic laminae as well as to the case of a single anisotropic lamina.

The composite reinforcement is normally in the form of continuous fibers, but short fibers, particulates, or spheres may be employed for reasons of cost, ease of fabrication, or toughness. The matrix is generally present to offer a stress transfer medium for the high-strength fibers. A large number of potential pairings of fiber and matrix are possible, with the most important fibers consisting of graphite, glass, metals, and ceramics, which are generally combined with matrices such as polymers, glasses, ceramics, metals, and intermetallic compounds. Given the wide variety of possible pairings of reinforcement and matrix, a composite may be tailored to the application, offering stiffness and strength according to the application requirements. A general discussion of the tailoring procedure is beyond the scope of the present treatment. We shall assume that a given composite material has been chosen, with its properties either directly measured or predicted from micromechanics models. The laminated plate analysis presented here assumes that the laminae are found as discrete units with a given orientation; this is the normal form found in aerospace laminates. However, composites such as injection-molded and sheet-molded polymer composites that are not formally layered may be approximated as a series of laminae whose thicknesses and reinforcements are influenced by molding conditions (Pipes et al., 1982; Fakirov and Fakirova, 1985). After fully describing the thermo-elastic properties and their transformations for a lamina, we shall combine the laminae into the structural laminate using both shear-deformable and classical plate theory models previously developed in Chapters 3 and 4.

## 10.1 ELEMENTS OF ANISOTROPIC ELASTICITY

Before dealing with the complexities of a laminated plate, we must understand the deformation response of a single anisotropic lamina under load. The similarities and differences of anisotropic and isotropic materials may be summarized as follows. Both the strain-displacement relations, in either linear (2.9) or nonlinear form (9.20), and the equations of motion (9.88)–(9.91) remain the same, since they are based on principles that are insensitive to the choice of material. The stress-strain relationship, however, will become more complex, and greater care must be taken. The general Hooke's law form of the stress-strain or constitutive relation contains 21 independent constants (see Section 2.3):

$$\tau_{ij} = C_{ijkl}\varepsilon_{kl} \tag{10.1}$$

In traditional notation, the unprimed coordinate system is reserved for "material coordinates," in which system the material exhibits its natural symmetries and is thus easiest to characterize. The use of the words *material coordinates* does not imply a convective coordinate system in the continuum mechanics sense; both unprimed (material) and primed coordinate systems are based on Lagrangian descriptions. Due to the tensorial character of the elements in (10.1), the constitutive relation in the primed system after a general rotation of orthogonal axes is expressed as

$$\tau_{p'q'} = C_{p'q'm'n'}\varepsilon_{m'n'} \tag{10.2}$$

where

$$C_{p'q'm'n'} = C_{ijkl}a_{p'i}a_{q'j}a_{m'k}a_{n'l}$$

Often the compliance relation is preferred to the stiffness relation, since the elements of the compliance tensor may be more easily related to traditional "engineering constants" such as Young's moduli, Poisson's ratios, and shear moduli:

$$\varepsilon_{m'n'} = S_{p'q'm'n'}\tau_{m'n'} \tag{10.3}$$

where

$$S_{p'q'm'n'} = S_{ijkl}a_{p'i}a_{q'j}q_{m'k}a_{n'l}$$

The transformation implied by (10.2) and (10.3) is frequently required since the primed or geometric axes that comprise the global axes of convenience for the plate may not be coincident with the axes of material symmetry.

The most general form of the stiffness tensor given by (10.2) has 21 independent elastic constants. Such a form is virtually never found in practical materials. Instead, there are material symmetries that serve to reduce the number of independent coefficients. We often use the notation of an "$n$-fold symmetry" about a specified axis. By this we mean that a positive rotation of $(2\pi/n)i$ about the axis, with $i$ integer from 0 to $n$, will give a physical form indistinguishable from the original form. The values $i = 0$ and $i = n$ give no additional information, since they represent the identity operator, but all other values of $i$ will supply information. The highest order of symmetry the structure possesses is used to describe the symmetry order. As an example, consider a twofold rotational symmetry about the $x_3$-axis as shown in Fig. 10.1. For this example, rotations of 0, $\pi$, and $2\pi$ about the $x_3$-axis return the

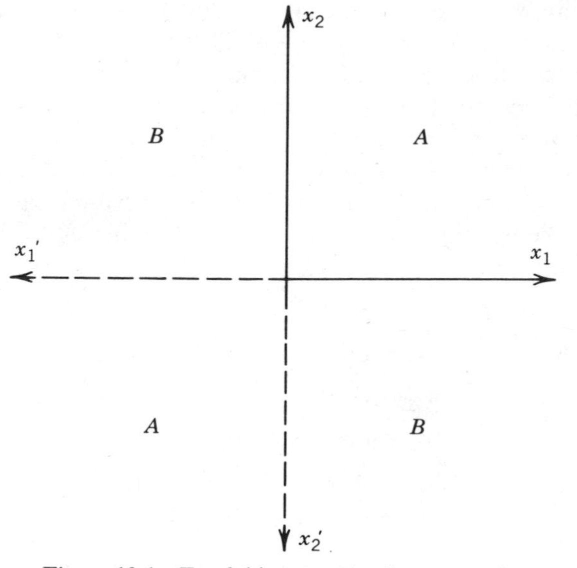


**Figure 10.1** Twofold symmetry about $x_3$-axis.

same physical form. The rotation matrix $a_{i'j}$ takes the following simple form for the rotation of angle $\pi$ about $x_3$: $a_{i'j} = \cos(\hat{\mathbf{e}}_{i'}, \hat{\mathbf{e}}_j)$:

$$[a_{i'j}] = \begin{bmatrix} -1 & 0 & 0 \\ 0 & -1 & 0 \\ 0 & 0 & 1 \end{bmatrix} \tag{10.4}$$

We may use the coordinate transformation equation (10.2) with (10.4) to find the stiffness properties in the primed coordinate system. By the physical symmetry observable in Fig. 10.1, we note that there is additional information available. Since the structure in the primed coordinate system is indistinguishable from the structure in the unprimed system, the properties must be invariant. In tensor notation, the elastic properties must be equal before and after the transformation:

$$C_{i'j'k'l'} = C_{ijkl} \quad \text{for } i = i', \quad j = j', \quad k = k', \quad l = l' \tag{10.5}$$

We may compare (10.5) and (10.2) in order to reduce the number of independent elastic constants present in $C_{ijkl}$. For example, consider $C_{1'1'2'3'} = a_{1'1}a_{1'1}a_{2'2}a_{3'3}C_{1123} = -C_{1123}$ from (10.2) and $C_{1'1'2'3'} = C_{1123}$ from (10.5); clearly, $C_{1123} = 0$. Many other symmetries are possible and are commonly found in composite laminae. A further discussion of this is postponed until Section 10.2, where a "contracted" notation is introduced.

## 10.2 CONTRACTED NOTATION

At this juncture, we choose to leave the traditional elasticity tensor notation, pursuing a contracted Voigt notation (Woldemar Voigt, 1850–1919). This is done in order to achieve a compact column matrix representation for stress and strain, and represents the standard notation for laminated plates. This contraction is not a true tensorial contraction of order but is rather a change of notation for convenience. This notation will be followed in all subsequent sections in this chapter, since it produces the most natural, least complicated form for the constitutive equations. We choose the following contracted notation:

$$\begin{aligned} \tau_{11} &= \sigma_1 & \varepsilon_{11} &= \varepsilon_1 \\ \tau_{22} &= \sigma_2 & \varepsilon_{22} &= \varepsilon_2 \\ \tau_{33} &= \sigma_3 & \varepsilon_{33} &= \varepsilon_3 \\ \tau_{23} &= \sigma_4 & 2\varepsilon_{23} &= \varepsilon_4 = \gamma_{23} \\ \tau_{13} &= \sigma_5 & 2\varepsilon_{13} &= \varepsilon_5 = \gamma_{13} \\ \tau_{12} &= \sigma_6 & 2\varepsilon_{12} &= \varepsilon_6 = \gamma_{12} \end{aligned} \tag{10.6}$$

We may form stiffness and compliance matrices from their associated tensors for use with (10.6) in a constitutive law. The relationship between the elements of the stiffness tensor and its associated matrix contraction is highlighted in Exercise 10.1:

$$\sigma_i = C_{ij}\varepsilon_j, \qquad i, j = 1, \ldots, 6 \tag{10.7}$$

or

$$\{\sigma\} = [C]\{\varepsilon\}$$

$$\varepsilon_i = S_{ij}\sigma_j, \qquad i, j = 1, \ldots, 6 \tag{10.8}$$

or

$$\{\varepsilon\} = [S]\{\sigma\}$$

The summation convention employed in (10.7) and (10.8) runs over the values 1 to 6 in the manner of a standard matrix product. Both the $C_{ij}$ and $S_{ij}$ matrices are symmetric (see Exercise 10.1), confirming that there are at most 21 independent elastic constants.

## 10.3 TRANSFORMATION OF CONSTITUTIVE LAWS

The transformation of stress, strain, and stiffness and compliance coefficients in the contracted system may be accomplished by a series of matrix multiplications. The elements of the rotational transformation matrix $a_{i'j}$ may be assigned as follows:

$$\begin{bmatrix} a_{1'1} & a_{1'2} & a_{1'3} \\ a_{2'1} & a_{2'2} & a_{2'3} \\ a_{3'1} & a_{3'2} & a_{3'3} \end{bmatrix} = \begin{bmatrix} m_1 & n_1 & p_1 \\ m_2 & n_2 & p_2 \\ m_3 & n_3 & p_3 \end{bmatrix} \tag{10.9}$$

where $a_{i'j} = \cos(\hat{\mathbf{e}}_{i'}, \hat{\mathbf{e}}_j)$. We define a transformation matrix $[T]$ for the stress components according to (10.10) and compare the results with the previously defined stress tensor transformation (2.25) to obtain the desired transformation (see Exercise 10.3):

$$\{\sigma'\} = [T]\{\sigma\} \tag{10.10}$$

where

$$[T] = \begin{bmatrix} m_1^2 & n_1^2 & p_1^2 & 2n_1p_1 & 2m_1p_1 & 2m_1n_1 \\ m_2^2 & n_2^2 & p_2^2 & 2n_2p_2 & 2m_2p_2 & 2m_2n_2 \\ m_3^2 & n_3^2 & p_3^2 & 2n_3p_3 & 2m_3p_3 & 2m_3n_3 \\ m_2m_3 & n_2n_3 & p_2p_3 & n_2p_3 + p_2n_3 & m_2p_3 + p_2m_3 & m_2n_3 + n_2m_3 \\ m_1m_3 & n_1n_3 & p_1p_3 & n_1p_3 + p_1n_3 & m_1p_3 + p_1m_3 & m_1n_3 + n_1m_3 \\ m_1m_2 & n_1n_2 & p_1p_2 & n_1p_2 + p_1n_2 & m_1p_2 + p_1m_2 & m_1n_2 + n_1m_2 \end{bmatrix} \tag{10.11}$$

In using the contracted notation, there is convenience of notation and manipulation, but the tensorial character of stress and strain is lost. There is a different transformation matrix $T^*$ for strain, which is found in a similar fashion to $T$ (see Exercise 10.3):

$$\{\varepsilon'\} = [T^*]\{\varepsilon\} \tag{10.12}$$

where

$$[T^*] = \begin{bmatrix} m_1^2 & n_1^2 & p_1^2 & n_1 p_1 & m_1 p_1 & m_1 n_1 \\ m_2^2 & n_2^2 & p_2^2 & n_2 p_2 & m_2 p_2 & m_2 n_2 \\ m_3^2 & n_3^2 & p_3^2 & n_3 p_3 & m_3 p_3 & m_3 n_3 \\ 2m_2 m_3 & 2n_2 n_3 & 2p_2 p_3 & n_2 p_3 + p_2 n_3 & m_2 p_3 + p_2 m_3 & m_2 n_3 + n_2 m_3 \\ 2m_1 m_3 & 2n_1 n_3 & 2p_1 p_3 & n_1 p_3 + p_1 n_3 & m_1 p_3 + p_1 m_3 & m_1 n_3 + n_1 m_3 \\ 2m_1 m_2 & 2n_1 n_2 & 2p_1 p_2 & n_1 p_2 + p_1 n_2 & m_1 p_2 + p_1 m_2 & m_1 n_2 + n_1 m_2 \end{bmatrix} \tag{10.13}$$

It may be shown that $T$ and $T^*$ possess a special relationship, namely,

$$[T]^{-1} = [T^*]^{\mathrm{T}}, \qquad [T^*]^{-1} = [T]^{\mathrm{T}} \tag{10.14}$$

A method to demonstrate (10.14) is to assume that it is true, calculate the matrix product $[T][T^*]^{\mathrm{T}}$, and use the orthogonality conditions $a_{p'i}a_{q'i} = \delta_{p'q'}$ with (10.9) to show that the product is the identity matrix. The proper constitutive law may thus be constructed for other coordinate systems. The elements of the original stiffness relation as defined by Hooke's law in the matrix notation has been stated in (10.7). Inverting the transformations given by (10.10) and (10.12), manipulating matrices, and using (10.14), the transformed stiffness matrix $[C']$ relating stress to strain in the transformed coordinate system may be found:

$$\{\sigma'\} = [C']\{\varepsilon'\} \tag{10.15a}$$

where

$$[C'] = [T][C][T]^{\mathrm{T}} \tag{10.15b}$$

A similar use of (10.8), (10.10), (10.12), and (10.14) will give the transformed compliance equation

$$\{\varepsilon'\} = [S']\{\sigma'\} \tag{10.16a}$$

**TABLE 10.1 Simplified Transformations for In-Plane Rotation about $x_1$-Axis**

$$[a_{i'j}] = \begin{bmatrix} 1 & 0 & 0 \\ 0 & \cos\theta & \sin\theta \\ 0 & -\sin\theta & \cos\theta \end{bmatrix}, \qquad m = \cos\theta, \quad n = \sin\theta \tag{10.17}$$

*A. Simplified Stress Transformation Matrix*

$$\{\sigma'\} = [T_1]\{\sigma\}$$

$$[T_1(\theta)] = \begin{bmatrix} 1 & 0 & 0 & 0 & 0 & 0 \\ 0 & m^2 & n^2 & 2mn & 0 & 0 \\ 0 & n^2 & m^2 & -2mn & 0 & 0 \\ 0 & -mn & mn & m^2 - n^2 & 0 & 0 \\ 0 & 0 & 0 & 0 & m & -n \\ 0 & 0 & 0 & 0 & n & m \end{bmatrix} \tag{10.18}$$

*B. Simplified Engineering Strain Transformation Matrix*

$$\{\varepsilon'\} = [T_1^*]\{\varepsilon\}$$

$$[T_1^*(\theta)] = \begin{bmatrix} 1 & 0 & 0 & 0 & 0 & 0 \\ 0 & m^2 & n^2 & mn & 0 & 0 \\ 0 & n^2 & m^2 & -mn & 0 & 0 \\ 0 & -2mn & 2mn & m^2 - n^2 & 0 & 0 \\ 0 & 0 & 0 & 0 & m & -n \\ 0 & 0 & 0 & 0 & n & m \end{bmatrix} \tag{10.19}$$

where

$$[S'] = [T^*][S][T^*]^{\mathrm{T}} \tag{10.16b}$$

Under normal circumstances, we are interested in only certain prescribed rotations, such as about the $x_1$-axis or about the $x_3$-axis. Transformations about the $x_1$-axis will be used to demonstrate reductions in the number of independent stiffness coefficients. Transformations about the $x_3$-axis will be referred to as in-plane rotations for a plate whose $x_3$-axis is in the plate thickness direction. The transformation matrices for these special cases are given in (10.18) and (10.19) and in (10.21) and (10.22) (see Tables 10.1 and 10.2). For a rotation about the $x_1$-axis, the elements of $a_{i'j}$ are related to the $m_i$, $n_i$, $p_i$ of (10.9) by (10.17). The similar representation for the rotation about the $x_3$-axis is given by (10.20). A pictorial representation for the rotation angles is given in Figs. 10.2 and 10.3.

**TABLE 10.2 Simplified Transformations for In-Plane Rotation about $x_3$-Axis**

$$[a_{i'j}] = \begin{bmatrix} \cos\phi & \sin\phi & 0 \\ -\sin\phi & \cos\phi & 0 \\ 0 & 0 & 1 \end{bmatrix}, \qquad m = \cos\phi, \quad n = \sin\phi \tag{10.20}$$

*A. Simplified Stress Transformation Matrix*

$$\{\sigma'\} = [T_3]\{\sigma\}$$

$$[T_3(\phi)] = \begin{bmatrix} m^2 & n^2 & 0 & 0 & 0 & 2mn \\ n^2 & m^2 & 0 & 0 & 0 & -2mn \\ 0 & 0 & 1 & 0 & 0 & 0 \\ 0 & 0 & 0 & m & -n & 0 \\ 0 & 0 & 0 & n & m & 0 \\ -mn & mn & 0 & 0 & 0 & m^2 - n^2 \end{bmatrix} \tag{10.21}$$

*B. Simplified Engineering Strain Transformation Matrix*

$$\{\varepsilon'\} = [T_3^*]\{\varepsilon\}$$

$$[T_3^*(\phi)] = \begin{bmatrix} m^2 & n^2 & 0 & 0 & 0 & mn \\ n^2 & m^2 & 0 & 0 & 0 & -mn \\ 0 & 0 & 1 & 0 & 0 & 0 \\ 0 & 0 & 0 & m & -n & 0 \\ 0 & 0 & 0 & n & m & 0 \\ -2mn & 2mn & 0 & 0 & 0 & m^2 - n^2 \end{bmatrix} \tag{10.22}$$

It should be noted that there is considerable variation in practice of the sign of the angle $\phi$ for in the in-plane rotations (10.20)–(10.22) as well as variation in the assignment of transformation definitions such as (10.10) and (10.12). It can be seen that the transformation equations define transformations for stress, strain, and the constitutive behavior. If the constitutive behavior is known in the $(x_1, x_2, x_3)$-axes (the material coordinate axes) and we wish to transform the constitutive law into the global $(x_1', x_2', x_3')$-axes of convenience for the plate, then the transformations defined here are the most straightforward. This is the system we shall pursue in the remainder of this chapter. On the other hand, if the applied stress or strain vector were known in the global plate axes, then an alternative transformation definition might be preferred, such as $\{\sigma\} = [T]\{\sigma'\}$, in contradiction to (10.10). This will lead to alternative definitions for the transformed constitutive laws, whose definitions will be similar to (10.15) and (10.16).

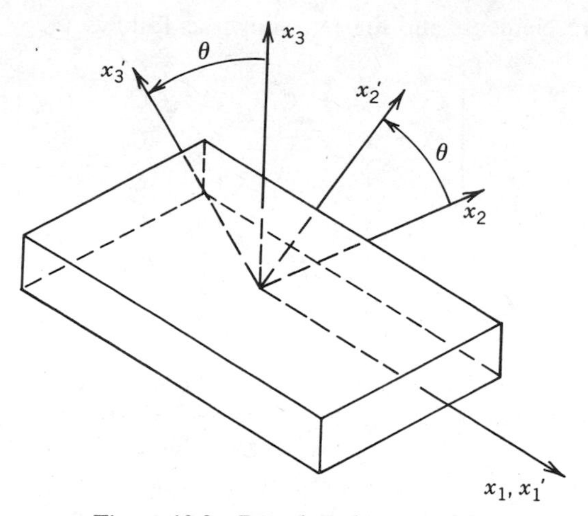


**Figure 10.2** Rotation about $x_1$-axis.

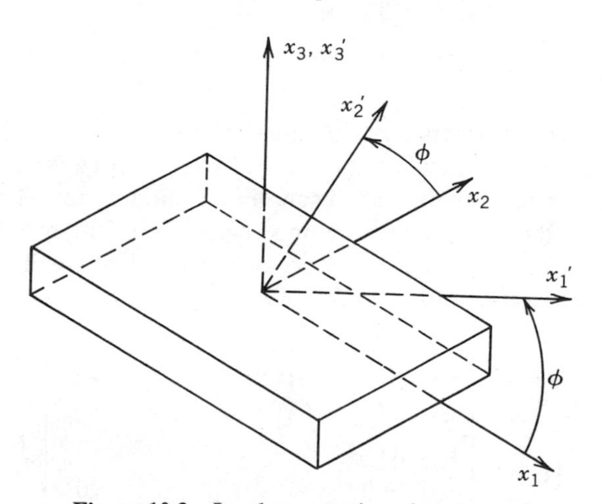


**Figure 10.3** In-plane rotation about $x_3$-axis.

## 10.4 MATERIAL SYMMETRIES

In order to decrease the number of independent elastic constants below 21, we must pursue the possible material symmetries and their effect on the number of independent elastic constants. This is important since the number of constants will determine the number of experiments that must be performed to fully characterize the material. Initially, the $C_{ij}$ matrix, although symmetric, contains no zero elements and may be shown as follows:

$$[C_{ij}] = \begin{bmatrix} C_{11} & C_{12} & C_{13} & C_{14} & C_{15} & C_{16} \\ C_{12} & C_{22} & C_{23} & C_{24} & C_{25} & C_{26} \\ C_{13} & C_{23} & C_{33} & C_{34} & C_{35} & C_{36} \\ C_{14} & C_{24} & C_{34} & C_{44} & C_{45} & C_{46} \\ C_{15} & C_{25} & C_{35} & C_{45} & C_{55} & C_{56} \\ C_{16} & C_{26} & C_{36} & C_{46} & C_{56} & C_{66} \end{bmatrix} \tag{10.23}$$

Consider the case of a twofold symmetry about the $x_3$-axis (see Fig. 10.1). This symmetry is generally the lowest order of symmetry that we shall experience and is referred to as a *monoclinic* symmetry. The transformation law (10.15b) relates the stiffness properties in the original and transformed axes; in addition, due to material symmetry, these properties must be invariant under the proper rotation:

$$[C'] = [T_3(\pi)][C][T_3(\pi)]^{\mathrm{T}} \tag{10.24a}$$

$$[C'] = [C] \tag{10.24b}$$

We perform the transformation calculation given by (10.24a) and compare the results with the invariance equation (10.24b). A series of equations relating the values of stiffness coefficients will result. For example, any elements of $[C]$ that are equal to their negative value must equal zero. The resultant form for $[C]$ is given in (10.24c), with the number of independent coefficients equaling $13 = 21 - 8$ (see Exercise 10.5):

$$[C] = \begin{bmatrix} C_{11} & C_{12} & C_{13} & 0 & 0 & C_{16} \\ C_{12} & C_{22} & C_{23} & 0 & 0 & C_{26} \\ C_{13} & C_{23} & C_{33} & 0 & 0 & C_{36} \\ 0 & 0 & 0 & C_{44} & C_{45} & 0 \\ 0 & 0 & 0 & C_{45} & C_{55} & 0 \\ C_{16} & C_{26} & C_{36} & 0 & 0 & C_{66} \end{bmatrix} \tag{10.24c}$$

If we enforce an additional twofold rotational symmetry about the $x_1$-axis (see Fig. 10.4), the resulting symmetry is referred to as *orthotropic* and possesses $9 = 13 - 4$ independent coefficients. Normally, any composite lamina of technological interest will have axes about which an orthotropic symmetry may be found, and this will be the assumed form for material symmetry for the remainder of this chapter. (For a discussion of laminae possessing monoclinic symmetry, see Exercise 10.8.) Pursuing the same logic for invariance and transformation of properties as in (10.24), the following form for an ortho-

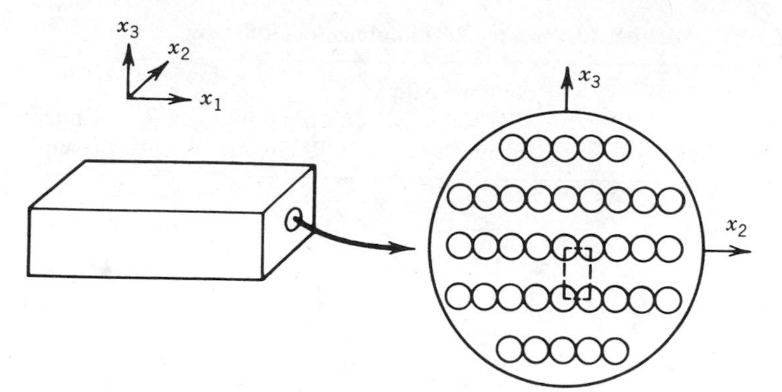


**Figure 10.4** Twofold symmetry about $x_1$-axis.

tropic stiffness matrix $C_{ij}$ results:

$$[C_{ij}] = \begin{bmatrix} C_{11} & C_{12} & C_{13} & 0 & 0 & 0 \\ C_{12} & C_{22} & C_{23} & 0 & 0 & 0 \\ C_{13} & C_{23} & C_{33} & 0 & 0 & 0 \\ 0 & 0 & 0 & C_{44} & 0 & 0 \\ 0 & 0 & 0 & 0 & C_{55} & 0 \\ 0 & 0 & 0 & 0 & 0 & C_{66} \end{bmatrix} \tag{10.25a}$$

Since all of these transformations may be carried out using compliance transformations, a similar form may be found for the orthotropic compliance matrix $[S_{ij}]$:

$$[S_{ij}] = \begin{bmatrix} S_{11} & S_{12} & S_{13} & 0 & 0 & 0 \\ S_{12} & S_{22} & S_{23} & 0 & 0 & 0 \\ S_{13} & S_{23} & S_{33} & 0 & 0 & 0 \\ 0 & 0 & 0 & S_{44} & 0 & 0 \\ 0 & 0 & 0 & 0 & S_{55} & 0 \\ 0 & 0 & 0 & 0 & 0 & S_{66} \end{bmatrix} \tag{10.25b}$$

In addition to the normally found orthotropic form given by (10.25), the results of several additional symmetries are given in Table 10.3 (see also Figs. 10.5–10.7). Since the fiber reinforcement direction is traditionally aligned with the $x_1$-axis, the simplifying relations are all derived from combining the transformation equation $[C'] = [T_1(\theta)][C][T_1(\theta)]^{\mathrm{T}}$ with the invariance equation $[C'] = [C]$ for values of the rotation angles $\theta$ appropriate to the physical symmetry.

The preceding simplifications refer directly to the orthotropic stiffness matrix (10.25). Two additional symmetries that are separate from Table 10.3

**TABLE 10.3 Additional Symmetry Relationships for Stiffness**

| Symmetry about $x_1$-Axis | Rotation Angle $\theta$ for Symmetry | Simplifying Relationships | Independent Elastic Constants |
|---|---|---|---|
| Square, Fig. 10.5 | $\left(\frac{2\pi}{4}\right)i$ | $C_{22} = C_{33}$, $C_{12} = C_{13}$, $C_{55} = C_{66}$, | 6 |
| Hexagonal, Fig. 10.6 | $\left(\frac{2\pi}{6}\right)i$ | $C_{22} = C_{33}$, $C_{12} = C_{13}$, $C_{55} = C_{66}$, $C_{44} = \frac{1}{2}(C_{22} - C_{23})$ | 5 |
| Random (transverse isotropy in $x_2$-$x_3$ plane), Fig. 10.7 | Any | $C_{22} = C_{33}$, $C_{12} = C_{13}$, $C_{55} = C_{66}$, $C_{44} = \frac{1}{2}(C_{22} - C_{23})$ | 5 |

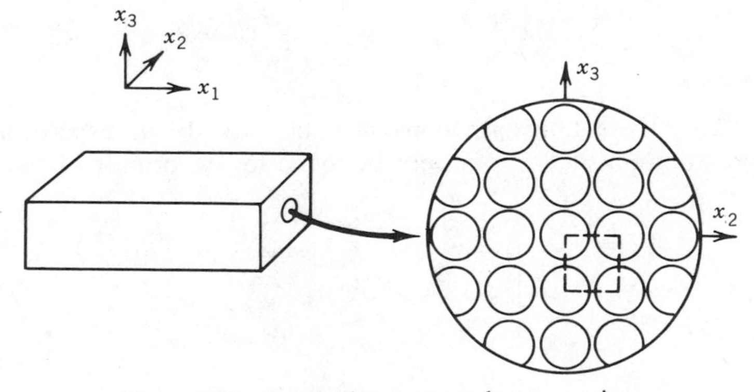


**Figure 10.5** Fourfold symmetry about $x_1$-axis.

may also be applied to the orthotropic form (10.25). For transverse isotropy in the $x_1$-$x_2$ plane (insensitivity to rotation about the $x_3$-axis), $C_{11} = C_{22}$, $C_{23} = C_{13}$, $C_{44} = C_{55}$, and $C_{66} = \frac{1}{2}(C_{11} - C_{12})$. For isotropic materials, $C_{11} = C_{22} = C_{33}$, $C_{12} = C_{13} = C_{23}$, and $C_{44} = C_{55} = C_{66} = \frac{1}{2}(C_{11} - C_{12})$. It should be noted that a series of relationships different from Table 10.3 will be found for the compliance matrix (10.25b) based on symmetry considerations. This is due to the difference in transformation equations between (10.15b) and (10.16b). Similar logic may be applied, however, to get a suitable series of relations as in Table 10.3 for compliances.

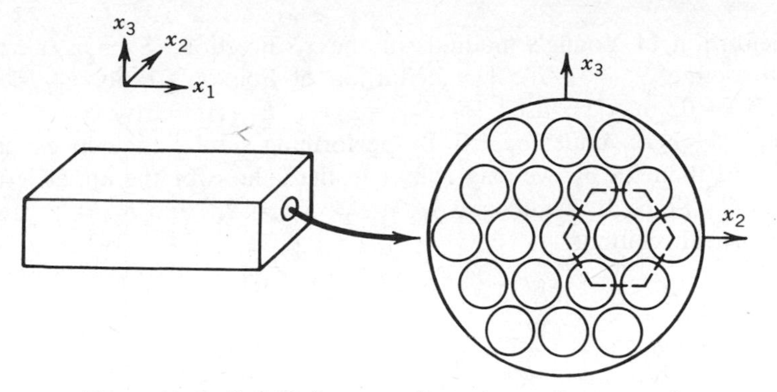


**Figure 10.6** Sixfold (hexagonal) symmetry about $x_1$-axis.

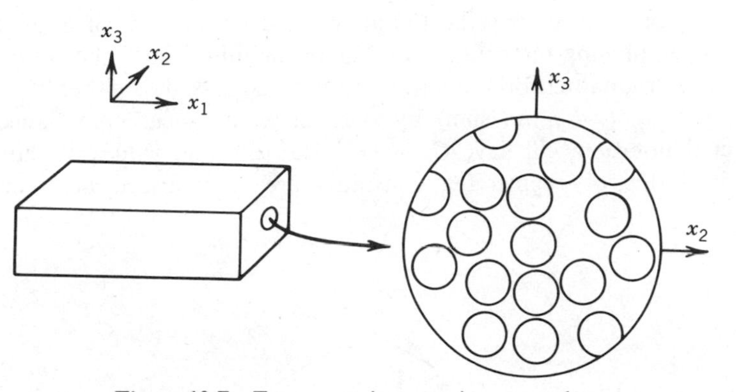


**Figure 10.7** Transverse isotropy in $x_2$-$x_3$ plane.

## 10.5 ENGINEERING CONSTANTS

The compliance equation (10.8) is especially suitable for considering how the elements of the compliance matrix may be related to engineering constants. The general definitions may be introduced for a Young modulus ($E$ is the ratio of normal stress to normal strain), shear modulus ($G$ is the ratio of shear stress to engineering shear strain), and Poisson's ratio ($\nu_{ij} = -\varepsilon_j/\varepsilon_i$ under $\sigma_i$). Consider an orthotropic material subject to a single nonzero stress component $\sigma_1$:

$$\begin{Bmatrix} \varepsilon_1 \\ \varepsilon_2 \\ \varepsilon_3 \\ \varepsilon_4 \\ \varepsilon_5 \\ \varepsilon_6 \end{Bmatrix} = \begin{Bmatrix} S_{11}\sigma_1 \\ S_{21}\sigma_1 \\ S_{31}\sigma_1 \\ 0 \\ 0 \\ 0 \end{Bmatrix}$$

The definition of Young's modulus in the $x_1$-direction, $E_1 = \sigma_1/\varepsilon_1$, may be used to define $S_{11} = 1/E_1$. The definition of Poisson's ratio, $\nu_{12} = -\varepsilon_2/\varepsilon_1$ under $\sigma_1 \neq 0$, may be used for $S_{21} = -\nu_{12}/E_1$; similarly, $S_{31} = -\nu_{13}/E_1$ with $\nu_{13} = -\varepsilon_3/\varepsilon_1$ under $\sigma_1 \neq 0$. By performing similar thought experiments for $\sigma_2$ and then for $\sigma_3$, we may complete the values for the upper left $3 \times 3$ block of $S_{ij}$. Since, by symmetry, $S_{12} = S_{21}$, $S_{13} = S_{31}$, and $S_{23} = S_{32}$, we have the reciprocal relations

$$\frac{\nu_{ij}}{E_i} = \frac{\nu_{ji}}{E_j}, \qquad i, j = 1, 2, 3 \tag{10.26}$$

The shear compliances may be found through application of a single shear stress and employing the definition of shear modulus. Since the shear compliance in the original tensorial notation may be expressed as either $2\varepsilon_{ij} = G_{ij}\tau_{ij}$ or $2\varepsilon_{ji} = G_{ji}\tau_{ji}$ ($i \neq j$, no sum) by symmetry, the shear compliance in the contracted notation will have $G_{ij} = G_{ji}$. The resulting values of the compliances in terms of engineering constants are summarized next, and their

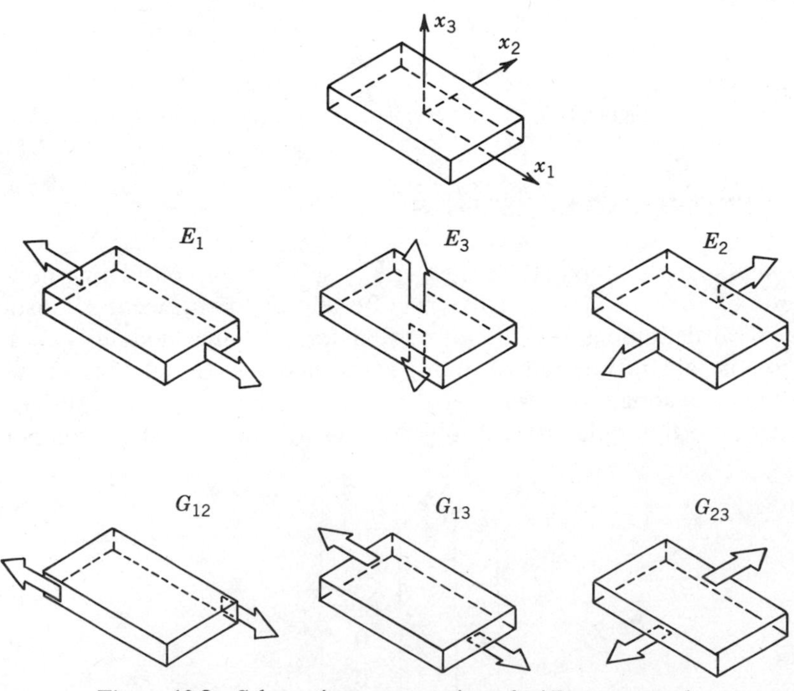


**Figure 10.8** Schematic representation of stiffness properties.

schematic stiffnesses are given in Fig. 10.8:

$$S_{ij} = \begin{cases} \dfrac{1}{E_i}, & i = j, \quad i, j = 1, 2, 3 \\ -\dfrac{\nu_{ij}}{E_i} & i \neq j, \quad i, j = 1, 2, 3 \end{cases} \tag{10.27}$$

$$S_{44} = \frac{1}{G_{23}}, \qquad S_{55} = \frac{1}{G_{13}}, \qquad S_{66} = \frac{1}{G_{12}}$$

The elements of the stiffness matrix $C_{ij}$ are less simply expressed in terms of engineering constants. They may be defined by the inverse of the compliance matrix,

$$[C] = [S]^{-1} \tag{10.28}$$

and are summarized as

$$\begin{aligned}
C_{11} &= E_1\left[1 - \frac{E_3}{E_2}\nu_{23}^2\right]D \\
C_{12} &= C_{21} = [E_2\nu_{12} + E_3\nu_{13}\nu_{23}]D \\
C_{13} &= C_{31} = E_3[\nu_{12}\nu_{23} + \nu_{13}]D \\
C_{22} &= E_2\left[1 - \frac{E_3}{E_1}\nu_{13}^2\right]D \\
C_{23} &= C_{32} = \frac{E_3}{E_1}[E_1\nu_{23} + E_2\nu_{12}\nu_{13}]D \\
C_{33} &= E_3\left[1 - \frac{E_2}{E_1}\nu_{12}^2\right]D \\
C_{44} &= G_{23} \\
C_{55} &= G_{13} \\
C_{66} &= G_{12}
\end{aligned} \tag{10.29}$$

with

$$D^{-1} = 1 - 2\frac{E_3}{E_1}\nu_{12}\nu_{23}\nu_{13} - \nu_{13}^2\frac{E_3}{E_1} - \nu_{23}^2\frac{E_3}{E_2} - \nu_{12}^2\frac{E_2}{E_1}$$

The requirements on the values of orthotropic compliances $S_{ij}$ are not the traditional requirements of isotropic materials. It may be possible, for example, to have a Poisson's ratio that exceeds $\frac{1}{2}$, which is thermodynamically forbidden for isotropic materials. The conditions that are satisfied are based on the positive definiteness of both the stiffness and compliance matrices, since the strain energy density must remain positive. The resulting requirements, expressed both in terms of $S_{ij}$ and $C_{ij}$ as well as in terms of engineering constants may be found elsewhere (Jones, 1975, Section 2.4).

## 10.6 STIFFNESS TRANSFORMATIONS AND REDUCED STIFFNESS COEFFICIENTS IN LAMINATION

A laminated plate will normally consist of laminae that possess potentially different in-plane rotation angles (rotations about the $x_3$-axis), and have been joined into a laminated plate whose thickness direction is along the $x_3$-axis. It is thus necessary to consider the lamina constitutive behavior derived in the previous section, to combine it with the usual plate assumptions and notation, and then to consider coordinate transformations. The usual assumption on the physical action of a plate is that it resists lateral and in-plane loads by bending, transverse shear stresses and in-plane action and not through block-like tension or compression in the plate thickness direction (Vinson and Chou, 1975). In agreement with traditional thin plate theory, the integrated resultant of the transverse normal stress $\sigma_3$ is set to zero. In addition, the normal stress $\sigma_3$ itself is assumed to be zero everywhere throughout the thickness, except in very thin boundary layers near the surface where it rises to match the surface pressures. This assumption is consistent with the generalized plane stress condition often assumed in plate analysis. This represents an improvement over classical theories that ignore both the transverse normal stress and transverse normal strain (e.g., Timoshenko and Woinowsky-Krieger, 1959), in which the Poisson ratio effects are generally ignored (although there are a few exceptions). This improvement is of importance in the case of laminated composites, in which the stiff fiber reinforcement is generally in-plane and the Poisson ratio effects from $\nu_{13}$, $\nu_{23}$ may be of greater importance.

This assumption of zero transverse normal stress has consequences for the orthotropic stiffness relationship (10.25a). Consider the case of zero transverse normal stress in each lamina within the plate,

$$\sigma_3 = 0 \tag{10.30}$$

and examine the effect on the constitutive equation (10.8). The normal strain $\varepsilon_z$ may be expressed in terms of in-plane normal stresses:

$$\varepsilon_3 = S_{13}\sigma_1 + S_{23}\sigma_2 \tag{10.31}$$

The normal strain may be dropped from further consideration since it is no longer independent and need not be carried in the compliance relationship.

The resultant constitutive relationship reduces to a compact 5 × 5 system:

$$\begin{Bmatrix} \varepsilon_1 \\ \varepsilon_2 \\ \varepsilon_4 \\ \varepsilon_5 \\ \varepsilon_6 \end{Bmatrix} = \begin{bmatrix} S_{11} & S_{12} & 0 & 0 & 0 \\ S_{12} & S_{22} & 0 & 0 & 0 \\ 0 & 0 & S_{44} & 0 & 0 \\ 0 & 0 & 0 & S_{55} & 0 \\ 0 & 0 & 0 & 0 & S_{66} \end{bmatrix} \begin{Bmatrix} \sigma_1 \\ \sigma_2 \\ \sigma_4 \\ \sigma_5 \\ \sigma_6 \end{Bmatrix} \tag{10.32}$$

The stiffness relationship may be obtained by inverting (10.32):

$$\begin{Bmatrix} \sigma_1 \\ \sigma_2 \\ \sigma_4 \\ \sigma_5 \\ \sigma_6 \end{Bmatrix} = \begin{bmatrix} Q_{11} & Q_{12} & 0 & 0 & 0 \\ Q_{12} & Q_{22} & 0 & 0 & 0 \\ 0 & 0 & Q_{44} & 0 & 0 \\ 0 & 0 & 0 & Q_{55} & 0 \\ 0 & 0 & 0 & 0 & Q_{66} \end{bmatrix} \begin{Bmatrix} \varepsilon_1 \\ \varepsilon_2 \\ \varepsilon_4 \\ \varepsilon_5 \\ \varepsilon_6 \end{Bmatrix} \tag{10.33}$$

The $Q_{ij}$ are referred to as elements of a "reduced" stiffness matrix.

The elements of the compliance matrix in terms of engineering constants are already known through (10.27). The elements of the reduced stiffness matrix $[Q]$ may be obtained by inverting (10.32). The results are

$$\begin{aligned}
Q_{11} &= \frac{S_{22}}{S_{11}S_{22} - S_{12}^2} = \frac{E_1}{1 - \nu_{12}\nu_{21}} \\
Q_{12} &= -\frac{S_{12}}{S_{11}S_{22} - S_{12}^2} = \frac{\nu_{12}E_2}{1 - \nu_{12}\nu_{21}} \\
Q_{22} &= \frac{S_{11}}{S_{11}S_{22} - S_{12}^2} = \frac{E_2}{1 - \nu_{12}\nu_{21}} \\
Q_{44} &= C_{44} = \frac{1}{S_{44}} = G_{23} \\
Q_{55} &= C_{55} = \frac{1}{S_{55}} = G_{13} \\
Q_{66} &= C_{66} = \frac{1}{S_{66}} = G_{12}
\end{aligned} \tag{10.34}$$

Notice that several out-of-plane elastic constants ($E_3$, $\nu_{13}$, $\nu_{23}$) have been eliminated from the constitutive relation. An alternative systematic derivation of (10.34) may be achieved as follows.

**Example:** Reduced Stiffness Coefficients: Show that the relationship

$$Q_{ij} = C_{ij} - \frac{C_{i3}C_{j3}}{C_{33}}, \qquad i, j = 1, 2, 4, 5, 6 \tag{a}$$

holds for an orthotropic material with zero transverse normal stress ($\sigma_3 = 0$). Using $\sigma_3 = 0$ and the stiffness relation (10.25a) with (10.7), we find that

$$\varepsilon_3 = \frac{-C_{13}\varepsilon_1 - C_{23}\varepsilon_2}{C_{33}} \tag{b}$$

Consider again the stiffness equation from (10.25a) with (10.7), locate the equation for $\sigma_1$, and employ (b):

$$\sigma_1 = \left( C_{11} - \frac{C_{13}^2}{C_{33}} \right) \varepsilon_1 + \left( C_{12} - \frac{C_{13}C_{23}}{C_{33}} \right) \varepsilon_2$$

but, from (10.33),

$$\sigma_1 \equiv Q_{11}\varepsilon_1 + Q_{12}\varepsilon_2 \tag{c}$$

This gives the values of $Q_{1j}$, $j = 1, 2, 4, 5, 6$. Similar consideration for the other stress components completes the exercise. It can be seen from (a) that the term *reduced stiffnesses* for $Q_{ij}$ confirms the fact that $Q_{ij} \le C_{ij}$.

The axes of material symmetry ($x_1$, $x_2$, $x_3$) are not necessarily aligned with the global or geometric plate axes of convenience $(x, y, z) = (x_1', x_2', x_3')$, so it becomes necessary to perform a rotation of axes about the $x_3$- or $z$-axis to obtain the required elastic properties. Rotation by an arbitrary angle about the $x_3$-axis may be accomplished using (10.21) and (10.22) with either (10.16b) for compliance or (10.15b) for stiffness. The resultant transformed compliance and stiffness matrices for use in the $(x, y, z)$-coordinate system become

$$\begin{Bmatrix} \varepsilon_x \\ \varepsilon_y \\ \varepsilon_z \\ \gamma_{yz} \\ \gamma_{xz} \\ \gamma_{xy} \end{Bmatrix} = [S'] \begin{Bmatrix} \sigma_x \\ \sigma_y \\ \sigma_z \\ \sigma_{yz} \\ \sigma_{xz} \\ \sigma_{xy} \end{Bmatrix} \tag{10.35a}$$

where

$$S'_{ij} = \begin{bmatrix} S'_{11} & S'_{12} & S'_{13} & 0 & 0 & S'_{16} \\ S'_{12} & S'_{22} & S'_{23} & 0 & 0 & S'_{26} \\ S'_{13} & S'_{23} & S'_{33} & 0 & 0 & S'_{36} \\ 0 & 0 & 0 & S'_{44} & S'_{45} & 0 \\ 0 & 0 & 0 & S'_{45} & S'_{55} & 0 \\ S'_{16} & S'_{26} & S'_{36} & 0 & 0 & S'_{66} \end{bmatrix} \tag{10.35b}$$

and

$$\begin{Bmatrix} \sigma_x \\ \sigma_y \\ \sigma_z \\ \sigma_{yz} \\ \sigma_{xz} \\ \sigma_{xy} \end{Bmatrix} = [C'] \begin{Bmatrix} \varepsilon_x \\ \varepsilon_y \\ \varepsilon_z \\ \gamma_{yz} \\ \gamma_{xz} \\ \gamma_{xy} \end{Bmatrix} \tag{10.36a}$$

where

$$C'_{ij} = \begin{bmatrix} C'_{11} & C'_{12} & C'_{13} & 0 & 0 & C'_{16} \\ C'_{12} & C'_{22} & C'_{23} & 0 & 0 & C'_{26} \\ C'_{13} & C'_{23} & C'_{33} & 0 & 0 & C'_{36} \\ 0 & 0 & 0 & C'_{44} & C'_{45} & 0 \\ 0 & 0 & 0 & C'_{45} & C'_{55} & 0 \\ C'_{16} & C'_{26} & C'_{36} & 0 & 0 & C'_{66} \end{bmatrix} \tag{10.36b}$$

Note that the orthotropic lamina has the potential for in-plane coupling of normal and shear effects in the geometric system and thus returns to the form of the monoclinic system given by (10.24c). Should we desire to implement the zero transverse normal stress assumption (10.30), a relationship similar to (a) of the reduced stiffness coefficients example will be obtained; if not, there is a direct equality of the elements of $C'_{ij}$ and $Q'_{ij}$:

$$\begin{Bmatrix} \sigma_x \\ \sigma_y \\ \sigma_{yz} \\ \sigma_{xz} \\ \sigma_{xy} \end{Bmatrix} = \begin{bmatrix} Q'_{11} & Q'_{12} & 0 & 0 & Q'_{16} \\ Q'_{12} & Q'_{22} & 0 & 0 & Q'_{26} \\ 0 & 0 & Q'_{44} & Q'_{45} & 0 \\ 0 & 0 & Q'_{45} & Q'_{55} & 0 \\ Q'_{16} & Q'_{26} & 0 & 0 & Q'_{66} \end{bmatrix} \begin{Bmatrix} \varepsilon_x \\ \varepsilon_y \\ \gamma_{yz} \\ \gamma_{xz} \\ \gamma_{xy} \end{Bmatrix} \tag{10.37}$$

with either

$$Q'_{ij} = C'_{ij} - \frac{C'_{i3}C'_{j3}}{C'_{33}} \qquad (\sigma_3 = 0), \qquad i, j = 1,2,4,5,6 \tag{10.38a}$$

or

$$Q'_{ij} = C'_{ij} \qquad (\sigma_3 \neq 0), \qquad i, j = 1,2,4,5,6 \tag{10.38b}$$

Equation (10.38a) may be verified with the techniques of the reduced stiffness coefficients example, with the provision that the computations for the coupling elements such as $Q'_{16}$ are no longer trivial. All subsequent plate stiffness terminology in this chapter will employ the notation $Q'_{ij}$ for stiffness properties, with the understanding that either (10.38a) or (10.38b) will be used. In accordance with the arguments of the beginning of this section, (10.38a) is the preferred form.

## 10.7 EXPANSION BEHAVIOR

The inclusion of expansion behavior is important since composites are routionely fabricated at a temperature different from their use temperature and often experience thermal excursions during a duty cycle, and since many polymer matrix composites can absorb water and other solvents. The inclusion of a variety of expansional strains in Duhamel-Neumann form may be considered in the constitutive equation. In contracted notation, (10.7) and (10.8) may be rewritten as

$$\varepsilon_i = S_{ij}\sigma_j + \varepsilon_i^E, \qquad i, j = 1,\ldots,6 \tag{10.39}$$

and

$$\sigma_i = C_{ij}\left(\varepsilon_j - \varepsilon_j^E\right), \qquad i, j = 1,\ldots,6 \tag{10.40}$$

where the expansional strain $\varepsilon_i^E = \varepsilon_i^T + \varepsilon_i^h + \cdots$ is composed of thermal strain, hygroscopic (swelling) strain, and other similar strain components. The compliance equation (10.39) is shown in more detail in what follows for completeness. There are no expansional shear strains in the unprimed material coordinates since symmetry conditions require that free expansion produce only normal strains in those coordinates, $\varepsilon_4^E = \varepsilon_5^E = \varepsilon_6^E = 0$:

$$\begin{Bmatrix} \varepsilon_1 \\ \varepsilon_2 \\ \varepsilon_3 \\ \varepsilon_4 \\ \varepsilon_5 \\ \varepsilon_6 \end{Bmatrix} = \begin{bmatrix} S_{11} & S_{12} & S_{13} & 0 & 0 & 0 \\ S_{12} & S_{22} & S_{23} & 0 & 0 & 0 \\ S_{13} & S_{23} & S_{33} & 0 & 0 & 0 \\ 0 & 0 & 0 & S_{44} & 0 & 0 \\ 0 & 0 & 0 & 0 & S_{55} & 0 \\ 0 & 0 & 0 & 0 & 0 & S_{66} \end{bmatrix} \begin{Bmatrix} \sigma_1 \\ \sigma_2 \\ \sigma_3 \\ \sigma_4 \\ \sigma_5 \\ \sigma_6 \end{Bmatrix} + \begin{Bmatrix} \varepsilon_1^E \\ \varepsilon_2^E \\ \varepsilon_3^E \\ 0 \\ 0 \\ 0 \end{Bmatrix} \tag{10.41}$$

The particular expansional strain chosen in this treatment will be thermal, with an assumed simple proportional behavior. Any other expansional strain may be considered using a similar analysis:

$$\varepsilon_i^E = \alpha_i T, \qquad i = 1, 2, 3 \tag{10.42}$$

where $T$ is the temperature above a reference state.

Consider an arbitrary rotation about the $x_3$- or $z$-axis, typical of a lamina in-plane rotation. The constitutive relationships may be expressed in a form similar to (10.35) and (10.36):

$$\begin{Bmatrix} \varepsilon_x \\ \varepsilon_y \\ \varepsilon_z \\ \gamma_{yz} \\ \gamma_{xz} \\ \gamma_{xy} \end{Bmatrix} = [S'] \begin{Bmatrix} \sigma_x \\ \sigma_y \\ \sigma_z \\ \sigma_{yz} \\ \sigma_{xz} \\ \sigma_{xy} \end{Bmatrix} + \begin{Bmatrix} \alpha_x T \\ \alpha_y T \\ \alpha_z T \\ 0 \\ 0 \\ \alpha_{xy} T \end{Bmatrix} \tag{10.43}$$

where $S'$ is given by (10.35b) and

$$\left[\alpha_x, \alpha_y, \alpha_z, 0, 0, \alpha_{xy}\right]^{\mathrm{T}} = [T_3^*][\alpha_1, \alpha_2, \alpha_3, 0, 0, 0]^{\mathrm{T}} \tag{10.44}$$

Similarly,

$$\begin{Bmatrix} \sigma_x \\ \sigma_y \\ \sigma_z \\ \sigma_{yz} \\ \sigma_{xz} \\ \sigma_{xy} \end{Bmatrix} = [C'] \begin{Bmatrix} \varepsilon_x - \alpha_x T \\ \varepsilon_y - \alpha_y T \\ \varepsilon_z - \alpha_z T \\ \gamma_{yz} \\ \gamma_{xz} \\ \gamma_{xy} - \alpha_{xy} T \end{Bmatrix} \tag{10.45}$$

where $[C']$ is given by (10.36b). Note that under this in-plane rotation, there are no transverse thermal expansion shear strains ($\alpha_{yz} = \alpha_{xz} = 0$). See Exercise 10.7.

In agreement with the reasoning of Section 10.6, we may discuss the effect of ignoring the transverse normal stress ($\sigma_3 = 0$) under nonisothermal conditions. The reduction of thermo-elastic coefficients may be accomplished as before. We may write a reduced stiffness constitutive relationship

$$\begin{Bmatrix} \sigma_x \\ \sigma_y \\ \sigma_{yz} \\ \sigma_{xz} \\ \sigma_{xy} \end{Bmatrix} = [Q'] \begin{Bmatrix} \varepsilon_x - \alpha_x T \\ \varepsilon_y - \alpha_y T \\ \gamma_{yz} \\ \gamma_{xz} \\ \gamma_{xy} - \alpha_{xy} T \end{Bmatrix} \tag{10.46}$$

where $Q'$ is given by either (10.38a) or (10.38b), with (10.38a) as the preferred form.

## 10.8 LAMINATE CONSTITUTIVE BEHAVIOR

In preparation for posing the differential equations of motion in terms of displacements for a laminated plate, we must obtain the integrated force and moment resultants. In order to accomplish this, we will propose a form for the displacement functions. The displacement functions are assumed to be linear in the thickness direction, so that plane sections remain plane [see (9.73)]. The deformations $(u_1, u_2, u_3) = (u, v, w)$ along the $(x, y, z)$-axes, respectively, are given by

$$\begin{aligned} u(x, y, z, t) &= u^0(x, y, t) + z\psi_x(x, y, t) \\ v(x, y, z, t) &= v^0(x, y, t) + z\psi_y(x, y, t) \\ w(x, y, z, t) &= w(x, y, t) \end{aligned} \tag{10.47}$$

The deformations $u^0$, $v^0$ represent in-plane deflections at the geometric midsurface. The plate thickness direction is along the $z$-axis, and thus the value of $w$ is referred to as the transverse deformation. The functions $\psi_x$, $\psi_y$ represent the small angle rotations of a scribe line that was originally perpendicular to the geometric midsurface. Equations (10.47) are the assumptions of a theory of nondeformable normals and are general enough either to permit the inclusion of transverse shear deformation $\gamma_{xz}$, $\gamma_{yz}$ or to simplify to a shear-indeformable classical plate theory (CPT) under the assumptions $(\psi_x, \psi_y) = (-w,_x, -w,_y)$. The strain-displacement relations take the form

$$\begin{aligned} \varepsilon_x &= \varepsilon_x^0 + z\kappa_x \\ \varepsilon_y &= \varepsilon_y^0 + z\kappa_y \\ \varepsilon_z &= 0 \\ \gamma_{yz} &= \psi_y + w,_y \\ \gamma_{xz} &= \psi_x + w,_x \\ \gamma_{xy} &= \gamma_{xy}^0 + z\kappa_{xy} \end{aligned} \tag{10.48a}$$

where

$$\begin{Bmatrix} \varepsilon_x^0 \\ \varepsilon_y^0 \\ \gamma_{xy}^0 \end{Bmatrix} = \begin{Bmatrix} u^0,_x \\ v^0,_y \\ u^0,_y + v^0,_x \end{Bmatrix} \tag{10.48b}$$

or

$$\begin{Bmatrix} \varepsilon_x^0 \\ \varepsilon_y^0 \\ \gamma_{xy}^0 \end{Bmatrix} = \begin{Bmatrix} u^0,_x + \frac{1}{2}(w,_x)^2 \\ v^0,_y + \frac{1}{2}(w,_y)^2 \\ u^0,_y + v^0,_x + w,_x w,_y \end{Bmatrix} \quad \text{(buckling or moderate displacement)} \tag{10.48c}$$

and

$$\begin{Bmatrix} \kappa_x \\ \kappa_y \\ \kappa_{xy} \end{Bmatrix} = \begin{Bmatrix} \psi_{x,x} \\ \psi_{y,y} \\ \psi_{x,y} + \psi_{y,x} \end{Bmatrix} \tag{10.48d}$$

The quantities $\varepsilon_x^0$, $\varepsilon_y^0$, $\gamma_{xy}^0$ represent midsurface strains, while the $\kappa_x$, $\kappa_y$, $\kappa_{xy}$ are simple curvatures. The curvatures $\kappa_x$, $\kappa_y$ are bending curvatures, while the $\kappa_{xy}$ curvature represents a twisting of the plate. Note that the nonlinear von Kármán formulation for the midsurface strains in the strain-displacement equations (10.48c) also given in (2.5) contains quadratic terms that should be included only under the conditions of buckling or moderately large displacement. There can be no derivation of buckling equations without the inclusion of these terms, which represent a coupling of in-plane and transverse displacements. [This coupling found by the use of (10.48a) with the nonlinear formulation (10.48c) may also be seen in (9.20).] Laminated plates possess an important difference from isotropic plates in that this coupling of displacements may still exist even under conditions without buckling or moderate displacement. That is, coupling can exist even when the strain-displacement equations are linearized into the form (10.48b), also given by (2.9), due to the laminate constitutive behavior. Unless we are considering buckling or moderate displacement, we shall assume that the linearized strain-displacement assumption (10.48b) is employed.

Define the force and moment resultants in the standard way (3.24):

$$\begin{aligned} (N_x, N_y, N_{xy}) &= \int_{-h/2}^{h/2} (\sigma_x, \sigma_y, \sigma_{xy})\, dz \\ (M_x, M_y, M_{xy}) &= \int_{-h/2}^{h/2} (\sigma_x, \sigma_y, \sigma_{xy})\, z\, dz \\ (Q_x, Q_y) &= \int_{-h/2}^{h/2} (\sigma_{xz}, \sigma_{yz})\, dz \end{aligned} \tag{10.49}$$

where $h$ is the plate thickness. Note that the resultants are given in the consistent contracted notation, which may be related to the tensor notation of Chapter 3. The sign convention for positive moment resultants $(M)$ and shear resultants $(Q)$ may be seen in Fig. 3.5. The direction of a positive force resultant $(N)$ follows similarly as the direction defined by the standard elasticity positive stress direction, and may be seen in Figs. 7.1 and 7.2.

$$\begin{aligned} N_x &= N_{xx}, \qquad N_y = N_{yy}, \qquad N_{xy} = N_{xy} \\ M_x &= M_{xx}, \qquad M_y = M_{yy}, \qquad M_{xy} = M_{xy} \end{aligned} \tag{10.50}$$

These definitions may be used with the constitutive equations (10.46) to define resultants in terms of derivatives of the displacement functions:

$$\begin{Bmatrix} N_x \\ N_y \\ N_{xy} \\ M_x \\ M_y \\ M_{xy} \end{Bmatrix} = \left[\begin{array}{ccc|ccc} A_{11} & A_{12} & A_{16} & B_{11} & B_{12} & B_{16} \\ A_{12} & A_{22} & A_{26} & B_{12} & B_{22} & B_{26} \\ A_{16} & A_{26} & A_{66} & B_{16} & B_{26} & B_{66} \\ \hline B_{11} & B_{12} & B_{16} & D_{11} & D_{12} & D_{16} \\ B_{12} & B_{22} & B_{26} & D_{12} & D_{22} & D_{26} \\ B_{16} & B_{26} & B_{66} & D_{16} & D_{26} & D_{66} \end{array}\right] \begin{Bmatrix} \varepsilon_x^0 \\ \varepsilon_y^0 \\ \gamma_{xy}^0 \\ \kappa_x \\ \kappa_y \\ \kappa_{xy} \end{Bmatrix} - \begin{Bmatrix} N_x^T \\ N_y^T \\ N_{xy}^T \\ M_x^T \\ M_y^T \\ M_{xy}^T \end{Bmatrix} \tag{10.51a}$$

or

$$\begin{Bmatrix} \{N\} \\ \{M\} \end{Bmatrix} = \left[\begin{array}{c|c} A & B \\ \hline B & D \end{array}\right] \begin{Bmatrix} \{\varepsilon^0\} \\ \{\kappa\} \end{Bmatrix} - \begin{Bmatrix} \{N^T\} \\ \{M^T\} \end{Bmatrix} \tag{10.51b}$$

where

$$\left.\begin{aligned} A_{ij} &= \sum_{k=1}^{N} Q_{ij}'^{(k)}(z_k - z_{k-1}) \\ B_{ij} &= \frac{1}{2} \sum_{k=1}^{N} Q_{ij}'^{(k)}(z_k^2 - z_{k-1}^2) \\ D_{ij} &= \frac{1}{3} \sum_{k=1}^{N} Q_{ij}'^{(k)}(z_k^3 - z_{k-1}^3) \end{aligned}\right\} \qquad (i, j = 1, 2, 6) \tag{10.52}$$

$$\left.\begin{aligned} N_i^T &= \int_{-h/2}^{h/2} Q_{ij}' \alpha_j T \, dz \\ M_i^T &= \int_{-h/2}^{h/2} Q_{ij}' \alpha_j T z \, dz \end{aligned}\right\} \qquad (i, j = 1, 2, 6) \tag{10.53}$$

and the thermal expansion coefficients are in the geometric or global axes. The

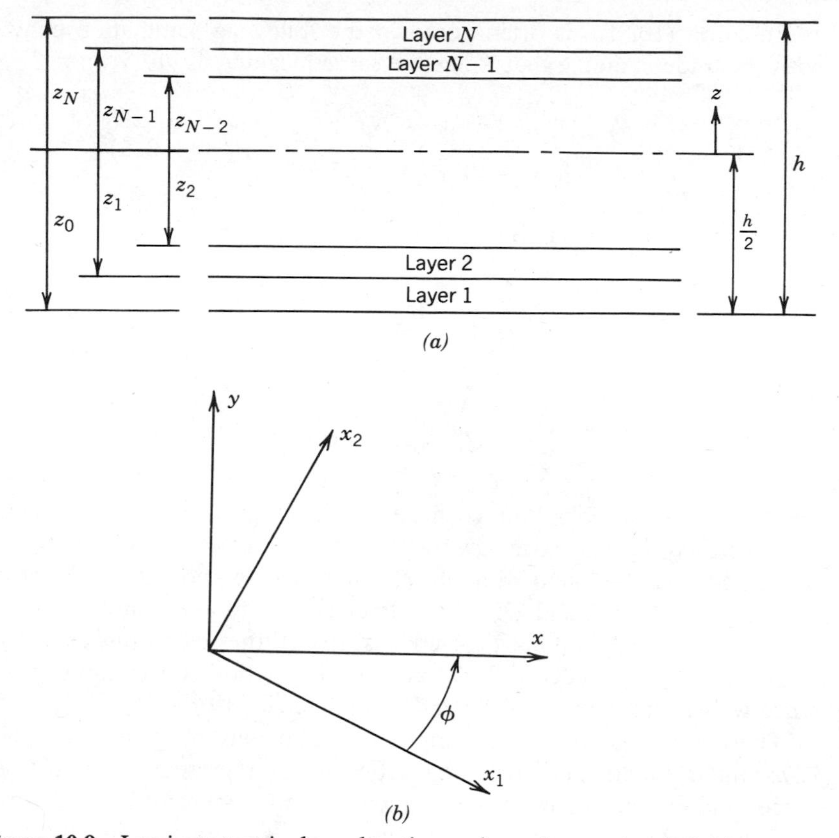


**Figure 10.9** Laminate terminology: location and angular rotation definitions.

standard terminology for lamina locations has been employed in (10.52), (10.53), and (10.56) and is shown in Fig. 10.9($a$). The $z_k$ values are signed numbers varying from $z_0 = -h/2$ to $z_N = h/2$ in a laminate of $N$ layers. The angle definition for the transformed stiffness $Q_{ij}'^{(k)}$ for lamina $k$ is as shown in Fig. 10.9($b$); $Q'_{ij}$ will be based on the transformations of Section 10.3. The laminate stacking sequence is the description of the angles $\phi$ for each lamina from the bottom to the top lamina. Several common notations are as follows:

$$[0/90]_s = [0/90/90/0] \quad (s \text{ denotes symmetry})$$

$$[0/90/0]_T = [0/90/0] \quad \begin{pmatrix} T \text{ confirms that listing is total} \\ \text{of all layers; may be omitted} \end{pmatrix}$$

$$[(\pm\theta)]_n = [(\pm\theta)_n] = [+\theta/-\theta/+\theta/-\theta/\cdots]$$

$$(n \text{ replicates of indicated unit})$$

Equation (10.51b) is often posed in the following summation convention, with the understanding that it has the same meaning as (10.51a):

$$\left.\begin{aligned} N_i &= A_{ij}\varepsilon_j^0 + B_{ij}\kappa_j - N_i^T \\ M_i &= B_{ij}\varepsilon_j^0 + D_{ij}\kappa_j - M_i^T \end{aligned}\right\} \qquad (i, j = 1, 2, 6) \tag{10.54}$$

The shear resultants may be similarly integrated,

$$\begin{Bmatrix} Q_y \\ Q_x \end{Bmatrix} = \begin{bmatrix} A_{44} & A_{45} \\ A_{45} & A_{55} \end{bmatrix} \begin{Bmatrix} \gamma_{yz} \\ \gamma_{xz} \end{Bmatrix} \tag{10.55}$$

where

$$A_{ij} = \sum_{k=1}^{N} K_{ij} Q_{ij}^{\prime(k)} (z_k - z_{k-1}), \qquad i, j = 4, 5 \tag{10.56}$$

and $K_{ij}$ are shear correction coefficients. Equations (10.51)–(10.56) constitute an integrated thermo-elastic law for the plate. All of the constants $A_{ij}$, $B_{ij}$, $D_{ij}$ for (10.54) may be found without requiring that we know the transverse shear stiffnesses $Q'_{44}$, $Q'_{45}$, and $Q'_{55}$. If we later choose to use a shear-indeformable plate theory, we do not need these transverse stiffnesses. In this event, we may work with a much reduced $3 \times 3$ system of equations for computing the $Q'_{ij}$, since we will not use (10.55) directly. See Exercise 10.10.

There are a variety of coupling effects that may be seen in the laminate constitutive equation (10.51). The coefficients $B_{ij}$ represent a coupling between forces and curvatures and between moments and extensions. It will be shown in Section 10.9 that nonzero $B_{ij}$ values produce characteristic equations with coupling between in-plane displacements and the transverse displacement. This coupling can occur even under conditions with neither buckling nor moderate displacement. The terms $A_{16}$, $A_{26}$ represent a coupling between extensional and shearing deformation and affect only the in-plane displacements. The terms $D_{16}$, $D_{26}$ represent coupling between bending and twisting and concern only the transverse displacement.

There are several important classes of laminate configuration that will result in important simplifications of the form of the laminate constitutive law (10.51). These simplifications depend mainly on the odd or even character of the stiffnesses $Q'_{ij}$ and the thermal expansion coefficients $\alpha_i$ as found in (10.46). The following can be shown for an orthotropic material [see Exercise 10.9 with (10.38a)]:

$$Q'_{ij}(\phi) = \begin{cases} -Q'_{ij}(-\phi), & (i, j) = (1,6), (2,6), (4,5) \\ Q'_{ij}(-\phi), & (i, j) = (1,1), (1,2), (2,2), (4,4), (5,5), (6,6) \end{cases} \tag{10.57}$$

$$\alpha_i(\phi) = \begin{cases} \alpha_i(-\phi), & i = x, y, z \\ -\alpha_i(-\phi), & i = xy \end{cases} \tag{10.58}$$

***Balanced Laminate*** For every lamina with orientation $\phi$, there exists a companion lamina of identical material and thickness at an orientation $-\phi$ somewhere within the laminate. Observing the definitions of $A_{ij}$ in (10.52), the elements are based on a simple thickness-weighted summation. The in-plane stretching-shearing coupling thus is not present:

$$A_{16} = A_{26} = 0 \quad \text{(balance)} \tag{10.59}$$

***Symmetric Laminate*** This laminate possesses an orientation symmetry with respect to the geometric midsurface. If a lamina at location $Z$ has orientation $\phi$, there is another lamina with identical orientation $\phi$ at $-Z$. The affected constants are the $B_{ij}$, which are sensitive to a mirror symmetry about the midsurface (see Exercise 10.12):

$$B_{ij} = 0, \qquad i, j = 1, 2, 6 \quad \text{(symmetry)} \tag{10.60}$$

***Antisymmetric Laminate*** For this configuration, the location of a layer with orientation $\phi$ at location $Z$ implies the existence of a layer with orientation $-\phi$ at location $-Z$. Some members of the class of balanced laminates belong to this class:

$$A_{16} = A_{26} = D_{16} = D_{26} = 0 \quad \text{(antisymmetry)} \tag{10.61a}$$

If the antisymmetric laminate is an angle-ply laminate, that is, if it is of the form $[\pm\theta]_{nT}$, there are additional simplifications:

$$B_{11} = B_{12} = B_{22} = B_{66} = 0 \quad \text{(antisymmetry, angle-ply)} \tag{10.61b}$$

**Example:** *Simplifications of Thermal Resultants.* Useful simplifications of the thermal terms of the general laminate constitutive relationship may be accomplished for certain laminates using (10.57) and (10.58). Consider the case of a uniform temperature field for a symmetric laminate. Using definition (10.53) for thermal moment resultants, we may write

$$M_i^T = T \sum_{k=1}^{N} Q_{ij}'^{(k)} \alpha_j^{(k)} \left( z_k^2 - z_{k-1}^2 \right), \qquad i, j = 1, 2, 6 \tag{a}$$

By considering the placement of layers, we can see that the product $Q_{ij}'^{(k)} \alpha_j^{(k)}$ is a constant vector for any pair of symmetric laminae, since they have the same in-plane rotation angle, while the term $z_k^2 - z_{k-1}^2$ will change sign between one element of the pair and the other. The terms of the summation (a) thus pairwise cancel for the symmetric laminae, and there can be no thermally induced moments:

$$M_i^T = 0, \qquad i = 1, 2, 6 \quad \text{(symmetry, constant temperature)} \tag{b}$$

We may similarly consider terms in the thermal shearing term $N_{xy}^T$ in a balanced laminate. The integrated laminate form of (10.53) is

$$N_{xy}^T = T \sum_{k=1}^{N} Q_{6j}^{\prime(k)} \alpha_j^{(k)} (z_k - z_{k-1}) \tag{c}$$

By observing the terms of the product $Q'_{6j}\alpha_j$, we determine that the product is a combination of even and odd functions of $\phi$ such that the product is an odd function. The thickness term $z_k - z_{k-1}$ is insensitive to lamina location and thus is an even function. The laminae may thus be considered as a series of balanced pairs, each pair of which will cancel:

$$N_{xy}^T = 0 \quad \text{(balance, constant temperature)} \tag{d}$$

## 10.9 CHARACTERISTIC EQUATIONS AND BOUNDARY CONDITIONS IN DISPLACEMENT FORM

### 10.9.1 Shear-Deformable Plate Theory

To derive the differential equations of motion for a shear-deformable theory of plates, we may employ Hamilton's principle for a general, simply connected laminated plate. We use the assumed displacement formulation given in (10.47), in which the "small" rotation vector $\psi = (\psi_x, \psi_y)$ is independent of the transverse deflection $w$. See (3.2). This shear-deformable theory is also referred to as a first-order, or improved, plate theory. The details of the use of Hamilton's principle may be found in Section 9.6. The resulting differential equations and natural boundary conditions (9.88)–(9.94) are repeated in what follows. Those terms dealing with the contribution of in-plane force resultants to the out-of-plane equilibrium condition are noted by an underscore. These underscored terms are important only for moderately large deflections or for buckling considerations and represent an inherent coupling of in-plane and transverse displacements provided by the nonlinear strain-displacement equations (10.48c). Unlike the case of an isotropic plate, laminated plates possess the potential for such coupling even if there is no buckling or moderate displacement, that is, even if the linearized strain-displacement equations (10.48b) are employed. Coupling may arise when we use the linearized strain-displacement equations due to the form of the laminate constitutive law (10.51). Note that the contracted notation represented by (10.50) is used in

(10.62)–(10.64):

$$
\begin{aligned}
N_{x,x} + N_{xy,y} &= I_1 \ddot{u}^0 + I_2 \ddot{\psi}_x \\
N_{xy,x} + N_{y,y} &= I_1 \ddot{v}^0 + I_2 \ddot{\psi}_y \\
S_{x,x} + S_{y,y} + p &= I_1 \ddot{w} \\
M_{x,x} + M_{xy,y} - Q_x &= I_3 \ddot{\psi}_x + I_2 \ddot{u}^0 \\
M_{xy,x} + M_{y,y} - Q_y &= I_3 \ddot{\psi}_y + I_2 \ddot{v}^0
\end{aligned}
\tag{10.62}
$$

where

$$(I_1, I_2, I_3) = \int_{-h/2}^{h/2} (1, z, z^2) \rho \, dz = \sum_{i=1}^{N} \rho^{(i)} \int_{z_{i-1}}^{z_i} (1, z, z^2) \, dz \tag{10.63a}$$

$$S_x = Q_x + \underline{N_x w_{,x} + N_{xy} w_{,y}} \qquad S_y = Q_y + \underline{N_{xy} w_{,x} + N_y w_{,y}} \tag{10.63b}$$

and the overdot represents $\partial/\partial t$. The quantities $I_1, I_2, I_3$ are the normal, coupled normal-rotatory, and rotatory inertia coefficients, respectively [see (9.78)]. The variational derivation also offers five admissible boundary condition pairs. One member of each pair must be specified on any boundary; $(n, s)$ represents the (normal, tangential) directions in the right-handed sense of Fig. 3.4:

$$
\begin{gathered}
(M_n, \psi_n), \qquad (M_{ns}, \psi_s), \qquad (S_n, w), \\
(N_n, u_n), \qquad (N_{ns}, u_s)
\end{gathered}
\tag{10.64}
$$

The differential equations (10.62) may be placed in terms of the generalized displacements by use of the strain-displacement equations (10.48) with the plate constitutive law (10.51). The strain-displacement equations are used in their linearized form (we use 10.48b, not 10.48c), since we are not interested in postbuckling displacements. The vector $\{f\}$ is a transverse forcing function, $\{b\}$ contains the terms which may be used to predict buckling, and $\{\theta\}$ contains the thermal terms:

$$[L]\{\Delta\} = \{f\} + \{b\} + \{\theta\} \tag{10.65}$$

where

$$\{\Delta\} = \left\{u^0, v^0, w, \psi_x, \psi_y\right\}^{\mathrm{T}} \tag{10.66a}$$

$$\{f\} = \{0, 0, p, 0, 0\}^{\mathrm{T}} \tag{10.66b}$$

$$\{\theta\} = \left\{N^T_{x,x} + N^T_{xy,y}, N^T_{xy,x} + N^T_{y,y}, 0, M^T_{x,x} + M^T_{xy,y}, M^T_{xy,x} + M^T_{y,y}\right\}^{\mathrm{T}} \tag{10.66c}$$

$$\{b\} = \left\{0, 0, (N_x w_{,x} + N_{xy} w_{,y})_{,x} + (N_{xy} w_{,x} + N_y w_{,y})_{,y}, 0, 0\right\}^{\mathrm{T}} \tag{10.66d}$$

and the operator $L_{ij}$ is given by

$$\begin{aligned}
L_{11} &= A_{11}d_{xx} + 2A_{16}d_{xy} + A_{66}d_{yy} - I_1 d_{tt} \\
L_{12} &= (A_{12} + A_{66})d_{xy} + A_{16}d_{xx} + A_{26}d_{yy} \\
L_{13} &= 0 \\
L_{14} &= B_{11}d_{xx} + 2B_{16}d_{xy} + B_{66}d_{yy} - I_2 d_{tt} \\
L_{15} &= (B_{12} + B_{66})d_{xy} + B_{16}d_{xx} + B_{26}d_{yy} = L_{24} \\
L_{22} &= 2A_{26}d_{xy} + A_{22}d_{yy} + A_{66}d_{xx} - I_1 d_{tt} \\
L_{23} &= 0 = L_{32} \\
L_{25} &= 2B_{26}d_{xy} + B_{22}d_{yy} + B_{66}d_{xx} - I_2 d_{tt} \\
L_{33} &= -A_{55}d_{xx} - 2A_{45}d_{xy} - A_{44}d_{yy} + I_1 d_{tt} \\
L_{34} &= -A_{55}d_x - A_{45}d_y \\
L_{35} &= -A_{45}d_x - A_{44}d_y \\
L_{44} &= D_{11}d_{xx} + 2D_{16}d_{xy} + D_{66}d_{yy} - A_{55} - I_3 d_{tt} \\
L_{45} &= (D_{12} + D_{66})d_{xy} + D_{16}d_{xx} + D_{26}d_{yy} - A_{45} = L_{54} \\
L_{55} &= 2D_{26}d_{xy} + D_{22}d_{yy} + D_{66}d_{xx} - A_{44} - I_3 d_{tt}
\end{aligned} \tag{10.66e}$$

with $L_{ij} = L_{ji}$ and the differential operator $d_{ij} = \partial^2/\partial x_i\, \partial x_j$. Note that the third equation of (10.62) has had all signs inverted for convenience.

### 10.9.2 Classical (Shear-Indeformable) Plate Theory

We may perform a similar derivation of differential equations of motion and boundary conditions for CPT. In CPT, the transverse shear deformation is assumed to be zero, and the displacement formulation (10.47) is simplified by the Kirchhoff-Love assumption:

$$(\psi_x, \psi_y) = (-w_{,x}, -w_{,y}) \tag{10.67a}$$

As a consequence, the plate curvatures given in (10.48d) may be given in terms of derivatives of the transverse displacement $w$:

$$\kappa_x = -w_{,xx}, \qquad \kappa_y = -w_{,yy}, \qquad \kappa_{xy} = -2w_{,xy} \tag{10.67b}$$

There are only three independent displacement variables: $u^0$, $v^0$, and $w$. The variational formulation is given in Section 9.8. Care must be exercised to realize that there can be no independent variation of rotation angles, so that $\delta\psi \equiv 0$. As a consequence, terms involving $I_3$ are zero. In addition, the coupling rotatory inertia terms involving $I_2$ are set to zero, in accordance with standard practice for thin plates. If this is not done, the displacement formulation of the governing equations in operator notation becomes nonsymmetric and thus formally non-self-adjoint (a most undesirable occurrence).

The differential equations of motion in terms of resultants are, for CPT [see (9.106)–(9.109)],

$$\begin{aligned} N_{x,x} + N_{xy,y} &= I_1 \ddot{u}^0 \\ N_{xy,x} + N_{y,y} &= I_1 \ddot{v}^0 \\ S_{x,x} + S_{y,y} + p &= I_1 \ddot{w} \end{aligned} \tag{10.68}$$

Additional relationships from the variational derivation are then required to eliminate the shear resultants from (10.68) as well as to describe the terms for buckling or moderate displacement, which are identified by an underscore:

$$\begin{aligned} S_x &= Q_x + \underline{N_x w_{,x} + N_{xy} w_{,y}} \\ S_y &= Q_y + \underline{N_{xy} w_{,x} + N_y w_{,y}} \end{aligned} \tag{10.69}$$

$$\begin{aligned} Q_x &= M_{x,x} + M_{xy,y} \\ Q_y &= M_{xy,x} + M_{y,y} \end{aligned} \tag{10.70}$$

These differential equations may be written in terms of displacements by employing the laminate constitutive relations (10.51), yielding a compact operator notation. As before, the lateral forcing function is given by $\{f\}$, the buckling terms by $\{b\}$, and the thermal terms by $\{\theta\}$:

$$\begin{bmatrix} L_{11} & L_{12} & L_{13} \\ L_{12} & L_{22} & L_{23} \\ L_{13} & L_{23} & L_{33} \end{bmatrix} \begin{Bmatrix} u^0 \\ v^0 \\ w \end{Bmatrix} = \{f\} + \{b\} + \{\theta\} \tag{10.71a}$$

where

$$\{f\} = \{0, 0, p\}^{\mathrm{T}} \tag{10.71b}$$

$$\{\theta\} = \left\{ N^T_{x,x} + N^T_{xy,y}, N^T_{xy,x} + N^T_{y,y}, 0 \right\}^{\mathrm{T}} \tag{10.71c}$$

$$\{b\} = \left\{ 0, 0, \left( N_x w_{,x} + N_{xy} w_{,y} \right)_{,x} + \left( N_{xy} w_{,x} + N_y w_{,y} \right)_{,y} \right\}^{\mathrm{T}} \tag{10.71d}$$

$$\begin{aligned} L_{11} &= A_{11} d_{xx} + 2A_{16} d_{xy} + A_{66} d_{yy} - I_1 d_{tt} \\ L_{12} &= A_{16} d_{xx} + (A_{12} + A_{66}) d_{xy} + A_{26} d_{yy} \\ L_{13} &= -B_{11} d_{xxx} - 3B_{16} d_{xxy} - (B_{12} + 2B_{66}) d_{xyy} - B_{26} d_{yyy} \\ L_{22} &= A_{66} d_{xx} + 2A_{26} d_{xy} + A_{22} d_{yy} - I_1 d_{tt} \\ L_{23} &= -B_{16} d_{xxx} - (B_{12} + 2B_{66}) d_{xxy} - 3B_{26} d_{xyy} - B_{22} d_{yyy} \\ L_{33} &= D_{11} d_{xxxx} + 4D_{16} d_{xxxy} + 2(D_{12} + 2D_{66}) d_{xxyy} + 4D_{26} d_{xyyy} \\ &\quad + D_{22} d_{yyyy} + I_1 d_{tt} \end{aligned} \tag{10.71e}$$

The four corresponding admissible boundary condition pairs, one member of which must be specified on any boundary, also arise from the variational derivation [see (9.109)]

$$\left(u_n^0, N_n\right), \qquad \left(u_s^0, N_{ns}\right), \qquad \left(w,_n, M_n\right), \qquad \left(w, S_n\right) \tag{10.72}$$

The quantity $S_n$ is the familiar Kirchhoff shear condition, modified for coupling effects produced by buckling or moderate displacement as indicated by the underlined terms:

$$S_n = Q_n + M_{ns,s} + \underline{N_n w,_n + N_{ns} w,_s} \tag{10.73}$$

The corner forces required in CPT have been discussed previously in Section 4.2 and in Exercise 9.17.

## 10.10 SOLUTIONS FOR CLASSICAL PLATE THEORY

In this section, we explore a variety of solutions for different classes of laminated plates. The solutions include problems of bending, buckling, and vibration, in which there may be coupling between the in-plane and transverse displacements. Emphasis is placed on analytical solutions, with summaries given to introduce appropriate approximation methods. The classes of plates are organized as to specially orthotropic, midplane symmetric, and general laminated plates.

### 10.10.1 Specially Orthotropic Plates

If a laminated plate is balanced and symmetric ($A_{16} = A_{26} = 0$, $B_{ij} = 0$) and also possesses no bending-twisting coupling ($D_{16} = D_{26} = 0$), it is said to be specially orthotropic. Examining the laminate constitutive law (10.51), we see that we have not only eliminated the coupling of in-plane and transverse displacement but have also uncoupled the in-plane extensional and shear responses as well as the transverse bending and twisting responses. This class of plates includes laminates composed of isotropic laminae as well as orthotropic layers with an in-plane rotation of 0° or 90°. For both of these types of laminates, each individual lamina has no bending-twisting coupling ($Q'_{16} = Q'_{26} = 0$). The operator displacement notation (10.71a) simplifies to an "orthotropic" operator with uncoupled transverse and in-plane displacements:

$$\left[L_{ij}\right] = \begin{bmatrix} L_{11} & L_{12} & 0 \\ L_{12} & L_{22} & 0 \\ 0 & 0 & L_{33} \end{bmatrix} \tag{10.74}$$

where

$$
\begin{aligned}
L_{11} &= A_{11}d_{xx} + A_{66}d_{yy} - I_1 d_{tt} \\
L_{12} &= (A_{12} + A_{66})d_{xy} \\
L_{22} &= A_{66}d_{xx} + A_{22}d_{yy} - I_1 d_{tt} \\
L_{33} &= D_{11}d_{xxxx} + 2(D_{12} + 2D_{66})d_{xxyy} + D_{22}d_{yyyy} + I_1 d_{tt}
\end{aligned}
$$

***Bending*** Consider a rectangular plate of dimensions $a, b$ along the $x$- and $y$-axes, respectively, simply supported along all four edges and subjected to a transverse loading $p(x, y)$. There is assumed to be no thermal expansion loading ($\{\theta\} = \{0\}$). The only portion of the operator form (10.74) that need be considered is $L_{33}w = p$ under static equilibrium:

$$D_{11}w,_{xxxx} + 2(D_{12} + 2D_{66})w,_{xxyy} + D_{22}w,_{yyyy} = p \qquad (10.75)$$

The boundary conditions are

$$
\begin{aligned}
\text{For } x = 0, a: \quad & w = 0 \\
& M_x = -D_{11}w,_{xx} - D_{12}w,_{yy} = 0 \\
\text{For } y = 0, b: \quad & w = 0 \\
& M_y = -D_{12}w,_{xx} - D_{22}w,_{yy} = 0
\end{aligned}
\qquad (10.76)
$$

Assuming that the loading may be represented by the double sine series expansion of (10.77), the displacements may be similarly represented using a Navier method. This solution is found by substituting the proposed form for the displacement (10.78) into the differential equation (10.75), employing the load expansion (10.77), and comparing like terms of the orthogonal functions $\sin(m\pi x/a)\sin(n\pi x/b)$. The boundary conditions are found to be identically satisfied.

$$p(x, y) = \sum_{m=1}^{\infty} \sum_{n=1}^{\infty} p_{mn} \sin\frac{m\pi x}{a} \sin\frac{n\pi y}{b} \qquad (10.77)$$

$$w = \sum_{m=1}^{\infty} \sum_{n=1}^{\infty} w_{mn} \sin\frac{m\pi x}{a} \sin\frac{n\pi y}{b} \qquad (10.78)$$

where

$$
\begin{aligned}
p_{mn} &= \frac{4}{ab}\int_0^b \int_0^a p(x, y) \sin\frac{m\pi x}{a} \sin\frac{n\pi y}{b}\, dx\, dy \\
w_{mn} &= \frac{1}{\pi^4} p_{mn} \left[ D_{11}\left(\frac{m}{a}\right)^4 + 2(D_{12} + 2D_{66})\left(\frac{mn}{ab}\right)^2 + D_{22}\left(\frac{n}{b}\right)^4 \right]^{-1}
\end{aligned}
$$

The case of a rectangular plate that is simply supported along two opposite edges may also be proposed in analytical form, although the values of the material constants play a role in determining the form of the solution. The differential equation is given by (10.75), and the boundary conditions are

$$y = 0, b, \qquad w = 0$$
$$M_y = -D_{12}w,_{xx} - D_{22}w,_{yy} = 0$$
$$x = 0, a \qquad \text{to be specified}$$

Consider the case of a uniform transverse load $p_0$. Following an approach by M. Lévy (see Section 5.1.2) a product solution is proposed:

$$w(x, y) = \sum_{n=1}^{\infty} X_n(x) \sin\frac{n\pi y}{b} \tag{10.79}$$

The transverse loading may be similarly expanded:

$$p(x, y) = p_0 = \sum_{n=1}^{\infty} p_n \sin\frac{n\pi y}{b} \tag{10.80}$$

where

$$p_n = \begin{cases} \dfrac{4p_0}{n\pi}, & n \text{ odd} \\ 0, & n \text{ even} \end{cases}$$

from

$$p_n = \frac{2}{b}\int_0^b p \sin\frac{n\pi y}{b}\, dy$$

Upon substitution of (10.79) into the differential equation (10.75) and comparing like terms of the orthogonal functions $\sin(n\pi y/b)$, we achieve an ordinary differential equation:

$$D_{11}X_n'''' - 2(D_{12} + 2D_{66})\left(\frac{n\pi}{b}\right)^2 X_n'' + D_{22}\left(\frac{n\pi}{b}\right)^4 X_n = p_n \tag{10.81}$$

All $p_n = 0$ for $n$ even; if the boundary conditions along the edges of $x = 0, a$ are homogeneous admissible boundary conditions, then all $X_n(x)$ are zero for $n$ even. The solution for the functions $X_n(x)$, $n$ odd, may be broken down into a homogeneous and a particular solution. The particular solutions to (10.81) are given by

$$X_{np}(x) = \frac{4b^4 p_0}{D_{22}n^5\pi^5} \qquad n \text{ odd} \tag{10.82}$$

whose contribution to the solution for displacement $w$ can be given as a simple polynomial that satisfies the boundary conditions on $y = 0, b$:

$$\sum_{n=1,3,5,\ldots}^{\infty} X_{np} \sin\frac{n\pi y}{b} = \frac{p_0}{24D_{22}}(y^4 - 2by^3 + b^3 y) \tag{10.83}$$

If we propose an exponential solution $X_n(x) = Ce^{n\pi sx/b}$ for the homogeneous part of (10.81), the form of the solution depends on the roots of the characteristic equation

$$D_{11}S^4 - 2(D_{12} + 2D_{66})S^2 + D_{22} = 0 \tag{10.84}$$

These roots may be real and unequal, real and equal, or complex; this will determine the form of the solution (Whitney, 1987). See Exercise 10.17. The type of boundary conditions on the edges $x = 0, a$ will determine the coefficients of the terms.

The solution for plates whose edges are not simply supported is not generally available in terms of elementary functions. Approximation methods based on energy principles may then be applied. Assume that the displacement may be represented by the form

$$w(x, y) = \sum_{i=1}^{M} \sum_{j=1}^{N} a_{ij} X_i(x) Y_j(y) \tag{10.85}$$

where the functions $X_i$, $Y_j$ are selected to satisfy the displacement boundary conditions if possible. If the functions do satisfy the boundary conditions, then they span the space, and the solution for $w(x, y)$ may be properly expanded in terms of the selected functions. Since only a finite number of terms is considered, a polynomial series or beam vibration mode series is often selected to provide appropriate functions. The principle of minimum potential energy is invoked in order to solve for the coefficients $a_{ij}$:

$$\frac{\partial W_p}{\partial a_{ij}} = 0, \qquad i = 1, \ldots, M, \quad j = 1, \ldots, N \tag{10.86}$$

where a suitable expression for the potential energy $W_p$ for a plate may be found elsewhere (Whitney, 1987).

**Example:** *Lamina Stress Calculation.* Having found the displacement field $w(x, y)$ for a specially orthotropic plate in bending by equations such as (10.78), (10.79), or (10.85), the determination of all strains and stresses may be made. The general procedure is to solve for the strain state throughout the thickness using the strain-displacement assumptions (10.48a), with the curvature $\{\kappa\}$ defined in terms of $w(x, y)$ by (10.67b). The midsurface strains $\{\varepsilon^0\}$

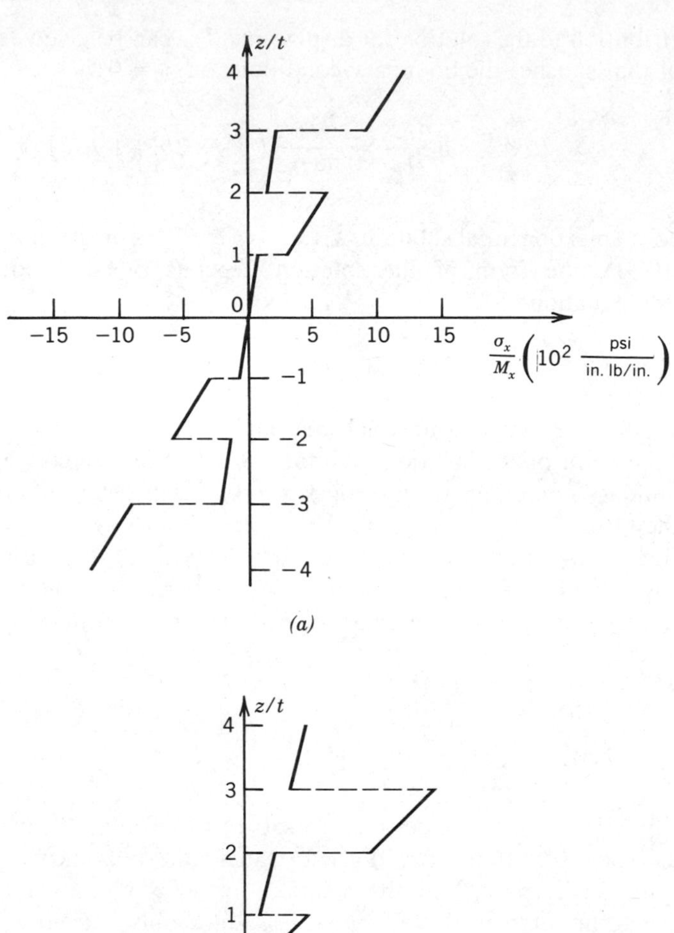


(a)

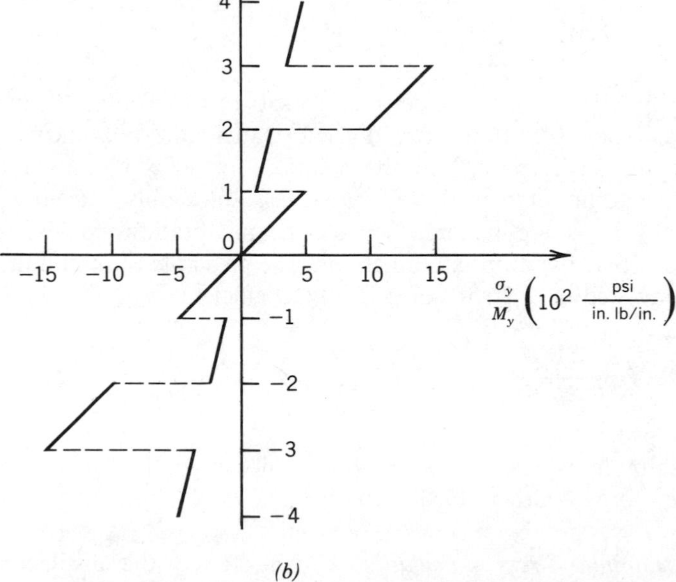


(b)

**Figure 10.10** Bending stress profiles for orthotropic plate.

are zero for the case of simple bending only. The lamina constitutive relations in the geometric coordinate system (10.46) may then be employed to solve for the stresses throughout the thickness.

We may also find that the bending moment resultants may be given directly in cases such as a mechanical characterization flexure test, and interesting results may be observed in calculating the stresses within the laminae under these bending loads. The results of a simple bending load $M_x$ is shown in Fig. 10.10($a$), and load $M_y$ is shown in Fig. 10.10($b$) for a $[(0/90)_2]_s$ laminate. The properties for this $E$ glass-epoxy laminate are

$$E_1 = 5.7 \times 10^6 \text{ psi}$$
$$E_2 = 1.4 \times 10^6 \text{ psi}$$
$$\nu_{12} = 0.25$$
$$G_{12} = 6.0 \times 10^5 \text{ psi}$$
$$t = 0.01 \text{ in.} \quad \text{(lamina thickness)}$$

These stress results are obtained from inverting the laminate constitutive relationship (10.51) to solve for the midplane strains $\{\varepsilon^0\}$ (zero in this case) and the curvatures $\{\kappa\}$, calculating the strains using the Kirchhoff assumption (10.48a) and employing the lamina constitutive relationship (10.46). Although the strain profile is linear, the stress profile is seen to be piecewise linear due to the stiffness changes between laminae. The laminate response is also sensitive to the orientation $M_x$ or $M_y$ of the bending load. This behavior is not seen with laminates of isotropic materials.

***Stability*** Consider the case of a simply supported rectangular plate subjected to uniform in-plane compressive loadings $N_x$ and $N_y$ as shown in Fig. 10.11:

$$(N_x, N_y, N_{xy}) = (-k_1 N_0, -k_2 N_0, 0) \quad \text{with } N_0 > 0 \tag{10.87}$$

The differential equation for this case is given from (10.74), with buckling

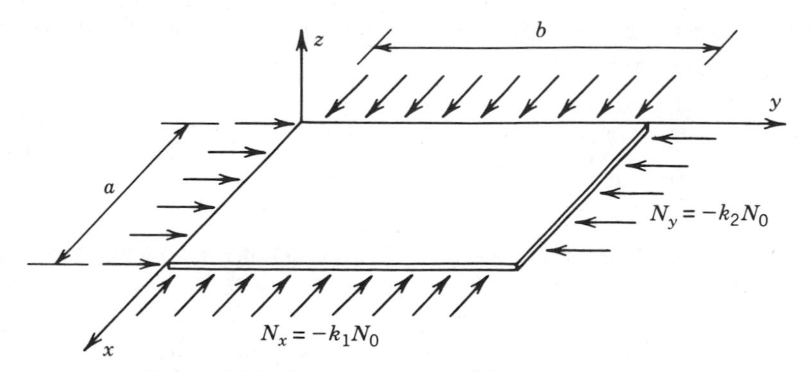


**Figure 10.11** Plate subjected to biaxial compression.

terms as shown in (10.71d):

$$D_{11}w,_{xxxx} + 2(D_{12} + 2D_{66})w,_{xxyy} + D_{22}w,_{yyyy}$$
$$= (-k_1N_0)w,_{xx} + (-k_2N_0)w,_{yy} \qquad (10.88)$$

The boundary conditions are as before (10.76). The possible solutions are of the double sine series form:

$$w_{mn} = a_{mn} \sin\frac{m\pi x}{a} \sin\frac{n\pi y}{b} \quad (\text{integers } m, n \geq 1) \qquad (10.89)$$

If we substitute this form into the differential equation (10.88) and choose the nontrivial solution ($a_{mn} \neq 0$), we may solve for the buckling load. The critical buckling load is defined as the lowest load that will cause buckling:

$$N_0 = \frac{\pi^2\left[D_{11}(m/a)^2 + 2(D_{12} + 2D_{66})(n/b)^2 + D_{22}(n/b)^4(a/m)^2\right]}{k_1 + k_2[an/bm]^2} \qquad (10.90)$$

We thus seek solutions for the values of $N_0$ that satisfy (10.90): these constitute the possible buckling loads, while (10.89) give the corresponding mode shapes.

Since the various $D_{ij}$ may take on a variety of relative numerical values for different laminated plates, it is not always clear how to determine the lowest, or critical, buckling load $N_{0,\mathrm{cr}}$ as a function of $m$ and $n$. For certain simple cases, however, this can be predicted. Consider the case $N_y = 0$ ($k_1 = 1$, $k_2 = 0$), a unidirectional compression. Examining (10.90), we see that $n = 1$ will supply the critical load, although the value of $m$ is yet to be determined. The behavior of the buckling load as a function of $m$ may be found by plotting the critical (lowest) buckling load as a function of plate aspect ratio. In particular, we are interested in the aspect ratio at the crossover points for different $m$ values according to

$$\frac{D_{11}}{D_{22}}\left(\frac{m}{R}\right)^2 + \left(\frac{R}{m}\right)^2 = \frac{D_{11}}{D_{22}}\left(\frac{m+1}{R}\right)^2 + \left(\frac{R}{m+1}\right)^2 \qquad (10.91)$$

where $R = a/b$ is the plate aspect ratio. Equation (10.91) would be solved first for $m = 1$ and then for increasing $m$ values.

**Example:** *Critical Buckling Load.* Following an example by Whitney (1987), we consider a $[0/90]_s$ graphite-epoxy laminate whose laminae possess orthotropic properties:

$$E_1 = 21 \times 10^6 \text{ psi}$$
$$E_2 = 1.5 \times 10^6 \text{ psi}$$
$$\nu_{12} = 0.3$$
$$G_{12} = 0.75 \times 10^6 \text{ psi}$$

The ratio of plate stiffness is given by

$$\frac{D_{11}}{D_{22}} = 4.71, \qquad \frac{D_{12} + 2D_{66}}{D_{22}} = 0.496$$

The buckling load equation (10.90) thus becomes

$$N_0 = \frac{\pi^2 D_{22}}{b^2}\left[4.71\left(\frac{m}{R}\right)^2 + 0.992 + \left(\frac{R}{m}\right)^2\right] \quad (m \text{ integer})$$

where $R = a/b$. Note that, for example, at $R = 3$ the buckling load in the second mode ($m = 2$) is lower than in the first mode ($m = 1$), so that $m = 2$ is the critical mode and gives the critical buckling load $N_{0,\text{cr}}$. That is,

$$4.71\left(\tfrac{2}{3}\right)^2 + \left(\tfrac{3}{2}\right)^2 < 4.71\left(\tfrac{1}{3}\right)^2 + \left(\tfrac{3}{1}\right)^2$$

By solving for the crossover points between modes as in (10.91) and by solving (10.90) within a given mode, a graph of buckling load (and mode) may be drawn as a function of plate aspect ratio. See Fig. 10.12. It should be noted that a figure of this nature must be constructed for each lamination stacking sequence under consideration even if the material is the same. This stands in contrast to an isotropic plate, where the relative values of the $D_{ij}$ are fixed for a given material.

The case of shear buckling is briefly summarized to show the need for approximation methods. Consider a simply supported plate with a uniform in-plane shear stress $N_{xy}$. The differential equation is given by

$$L_{\text{sb}}(w) = 0 \tag{10.92}$$

where the $L_{\text{sb}}$ operator is

$$L_{\text{sb}} = D_{11} d_{xxxx} + 2(D_{12} + 2D_{66}) d_{xxyy} + D_{22} d_{yyyy} - 2N_{xy} d_{xy}$$

and the boundary conditions are given by (10.76). A double sine series expansion for $w(x, y)$ such as (10.78) is found to satisfy the boundary

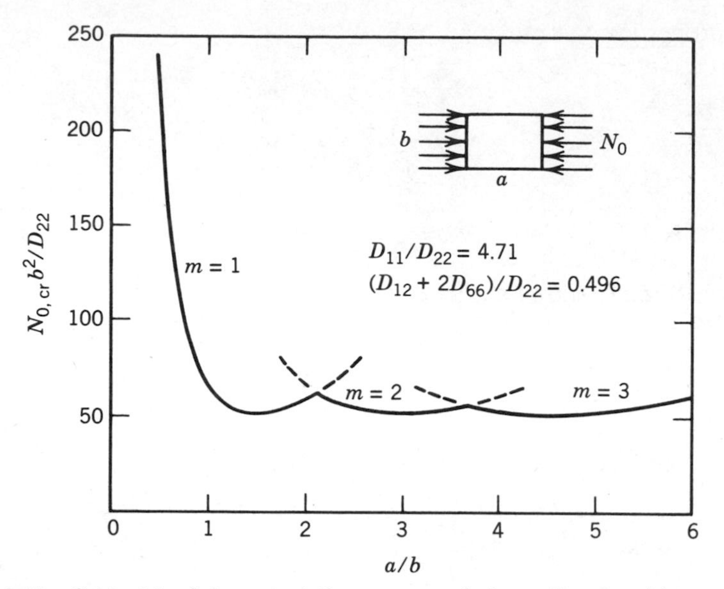


**Figure 10.12** Critical load for uniaxially compressed plate. (Reprinted by permission from K. T. Kedward, J. M. Whitney, *Composites Design Encyclopedia*, Vol. 5, Center for Composite Materials, University of Delaware, 1984, p. 39.)

conditions but not the differential equation. Because there is no solution available for this system, the Galerkin method may be applied (Whitney, 1987). A double sine series of the form (10.78) but with a finite number of terms is employed, and the Galerkin formulation takes the form

$$\int_0^b \int_0^a L_{\mathrm{sb}}(w) w \, dx \, dy = 0 \tag{10.93}$$

A series of algebraic equations will result from this formulation, which may be placed in an eigenvalue formulation to solve for the critical shear buckling load.

***Free Vibration*** In this section, we consider the free vibration of a simply supported rectangular plate. For this case, the differential equation (10.71) simplifies to

$$D_{11} w,_{xxxx} + 2(D_{12} + 2D_{66}) w,_{xxyy} + D_{22} w,_{yyyy} + I_1 w,_{tt} = 0 \tag{10.94}$$

The solution may be sought as the sum of products of a spatial function and a harmonic time function:

$$w(x, y, t) = \sum_{n=1}^{\infty} \sum_{m=1}^{\infty} w_{mn}(x, y) e^{i\Omega_{mn} t} \tag{10.95}$$

where $\Omega_{mn}$ is the natural vibration frequency of mode $mn$ and

$$w_{mn}(x, y) = A_{mn} \sin\frac{m\pi x}{a} \sin\frac{n\pi y}{b} \tag{10.96}$$

The displacement function $w_{mn}$ in (10.96) satisfies the boundary conditions (10.76), so that $w(x, y, t)$ satisfies the boundary conditions at all times.

Substitution of the proposed solution (10.95) into the differential equation and comparison of like terms of the orthogonal functions $\sin(m\pi x/a)\sin(n\pi y/b)$ will result in the following form for the natural frequencies:

$$\Omega_{mn}^2 = \frac{\pi^4}{b^4 I_1}\left[D_{11}\left(\frac{mb}{a}\right)^4 + 2(D_{12} + 2D_{66})\left(\frac{mnb}{a}\right)^2 + D_{22}n^4\right] \tag{10.97}$$

Unlike the buckling case, the order of solutions for the vibration frequencies given by (10.97) is clear. The fundamental frequency for the plate is $\Omega_{11}$, which, for the case of an isotropic plate, can be shown to be

$$\Omega_{11} = \frac{\pi^2}{b^2}\left(\frac{D}{I_1}\right)^{1/2}\left[\left(\frac{b}{a}\right)^2 + 1\right] \tag{10.98}$$

where $D = Eh^3/12(1 - \nu^2)$, $I_1 = \rho h$, which matches the results of (6.19). See Exercise 10.15.

### 10.10.2 Midplane Symmetric Plates

For the class of laminates that are symmetric with respect to the midplane, there is no coupling of in-plane and transverse displacements according to (10.71) since all $B_{ij} = 0$. However, the bending-twisting coupling terms $D_{16}$, $D_{26}$ present in (10.71e) are not necessarily zero, even in the case of a balanced symmetric laminate. This is confirmed by an examination of (10.52), in which the $Q'_{16}$, $Q'_{26}$ terms are greatly modified by the laminate stacking sequence. The differential operator notation (10.71) gives the following simplification:

$$[L_{ij}] = \begin{bmatrix} L_{11} & L_{12} & 0 \\ L_{12} & L_{22} & 0 \\ 0 & 0 & L_{33} \end{bmatrix} \tag{10.99}$$

where

$$\begin{aligned}
L_{11} &= A_{11}d_{xx} + 2A_{16}d_{xy} + A_{66}d_{yy} - I_1 d_{tt} \\
L_{12} &= A_{16}d_{xx} + (A_{12} + A_{66})d_{xy} + A_{26}d_{yy} \\
L_{22} &= A_{66}d_{xx} + 2A_{26}d_{xy} + A_{22}d_{yy} - I_1 d_{tt} \\
L_{33} &= D_{11}d_{xxxx} + 4D_{16}d_{xxxy} + 2(D_{12} + 2D_{66})d_{xxyy} \\
&\quad + 4D_{26}d_{xyyy} + D_{22}d_{yyyy} + I_1 d_{tt}
\end{aligned}$$

We consider a simply supported rectangular plate to gain insight into the problems encountered with previously presented solution techniques. The differential equation for $w$ is given by (10.99) under static loading,

$$D_{11}w,_{xxxx} + 4D_{16}w,_{xxxy} + 2(D_{12} + 2D_{66})w,_{xxyy} + 4D_{26}w,_{xyyy} + D_{22}w,_{yyyy} = p \tag{10.100}$$

and the boundary conditions are given by

$$\begin{aligned} x = 0, a: \quad & w = 0 \\ & M_x = -D_{11}w,_{xx} - D_{12}w,_{yy} - 2D_{16}w,_{xy} = 0 \\ y = 0, b: \quad & w = 0 \\ & M_y = -D_{12}w,_{xx} - D_{22}w,_{yy} - 2D_{26}w,_{xy} = 0 \end{aligned} \tag{10.101}$$

The usual approach of the assumption of a double sine series for transverse displacement will lead to difficulties. When the sine series is substituted into the differential equation, the odd derivative terms such as $w,_{xyyy}$ produce both sine and cosine terms in the final result. A comparison against like terms of a sine series expansion of the loading function can no longer be made. In addition, the boundary conditions are no longer satisfied due to the coupling terms $D_{16}$, $D_{26}$.

The solutions for this class of laminates are of considerable technological significance, since a bending-twisting coupling may be of importance as a control element in "aeroelastic tailoring" (Shirk et al., 1986). Solutions are usually sought using an approximation method. Examples include the Rayleigh-Ritz and Kantorovich methods for natural frequency determination (Jensen and Crawley, 1984), the Ritz method using the minimum potential energy principle for transverse deflection (Whitney, 1987), and the Galerkin method for buckling (Whitney, 1987). Considerable physical insight is often required in these approximation methods as to the form of the displacement functions, since twisting warpage must be accounted for.

### 10.10.3 General Laminated Plates

In this class of laminates, the in-plane and transverse displacements are fully coupled according to (10.71). This can give rise to extremely unusual mechanical behavior. For example, such general plates can bend and twist under an in-plane load and can shear in-plane when subjected to a transverse load. For isotropic plates, the in-plane and transverse displacements are only coupled under conditions of moderate transverse deflections or buckling in which the nonlinear strain-displacement relations (10.48a) with (10.48c) are employed.

For general laminated plates, this coupling may occur under any loading, no matter how small, for which the linearized strain-displacement relations (10.48a) with (10.48b) are appropriate. In this case, the coupling arises from the constitutive law (10.51). It is this coupling of the differential equations as well as the prescription of suitable boundary conditions that make solutions particularly difficult.

Solutions do exist for certain carefully defined problems, however. A solution is first demonstrated for transverse loading of an antisymmetric angle-ply laminate (Whitney and Leissa, 1969). The reduction of the plate stiffness terms is given by (10.61): $B_{11} = B_{12} = B_{22} = B_{66} = A_{16} = A_{26} = D_{16} = D_{26} = 0$. It is assumed that the transverse load may be expanded in terms of a double sine series of the form (10.77). If the simple support boundary conditions are of a special hinged type referred to as S3, then the displacement formulation may be posed in terms of elementary functions. These boundary conditions are those of smooth pins, which permit tangential displacements at the boundaries but forbid normal displacements in-plane at the boundaries:

$$
\begin{aligned}
x = 0, a: \quad & u^0 = 0 \\
& w = 0 \\
& N_{xy} = A_{66}\left(u^0_{,y} + v^0_{,x}\right) - B_{16}w_{,xx} - B_{26}w_{,yy} = 0 \\
& M_x = B_{16}\left(u^0_{,y} + v^0_{,x}\right) - D_{11}w_{,xx} - D_{12}w_{,yy} = 0 \\
y = 0, b: \quad & v^0 = 0 \\
& N_{xy} = A_{66}\left(u^0_{,y} + v^0_{,x}\right) - B_{16}w_{,xx} - B_{26}w_{,yy} = 0 \\
& w = 0 \\
& M_y = B_{26}\left(u^0_{,y} + v^0_{,x}\right) - D_{12}w_{,xx} - D_{22}w_{,yy} = 0
\end{aligned}
\tag{10.102}
$$

The form of the displacements is given by

$$
\begin{Bmatrix} u^0 \\ v^0 \\ w \end{Bmatrix} = \begin{Bmatrix} \sum_{m=1}^{\infty} \sum_{n=1}^{\infty} a_{mn} \sin\frac{m\pi x}{a} \cos\frac{n\pi y}{b} \\ \sum_{m=1}^{\infty} \sum_{n=1}^{\infty} b_{mn} \cos\frac{m\pi x}{a} \sin\frac{n\pi y}{b} \\ \sum_{m=1}^{\infty} \sum_{n=1}^{\infty} c_{mn} \sin\frac{m\pi x}{a} \sin\frac{n\pi y}{b} \end{Bmatrix} \tag{10.103}
$$

These displacements satisfy the differential equation and boundary conditions and so represent the exact solution. The displacement coefficients $a_{mn}, b_{mn}, c_{mn}$ may be found in terms of the load expansion coefficients $p_{mn}$ and the plate integrated elastic properties $A_{ij}, B_{ij}, D_{ij}$ by substitution of the proposed solution (10.103) into the differential equations (10.71). For the class of laminates with these boundary conditions, the free vibration problem may be solved to determine the natural frequencies of different modes by using the form

$$\{u, v, w\} = \sum_{m=1}^{\infty} \sum_{n=1}^{\infty} \{U_{mn}, V_{mn}, W_{mn}\} e^{i\Omega_{mn}t} \tag{10.104}$$

where $U_{mn}, V_{mn}, W_{mn}$ are the spatial function terms represented in (10.103).

The case of a cross-ply laminated plate with an even number of alternating layers also admits solutions to transverse loading and vibration. These laminates have the stacking sequence $[0/90]_{nT}$. The differential equations (10.71) will be simplified by the requirement that

$$\begin{aligned} &A_{16} = A_{26} = B_{12} = B_{16} = B_{26} = D_{16} = D_{26} = 0 \\ &A_{22} = A_{11} \\ &B_{22} = -B_{11} \\ &D_{22} = D_{11} \end{aligned} \tag{10.105}$$

The proposed simple support boundary conditions S2 allow normal in-plane displacements but prevent tangential displacements at the boundaries.

$$\begin{aligned} x = 0, a: \quad & w = 0 \\ & v^0 = 0 \\ & M_x = B_{11}u^0_{,x} - D_{11}w_{,xx} - D_{12}w_{,yy} = 0 \\ & N_x = A_{11}u^0_{,x} + A_{12}v^0_{,y} - B_{11}w_{,xx} = 0 \end{aligned} \tag{10.106a}$$

$$\begin{aligned} y = 0, b: \quad & w = 0 \\ & u^0 = 0 \\ & M_y = -B_{11}v^0_{,y} - D_{12}w_{,xx} - D_{11}w_{,yy} = 0 \\ & N_y = A_{12}u^0_{,x} + A_{11}v^0_{,y} + B_{11}w_{,yy} = 0 \end{aligned} \tag{10.106b}$$

The following set of displacement functions will identically satisfy the differential equations and boundary conditions and so constitute the unique solution:

$$\begin{Bmatrix} u^0 \\ \\ v^0 \\ \\ w \end{Bmatrix} = \begin{Bmatrix} \sum_{m=1}^{\infty} \sum_{n=1}^{\infty} a_{mn} \cos\frac{m\pi x}{a} \sin\frac{n\pi y}{b} \\ \sum_{m=1}^{\infty} \sum_{n=1}^{\infty} b_{mn} \sin\frac{m\pi x}{a} \cos\frac{n\pi y}{b} \\ \sum_{m=1}^{\infty} \sum_{n=1}^{\infty} c_{mn} \sin\frac{m\pi x}{a} \sin\frac{n\pi y}{b} \end{Bmatrix} \tag{10.107}$$

Provided the transverse load may be expanded in the double sine series (10.77), the solution for displacements may be found.

General laminated plates subjected to thermal loading offer another case of unusual behavior. Consider a free plate with no applied mechanical loads and subjected to a uniform temperature excursion $T$. By inspection of (10.54), the laminated constitutive equation can be placed in the general form

$$\begin{Bmatrix} \{N^T\} \\ \{M^T\} \end{Bmatrix} = [H]^{-1} \begin{Bmatrix} \{\varepsilon^0\} \\ \{\kappa\} \end{Bmatrix} \tag{10.108}$$

where

$$[H]^{-1} = \begin{bmatrix} A & B \\ B & D \end{bmatrix}$$

This equation may be inverted to solve for the midplane strains and curvatures:

$$\begin{Bmatrix} \{\varepsilon^0\} \\ \{\kappa\} \end{Bmatrix} = [H] \begin{Bmatrix} \{N^T\} \\ \{M^T\} \end{Bmatrix} \tag{10.109}$$

It is worth noting that the right side of (10.109) is simply a vector of constants, that is,

$$\begin{Bmatrix} \{\varepsilon^0\} \\ \{\kappa\} \end{Bmatrix} = C_i, \qquad i = 1, \ldots, 6 \tag{10.110}$$

The general solution for transverse deflection $w$ may be found by considering the $i = 4, 5, 6$ terms in (10.110):

$$w(x, y) = k_1 y^2 + k_2 y + k_3 x^2 + k_4 x + k_5 xy + k_6$$

The application of the derivatives implied by the $\kappa_i$ terms in (10.110), in addition to the elimination of the rigid body rotation linear terms and the arbitrary constant $k_6$, yields the displacement form

$$w(x, y) = -\tfrac{1}{2}C_4x^2 - \tfrac{1}{2}C_5y^2 - \tfrac{1}{2}C_6xy \tag{10.111}$$

The terms $C_4$ and $C_5$ represent bending curvatures, while the $C_6$ term represents a twisting of the laminate.

We consider two cases where this thermal loading will produce warpage. The material chosen is a graphite-epoxy laminate whose lamina properties are given by

$$E_1 = 20.0 \times 10^6 \text{ psi}$$
$$E_2 = 1.4 \times 10^6 \text{ psi}$$
$$\nu_{12} = 0.31$$
$$G_{12} = 6.0 \times 10^5 \text{ psi}$$
$$\alpha_1 = -2.5 \times 10^{-7}/°\text{F}$$
$$\alpha_2 = 1.4 \times 10^{-5}/°\text{F}$$

**Case 1)** Choose a $[0/90]_T$ laminate whose twisting term $C_6 = 0$. This can be shown from the consideration of the orthotropic forms of the $[A]$, $[B]$, and $[D]$ matrices and how this affects the form of $[H]$. The curvature terms of (10.110) are $\{C_4, C_5, C_6\}^T = \{1.09 \times 10^{-3}, -1.09 \times 10^{-3}, 0\}^T$ per degree Fahrenheit. The displacement is an anticlastic bending, with a downward curvature in the $x$-direction in which the 0° lamina is stiffer and upward curvature in the $y$-direction in which the 90° lamina is stiffer. Since $C_4 = -C_5$, the curvatures $\kappa_x$, $\kappa_y$ are of opposite sign; here, $\kappa_x = -\kappa_y$. See the figure for Exercise 3.35 for a visualization of anticlastic bending. Such a curvature condition is not possible for a laminate composed of isotropic laminae, where the stiffness and thermal expansion behavior is independent of the direction chosen. In that case, the bending curvatures $\kappa_x$, $\kappa_y$ have the same sign, and the deformation is referred to as "synclastic" bending.

**Case 2)** Choose a graphite-epoxy laminate of the form $[\phi/-\phi]_T$, which is expected to have a twisting deformation $C_6 \neq 0$. This twisting warpage is due to the presence of the shearing-normal coupling stiffnesses $Q'_{16}$, $Q'_{26}$, which result in a fully populated $[H]$ matrix. (Such a warpage cannot occur for isotropic laminae, since their coupling terms $Q'_{16}$, $Q'_{26}$ are identically zero.) The top and bottom layers, if free, would seek to shear in opposite directions. Since the layers are constrained in a laminate, they will twist. For $\phi = 30°$, the curvature values are $\{C_4, C_5, C_6\}^T = \{0, 0, -1.9 \times$

$10^{-3}\}^T$ per degree Fahrenheit, indicating that there is only a twisting curvature.

It is interesting to note that virtually every author has stated that a general laminated plate must exhibit bending or twisting warpage during a temperature excursion. This is, however, not true. General laminated plates with useful aero-elastic coupling may be constructed which do not warp during a temperature excursion, as has been demonstrated by Winckler (1985).

## REFERENCES FOR CHAPTER 10

S. Fakirov and C. Fakirova, Direct Determination of the Orientation of Short Glass Fibers in an Injection-Molded Poly (Ethylene Terephthalate) System, *Polym. Compos.* **6**, 41 (1985).

D. W. Jensen and E. F. Crawley, Frequency Determination Techniques for Cantilevered Plates with Bending-Torsion Coupling, *AIAA J.* **22**, 415 (1985).

R. M. Jones, Mechanics of Composite Materials, McGraw-Hill/Scripta, New York, 1975.

R. L. McCullough, Micromechanical Materials Modeling, in University of Delaware Composites Design Guide, Newark, DE, 1982.

R. B. Pipes, R. L. McCullough, and D. G. Taggart, Behavior of Discontinuous Fiber Composites: Fiber Orientation, *Polym. Compos.* **3**, 34 (1982).

M. H. Shirk, T. J. Hertz, and T. A. Weisshaar, Aeroelastic Tailoring-Theory, Practice, and Promise, *J. Aircraft* **23**, 6 (1986).

S. Timoshenko and S. Woinowsky-Krieger, *Theory of Plates and Shells*, McGraw-Hill, New York, 1959.

J. R. Vinson and T.-W. Chou, *Composite Materials and Their Use in Structures*, Applied Science, London, 1975.

J. M. Whitney, *Structural Analysis of Laminated Anisotropic Plates*, Technomic, Lancaster, PA, 1987.

J. M. Whitney and J. E. Ashton, Effect of Environment on the Elastic Response of Layered Composite Plates, *AIAA J.* **9**, 1708 (1971).

J. M. Whitney and A. W. Leissa, Analysis of Heterogeneous Anisotropic Plates, *J. Appl. Mechan.*, **36**, 261 (1969).

S. J. Winckler, *Hygrothermally Curvature Stable Laminates with Tension–Torsion Coupling*, J. Am. Helicopter Soc. **30**, 56 (1985).

## EXERCISES

**10.1** Assuming that we use the contraction implied by (10.6), the values of $C_{ijkl}$ must be related to the values of $C_{pq}$. Show that $C_{ijkl} = C_{pq}$

according to the rule

| $ij$ or $kl$ | $p$ or $q$ |
|---|---|
| 11 | 1 |
| 22 | 2 |
| 33 | 3 |
| 23 | 4 |
| 13 | 5 |
| 12 | 6 |

Consider the expansion for $\tau_{ij}$ in (10.1) and the expansion for $\sigma_p$ in (10.7); then write and compare terms using (10.6).

**10.2** A simple proof that $C_{ijkl} = C_{klij}$ may be pursued as follows: For a hyperelastic solid, a strain energy density $U$ may be defined as

$$U = U_0 + \tfrac{1}{2}C_{ijkl}\varepsilon_{ij}\varepsilon_{kl}$$

Thus,

$$\begin{aligned}\frac{\partial U}{\partial \varepsilon_{mn}} &= \tfrac{1}{2}C_{ijkl}\left\{\varepsilon_{ij}\frac{\partial \varepsilon_{kl}}{\partial \varepsilon_{mn}} + \varepsilon_{kl}\frac{\partial \varepsilon_{ij}}{\partial \varepsilon_{mn}}\right\} \\ &= \tfrac{1}{2}C_{ijkl}\{\varepsilon_{ij}\delta_{km}\delta_{ln} + \varepsilon_{kl}\delta_{im}\delta_{jn}\} \\ &= \tfrac{1}{2}\{C_{ijmn}\varepsilon_{ij} + C_{mnkl}\varepsilon_{kl}\} = \tfrac{1}{2}\{C_{ijmn}\varepsilon_{ij} + C_{mnij}\varepsilon_{ij}\}\end{aligned}$$

But $\sigma_{mn} \equiv \partial U/\partial \varepsilon_{mn} = C_{mnij}\varepsilon_{ij}$, and therefore, $C_{ijmn} = C_{mnij}$. Using a summation convention for $C_{ij}$ and $\delta_{ij}$ with $i$ and $j$ defined from 1 to 6, give a similar proof that $C_{ij} = C_{ji}$ using the strain energy definition $U = U_0 + \frac{1}{2}C_{ij}\varepsilon_i\varepsilon_j$, noting that $\sigma_m = \partial U/\partial \varepsilon_m$.

**10.3** By performing an expansion for $\tau_{1'1'}$ using (2.25) and $\sigma_1'$ using (10.10) and comparing like terms, derive the first row of the transformation matrix $[T]$ of (10.11). Perform a similar derivation for the first row of the strain transformation matrix $[T^*]$.

**10.4** Demonstrate that $[S'] = [T^*][S][T^*]^T$ as in (10.16b).

**10.5** Perform the matrix multiplication implied by (10.24a) and verify that a twofold symmetry about the $x_3$-axis will reduce the form of $C_{ij}$ from that of (10.23) to that given in (10.24c).

**10.6** It is often stated that twofold (orthotropic) symmetry with respect to two perpendicular axes implies symmetry with respect to a third perpendicular axis. First draw a figure to demonstrate that the rotation matrix $[T_2]$ is

$$[a'_{ij}] = \begin{bmatrix} \cos\alpha & 0 & -\sin\alpha \\ 0 & 1 & 0 \\ \sin\alpha & 0 & \cos\alpha \end{bmatrix}$$

where $\alpha$ is the positive rotation angle about the $x_2$-axis. Perform an expansion for the elements of $\sigma_{i'j'} = \sigma_{kl}a_{i'k}a_{j'l}$ and, using $\sigma_i' = T_{ij}\sigma_j$, rearrange to form $[T_2(\alpha)]$. Form $[T^*_{(\alpha)}]$ by careful manipulation of $T$, noting that $\{\sigma_1, \sigma_2, \sigma_3, \sigma_4, \sigma_5, \sigma_6\}^T$ and $\{\varepsilon_1, \varepsilon_2, \varepsilon_3, \varepsilon_{23}, \varepsilon_{13}, \varepsilon_{12}\}^T$ will transform identically using $[T]$. Using the fact that $[T_2^*]^{-1} = [T_2]^T$, perform the twofold symmetry operation around the $x_2$-axis using the orthotropic matrix (10.25) and show that it yields no more information.

**10.7** Prove (10.44) for thermal expansion coefficients in the rotated coordinate system. Start with (10.43), use the transformation equations to yield a form similar to (10.41), and compare terms. Comment on the odd or even functional character of $\alpha_x, \alpha_y, \alpha_{xy}$ with $\phi$.

**10.8** Assume that a lamina possesses only a monoclinic symmetry, as given by (10.24c). Show that the form of the stiffness relationship will look identical to (10.36b) after an in-plane rotation. Follow this through to indicate that the equations of generalized plane stress (10.38a) and the laminate constitutive behavior (10.51) do not change in form.

**10.9** Use the orthotropic stiffness matrix $[C_{ij}]$ and the in-plane rotation transformation (10.15b) with (10.21) to verify the following values of $C'_{ij}$ in terms of $C_{ij}$:

$$
\begin{aligned}
C'_{11} &= C_{11}m^4 + 2(C_{12} + 2C_{66})m^2n^2 + C_{22}n^4 \\
C'_{12} &= C_{12}(m^4 + n^4) + (C_{11} + C_{22} - 4C_{66})m^2n^2 \\
C'_{13} &= C_{13}m^2 + C_{23}n^2 \\
C'_{16} &= -\left[(C_{11} - C_{12} - 2C_{66})m^2 - (C_{22} - C_{12} - 2C_{66})n^2\right]mn \\
C'_{22} &= C_{22}m^4 + 2(C_{12} + 2C_{66})m^2n^2 + C_{11}n^4 \\
C'_{23} &= C_{23}m^2 + C_{13}n^2 \\
C'_{26} &= -\left[(C_{22} - C_{12} - 2C_{66})m^2 - (C_{11} - C_{12} - 2C_{66})n^2\right]mn \\
C'_{33} &= C_{33} \\
C'_{36} &= -(C_{13} - C_{23})mn \\
C'_{44} &= C_{44}m^2 + C_{55}n^2 \\
C'_{45} &= -(C_{55} - C_{44})mn \\
C'_{55} &= C_{55}m^2 + C_{44}n^2 \\
C'_{66} &= (C_{11} + C_{22} - 2C_{12})m^2n^2 + C_{66}(m^2 - n^2)^2
\end{aligned}
$$

$$m = \cos\phi, \qquad n = \sin\phi$$

Show that the $C'_{ij}$ are invariant with respect to the in-plane rotation angle for an isotropic material using the definitions (10.29), with the simplifications $E_i = E$, $\nu_{ij} = \nu$, and $G = E/2(1+\nu)$, and the result from Section 10.4 that $C_{44} = \frac{1}{2}(C_{11} - C_{12})$.

**10.10** Examining the stiffness relation (10.36) for the orthotropic lamina under an in-plane rotation, it can be noticed that there is no coupling between the in-plane stresses/strains and the transverse shear strains/stresses. Many authors use this fact to accomplish a simplification for the form of $Q'_{ij}$ and $Q_{ij}$, $i, j = 1, 2, 6$. Form a reduced transformation matrix $[T_{3r}]$ for $\{\sigma_1, \sigma_2, \sigma_6\}$ and a reduced matrix $[T^*_{3r}]$ for $\{\varepsilon_1, \varepsilon_2, \varepsilon_6\}$ based on (10.21) and (10.22). Starting from $\{\sigma_1, \sigma_2, \sigma_6\}^{\mathrm{T}} = [Q]\{\varepsilon_1, \varepsilon_2, \varepsilon_6\}^{\mathrm{T}}$, where

$$[Q] = \begin{bmatrix} Q_{11} & Q_{12} & 0 \\ Q_{12} & Q_{22} & 0 \\ 0 & 0 & Q_{66} \end{bmatrix}$$

confirm the following values for $Q'_{ij}$, $i, j = 1, 2, 6$, for use in (10.52):

$$Q'_{11} = Q_{11}m^4 + 2(Q_{12} + 2Q_{66})m^2n^2 + Q_{22}n^4$$
$$Q'_{12} = (Q_{11} + Q_{22} - 4Q_{66})m^2n^2 + Q_{12}(m^4 + n^4)$$
$$Q'_{16} = -(Q_{11} - Q_{12} - 2Q_{66})m^3n - (Q_{12} - Q_{22} + 2Q_{66})mn^3$$
$$Q'_{22} = Q_{11}n^4 + 2(Q_{12} + 2Q_{66})m^2n^2 + Q_{22}m^4$$
$$Q'_{26} = -(Q_{11} - Q_{12} - 2Q_{66})mn^3 - (Q_{12} - Q_{22} + 2Q_{66})m^3n$$
$$Q'_{66} = (Q_{11} + Q_{22} - 2Q_{12} - 2Q_{66})m^2n^2 + Q_{66}(m^4 + n^4)$$
$$m = \cos\phi, \qquad n = \sin\phi$$

**10.11** Consider the even or odd functional nature of the elements of $Q'_{ij}(\phi)$ for $i, j = 1, 2, 6$ from Exercise 10.10. By considering the pairing of balancing layers, show that $A_{16} = A_{26} = 0$ for a balanced laminate. Why is having balanced layers not sufficient to enforce $D_{16} = D_{26} = 0$?

**10.12** Prove that all $B_{ij} = 0$ for a symmetric laminate by considering a systematic pairing of the symmetric layers.

**10.13** Prove that the plate coupling stiffnesses $A_{16} = A_{26} = B_{16} = B_{26} = D_{16} = D_{26} = 0$ for a plate that contains only laminae oriented at 0° and 90°.

**10.14** Verify the first row of the operator $L_{1j}$, $j = 1, \ldots, 5$ as given in (10.66e) using equilibrium and laminate constitute equations.

**10.15** For a laminate composed of one layer of an isotropic material, show that

$$A_{11} = A_{22} = \frac{Eh}{1 - \nu^2}$$

$$A_{66} = \frac{1 - \nu}{2} A_{11}$$

$$A_{12} = \nu A_{11}$$

$$A_{16} = A_{26} = 0$$

$$B_{ij} = 0, \qquad i, j = 1, 2, 6$$

$$D_{ij} = \frac{h^2}{12} A_{ij}, \qquad i, j = 1, 2, 6$$

**10.16** For a laminate composed of one layer of an isotropic material, show that the final equation of motion (10.71) will reduce to

$$(1 + \nu)\left(u^0_{,xx} + v^0_{,xy}\right) + (1 - \nu)\nabla^2 u^0 = 0$$

$$(1 + \nu)\left(u^0_{,xy} + v^0_{,yy}\right) + (1 - \nu)\nabla^2 v^0 = 0$$

$$D\nabla^4 w = p$$

for the static case with no buckling or thermal terms.

**10.17** If the roots of (10.84) are real and unequal, taking the form $\pm S_1, \pm S_2, (S_1, S_2 > 0)$, show that the form of the solution for a specially orthotropic plate with two edges simply supported is given by (10.79) with

$$X_n(x) = A_n \cosh\frac{n\pi S_1 x}{b} + B_n \sinh\frac{n\pi S_1 x}{b} + C_n \cosh\frac{n\pi S_2 x}{b} + D_n \sinh\frac{n\pi S_2 x}{b} + \frac{4b^4 p_0}{D_{22} n^5 \pi^5}$$

Give the conditions on the $D_{ij}$ under which the roots will be real and unequal. Show that the roots will always be real and equal for an isotropic plate using the results of Exercise 10.15.

**10.18** Show that the assumed displacement forms (10.103) for an antisymmetric angle-ply laminated plate will satisfy the boundary conditions (10.102).

# BIBLIOGRAPHY

## CARTESIAN TENSORS


A. I. Borisenko and I. E. Tarapov, *Vector and Tensor Analysis with Applications*, Prentice-Hall, Englewood Cliffs, NJ, 1968.

D. E. Bourne and P. C. Kendall, *Vector Analysis and Cartesian Tensors*, 2nd ed., Academic, New York, 1977.

A. M. Goodbody, *Cartesian Tensors*, Ellis Horwood, Chichester, 1982.

H. Jeffreys, *Cartesian Tensors*, Cambridge University Press, London, 1931.

J. G. Simmonds, *A Brief on Tensor Analysis*, Springer-Verlag, New York, 1982.


## PLATES


C. Y. Chia, *Nonlinear Analysis of Plates*, McGraw-Hill, New York, 1980.

H. L. Cox, *The Buckling of Plates and Shells*, Macmillan, New York, 1963.

L. H. Donnell, *Beams, Plates and Shells*, McGraw-Hill, New York, 1976.

K. Girkmann, *Flächentragwerke*, 6th ed., Springer-Verlag, Vienna, 1963.

A. W. Leissa, *Vibration of Plates*, U.S. Government Printing Office, Washington, DC, 1969.

S. G. Lekhnitskii, *Anisotropic Plates*, 2nd ed., Gordon & Breach, New York, 1968.

S. Lukasiewicz, *Local Loads in Plates and Shells*, Noordhoff, Leyden, 1979.

E. H. Mansfield, *The Bending and Stretching of Plates*, Macmillan, New York, 1964.

K. Marguerre and H.-T. Woernle, *Elastic Plates*, Blaisdell, Waltham, MA, 1969.

R. D. Mindlin, *An Introduction to the Mathematical Theory of Vibrations of Elastic Plates*, U.S. Army Signal Corps Engineering Laboratories, Fort Monmouth, NJ, 1955.

A. Nadai, *Elastische Platten*, Springer-Verlag, Berlin, 1925.

V. Panc, *Theories of Elastic Plates*, Noordhoff, Leyden, 1975.

W. Soedel, *Vibrations of Shells and Plates*, Marcell Dekker, New York, 1981.

I. S. Sokolnikoff, *Mathematical Theory of Elasticity*, Chapter VI (Theory of Thin Plates), Notes of the Summer Session 1941, Brown University, Providence, RI, 1941.

J. J. Stoker, *Bending and Buckling of Elastic Plates*, New York University, Summer Session, 1941.

S. Timoshenko and S. Woinowsky-Krieger, *Theory of Plates and Shells*, McGraw-Hill, New York, 1959.

A. C. Ugural, *Stresses in Plates and Shells*, McGraw-Hill, New York, 1981.

V. Z. Vlasov and N. N. Leont'ev, *Beams, Plates and Shells on Elastic Foundations* (translated from the Russian), U.S. Department of Commerce, Washington, DC, 1966.

A. S. Volmir, *Gibkie plastinkii obolochki* (*Flexible Plates and Shells*), Gos: Izdvo Techniko-Teoret Lyt-ry, Moscow, 1956. See also AFFDL-TR-66-216, Air Force Flight Dynamics Laboratory, Wright-Patterson Air Force Base, Ohio, 1967.

M. D. Waller, *Chladni Plates*, Staples, London, 1960.

M. D. Waller, *Chladni Figures*, G. Bell, London, 1961.

## ELASTICITY THEORY

Yu. A. Amenzade, *Theory of Elasticity*, Mir, Moscow, 1979.

B. M. Fraeijs de Veubeke, *A Course in Elasticity*, Springer-Verlag, New York, 1979.

L. D. Landau and E. M. Lifshitz, *Theory of Elasticity*, 3rd ed., Pergamon, Oxford, 1986.

H. Leipholz, *Einführung in die Elastizitätsheorie*, G. Braun, Karlsruhe, 1968.

A. E. H. Love, *Theory of Elasticity*, 4th ed., Dover, New York, 1944.

V. Z. Parton and P. I. Perlin, *Mathematical Methods of the Theory of Elasticity*, Vols. I and II, Mir, Moscow, 1984.

H. Reismann and P. S. Pawlik, *Elastokinetics*, West, St. Paul, Minnesota, 1974.

H. Reismann and P. S. Pawlik, *Elasticity-Theory and Applications*, Wiley, New York, 1980.

## VARIATIONAL METHODS

H. L. Langhaar, *Energy Methods in Applied Mechanics*, Wiley, New York, 1962.

J. N. Reddy, *Energy and Variational Methods in Applied Mechanics*, Wiley, 1984.

T. H. Richards, *Energy Methods in Stress Analysis*, Ellis Horwood, Chichester, 1977.

K. Washizu, *Variational Methods in Elasticity and Plasticity*, 2nd ed., Pergamon, Oxford, 1975.

## FINITE DIFFERENCE METHODS

R. L. Burden, J. D. Faires, and A. C. Reynolds, *Numerical Analysis*, 2nd ed., Prindle, Weber and Schmidt, Boston, 1978.

S. H. Crandall, *Engineering Analysis*, McGraw-Hill, New York, 1956.

J. Dankert, *Numerische Methoden Der Mechanik*, Springer-Verlag, Wien, 1977.

V. F. D'yachenko, *Basic Computational Mathematics*, Mir, Moscow, 1979.

V. Vemuri and W. J. Karplus, *Digital Computer Treatment of Partial Differential Equations*, Prentice-Hall, Englewood Cliffs, NJ, 1981.

## FINITE ELEMENT METHODS

K.-J. Bathe, *Finite Element Procedures in Engineering Analysis*, Prentice-Hall, Englewood Cliffs, NJ, 1982.

S. S. Rao, *The Finite Element Method in Engineering*, Pergamon, Oxford, 1982.

L. J. Segerlind, *Applied Finite Element Analysis*, Wiley, New York, 1976.

I. H. Shames and C. L. Dym, *Energy and Finite Element Methods in Structural Mechanics*, McGraw-Hill, New York, 1985.

R. Vichnevetsky, *Computer Methods for Partial Differential Equations*, Vol. I, Prentice-Hall, Englewood Cliffs, NJ, 1981.

O. C. Zienkiewicz, *The Finite Element Method*, McGraw-Hill, London, 1977.

## LAMINATED PLATES

B. D. Agarwal and L. J. Broutman, *Analysis and Performance of Fiber Composites*, Wiley, New York, 1980.

S. A. Ambartsumyan, *Theory of Anisotropic Plates* (translation from the Russian), Technomic, Stamford, CT, 1970.

J. E. Ashton and J. M. Whitney, *Theory of Laminated Plates*, Technomic, Stamford, CT, 1970.

R. M. Christensen, *Mechanics of Composite Materials*, Wiley, New York, 1979.

J. C. Halpin, *Revised Primer on Composite Materials*, Technomic, Lancaster, PA, 1984.

R. F. S. Hearmon, *Applied Anisotropic Elasticity*, Oxford University Press, London, 1961.

R. M. Jones, *Mechanics of Composite Materials*, McGraw-Hill, New York, 1975.

A. W. Leissa, Buckling of Laminated Composite Plates and Shell Panels, AFWAL-Technical Report 85-3069, Wright-Patterson Air Force Base, Ohio, 1985.

S. G. Lekhnitskii, *Anisotropic Plates*, (S. W. Tsai and T. Cheron, transl.) Gordon and Breach, New York, 1968.

S. W. Tsai and H. T. Hahn, *Introduction to Composite Materials*, Technomic, Westport, CT, 1980.

J. R. Vinson and T.-W. Chou, *Composite Materials and Their Use in Structures*, Applied Science, London, 1975.

J. R. Vinson and R. L. Sierakowski, *The Behavior of Structures Composed of Composite Materials*, Nijhoff, Boston, 1986.

J. M. Whitney, *Structural Analysis of Laminated Anisotropic Plates*, Technomic, Lancaster, PA, 1987.

# AUTHOR INDEX

NOTE: Author index covers Chapters 1 through 10, but does not include references.

# SUBJECT INDEX